MORPHOLOGY, PHYSIOLOGY, AND BEHAVIORAL BIOLOGY OF TICKS

ELLIS HORWOOD SERIES IN ACAROLOGY

Series Editor: CLIVE BOWMAN, Ministry of Agriculture, Fisheries and Food, Slough Laboratory, Berkshire

Mites and tics are a neglected arthropod group of world-wide economic importance. This young series draws together many of the separate research and applied efforts on an international scale.

ACAROLOGY VI: Volumes 1 and 2
Editors: D. A. GRIFFITHS, Head of Storage Pests Department, and C. E. BOWMAN, Acarologist, both of Ministry of Agriculture, Fisheries and Food, Slough Laboratory, Berkshire

MORPHOLOGY, PHYSIOLOGY AND BEHAVIORAL BIOLOGY OF TICKS
Editors: J. R. SAUER and J. A. HAIR, Department of Entomology, Oklahoma State University, Oklahoma, USA

ECOLOGY OF SOIL ANIMALS, Second Edition
J. A. WALLWORK, Department of Zoology, Westfield College, University of London

MORPHOLOGY, PHYSIOLOGY, AND BEHAVIORAL BIOLOGY OF TICKS

Editors:
JOHN R. SAUER
Professor of Entomology
and
J. ALEXANDER HAIR
Regents Professor of Entomology
both of Oklahoma State University, Stillwater, USA.

ELLIS HORWOOD LIMITED
Publishers · Chichester

Halsted Press: a division of
JOHN WILEY & SONS
New York · Chichester · Brisbane · Toronto

First published in 1986 by
ELLIS HORWOOD LIMITED
Market Cross House, Cooper Street,
Chichester, West Sussex, PO19 1EB, England
The publisher's colophon is reproduced from James Gillison's drawing of the ancient Market Cross, Chichester.

Distributors:
Australia and New Zealand:
JACARANDA WILEY LIMITED
GPO Box 859, Brisbane, Queensland 4001, Australia
Canada:
JOHN WILEY & SONS CANADA LIMITED
22 Worcester Road, Rexdale, Ontario, Canada
Europe and Africa:
JOHN WILEY & SONS LIMITED
Baffins Lane, Chichester, West Sussex, England
North and South America and the rest of the world:
Halsted Press: a division of
JOHN WILEY & SONS
605 Third Avenue, New York, NY 10158, USA

British Library Cataloguing in Publication Data
Morphology, physiology, and behavioral biology of ticks. —
(Ellis Horwood series in acarology)
1. Ticks
I. Sauer, John R. II. Hair, J. Alexander
595.4′2 QL458.2.I9

ISBN 0–85312–841–3 (Ellis Horwood Limited)
ISBN 0–470–20327–7 (Halsted Press)

Phototypeset in Times by Ellis Horwood Limited
Printed in Great Britain by R.J. Acford, Chichester

Table of Contents

Foreword

Recent advances in our knowledge of the biology of ticks in relation to other animals are authoritatively documented in this book. Companion volumes contributing to the consolidation of this knowledge are *Physiology of ticks* [edited by F. D. Obenchain and R. Galun, published in 1983 (dated 1982), and covering work and literature up to 1980 or 1981], *An atlas of ixodid tick ultrastructure* [by Yu, S. Balashov, published in Russian in 1979, English translation published by the Entomological Society of America (ESA) in 1983], and *Bloodsucking ticks . . . vectors of diseases of man and animals* [by Yu, S. Balashov, published in Russian in 1968, English translation published by the ESA in 1972].

During the past two to four decades, there have been equally significant advances in our knowledge of tick biosystematics, tickborne disease epidemiology, and the kinds and properties of tickborne disease agents. However, this information has not yet been compiled into one or more review volumes.

This book on the morphology, physiology, and behavioral biology of ticks lucidly illustrates how the most sophisticated modern tools and technology are being utilized in North America, Australia, Europe, and Africa to elucidate the internal functions of ticks, their interrelationships with hosts, their adaptability to different natural and man-made environments, and the biological properties which permit them to serve as reservoirs and vectors of a variety of agents infecting man, livestock, and wildlife. The results of multidisciplinary studies such as these lead to the designing of more effective integrated methods for preventing and controlling the tickborne diseases that take such a tremendous toll of health and wealth in the developed and developing regions of the world. Modern man has brought many of these medical and veterinary problems upon himself; now he is well on the way to alleviating them.

Eight of the first authors of chapters in this book also contributed chapters to the 1982/3 volume on *Physiology of ticks*. The present updating and complementing of their earlier reviews demonstrate the vigor and vitality of contemporary research in these fields. The other first authors of chapters herein, and their co-authors, present important data and conclusions which diversify and extend the multifaceted knowledge of how molecules, cells, glands, and organs function during the on-host and off-host periods of tick life.

Each of the 'new' chapters on tick antigens, sense organs, perception and response to sounds, etc., adds significantly to understanding tick–host–environmental interactions and their roles in determining the degree of success of the individual tick species as a parasite and as a vector.

This 'state-of-the-art' book should stimulate worldwide research to enhance biomedical knowledge of ticks and other bloodsucking arthropods and to apply this knowledge for the good of mankind.

Harry Hoogstraal

Preface

This volume contains contributions from two symposia held during August 1984 at the XVII International Congress of Entomology in Hamburg, West Germany and organized by the German Society for General and Applied Entomology. Additional manuscripts were solicited from investigators unable to attend the Congress.

Ticks are perhaps the most important arthropod vectors of pathogens and toxic agents of man and amimals. At the same time, and in comparison to many groups of arthropods, relatively few scientists are investigating basic tick biology and physiology. This volume therefore takes on added significance since it incorporates contributions from many of the active tick research centers around the world. It is unfortunate that in recent years we have observed a decrease in fundamental tick research in some parts of the world because of political changes and subsequent decrease in financial support. If this volume aids in the restoration of deemphasized efforts or increases interest in tick biology and physiology, it will have achieved a measure of success.

The XVII International Congress provided an opportunity for scientists who were investigating fundamental tick morphology, physiology, and biology to meet and discuss recent developments in their respective fields. It is perhaps equally important to provide this published account of those proceedings because tick research has characteristically been published in a variety of journals.

We adhere to the philosophy that an understanding of the basic biology of the organism is required before rational solutions to pest management are posssible. Some avenues of research described herein provide a basis for future research and possible breakthroughs in the practical management of ticks. We are hopeful that other scientists will discover that ticks provide a fascinating challenge for future basic biological research.

The editors thank Dr Harry Hoogstraal for graciously consenting to write the Forword, and the contributors for timely submission of manuscripts. We also thank Linda Hope for secretarial help above the call of duty. A special thank you is extended to Mr. James S. Tucker and Drs. Janis L. McSwain and Richard C. Essenberg who provided considerable skill in helping proof read the final manuscript.

John R. Sauer
Jakie A. Hair

Directory of Contributors

AESCHLIMANN, André
Institut de Zoologie, Université de Neuchâtel, Chantermerle 22, CH-2000 Neuchâtel 7, Switzerland
AGBEDE, R. I. S.
Department of Veterinary Parasitology and Entomology, Ahmadu Bello University, Zaria, Nigeria, Africa
ASKENASE. Philip W.
Department of Epidemiology and Public Health, Yale University School of Medicine, New Haven, Connecticut, USA
BARKER, R. W.
Department of Entomology, Oklahoma State University, Stillwater, Oklahoma, USA
BARNARD, Donald R.
Lone Star Tick Research Laboratory, ARS, USDA, Poteau, Oklahoma, USA
BASSAL, Taha T. M.
Entomology Department, Faculty of Science, Cairo University, Giza, Egypt
BINNINGTON, Keith
Division of Entomology, Commonwealth Scientific and Industrial Research Organization, Canberra, Australia
BEN-YAKIR, David
Department of Entomology, Oklahoma State University, Stillwater, Oklahoma, USA and The Juvin Center for the Study of Infectious and Tropical Diseases, The Hebrew University, Hadassah Medical School, P.O. Box 1172, Jerusalem, 91010, Israel
BOWMAN, Jerry L.
Department of Entomology, Oklahoma State University, Stillwater, Oklahoma, USA
BROWN, Stephen J.
Department of Pathobiology, College of Veterinary Medicine, University of Illinois, Urbana, Illinois, USA

BULL, C. M.
School of Biological Sciences, Flinders University of South Australia, Bedford Park, South Australia

CONNAT, Jean-Louis (J. L.)
Institut de Zoologie, Université de Neuchâtel, Chantermerle 22, CH-2000, Neuchâtel 7, Switzerland

COONS, Lewis B.
Department of Biology, Memphis State University, Memphis, Tennessee, USA

DIEHL, Peter A.
Institut de Zoologie, Université de Neuchâtel, Chantermerle 22, CH-2000, Neuchâtel 7, Switzerland

DOTSON, E.
Institut de Zoologie, Université de Neuchâtel, Chantermerle 22, CH-2000, Neuchâtel 7, Switzerland

DUCOMMUN, Jacques
Institut de Zoologie, Université de Neuchâtel, Chantermerle 22, CH-2000, Neuchâtel 7, Switzerland

ESSENBERG, R. C.
Department of Biochemistry, Oklahoma State University, Stillwater, Oklahoma, USA

FAWCETT, Don W.
Department of Anatomy and Cellular Biology, Harvard Medical School, Boston, Massachusetts, USA

FELDMAN-MUHSAM, B.
Department of Medical Entomology, Hebrew University, Hadassah Medical School, Jerusalem, Israel

FOX, J. Carl
Department of Veterinary Parasitology, Microbiology and Public Health, College of Veterinary Medicine, Oklahoma State University, Stillwater, Oklahoma, USA

HAIR, Jakie A.
Department of Entomology, Oklahoma State University, Stillwater, Oklahoma, USA

HESS, E.
Institut de Zoologie, Université de Neuchâtel, Chantermerle 22, CH-2000, Neuchâtel 7, Switzerland

HOMER, John T.
Department of Veterinary Parasitology, Microbiology and Public Health, College of Veterinary Medicine, Oklahoma State University, Stillwater, Oklahoma, USA

HOYTE, H. M. D.
Department of Parasitology, University of Queensland, St. Lucia, Queensland 4067, Australia

KAUFMAN, W. Reuben
Department of Zoology, University of Alberta, Edmonton, Canada

KEMP, D. H.

CSIRO, Division of Tropical Animal Science, Long Pocket Laboratories, Private Bag No. 3, P.O. Indooroopilly, Queensland 4068, Australia

KOCAN, Katherine M.
Department of Veterinary Pathology, Oklahoma State University, Stillwater, Oklahoma, USA

MANE, Suresh D.
Department of Entomology, Oklahoma State University, Stillwater, Oklahoma, USA

NEEDHAM, Glen R.
Department of Entomology, Ohio State University, Columbus, Ohio, USA

OLIVER, James H. Jr.
Institute of Arthropodology and Parasitology and Department of Biology, Georgia Southern College, Statesboro, Georgia, USA

RECHAV, Y.
Department of Biological Sciences, Medical University of Southern Africa, P.O. Medunsa 0204, South Africa

ROSELL-DAVIS, Rosemarie
Department of Biology, Memphis State University, Memphis, Tennessee, USA

SAUER, John R.
Department of Entomology, Oklahoma State University, Stillwater, Oklahoma, USA

SCHMIDT, Stephen P.
Union Carbide Agricultural Products Co. Inc., Research Triangle Park, North Carolina, USA

SONENSHINE, Daniel E.
Department of Biological Sciences, Old Dominion University, Norfolk, Virginia, USA

STONE, Bernard F.
CSIRO, Division of Tropical Animal Science, Long Pocket Laboratories, Private Bag #3, P.O. Indooroopilly, Queensland 4068 Australia

TARNOWSKI, Betty I.
Department of Biology, Memphis State University, Memphis, Tennessee, USA

TEEL, Pete D.
Department of Entomology, Texas A&M University, College Station, Texas, USA

VLIMANT, M.
Institut de Zoologie, Université de Neuchâtel, Chantermerle 22, CH-2000, Neuchâtel 7, Switzerland

VOIGHT, Wolf P.
International Laboratory for Research on Animal Diseases, Nairobi, Kenya

ZEEDERBERG, M. E.
Department of Anatomy, Medical University of Southern Africa, P.O. Medunsa 0204, South Africa

Part 1
Structure, function

1

The cell biology of the ixodid tick salivary gland

Don W. Fawcett, Keith Binnington, and Wolf P. Voigt

1.1 INTRODUCTION

Interest in the salivary gland of ixodid ticks relates mainly to its role as a site of development of infectious stages of protozoan parasites that cause disease of livestock throughout the world. However, the gland also merits intensive study by cell and developmental biologists for the remarkable and complex cytological transformations necessary to meet the tick's changing physiological requirements.

1.2 GENERAL ANATOMY

The organization of the salivary gland has been well described in earlier light microscopic studies on *Rhipicephalus appendiculatus* (Till 1961) and on *Boophilus microplus* Binnington 1978), and these observations have been extended recently to the electron microscopic level (Coons & Roshdy 1973, Megaw & Beadle 1979, Fawcett *et al*. 1981a, b, Krolak *et al*. 1982, Walker *et al*. 1985).

The paired salivary glands in the adult female of *Rhipicephalus appendic-*

ulatus each consist of about 1400 acini of three types (I, II, III) and in the male about 1350 acini of four types (I, II, III, IV) (Walker *et al*. 1985). The type I acini, which are thought to be osmoregulatory, are situated along the main duct, while the remaining acini, containing cells with secretory granules, are distributed along a ramifying system of intralobular ducts. Of the granule-containing acini, the type III tend to be most numerous, occupying the terminal branches of the ducts at the periphery of the gland. These and acinar types II and IV all conform to a common architectural pattern with large pyramidal secretory cells and intervening slender interstitial cells arranged radially around a small lumen (Fig. 1.1). Electron microscopic studies have shown that interstitial cells alternate with the secretory cells and occur in two tiers — *adlumenal interstitial cells* bordering on the lumen and surrounding the narrow apices of the glandular cells, and the *ablumenal interstitial cells* in contact with the bases of the adlumenal cells centrally and with the basal lamina of the acinus at the periphery (Fawcett *et al*. 1981b). A bicuspid valve is present in the acinar duct connecting the lumen of the acinus with the intralobular duct. The valve and the chitinous lining of this acinar duct are associated with interdigitating *neck cells* which have ultrastructural features in common with the adjoining adlumenal interstitial cells and the cells of the intralobular duct.

The nervous control of salivation has attracted considerable interest. The salivary nerve is a branch of a pedipalpal nerve (Megaw 1977). Where it lies free in the haemocoel it is ensheathed in a thick external lamina which sends extensions into the interstices between large axons of variable cross-sectional outline. The most conspicuous feature of some axons is the occurrence of dense neurosecretory granules in their axoplasm. Upon reaching the gland, the nerve runs along the main duct. More distally its branches become incorporated in the wall of the lobular ducts, with the external lamina of the nerve becoming continuous with the basal lamina of the duct epithelium (Fig. 1.2).† The nerve ramifies with the duct, giving off small lateral branches to each acinus. These enter the granule-containing acini along the acinar duct, terminating in synapses on the glandular cells near the lumen. Nerve endings have also been observed in contact with cytoplasmic processes of the peripheral cells of acinus I (Coons & Roshdy 1973, Megaw 1977). The axoplasm of the nerve endings is crowded with vesicular structures of three kinds — large vesicles with a granular content of moderate density; smaller vesicles with a very dense homogeneous content; and large numbers of small non-granular vesicles (Fig. 1.3). The non-granular vesicles are clustered around paramembranous dense configurations which are present in central, peripheral, and neuroendocrine terminals of both ticks and other arthropods (Binnington & Lane 1982). The function of these 'synaptic vesicles' is conjectural. While they may contain neurotransmitter, there is evidence that in vertebrate neuroendocrine terminals they function in calcium regulation (discussed by Binnington & Lane 1982). In vertebrates there is good evidence that small electron-dense

† Figs. 1.2–1.22 are electron micrographs of the salivary gland of *R. appendiculatus*.

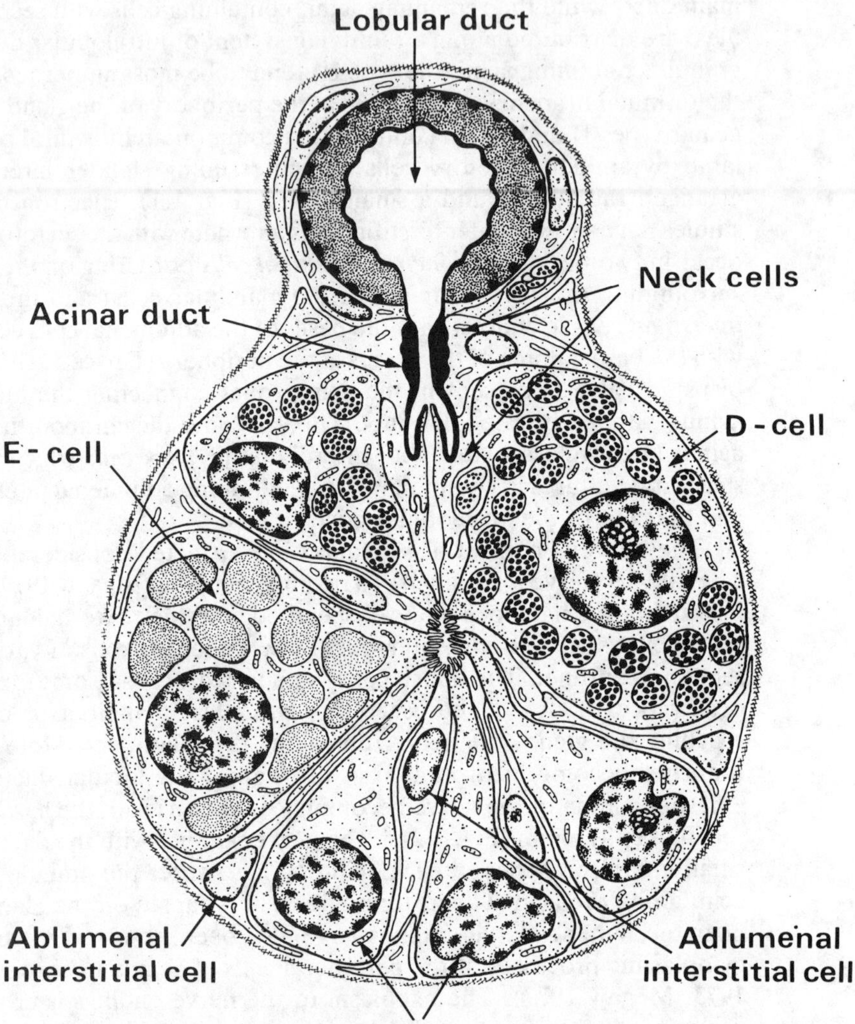

1

Fig. 1.1 — Schematic representation of the structure of the type III acinus. The type II and type IV acini have a similar organization but different glandular cell types. (After Fawcett *et al.* 1981a).

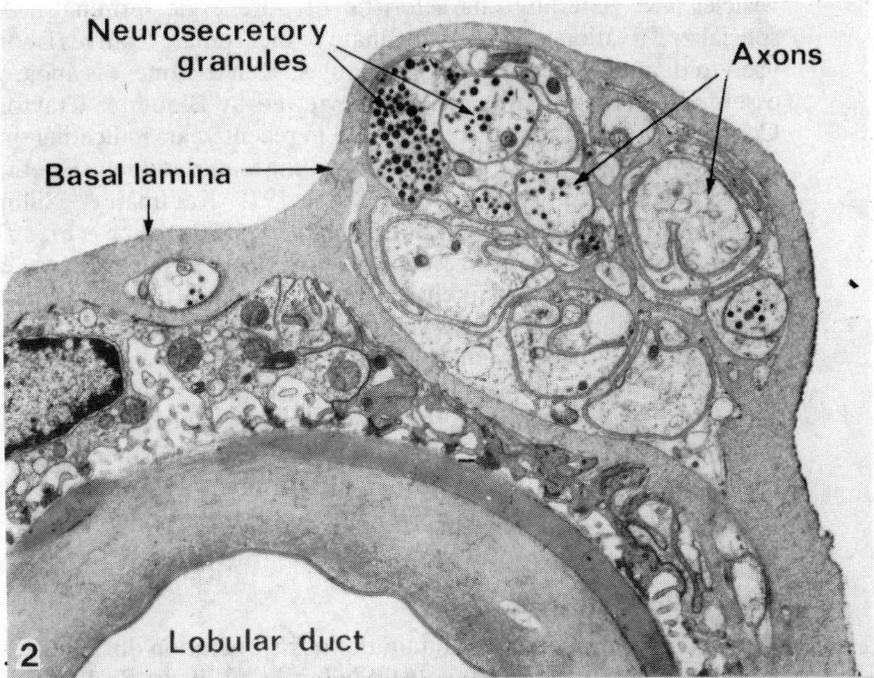

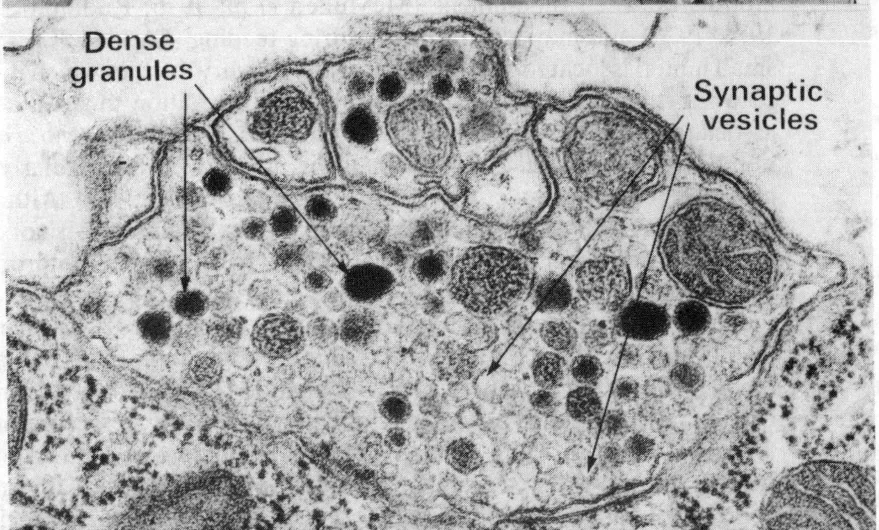

Fig. 1.2 — Portion of a lobular duct and the accompanying nerve. Some of the axons containing neurosecretory granules. Fig. 1.3 — A nerve ending on a d-cell of acinus III. In addition to small non-granular, 'synaptic' vesicles, the axoplasm contains a few large vesicles with a moderately dense content and smaller very dense neurosecretory granules.

vesicles are generally characteristic of adrenergic terminals. Although specialized fixation with permanganate or dichromate fixatives is sometimes used to demonstrate these small granules, such treatment is apparently not essential to reveal their presence (reviewed by Bloom & Crayton 1972). Compelling evidence has accumulated in recent years indicating that dopamine is the principal transmitter controlling water and ion secretion in the tick salivary gland (Binnington & Stone 1977, Kaufman & Phillips 1973, Megaw & Robertson 1974, Kaufmann 1976, 1979, Sauer *et al.* 1974), and this is consistent with the presence of small dense granules. Large electron-dense granules are more likely to contain peptides. In view of the range of functions attributed to the tick salivary gland, it seems reasonable to expect that a number of different transmitters and/or neuromodulators may be needed to control the development and secretory activities of the various cell types. Neural control of the different forms of salivary secretion in the tick certainly provides an interesting subject for further research. One approach which should prove profitable is immunocytochemical analysis of the contents of the large electron-dense axoplasmic granules.

1.3 TYPE I ACINUS

This acinus is thought to maintain the tick's hydration during the weeks or months between bloodmeals (McMullen *et al.* 1976, Rudolph & Knülle 1974, 1978) and may also function during feeding (Barker *et al.* 1984). Its small lumen is located at the pole near the salivary duct, to which it is joined by a short acinar duct which has no valve. In addition to small neck cells associated with the duct, the acinus consists of a single large *central cell* and a number of *peripheral cells* that interdigitate with the irregularly shaped central cell and extend outward to the basal lamina (Fig. 1.4). Although the peripheral cells are often called 'pyramidal cells,' their shape is not obvious, for their lateral cell boundaries are difficult to identify in the profusion of radially oriented membranes in the basal labyrinth of the acinus. The bulk of their volume is made up of a myriad of slender basal processes which extend, from a small perikaryon adjacent to the central cell, to the basal lamina at the periphery. Except where they are expanded to accommodate mitochondria, or lipid droplets, these foliate or lamellar basal processes are only 30 to 50 nm in thickness (Fig. 1.5), and in the aggregate they present an enormous surface area to the extracellular compartment of the basal labyrinth. The central cell and the peripheral cells contain an abundance of mitochondria, an inconspicuous Golgi complex, and little or no endoplasmic reticulum. The cytoplasm is poor in ribosomes but may contain widely dispersed alpha and beta particles of glycogen. No secretory granules are present, but small dense lysosome-like bodies of varying shape are common. Lipid droplets are numerous, both in the perikaryon and in the foliate basal processes (Figs. 1.4 and 1.5). This acinus has been reported to show little change in its structure in the course of the bloodmeal other than some diminution in the number of lipid droplets (Binnington 1978, Walker *et al.* 1985). But in recent studies on *Amblyomma americanum* (Barker *et al.* 1984, Needham *et*

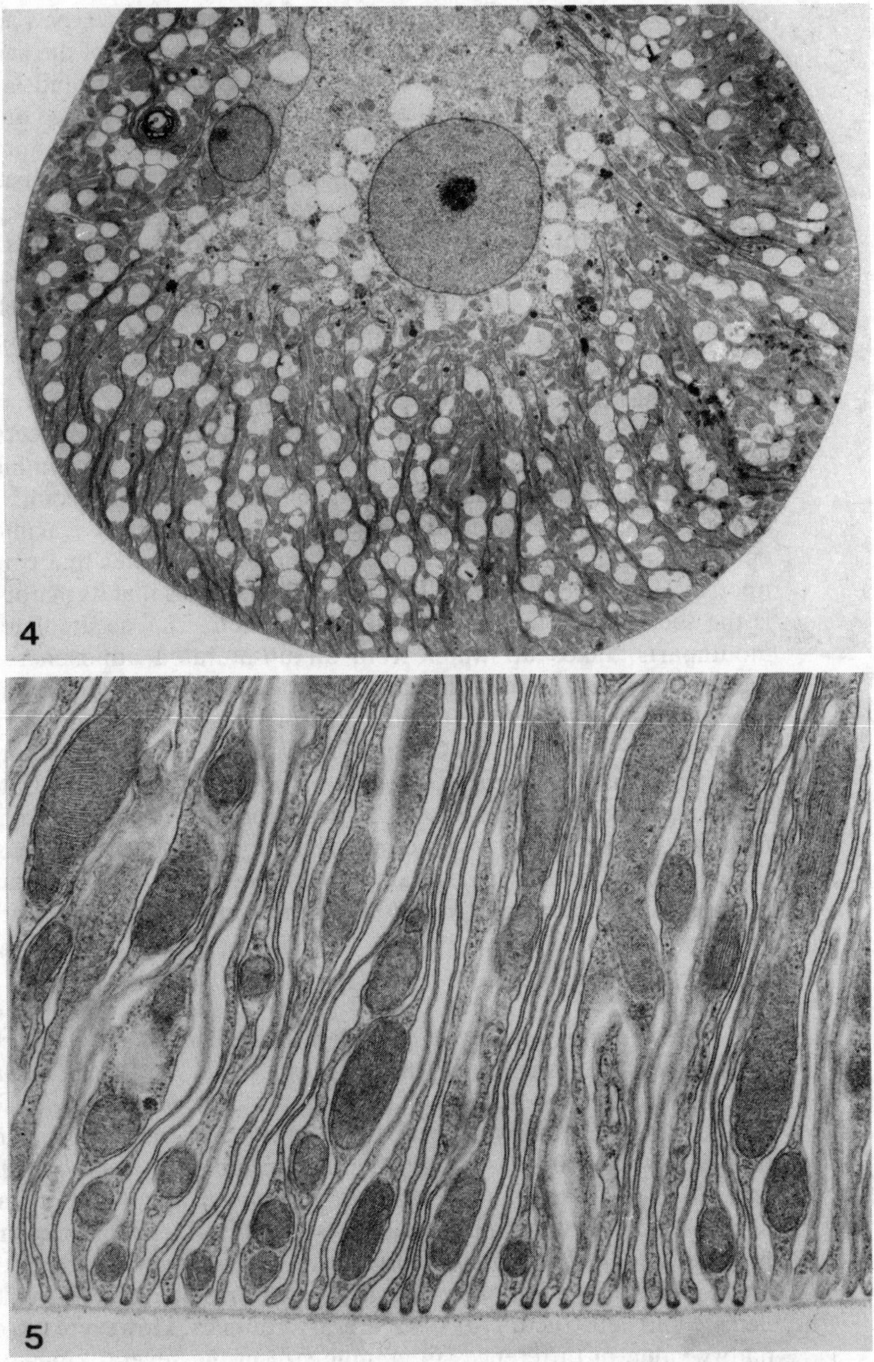

Fig. 1.4 — Low-power micrograph of a typical type I acinus showing the central cell with its large euchromatic nucleus. The smaller nucleus of one of the peripheral cells is seen. Numerous lipid droplets are present in the cell body and throughout the basal labyrinth of the acinus. Fig. 1.5 — A small area at the periphery of a type I acinus showing the extremely thin laminar cell processes and intervening extracellular clefts that form the basal labyrinth. The basal lamina of the epithelium is at the bottom of the figure.

al. 1983), significant changes in the mean diameter of the type I acini were found during feeding. Accompanying the change in size of the acini in the first few days of feeding there was a rapid depletion of lipid inclusions; changes in mitochondrial structure; and evidence of increased autophagic activity in the central cell. In the stage of rapid feeding immediately preceding repletion, the size of the acini and the appearance of the organelles returned to the state normally found in unfed ticks. It was speculated that the early disappearance of cytoplasmic fat droplets might be associated with the contribution by type I acini of a lipid component to the cement (Barker *et al.* 1984). However, bulk release of lipid from cells occurs only by the unique form of apocrine secretion observed in the mammalian mammary gland or by holocrine secretion, as in sebaceous glands. In the type I acinus there is no cytological evidence for either of these modes of secretion. It seems more likely that the lipid is not depleted by secretion but is consumed intracellularly as a source of energy for a transient increase in water and electrolyte transport during the early stages of feeding.

The striking ultrastructural resemblance of the type I acinus to the epithelium of the salt gland of marine birds and reptiles first commented upon by Balashov (1972) is consistent with the thesis that its major function is the secretion of a concentrated salt solution. This accumulates on the mouthparts, takes up water from an unsaturated atmosphere, and is ingested by the tick to maintain its hydration during the questing period (Rudolph & Knülle 1974). Further support for this view has recently been published by Needham & Coons (1984); these authors found significant ultrastructural differences in acini I from dehydrated and rehydrated ticks. The most striking change was in the mitochondria which were transformed from an orthodox form found during slow respiration to a condensed form thought to indicate fast respiration. Another indication that energy was being consumed during rehydration was the decrease in the amount of lipid.

1.4 TYPE II ACINUS

The type II acinus presents a bewildering diversity of glandular cell types designated by Binnington (1978) as a, b, c_1, c_2, c_3, and c_4. In *B. microplus* their identification at the light microscope level was largely based upon differences in size, staining properties, and the histochemical reactions of their secretory granules. In a further light microscopical study, Binnington *et al.* (1983) reported that the gland of *R. appendiculatus* was similar to that of *Boophilus* with 6 granular cell types in acinus II. Electron micrographs provide details of cytoplasmic organization beyond the reach of the light-microscope and additional identifying features in the relative electron density and substructure of the secretory granules. However, the discriminatory value of differences in granule size in the several kinds of cells is compromised somewhat in electron micrographs because ultra-thin sections transect only a fraction of the granules equatorially. The granule profiles therefore present a broader range of diameters than in histological sections where the great majority of the granules are included intact. Therefore in comparing cell types visually one is obliged to rely upon diameters of the largest granule profiles. This involves an obvious bias in its assumption that there is little actual variation in granule size within the same cell. But for simply identifying cell types in electron micrographs, elimination of this bias by application of stereological formulae is surely unnecessary.

In describing the ultrastructure of the type II acinus of *R. appendiculatus* one would like to avoid creating further terminological confusion; but, in some cases, it is difficult to relate ultrastructural appearance to the accepted staining criteria for cell identifications with the light-microscope. We have attempted to adhere to Binnington's classification, but have some reservations as to the degree of correspondence in our identifications of the cell types. The illustrations of the c-series (Figs. 1.9–1.12) are shown at the same magnification and printed on the same grade of photographic paper to facilitate comparison of granule size and relative electron density. It is hoped that this may prove helpful to others in distinguishing the cell types in electron micrographs.

The a-cells are situated in the hilus of the acinus and are easily identified by their large compound granules which have subunits visible with the light-microscope (Fig. 1.6). The a-cells of acinus II and d-cells of acinus III are similar in the structure of their granules (Figs. 1.6–1.13); and both are negative with the periodic acid-Schiff reaction (PAS) for carbohydrates (Binnington 1978, Walker *et al.* 1985). Both are believed to secrete a component of the cement in the attachment cone. In electron micrographs the large (up to 3 μm) membrane-bounded secretory granules are composed of round or oval subunits (0.5 μm) of relatively low density embedded in a matrix of slightly greater density. For reasons that remain unclear, the relative density of subunits and the surrounding matrix is variable. In the a-cell the subunits are often pale and the matrix darker, whereas in the d-cell the subunits are usually very dense and the matrix electronlucent (Fig. 1.14). These variations in appearance may occur in both cell types and may be vagaries of fixation and dehydration.

The b-cells are usually adjacent to the a-cell. Their secretory granules stain intensely with PAS (Binnington 1978). In electron micrographs their large granules (up to 2 μm) exhibit a bizarre inhomogeneity, with components of contrasting electron density. The dominant component of the granule is quite dense, but in some cells there is a crescentic or elliptical polar region of lower density (Fig. 1.8) while in other cells the pale component follows a vermiform course throughout the interior of the granule (Fig. 1.7). In the cytoplasm the endoplasmic reticulum is extensive and its cisternae are often moderately distended with a flocculent material of low density.

The c_1-cells are located in the fundus of the acinus. Their granules are small, PAS positive, have some affinity for basic dyes in histological sections, and exhibit more histochemical esterase activity than the other cells of the acinus (Binnington 1978, Walker *et al.* 1985). In electron micrographs the granules (up to 0.75 μm) are of lower density than those of any other member of the c-series, and their density may vary considerably

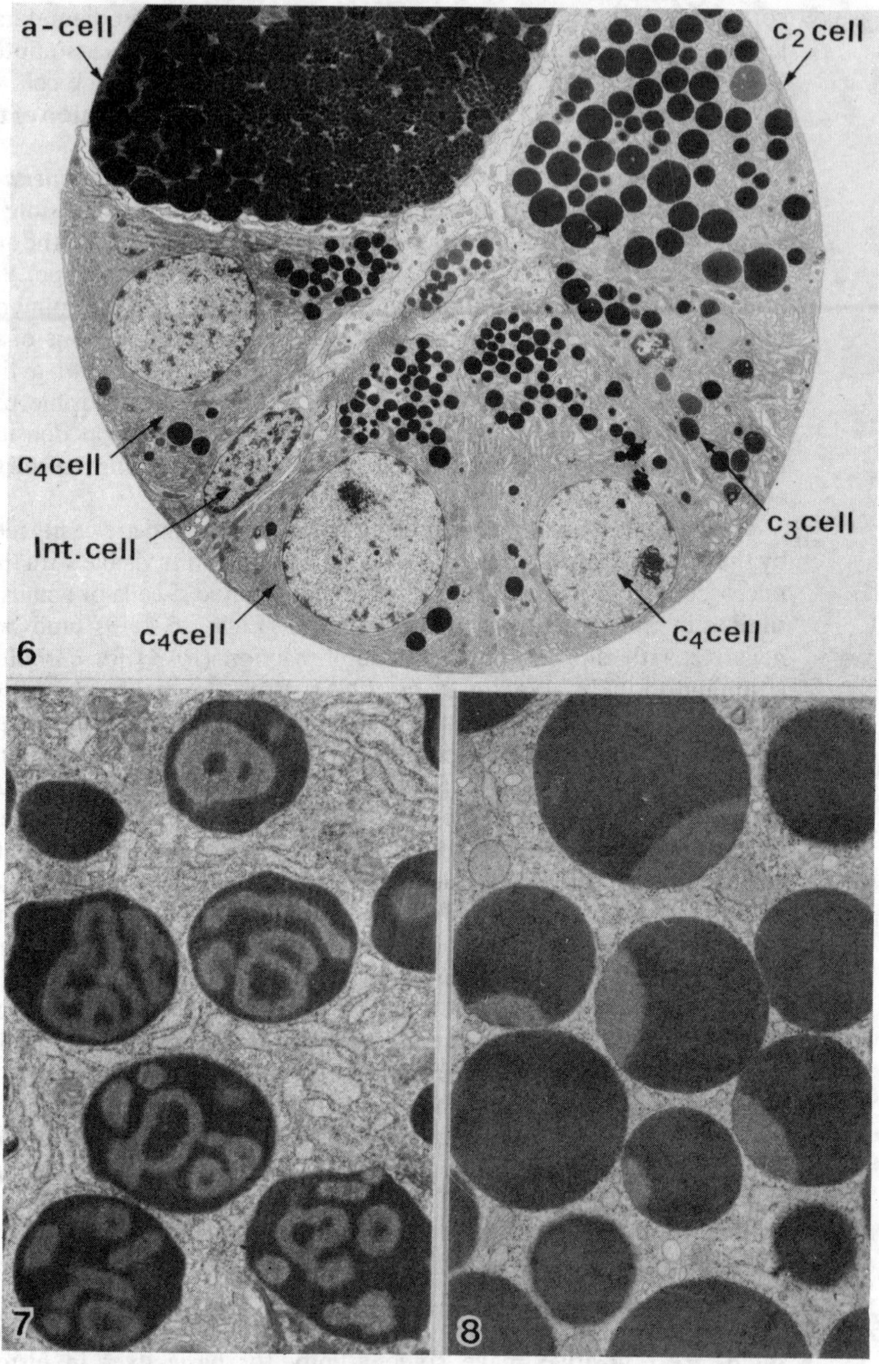

Fig. 1.6 — Low-power micrograph of a type II acinus from a male tick two days on its mammalian host. Cell types a, c_2, c_3, and c_4 are labelled. Types b and c_1 are not included in this plane of section. Fig. 1.7 — The granules of the b-cell of acinus II exhibit various patterns of segregation of two components of differing density. Here the paler component is seen as vermiform profiles in the dense substance of the granule. Fig. 1.8 — In other b-cells the less dense component forms a polar cap on the granule.

within the same cell (Fig. 1.9). These cells hypertrophy in the course of feeding and may have a very large nucleus with multiple nucleoli.

The c_2-cell is commonly situated either in the polar or equatorial region of the acinus adjacent to an a-cell. It has large secretory granules with a pale pink PAS staining reaction. In electron micrographs the granules (up to 2 μm) are spherical but their outline may show a fine scalloping. Their content is homogeneous and moderately dense (Fig. 1.10) The cisternae of the endoplasmic reticulum are often considerably distended.

The c_3-cell has granules that are slightly smaller than those of c_2 and stain more intensely with PAS. In electron micrographs they are spherical (up to 1.7 μm) smooth-contoured and distinctly more dense than the c_2 granules (Fig. 1.11).

C_4-cells are interspersed among the c_1-cells at the fundus of the acinus and have small granules of the same size as those of c_1 but they stain more intensely with PAS, and in electron micrographs the granules (up to 0.75–1 μm) are much more electron-dense (Fig. 1.12).

In the type II acinus of unfed adult females the interstitial cells conform in all respects to previous descriptions of these cells in acinus III (Fawcett *et al*. 1981a, b). In the course of feeding, the ablumenal interstitial cells enlarge somewhat and acquire numerous slender processes that form a basal labyrinth of limited extent. Thus it appears that the type II acinus acquires some capacity for fluid and electrolyte transport and makes a contribution to the copious salivary secretion in the rapid stage of feeding.

The kinetes of the protozoan parasite *Theileria parva* preferentially enter the e-cell of acinus III (Fawcett *et al*. 1982), but in rare instances infected type II acini may occur. In examination of nearly a hundred infected salivary glands, only two examples of infected type II acini have been found.

Acinus II of female *Ixodes holocyclus,* an ixodid which does not secrete cement, is quite different from that of *Dermacentor* (Coons & Roshdy 1973), *Boophilus*, and *Rhipicephalus* in containing only two cell types (Binnington & Stone 1977). Acinus III is also less complex than that of other ixodids and in the male there is only one granular acinus with a single cell type. The ultrastructure of the granular acini of *I. holocyclus* is described in Chapter 4.

1.5 TYPE III ACINUS

The type III acinus has been more thoroughly studied than the other acini because it is the principal site of sporogony of *Theileria parva*, the protozoan parasite causing African East Coast fever. It is also of interest to cell biologists and tick physiologists because of the rapid structural and functional transformations that occur in certain of its cells during feeding — a transformation believed to enable the tick to concentrate the bloodmeal by injecting a copious saliva into the host (Tatchell 1967).

The acinus contains three glandular cell types — cells d, e, and f (Binnington 1978, Fawcett *et al*. 1981a). The d-cells around the hilus of the acinus have secretory granules (up to 3 μm) with large subunits apparently

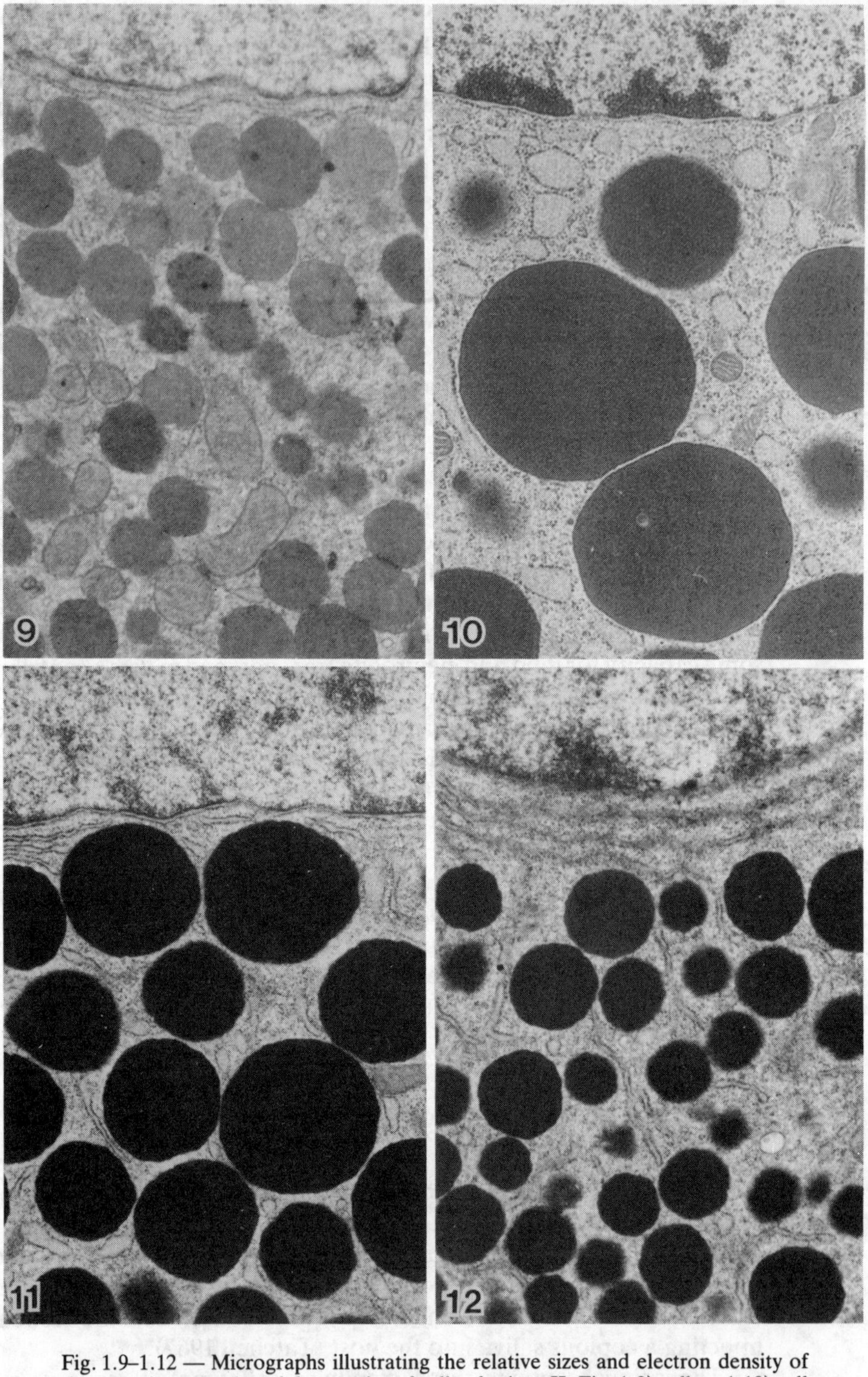

Fig. 1.9–1.12 — Micrographs illustrating the relative sizes and electron density of the secretion granules of the c-series of cells of acinus II. Fig. 1.9) cell c_1; 1.10) cell c_2; 1.11) cell c_3 and 1.12) c_4. The micrographs are printed at same magnification (12500×) and on the same grade of photographic paper to facilitate comparison.

identical to those of the a-cells of acinus II (Figs. 1.13, 1.14). The density of the subunits is inconsistent even in otherwise well preserved material. They are usually electron-dense, but cells are frequently observed in which the subunits are of lower density than the granule matrix. It is not clear whether these differences represent maturational changes in the granules or are attributable to variations in specimen preparation. Secretory product accumulates during questing, and, in the unfed tick, very large numbers of granules are found throughout the d-cell. The intervening cytoplasm is rich in mitochondria and crowded with narrow cisternal profiles of endoplasmic reticulum. A Golgi complex is rarely observed. The d-cells are believed to secrete the cement of the attachment cone. At the onset of feeding in females there is a marked depletion of d-cell granules, but some remain. There is little cytological evidence of continuing synthetic activity, and the cells diminish in size as feeding proceeds and undergo progressive involution. In males which may remain on the host for several weeks there is less depletion of d-cell granules, and the cells continue to be synthetically active.

The e-cells are the dominant cell type in the type III acinus, and, in the unfed tick, are filled with the largest secretory granules in the gland (up to 4 μm). As in other protein secreting glandular cells, parallel arrays of endoplasmic reticulum cisternae occupy much of the cytoplasm (Fig. 1.15). The secretion granules are of relatively low density and composed of closely packed small subunits (60–100 nm). Randomly scattered among these are small numbers of denser granules 100–500 nm in diameter (Figs. 1.15, 1.16). The e-cells are thought to be involved in the elaboration of a component of the cement, and, like the d-cells, secrete during attachment and subsequently regress. In males, however, the d- and e-cells remain active and provide an opportunity to study the role of the organelles in the formation of the secretory granules.

Correlated electron-microscopic and biochemical studies in other protein secreting glands have led to a widely accepted interpretation of the pathway for protein synthesis and transport common to most cells. Protein synthesis takes place on ribosomes with the nascent polypeptide chains coming off vectorially and traversing the underlying membrane to accumulate in the lumen of the endoplasmic reticulum. The product is then channeled through this reticular system to the Golgi region. Vesicles are budded off from the reticulum, carrying small quanta of the intermediate product to the stacks of Golgi cisternae. There, carbohydrate side-chains are added by glycosyltransferases, and the definitive glycoprotein is concentrated and packaged in secretory granules. Few variations on this theme have been described, and the Golgi complex is generally considered to be an indispensable organelle for secretory cells — essential for the formation of secretory granules and for their enclosure in a membrane capable of fusion with the plasma membrane in exocytosis. All of the secretory cells of the tick salivary gland, except the d- and e-cell of acinus III, conform to this description. In actively synthesizing cells of these two types the usual stacks of Golgi cisternae are not found. Instead the subunits of the e-cell granules are formed in the cisternae of the endoplasmic reticulum and released

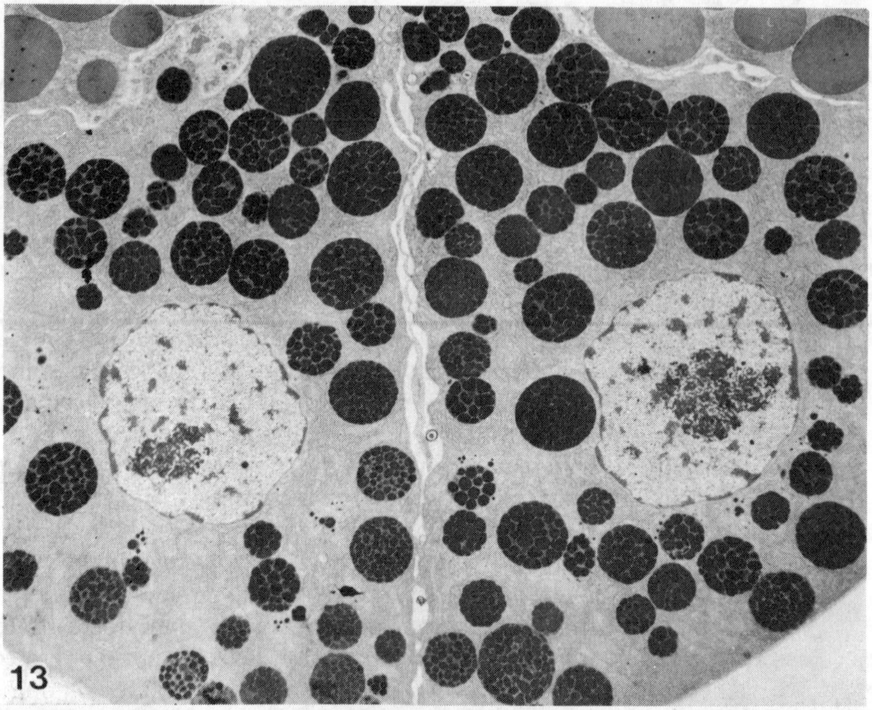

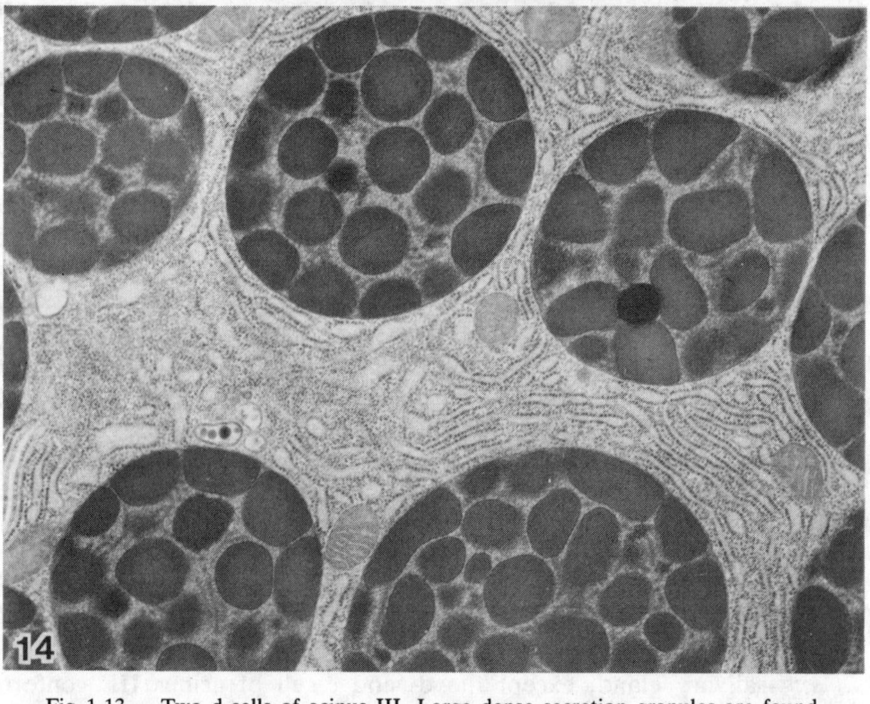

Fig. 1.13 — Two d-cells of acinus III. Large dense secretion granules are found throughout the cytoplasm. Fig. 1.14 — At higher magnification the compound granules of the a-cell of acinus II and the d-cell of acinus III have dense subunits in a less dense matrix. Not infrequently however the subunits are less dense than the matrix. This may be a consequence of inadequate fixation or extraction during specimen preparation.

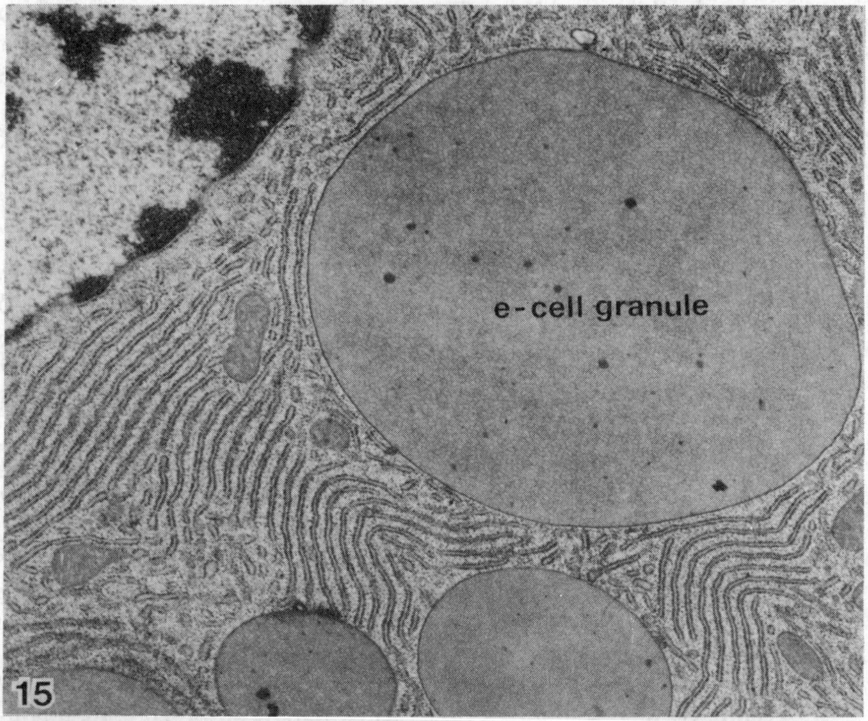

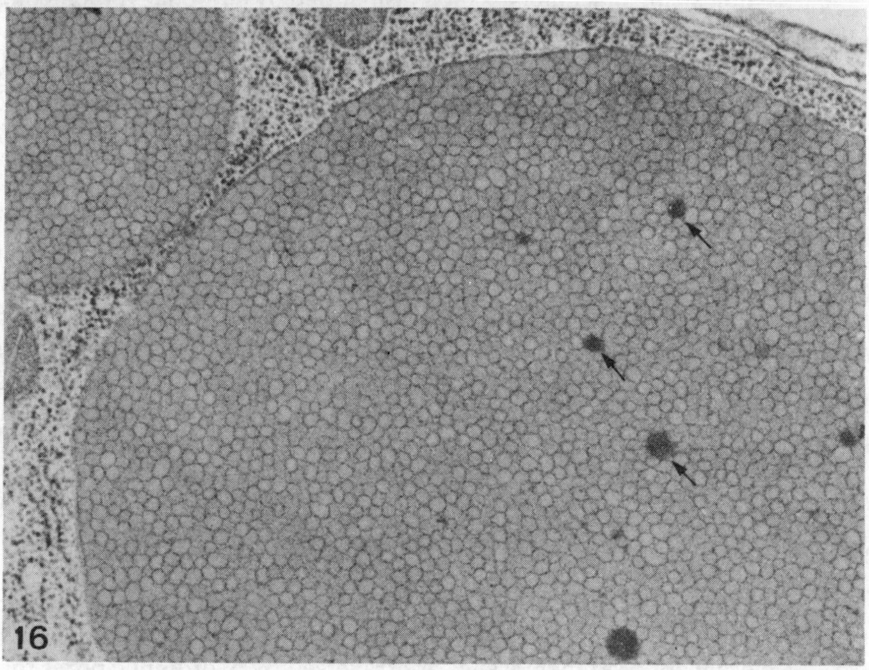

Fig. 1.15 — The cytoplasm of the e-cell of acinus III is permeated by cisternae of endoplasmic reticulum in close parallel array. The secretion granules are very large and of relatively low electron density. Fig. 1.16 — At high magnification the e-cell granules are found to be composed of closely packed small subunits. Scattered throughout are granules of a second component that is denser or more osmiophilic (at arrows).

individually in smooth-surfaced vesicles. The latter fuse with one another to produce membrane-bounded aggregations that are progressively enlarged until they attain the size of the definitive secretory granule. The same mode of granule formation is observed in the d-cell (Fawcett, unpublished observations). Thus, in these cells, the Golgi complex is bypassed. It is of interest that of the nine secretory cell types in the gland, the d- and the e-cells are the only ones with granules that are negative with the PAS reaction and evidently contain no carbohydrate. Their novel mode of secretory granule formation has two implications of interest to cell biologists. One is that the stacks of Golgi cisternae in other cell types are needed principally for the post-translational glycosylation of glycoprotein secretory products. The other is that the endoplasmic reticulum, and the transport vesicles derived from it, have the capacity to concentrate the product without participation of a Golgi complex.

The f-cells are clustered at the fundus of acinus III. In unfed ticks they are slender, undifferentiated cells with a cytoplasm rich in free ribosomes, containing few membranous organelles and no secretory granules (Fig. 1.17). Upon attachment of the tick to its mammalian host, the f-cells rapidly enlarge, and differentiate into typical glycoprotein secreting glandular cells with an extensive endoplasmic reticulum, and a prominent Golgi complex forming dense secretory granules (Fig. 1.18).

The f-cells secrete actively for only about two days. Their protein synthesizing machinery is then dismantled by autophagy; their complement of mitochondria rapidly increases; and they extend numerous lateral cell processes which establish close contacts with the intervening adlumenal interstitial cells (Fig. 1.19). From day 4 to 6 these remarkable changes in cell configuration continue until the body of each f-cell is surrounded by an intricate filigree of slender cell processes forming the adlumenal portion of a very elaborate basal labyrinth (Fawcett *et al.* 1981b).

Concurrently with these changes in the f-cells, the ablumenal interstitial cells hypertrophy, and undergo a striking proliferation of their mitochondria. These organelles come to reside in slender branching cell processes that resemble those of the f-cell and form the peripheral portion of the basal labyrinth (Fig. 1.20). The adlumenal interstitial cells increase in size and elongate, extending further toward the periphery of the acinus. Large numbers of processes of both the f-cells and the ablumenal interstitial cells terminate in contact with the membrane of the adlumenal interstitial cells. These remarkable cell transformations, occurring from the second to the fifth day of feeding, convert the glandular epithelium of acinus III to an epithelium of unparalleled complexity (Figs. 1.21, 1.23) specialized for fluid transport (Fawcett *et al.* 1981b).

Adjacent cells of most epithelia in vertebrates and invertebrates are joined by communicating or gap-junctions — small plaques where the closely apposed membranes are traversed by hydrophilic pores that permit the passage of ions and small molecules from cell to cell. These are the structural basis for the electrical and metabolic coupling of epithelial cells that share a common function. In the salivary acini of fasting ticks such

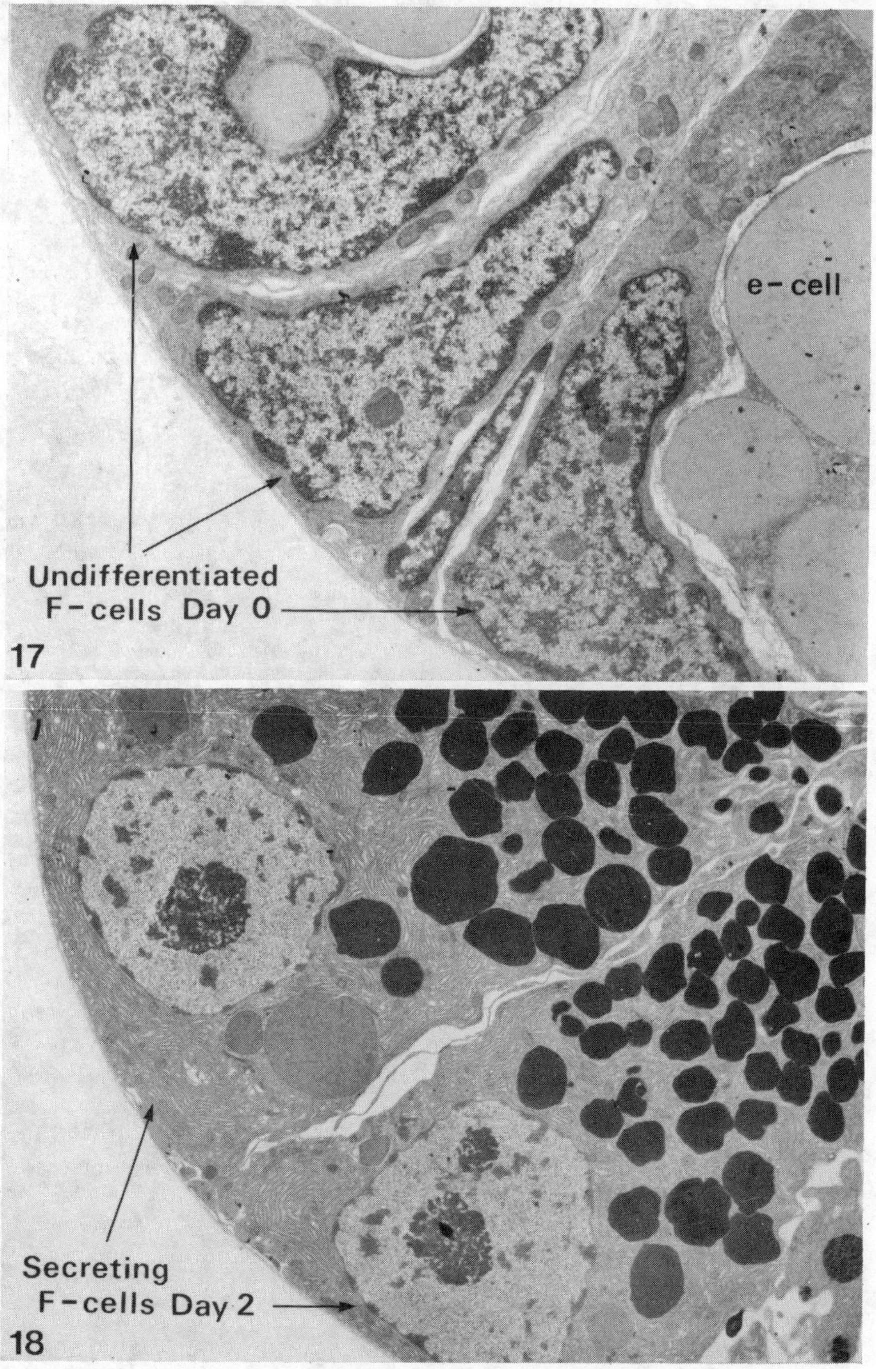

Fig. 1.17 — In the unfed tick the f-cells of acinus III are grouped at the fundus and are undifferentiated. Fig. 1.18 — After attachment of the tick to its mammalian host the f-cells rapidly enlarge and differentiate into glandular cells which are actively synthesizing and secreting glycoprotein on day 2.

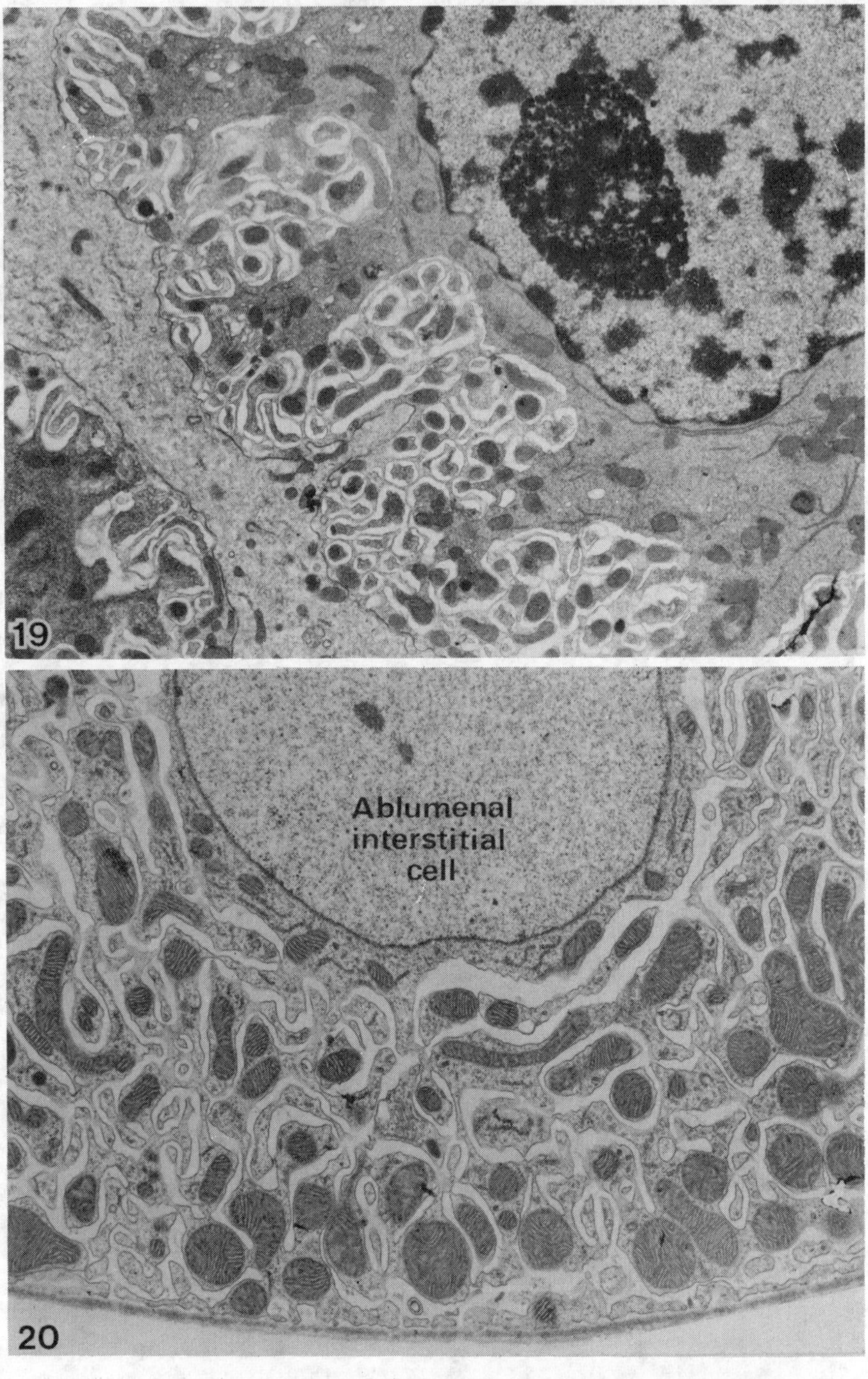

Fig. 1.19 — After secreting for only two days the cytomembranes and residual granules are eliminated by autophagy and the f-cells are transformed into cells specialized for transport as shown here. The slender lateral processes of a transformed f-cell at the upper right are in close contact with an adlumenal interstitial cell crossing the figure at the lower left. Day 4. Fig. 1.20 — The ablumenal interstitial cells also hypertrophy and extend numerous branching proceses which form the peripheral portion of an elaborate basal labyrinth. Day 5.

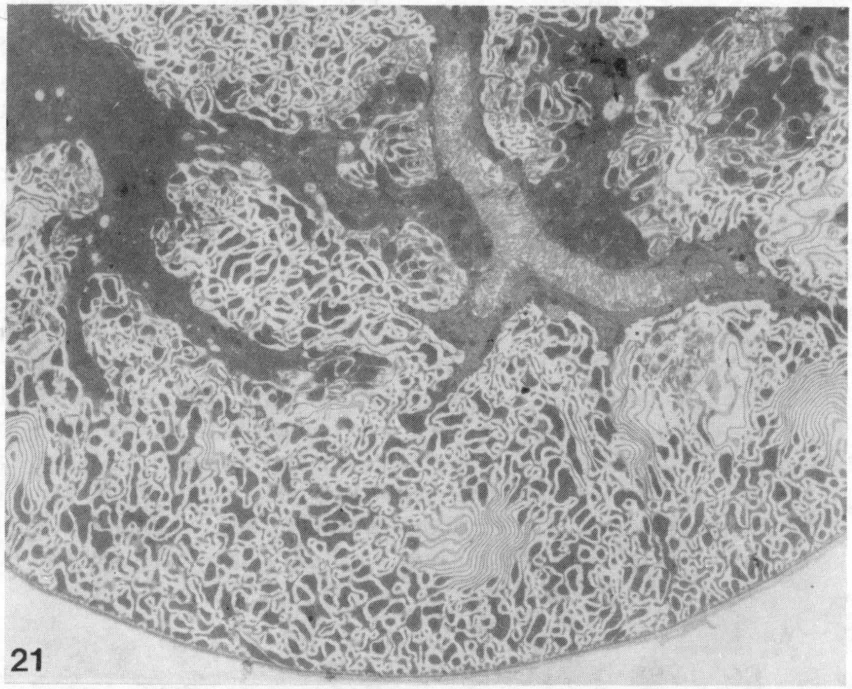

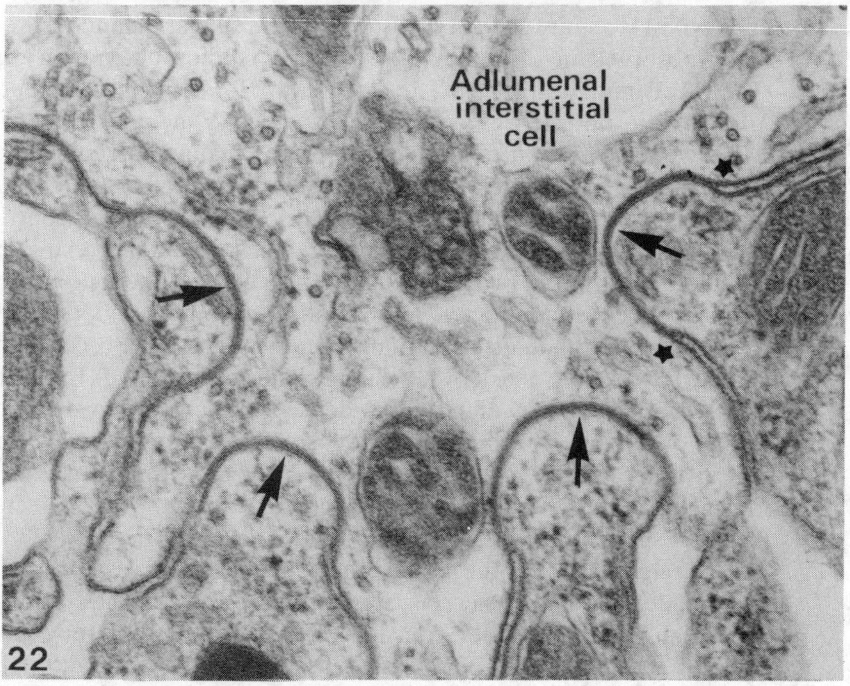

Fig. 1.21 — By day 7 of feeding the basal labyrinth occupies most of the volume of the acinus as shown in the low-power micrograph. Fig. 1.22 — At sites of contact of processes of the transformed f-cells with the adlumenal interstitial cells, the apposing membranes form curved communicating or gap junctions (at arrows). The convergence of the membrane at the periphery of the junctional plaque can be seen at the asterisks. Day 5.

junctions are only very rarely observed (Megaw & Beadle 1979), but in the course of specialization of the type III acinus for fluid transport they appear in great numbers at sites of contact of the processes of the transformed f-cells and ablumenal interstitial cells with the columnar adlumenal interstitial cells (Fig. 1.22). During chemical fixation for electron microscopy, these junctions, which are flat plaques *in vivo,* become curved, and in micrographs, present a characteristic ball-in-socket configuration that facilitates their recognition even in images of low magnification. They steadily increase in number from day 4 to day 6 of feeding (Fawcett *et al.* 1981b). In the low-power micrograph in Fig. 1.23, about 50 arc marked by asterisks along the slender body of a single adlumenal interstitial cell. With so many on one cell in a single plane of section, the total number of these low-resistance intercellular junctions in the type III acini would be astronomical during the stage of rapid feeding. Only a few gap-junctions suffice to maintain electrical and metabolic coupling of epithelial cells normally. We believe that their formation in such profusion in the tick salivary gand is an essential component of the specialization of this epithelium and is a unique example of gap-junctions providing for massive fluxes of solute in transcellular fluid transport (Fawcett *et al.* 1981b).

1.6 TYPE IV ACINUS

The type IV acinus which occurs only in the salivary gland of the male has been observed in some cement-secreting ixodids but not in others (see Table 1, Binnington 1978). The explanation for these inconsistent results is probably the undifferentiated nature of acinus IV in unfed males (Fig. 1.24). After *R. appendiculatus* has been on its mammalian host for a day or more, its glandular cells then enlarge and differentiate for protein synthesis. By day three, large electron-dense secretory granules are present in limited number in their apical cytoplasm (Fig. 1.25). Glandular cell hypertrophy and accumulation of product progress rapidly until the type IV acini are as large or larger than types II and III and the cells are filled with secretion granules. The granules are initially spherical and have a central region of somewhat greater density than the periphery (Fig. 1.26). Later, this zonation becomes more evident with a dense spherical region surrounded by a broad outer zone of relatively low density. This outer region is evidently of softer consistency, for where the granules are closely packed they are mutually deformed, becoming polygonal, while the dense central zone remains spherical (Fig. 1.27).

The significance of the sexual dimorphism of the tick salivary gland is not well understood. The secretion of the type IV acinus is believed to be essential for the smooth transfer of the spermatophore to the genital aperture of the female (Feldman-Muhsam *et al.* 1970). Intense salivation of male ticks during copulation led to the suggestion that a glycoprotein of the saliva serves to overcome the stickiness of the newly extruded spermatophore and facilitates its transfer without accidental adherence to the cuticle of the male or female. Considered teleologically, the evolution of a non-

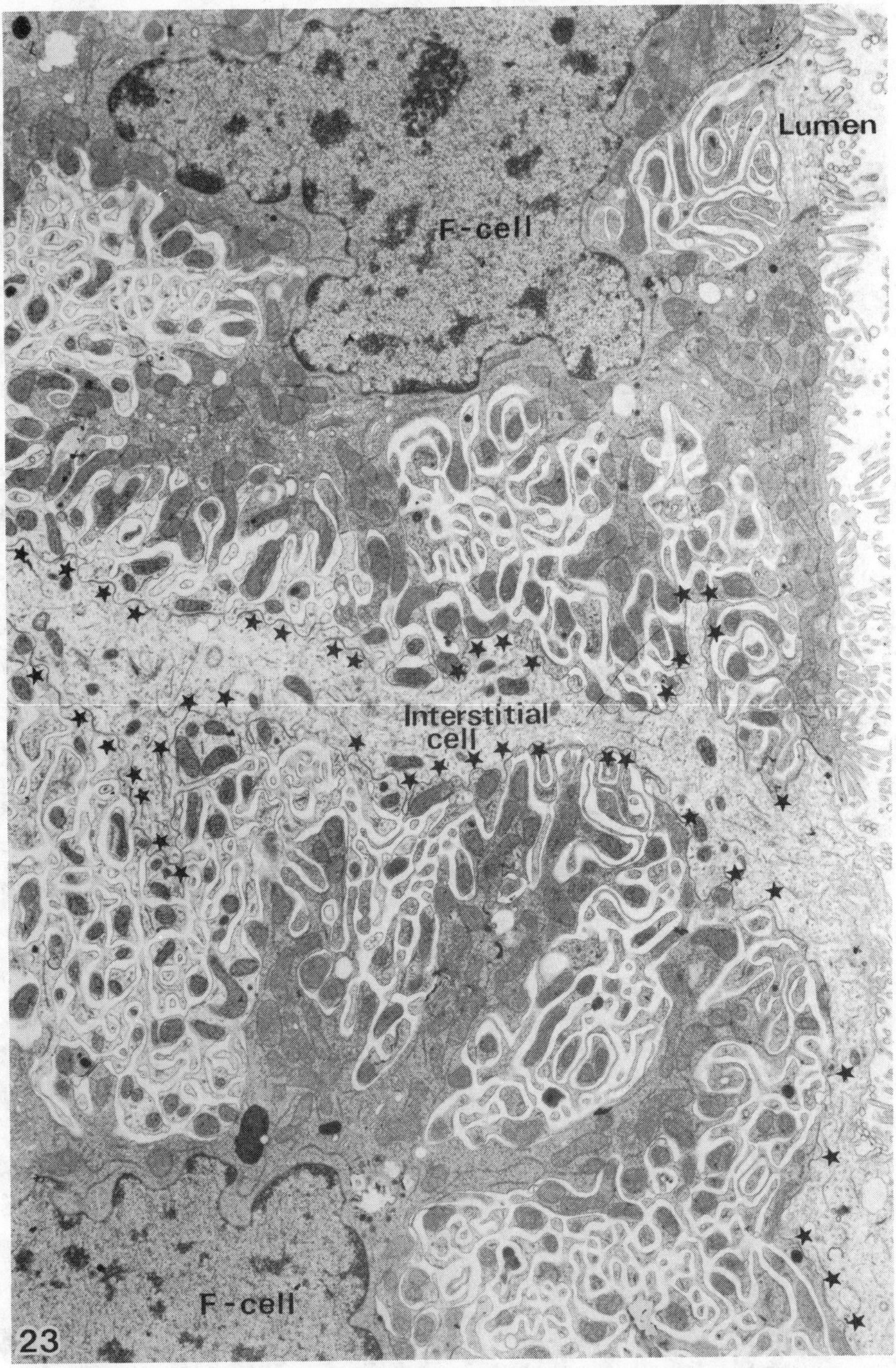

Fig. 1.23 — The number of communicating junctions between cell processes of the basal labyrinth and the columnar adlumenal interstitial cells has no parallel in cell biology. Two transformed f-cells are shown in this micrograph with their processes abutting the interposed adlumenal interstitial cell. At higher enlargement over 50 gap junctions were identified in this field. The location of some of these is indicated by the stars on the figure. Day 5.

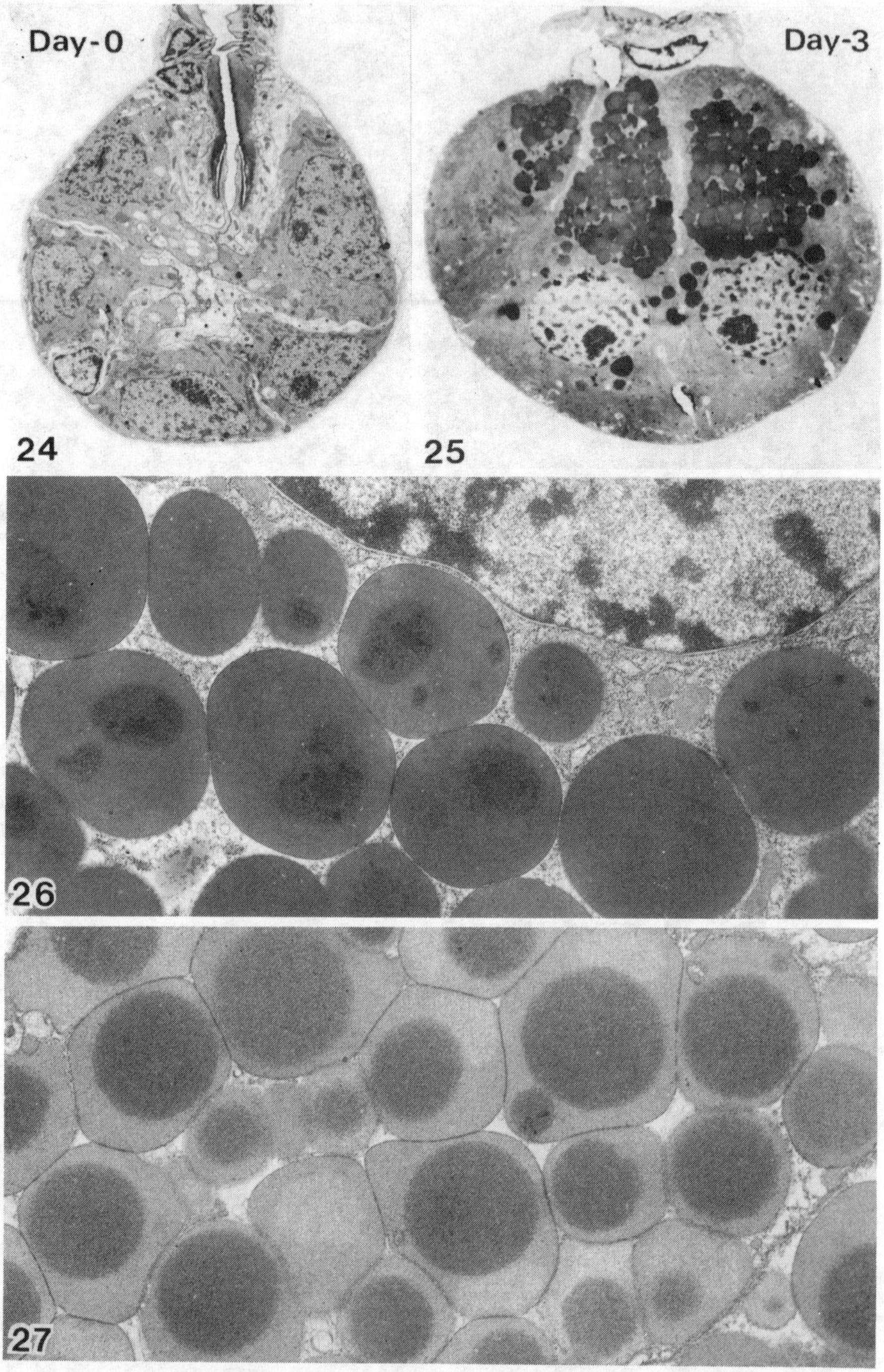

Figs. 1.24, 1.25 — The type IV acinus of the unfed male is small and its cells undifferentiated. The cells are rapidly activated, and by day 3 contain numerous secretion granules. The acinus continues to enlarge attaining a size equal to or greater than the other granular acini. Figs. 1.26, 1.27 — The electron-dense secretion granules of acinus IV commonly have a central region of slightly greater density (1.26). Under some conditions of specimen preparation there is a more hydrated peripheral zone of low density, and the closely packed granules become polygonal from mutual deformation (1.27).

sticky spermatophore would seem to have been a more direct solution to this problem than to evolve a separate salivary acinus to overcome its undesirable adhesive properties. Further studies are needed to explore the possibility that a component of the male saliva may affect the behaviour of the female or may have some role in post-copulatory maturation and capacitation of the spermatozoa.

1.7 CONCLUDING COMMENT

The tick salivary gland is a remarkably versatile organ. Its type I acini may have an important role in osmoregulation in the fasting state, and the glandular type II and III acini probably secrete numerous products needed for feeding. These include cement for the attachment cone; anticoagulant to keep the host blood fluid; vasoactive agents to increase the flow of the host's blood to the attachment site; pharmacologically active substances that induce the local release of chemical mediators of inflammation in the host; and enzymes whose functions have yet to be defined. Salivary products are also antigenic and induce tick resistance in the host, and in some species saliva contains a paralyzing toxin. The challenging task for future investigators will be the isolation and characterization of these substances and their localization to specific cell types. This should be facilitated by the recent morphological studies that have provided unambiguous ultrastructural criteria for the identification of the ten or more cell types in the tick salivary gland. As antibodies to purified constituents of tick saliva become available, immunoelectron microscopy using protein A-colloidal gold conjugates will permit precise localization of the antigens at ther cellular and subcellular levels. Identification of the cell surface characteristics of the secretory cells selected for invasion by protozoan parasites (*Theileria* and *Babesia*) will provide new information on the life cycle of these important disease agents.

REFERENCES

Balashov, Y. S. (1972). Bloodsucking ticks (Ixodoidea). Vectors of diseases of man and animals. *Misc. Publ. Entomol. Soc. Am.* **8**: 161–376.

Barker, D. M., Ownby, C. L., Krolak, J. M., Claypool, P. L. & Sauer, J. R. (1984). The effects of attachment, feeding and mating on the morphology of the type 1 alveolus of salivary glands of the lone star tick, *Amblyomma americanum. J. Parasitol.* **70**: 99–113

Binnington, K. D. (1978). Sequential changes in salivary gland structure during attachment and feeding of the cattle tick, *Boophilus microplus. Int. J. Parasitol.* **8**: 97–115.

Binnington, K. D. & Lane, N. J. (1982). Presence of T-bars, intramembranous particle arrays and exocytotic profiles in neuroendocrine terminals of an insect. *Tissue and Cell* **14**: 463–474.

Binnington, K. D. & Stone, B. F. (1978). Distribution of catecholamines in the nervous system of the cattle tick, *Boophilus microplus* Canestrini. *Comp. Biochem. Physiol.* **58C**: 21–28.

Binnington, K. C., Young, A. S. & Obenchain, F. D. (1983). Morphology of normal and *Theileria parva*-infected salivary glands of *Rhipicephalus appendiculatus* — (Acari: Ixodoidea). *J. Parisitol.* **69:** 421–424.

Bloom, F. E. & Crayton, J. W. (1972). Electron microscopic localization of biogenic amines. In: Methods in investigative and diagnostic endocrinology. *The thyroid and biogenic amines.* (J. E. Rall & J. Koplin, eds.), Amsterdam, Elsevicr/North-Holland.

Coons, L. B. & Roshdy, M. A. (1973). Fine structure of the salivary glands of unfed *Dermacentor variabilis* (Say) (Ixodidae, *Ixoidea*). *J. Parasitol.* **59**: 900–912.

Fawcett, D. W., Doxsey, S. J. & Büscher, G. (1981a). Salivary gland of the tick vector (*Rhipicephalus appendiculatus*) of East Coast fever. I. Ultrastructure of the type III acinus. *Tissue and Cell* **13**: 209–230.

Fawcett, D. W., Doxsey, S. J. & Büscher, G. (1981b). Salivary gland of the tick vector (*Rhipicephalus appendiculatus*) of East Coast fever. II. Cellular basis for fluid secretion in the type III acinus. *Tissue and Cell* **13**: 231–253.

Fawcett, D. W., Doxsey, S. J. & Büscher, G. (1982). Salivary gland of the tick vector (*Rhipicephalus appendiculatus*) of East Coast fever. IV. Cell type selectivity and host cell responses to *Theileria parva. Tissue and Cell* **14**: 397–414.

Feldman-Muhsam, B., Borut, S. & Saliternik-Givant, S. (1970). Salivary secretion of the male tick during copulation. *J. Insect Physiol.* **16**: 1945–1949.

Kaufman, W. R. (1976). The influence of various factors on secretion by *in vitro* salivary glands of ixodid ticks. *J. Exp. Biol.* **64**: 727–742.

Kaufman, W. R. (1979). Control of salivary fluid secretion in ixodid ticks. In: *Recent Advances in Acarology* (J. G. Rodriguez, ed.), Vol. 1, pp. 357–363.

Kaufman, W. R. & Phillips, J. E. (1973). Ion and water balance in the ixodid tick *Dermacentor andersoni.* II. Mechanism and control of salivary secretion. *J. Exp. Biol.* **58**: 537–547.

Krolak, J. M., Ownby, C. L. & Sauer, J. R. (1982). Alveolar structure of salivary glands of the lone star tick, *Amblyomma americanum* (L.): unfed females. *J. Parasitol.* **68**: 61–82.

McMullen, H. L., Sauer, J. R. & Burton, R. L. (1976). Possible role in uptake of water vapour by ixodid tick salivary glands. *J. Insect Physiol.* **22**: 1281–1285.

Megaw, M. W. J. (1977). The innervation of the salivary gland of the tick *Boophilus microplus. Cell Tiss. Res.* **184**: 551–558.

Megaw, M. W. & Beadle, D. J. (1979). Structure and function of the salivary glands of the tick, *Boophilus microplus, Canestrini (Acrina: Ixodidae). Int. J. Insect. Morphol. and Embryol.* **8**: 67–83.

Megaw, M. W. J. & Robertson, H. A. (1974). Dopamine and noradrenaline in the salivary glands and brain of the tick, *Boophilus microplus:* effect of reserpine. *Experientia* **30**: 1261–1262.

Needham, G. R., Greenwald, L., Freda, T. J. & Coons, L. B. (1983). The ultrastructure of the tick (*Amblyomma americanum*) salivary gland acini responsible for secretion of the hygroscopic fluid and localization of Na^+/K^+–ATPase. (Abstr.) *Am. Zool.* **23:** 1013.

Rudolph, D. & Knülle, W. (1974). Site and mechanism of water vapour uptake from the atmosphere in ixodid ticks. *Nature* **149**: 84–85.

Rudolph, D. & Knülle, W. (1978). Uptake of water vapour from air: process, site, and mechanism in ticks. In: Comparative Physiology: *Water, ions and fluid mechanics* (K. Schmidt-Nielsen, L. Bolio & S. H. P. Maddrell, eds.), Cambridge University Press. pp. 97–113.

Sauer, J. R., Frick, J. H. & Hair, J. A. (1974). Control of ^{36}Cl uptake by isolated salivary glands of the lone star tick. *J. Insect Physiol.* **20**: 1771–1778.

Tatchell, R. J. (1967). Salivary secretion in the cattle tick as a means of water elimination. *Nature* **213**: 940–941.

Till, W. M. (1961). A contribution to the anatomy and histology of the brown ear tick, *Rhipicephalus appendiculatus,* Neumann, *Mem. Ent. Soc. S. Afr.* 6 Apr. 1961, 124 pp.

Walker, A. R., Fletcher, J. D. & Gill, H. S. (1985). Structural and histochemical changes in the salivary glands of *Rhipicephalus appendiculatus* during feeding. *Internat. J. Parasitol.* (in press).

2

Salivary Gland Degeneration in the Female Tick, *Amblyomma hebraeum* Koch (Acari: Ixodidae)

W. Reuben Kaufman

2.1 INTRODUCTION

In most bloodsucking arthropods, the act of feeding heralds a host of physiological changes which require nutrients of the bloodmeal. Adult ixodid ticks feed to satiation in 7–14 days. Males imbibe a rather small volume of blood (the fed-to-unfed weight ratio rarely exceeds 1.5), but can reattach and feed intermittently. The major events promoted in males by feeding are (1) secretion of aggregation and attachment pheromones (Sonenshine *et al.* 1982), (2) enlargement of the testes and accessory glands, and (3) spermatogenesis (Oliver 1982).

In females, the fed-to-unfed weight ratio at full engorgement can exceed 100. During feeding, there is growth and development of most tissues, for example: cuticle, salivary glands, genital tract, Genè's organ, and even neural tissue. We are largely ignorant of the factors controlling these developmental processes. In the case of the salivary glands, however, a hormone is implicated (Coons & Kaufman, in preparation).

In the general physiology of most organisms, the salivary glands do not play a very glamorous role. Saliva is usually associated with the lubrication of solid food, the initiation of certain digestive processes, and perhaps a certain degree of bacteriostasis. In ixodid ticks, however, the salivary glands play a vital role in ionic and osmotic regulation, in addition to carrying out many other functions. Because they also transmit pathogenic agents and contribute most of the antigens that stimulate the immunological defense mechanisms of the host, the salivary glands have attracted much attention by acarologists (for recent reviews on tick salivary glands see Sauer 1977, Wikel & Allen 1982, Kaufman & Sauer 1982, Wikel 1983, Kaufman 1983).

2.2 DEVELOPMENT AND AUTOLYSIS OF TICK TISSUES

Following repletion, larvae and nymphs enter a quiescent period, during which there occur development and moulting to the succeeding stage. In the adult female, resources of the blood meal are largely devoted to egg production. In *Amblyomma hebraeum* Koch, oviposition commences within 11 days post-repletion at 27°C. The tick is spent about 20 days later, and death ensues shortly thereafter. During vitellogenesis and oviposition, at least 2 tissues suffer autolysis: the endocuticle (which had been synthesized only a few days earlier, Lees 1952) and the salivary glands.

In her detailed monograph on the anatomy of *Rhipicephalus appendiculatus,* Till (1961) reported that histological disruption of the salivary glands occurred within just a few days following the female's detachment from the host. Kaufman (1976) demonstrated that the isolated salivary glands of engorged *Dermacentor andersoni* secreted fluid at only 25% the rate of salivary glands taken from ticks at the end of the slow phase of engorgement. I suggested that this was a physiological manifestation of the impending histological autolysis. In *A. hebraeum* also, there is a marked loss of secretory response with time post-repletion, to drugs administered *in vivo* (Harris & Kaufman 1981).

Most prominent among the ultrastructural changes visible during salivary gland autolysis, is the appearance of numerous autophagic vacuoles in the fluid secretory labyrinth of the type III acinus within the first day post-repletion (Harris & Kaufman 1981). Autophagic vacuoles do not usually appear in the salivary glands of small partially-fed ticks for at least 4 or 5 days following forcible removal from the host (Harris & Kaufman 1981, 1984).

When we transplanted salivary glands from small partially-fed ticks into the haemolymph of replete specimens, autophagic vacuoles appeared in the donor tissue within 48 h. Such was not the case in control experiments (salivary glands of small partially-fed ticks residing in the haemocoels of other small partially-fed ticks). We thus suggested that post-prandial salivary gland degeneration was controlled by a humoral agent, which we call 'tick salivary gland degeneration factor' (TSGDF).

Since 1981, my laboratory has been studying some properties of ths new hormonal system in *A. hebraeum.* I shall now discuss, in turn, the time course of salivary gland degeneration, the neural pathways mediating the release of TSGDF, the role of insemination in the degeneration process, and the chemical nature of the hormone. In all of the experiments reported below, we have exploited a modified *in vitro* assay to assess the degree of salivary gland degeneration (described fully in Harris & Kaufman 1984). In brief, we measure net weight increase in isolated salivary glands bathed in a supra-maximal concentration of dopamine. Dopamine is a potent agonist of salivary fluid secretion, and is the putative transmitter substance at the neuroglandular junction (Kaufman, 1983). We propose that the loss of fluid secretory competence under the conditions of this assay can be directly correlated with prior exposure to TSGDF.

2.3 TIME COURSE OF SALIVARY GLAND DEGENERATION IN NORMAL TICKS (from Harris & Kaufman 1984)

The following definitions apply in this chapter. Ticks which have attained a weight above 1.0 g and which have detached spontaneously are termed 'engorged' or 'replete'. Ticks which have fed to a weight of 0.18 to 0.29 g are termed 'small partially fed'. Ticks which have fed to between 0.40 and 1.0 g are termed 'large partially fed'.

Engorged ticks lose over 95% of fluid secretory competence within 3–4 days. As mentioned earlier, this loss of secretory function correlates with the proliferation of autophagic vacuoles in the type III acini. To our surprise, small partially fed ticks also lost a great deal (75%) of their fluid secretory competence during 4–5 days post-removal from the host. There was no further loss of function, however, over the succeeding 10 days. Moreover, loss of function in small ticks was not correlated with the presence of autophagic vacuoles in the fluid secretory labyrinth. Also, loss of function in small ticks was reversed by simply allowing the latter to reattach and resume feeding. Unfortunately, one cannot coax replete females to resume feeding. However, considering the histological disruption of the salivary glands, it is most unlikely that salivary gland function could likewise be reversed in replete ticks. At the end of this chapter, I shall return to consider the biochemical lesions which may be responsible for the reversible loss of secretory function in small partially fed ticks.

2.4 NEURAL INVOLVEMENT IN SALIVARY GLAND DEGENERATION (from Harris & Kaufman 1984)

Although it is ultimately the humoral factor, TSGDF, which promotes salivary gland degeneration, we have recently shown that it is possible to prevent autolysis in large partially fed ticks by interrupting certain neural pathways.† We tested salivary gland function *in vitro* 4 days following

† For reasons discussed in Harris & Kaufman (1984) we used large partially fed ticks rather than replete ticks for these experiments. The glands of large partially fed ticks degenerate to the same extent as replete ticks within 4 days post-removal.

bilateral ablation of various groups of nerves, and compared it to the function of glands taken from appropriate sham-operated controls. Salivary gland degeneration was inhibited only by cutting all of the opisthosomal nerves. Unfortunatley, we have not yet refined our surgical technique so as to reliably denervate single nerve trunks bilaterally.

We interpret our results as follows. When the tick feeds to a critical size (corresponding to about 0.3–0.4 g in *A. hebraeum*), stretch receptors in the abdominal dorso-ventral muscles are activated. Signals from these stretch receptors pass along certain opisthosomal nerves to the central nervous system. Where the information passes from here is not known. Ultimately, there results the release of TSGDF from its as yet unidentified site of production. Thus, cutting the opisthosomal nerves blocks salivary gland degeneration by interfering with afferent signals from stretch receptors.

2.5 ROLE OF COPULATION IN SALIVARY GLAND DEGENERATION (from Harris & Kaufman 1984)

In most ixodid ticks, copulation (which usually occurs on the host) must take place before the female will engorge to repletion. Hence, mating would influence salivary gland degeneration by virtue of its effect on feeding. We have recently discovered, however, that transfer of the spermatophore also exerts a much more direct influence on autolysis. Although only a small percentage of virgin *A. hebraeum* will feed above 0.40 g, such large virgins do not suffer salivary gland degeneration (Kaufman 1983). Moreover, mated females above the critical weight can be surgically deprived of the seminal receptacle. Such females also do not suffer salivary gland degeneration. If, on the other hand, the excised seminal receptacle is left in the haemocoel, autolysis proceeds normally. Finally, a number of large partially fed females were deprived of their seminal receptacles, and an extract of male genital tract or male salivary glands was injected into their haemocoel. The genital tract extract, but not the salivary gland extract, was able to restore salivary gland degeneration. All the above can be interpreted as follows:

> Two chemicals are required for salivary gland degeneration; (1) TSGDF is the hormone ultimatcly responsible for autolysis, and presumably acts directly on the salivary gland epithelium. (2) A factor produced in the male reproductive tract is transferred to the female's seminal receptacle during copulation, from there it gains access to the haemocoel, and somehow renders TSGDF effective. There appear to be at least two ways whereby the male factor could act. It could 'prime' the salivary glands to become responsive to TSGDF, for example, by stimulating the synthesis of TSGDF receptors in the salivary gland. Or, it might represent an essential intermediate in the pathway leading to the synthesis or release of TSGDF. For the moment I favour the latter hypothesis. Below, I shall present evidence that TSGDF may be an ecdysteroid. The fact that ecdysteroids are effective in triggering

salivary gland degeneration in virgin as well as mated females (Kaufman, in preparation) indicates that the male factor is not required for imparting TSGDF sensitivity to salivary glands. We still do not know the chemical nature of the male factor nor the precise role it plays in promoting the appearance of TSGDF.

2.6 ECDYSTEROIDS: PUTATIVE CANDIDATES FOR TSGDF (from Harris and Kaufman 1985)

Circumstantial evidence led us to test ecdysteroids as putative candidates for TSGDF. The fact that resorption of endocuticle occurs during egg production suggests that ecdysteroid titres may be elevated in engorged ticks. The latter has in fact been demonstrated in *A. hebraeum* by Dr. P. A. Diehl, Université de Neuchâtel, Switzerland (personal communication) using a radioimmunoassay for ecdysteroids. Haemolymph ecdysteroid concentration rises from about 10 ng/ml in freshly gorged specimens to over 300 ng/ml in females about to oviposit. Finally, ecdysteroids are known to mediate a number of autolytic processes in insects (e.g. Lockshin & Williams 1964, Lockshin 1969, van PeltVerkuil 1979).

Recently, we have demonstrated that ecdysone and ecdysterone are able to cause salivary gland degeneration in small partially fed ticks. Single or double injections of ecdysterone are not very effective. However, if one infuses the hormone constantly over 24 h, relatively small amounts (about 300 ng/g body weight) can induce the loss of secretory competence as measured 3 days later.

Salivary glands set up in organ culture (the 'backless tick explant technique' of Bell 1980) also suffer salivary gland degeneration following 4 days exposure to as little as 15 ng ecdysone plus 15 ng ecdysterone/ml culture mdium. When tested at 100 ng total ecdysteroid/ml, ecdysone alone, ecdysterone alone, or a mixture of the 2, were equieffective. Note that these concentrations are well within the physiological range of haemolymph titres mentioned above.

Ecdysteroids are believed to play an important (but as yet undefined) role in vitellogenesis in ticks (Diehl *et al.* 1982). It thus appears that a putative vitellogenic hormone also regulates the autolytic processes occurring prior to and during egg production.

2.7 REVERSIBLE LOSS OF SECRETORY FUNCTION IN SMALL PARTIALLY FED TICKS

Before concluding, I wish to briefly consider further the reversible loss of secretory competence in small partially fed females following forcible removal from the host.

It has been well established by Sauer and his colleagues (Needham & Sauer 1979) that catecholamine-stimulated fluid secretion is mediated by the intracellular second messenger, cyclic 3′:5′-adenosine monophosphate (cyclic AMP; cAMP). Schmidt *et al.* (1981, 1982) have shown that the pharmacological properties of the adenylate cyclase system in salivary gland homogenates of *A. americanum* accord well with the response of isolated salivary glands to various agonist and antagonist drugs.

Hormone-stimulated adenylate cyclases are assembled from several multi-subunit proteins (Ross & Gilman 1980, Seamon & Daly 1981): (a) the cell surface receptor for the hormone or neurotransmitter, is linked to (b) a guanine nucleotide binding protein (G/F protein) which in turn is loosely associated with (c) the catalytic site — the component of the system which actually effects the conversion of ATP to cAMP. The hormone binds to the cell-surface receptor; NaF and cholera toxin (general stimulants of most adenylate cyclase systems) interact with the G/F protein, and the drug forskolin stimulates the catalytic site directly (Fig. 2.1).

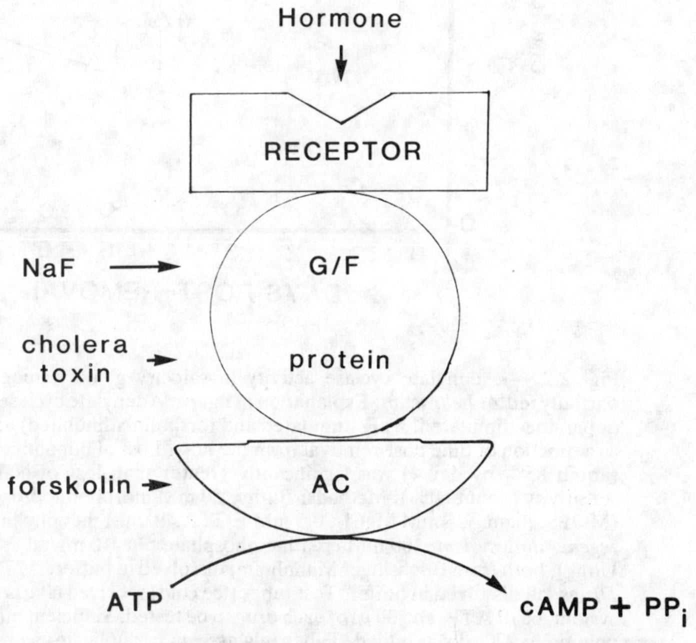

Fig. 2.1 — A schematic model showing the components of hormone-sensitive adenylate cyclases. A number of agents which stimulate adenylate cyclase activity in most cells are indicated, as are their sites of action. (AC) adenylate cyclase.

In Fig. 2.2, I present basal, dopamine-stimulated, NaF-stimulated, and forskolin-stimulated adenylate cyclase activities in salivary glands from small partially fed *A. hebraeum* removed from the host for up to 7 days. Over this time, dopamine-sensitive activity declined by 85%, whereas NaF− and forskolin-sensitive declined by 60%. The loss of forskolin sensitivity suggests that there might be a lesion in the catalytic subunit of the system. However, the larger decline in dopamine-sensitive activity suggests,

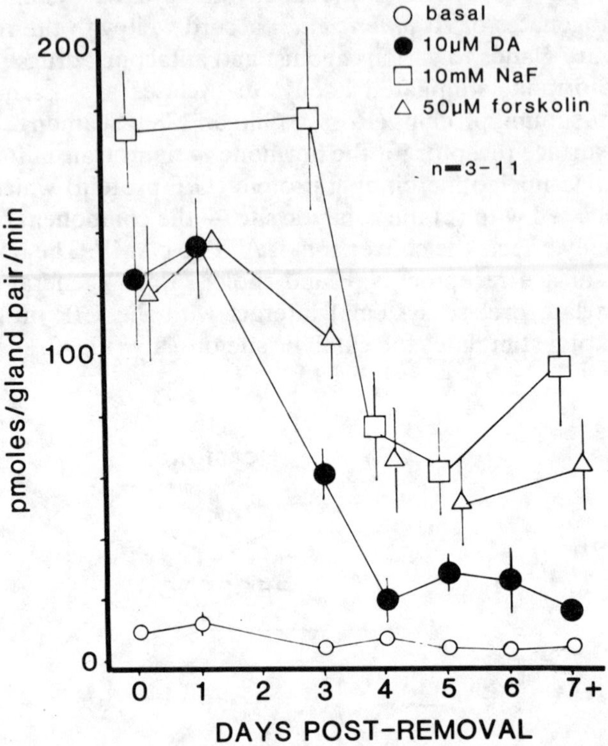

Fig. 2.2 — Adenylate cyclase activity in salivary gland homogenates of small, partially fed *A. hebraeum*. **Explanation of figure:** Adenylate cyclase activities (basal, dopamine-stimulated, NaF-stimulated and forskolin-stimulated) are plotted (±SE) as a function of time post-removal from the host. Loss of dopamine (DA) sensitivity (about 85% by day 4) was significantly greater than loss of NaF− or forskolin-sensitivity (about 60%). **Methods:** *Buffer:* 50 mM morpholinopropane sulfonic acid (MOPS; Sigma), 5 mM $MgCl_2$, 0.3 mM EGTA, 10 mM theophylline, pH 7.2. *ATP-regenerating system:* 100 mM creatine phosphate plus 10 mg/ml creatine kinase (25 U/mg), both from Boehringer Mannheim, dissolved in buffer. *ATP:* 1 mM in buffer. *Drugs:* all dissolved in buffer. Test tubes (ice cold) received 60 μl ATP regenerating system, 60 μl ATP, and 30 μl of each drug to be tested. Sufficient buffer to bring total volume to 270 μl was added. Tubes (always run in duplicate) were preincubated at 30°C for 2–3 minutes prior to beginning reaction with 30 μl enzyme. The reaction was stopped after 3 minutes by placing tubes in a boiling water bath for 2 minutes. Tubes were centrifuged for 5–10 minutes to remove precipitate, and the supernatant stored frozen until cAMP assay was performed. The cAMP assay was essentially that of Brown *et al.* (1971), except that a Millipore Filtration system was used to separate bound from free radio-labelled cAMP (New England Nuclear, NET-275, adenosine, 3′,5′-cyclic phosphate, ammonium salt, [2, $8-^3H$-**, 30–50 Ci/mmole). Radioactivity was measured on a Beckman LS9000 Liquid Scintillation Counter. Standard curves were prepared using authentic cAMP (Sigma). Specificity of this assay for cAMP was confirmed by the testing of several authentic nucelotides (cGMP; 2′:3′-cAMP; 5′-CMP; 5′-AMP; ADP; ATP; CTP; GTP; ITP). Although the binding protein cross-reacted with cGMP to a small degree (sensitivity was about 10^{-2} that of cAMP), there was essentially no recognition of the other nucleotides at 5000 times the concentration of cold cAMP which caused a 50% displacement of radiolabelled cAMP from the binding protein.

in addition to the latter, the uncoupling of the dopamine receptor from the system, or perhaps a reduction in the number of dopamine-receptors itself (Creese *et al.* 1981). This phenomenon clearly justifies more attention in the future.

ACKNOWLEDGEMENTS

Most of the work from my laboratory was supported by the Natural Sciences and Engineering Research Council of Canada (NSERC) and by the Canadian National Sportsmen's Fund. Attendance at the Congress was made possible by a generous travel grant from the Alberta Heritage Foundation for Medical Research.

REFERENCES

Bell, L. J. (1980). Organ culture of *Rhipicephalus appendiculatus* with maturation of *Theileria parva* in tick salivary glands *in vitro*. *Acta Tropica* **37**: 319–325.

Brown, B. L., Albano, J. D. M., Ekins, R. P., & Sgherzi, A. M. (1971). A simple and sensitive saturation assay method for the measurement of adenosine 3′:5′-cyclic monosphosphate. *Biochem. J.* **121**: 561–562.

Creese, I., Sibley, D. R., Leff, S., & Hamblin, M. (1981). Dopamine receptors: subtypes, localization and regulation. *Federation Proceedings* **40**: 147–152.

Diehl, P. A., Aeschlimann, A., & Obenchain, F. D. (1982). Tick reproduction: Oogenesis and oviposition. In: *Physiology of ticks* (F. D. Obenchain, & R. Galun, eds.) Pergamon Press, pp. 277–350.

Harris, R. A. & Kaufman, W. R. (1981). Hormonal control of salivary gland degeneration in the ixodid tick *Amblyomma hebraeum*. *J. Insect Physiol.* **27**: 241–248.

Harris, R. A. & Kaufman, W. R. (1984). Neural involvement in the control of salivary gland degeneration in the ixodid tick, *Amblyomma hebraeum* Koch. *J. exp. Biol.* **109**: 281–290.

Harris, R. A. & Kaufman, W. R. (1985). Ecdysteroids: possible candidates for the hormone which triggers salivary gland degeneration in the ixodid tick *Amblyomma hebraeum*. *Experienta* **41:** 740–742

Kaufman, W. (1976). The influence of various factors on fluid secretion by *in vitro* salivary glands of ixodid ticks. *J. Exp. Biol.* **64**: 727–742.

Kaufman, W. R. (1983). The function of tick salivary glands. In: *Current Topics in Vector Research,* Vol. 1 (K. F. Harris, ed.) Praeger Scientific, New York, pp. 215–247.

Kaufman, W. R. & Sauer, J. R. (1982). Ion and water balance in feeding ticks: Mechanisms of tick excretion. In: *Physiology of Ticks* (F. D. Obenchain & R. Galun, eds.) Pergamon Press, pp. 213–244.

Lees, A. D. (1952). The role of cuticle growth in the feeding process of ticks. *Proc. Zool. Soc. London* **121**: 759–772.

Lockshin, R. A. (1969). Lysosomes in Insects. In: *Lysosomes in Biology* and Pathology Vol. 1 (J. T. Dingle & H. B. Fell, eds.) North-Holland (Amsterdam and London), pp. 363–391.

Lockshin, R. A. & Williams, C. M. (1964). Programmed cell death — II. Endocrine potentiation of the breakdown of the intersegmental muscles of silkworms. *J. Insect Physiol.* **10**: 643–649.

Needham, G. & Sauer, J. R. (1979). Dynamics of calcium, catecholamines and cyclic AMP in control of salivary fluid secretion by female ixodid ticks, In: *Recent Advances in Acarology* **1**: 413–419. (J. G. Rodriguez, Ed.), Academic Press, New York, San Francisco, London.

Oliver, J. H., Jr. (1982). Tick Reproduction: Sperm development and cytogenetics. In: *Physiology of ticks* (F. D. Obenchain & R. Galun, eds.) Pergamon Press, pp.245–275.

Ross, E. M. & Gilman, A. G. (1980). Biochemical properties of hormone-sensitive adenylate cyclase. *Ann. Rev. Biochem.* **49**: 533–564.

Sauer, J. R., (1977) Acarine salivary glands-physiological relationships. *J. Med. Ent.* **14**: 1–9.

Schmidt, S. P., Essenberg, R. C., & Sauer, J. R. (1981). Evidence for a D_1 dopamine receptor in the salivary glands of Amblyomma americanum (L.). *J. Cyclic Nucleotide Research* **7**: 375–384.

Schmidt, S. P., Essenberg, R. C., & Sauer, J. R. (1982). A dopamine sensitive adenylate cyclase in the salivary glands of *Amblyomma americanum* (L.). *Comp. Biochem. Physiol.* **72C**: 9–14.

Seamon, K. B. & Daly, J. W. (1981). Forskolin: a unique diterpene activator of cyclic AMP-generating systems. *J. Cyclic Nucleotide Res.* **7**: 201–224.

Sonenshine, D. E., Silverstein, R. M., & Rechav, Y. (1982). Tick pheromone mechanisms. In: *Physiology of ticks* (F. D. Obenchain & R. Galun, eds.) Pergamon Press, pp. 439–468.

Till, W. M. (1961). A contribution to the anatomy and histology of the brown ear tick, *Rhipicephalus appendiculatus* (Neumann). *Mem. Ent. Soc. Sth. Afr.* **6,** April 1961. pp. 124.

van Pelt-Verkuil, E. (1979). Hormone mediated induction of acid phosphatase activity in the fat body of *Calliphora erythrocephala* prior to metamorphosis. *J. Insect Physiol.* **25**: 965–973.

Wikel, S. K., (1983) Host immunity to tick bite. In: *Current Topics in Vector Research,* Vol. 1 (K. F. Harris, ed.) Praeger Scientific, New York, pp. 249–269.

Wikel, S. K. & Allen, J. R. (1982). Immunological basis of host resistance to ticks. In: *Physiology of ticks* (F. D. Obenchain & R. Galun, eds.) Pergamon Press, pp. 169–196.

3

Molecular basis for salivary fluid secretion in ixodid ticks

John R. Sauer, Suresh D. Mane, S. P. Schmidt and R. C. Essenberg

3.1 INTRODUCTION

The salivary glands of ixodid ticks perform various functions in both the parasitic and non-parasitic phases of the tick's life cycle. The glands of most feeding ixodid ticks secrete a cement-like substance to insure firm attachment of the mouthparts to the host (Moorhouse 1969) and the feeding lesion associated with tick feeding is caused by host reaction to materials in tick saliva (see Brown & Askenase, Chapter 13). The salivary glands are believed to secrete hygroscopic material that is important in enabling the unfed tick to absorb water from unsaturated air (Rudolph & Knülle 1974, 1978, McMullen *et al.*, 1976). Needham and Teel review this topic in Chapter 5.

The salivary glands also serve as important organs of fluid excretion during tick feeding. As the bloodmeal is ingested, excess ions and water are moved across the gut epithelium into the hemocoel and secreted back to the host via the salivary glands (Tatchell 1967, 1969, Sauer 1977, Guenther *et al.* 1980). An active transport of chloride and sodium may be important driving forces for the fluid secretory process. The glands secrete fluid *in vitro* after a low concentration of dopamine is added to a medium bathing the gland (Kaufman 1976, Sauer *et al.* 1979) and *in vivo* following injection of dopamine into the hemocoel of partially fed females (Hsu & Sauer 1975). These results, together with morphological information (Megaw 1977) suggest that the glands are controlled by nerves. Overall nervous control of tick salivary glands has been reviewed by Kaufman (1979).

Other evidence indicates that dopamine is a neurotransmitter at the neuroeffector junction (Kaufman 1976, 1977, Megaw & Robertson 1974,

Binnington & Stone 1977). During stimulation of cells, an adenylate cyclase is activated causing an increase in cellular cyclic AMP (cAMP) which functions as an intracellular 'second messenger' mediating fluid secretion (Schmidt *et al.* 1981, 1982, Hume *et al.* 1984). Deletion of calcium or addition of the calcium antagonist verapamil to the bathing medium during experiments with isolated glands inhibits dopamine- and cAMP-stimulated fluid secretion indicating that Ca^{2+} too is an important factor mediating fluid secretion (Needham & Sauer 1979a,b).

Our purpose here is to examine these events in more detail and, where possible, to draw general conclusions about biochemical and anatomical changes to draw attention to gaps in our understanding of factors that mediate fluid secretion in feeding female ixodid ticks.

3.2 CHANGES IN SALIVARY GLANDS DURING TICK FEEDING

3.2.1 Anatomical

The paired salivary glands of ixodid ticks consist of clusters of alveoli running about two-thirds the length of the tick (Balashov 1972). The glands of females consist of three kinds of alveoli (acini) and multiple types of cells. Salivary gland structure is treated in detail by Fawcett *et al.* (Chapter 1), and only a brief description of the major changes observed in the salivary glands of female *Amblyomma americanum* (L.) during feeding will be presented here. As in other species, Type I alveoli are without valves, consist of 13–14 cells and are confined to the anterior region of the glands. These alveoli have peripheral 'pyramidal' cells with tortuous plasma membrane infoldings and closely associated mitochondria. A relatively large, 'central' cell occupies most of the alveolar midsection. Type II alveoli extend through the anterior half of the gland, open into the main duct or its principal branches, and contain several types of granule-secreting cells. Type III alveoli also contain granule-secreting cells and are found primarily in the posterior half of the glands and in the distal regions of the anterior branches. Type II alveoli have two complex granular cells (granules with subunits) whereas Type III have only one. Thin ablumenal interstitial cells separate adjacent granular cells in both alveolar Types II and III and only one adlumenal interstitial cell, with myoepithelia-like features lines the lumen of both alveolar types. In comparison to Types II and III, relatively few changes occur in Type I alveoli during tick feeding. Significant ultrastructural changes were observed in the central and pyramidal cells during feeding, but not in other cell types (Barker *et al.* 1984).

Cytological changes were already apparent in granular cells in Types II and III alveoli in salivary glands within 12 h post-infestation of the tick on the host (Claypool *et al.* 1986). Autophagic activity produced larger subunits within the complex granular cells. Granules in simple granular cells were in different stages of lysis ranging from homogeneously stained granules to granules that stained light and dark with uranyl acetate.

Significant cytological changes were observed in all Types II and III alveolar cells by 48 h of tick feeding. An extensive network of rough endoplasmic reticulum was present in these cells, some of which appeared swollen, suggesting increased protein synthesis by the cell. Considerable membrane interdigitation and proliferation between the ablumenal intersitial cell (ABI) and the lightly stained, simple granular cells was seen at this stage of tick feeding.

Types II and III alveoli from ticks weighing 100–500 mg were considerably larger, with greatly enlarged lumens surrounded by abundant microvilli. An extensive labyrinth of interdigitating membranes derived from the depleted lightly staining granular and ABI cells was seen in the basal regions of the alveolus, and granular material formerly seen in the granular cells was mostly absent. Numerous mitochondria were observed closely associated with the extensive membranous network, suggesting a transport function.

3.2.2 Molecular basis for salivary fluid secretion during tick feeding

3.2.2.1 *Adenylate cylase and effects of drugs on fluid secretion*

A dopamine-sensitive adenylate cyclase is present in tick salivary glands (Schmidt *et al.* 1982) with a K_m for ATP slightly lower than that reported in vertebrates (Perkins 1973; Beaumont *et al.* 1979) and insects (Clement-Cormier *et al.* 1975, Filburn & Wyatt 1976). The concentration of dopamine required for half-maximal stimulation of the enzyme (4.2×10^{-7} M) was shown to be the same as the concentration required to stimulate the half-maximal rate of fluid secretion by isolated salivary glands of *A. americanum*, supporting the contention that this enzyme is involved in fluid secretion.

Adenylate cyclase from the glands of *A. americanum* can be stimulated by several other derivatives of phenylethylamine including noradrenaline, adrenaline, and isoproterenol (Fig. 3.1). These same drugs potentiate fluid secretion by isolated glands in the same order of effectiveness. Octopamine and L-DOPA had no effect on basal adenylate cyclase activity. Dopamine had the highest potency and lowest K_a, 0.4 μM; followed by adrenaline and noradrenaline, 25 μM; and isoproterenol 0.15 mM. On the basis of effects of combinations of catecholamines, and effects of known α- and β- adrenergic and dopamine receptor antagonists on dopamine-stimulated activity, Schmidt *et al.* (1981) suggested that the salivary gland adenylate cyclase of *A. americanum* has a D_1 type dopamine receptor. A feature common to this type of receptor is linkage to an adenylate cyclase second messenger system (Kebabian & Calne 1979, Seeman, 1981).

Because salivary glands from unfed females exhibit little ability to secrete fluid (Sauer *et al.* 1979), adenylate cyclase activity could not be measured in salivary glands of unfed *A. americanum* females by an assay sensitive to 4 pmoles cAMP formed/min/mg protein (Schramke *et al.* 1984). However, activity increased in glands after tick attachment and continued to increase as the tick progressed through feeding, attaining a maximum activity at a tick weight of approx. 200 mg. Activity declined slightly thereafter as the tick progressed through the later, more rapid phase of feeding.

Of further significance is the fact that the average adenylate cyclase activity in the glands of ticks reared on hosts which had been exposed

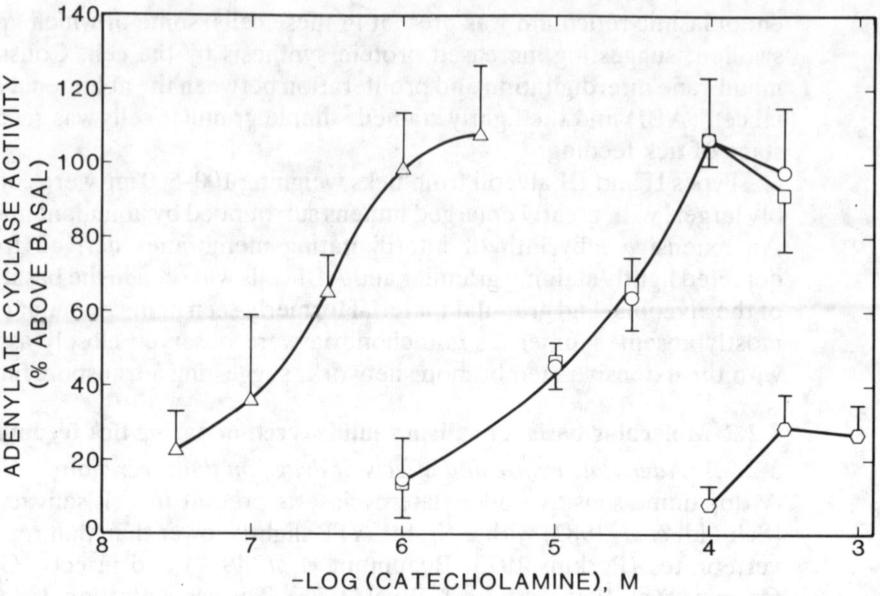

Fig. 3.1—The activation of dopamine-sensitive adenylate cyclase by catecholamines in homogenates of tick salivary glands. The agonists used were dopamine (△), adrenaline (□), noradrenaline (○) and isoproterenol (⬡). The basal adenylate cyclase activity was 37 ± 5.2 pmole cAMP formed/mg protein/min. From Schmidt *et al.* 1981 with permission.

previously to tick infestation in the laboratory was reduced as was the engorgement weight of ticks which had fed on the same hosts (Schramke *et al.* 1984). It is well known that previous host exposure to ticks affects the feeding and final engorgement weights of ticks. McGowen *et al.* (1981) have shown that on cattle immunized with extracts of *A. americanum* (L.) there is a significant reduction in the adult tick engorgement weight.

The adenylate cyclase complex in all cells studied to date consists of receptor, guanine nucleotide dependent and catalytic subunits (Ross & Gilman 1980). Various endogeneous factors such as GTP modulate activity of adenylate cyclase in many different systems (Beaumont *et al.* 1979, MacNeil *et al.* 1980). Basal adenylate cyclase activity of a washed particulate fraction of tick salivary glands can be stimulated by GTP and the non-hydrolyzable GTP analogue, Gpp(NH)p (Fig. 3.2)

Drug effects on fluid secretion by isolated glands are much more complicated than drug effects on adenylate cyclase in broken cells (Kaufman 1977, Wong & Kaufman 1981). Dopamine (DA) receptors in other animals consist of two sub-types, D_1 and D_2 (Kebabian & Calne 1979). Kaufman's group found that the stimulatory effect of 1 µM DA on tick salivary gland fluid secretion was not blocked by 1 mM chloropromazine or (+)-flupenthixol, both inhibitors of D_1 receptors. These results contrast with those where chlorpromazine and other D_1 receptor inhibitors block dopamine-stimulated adenylate cyclase in tick salivary glands (Schmidt *et al.* 1981).

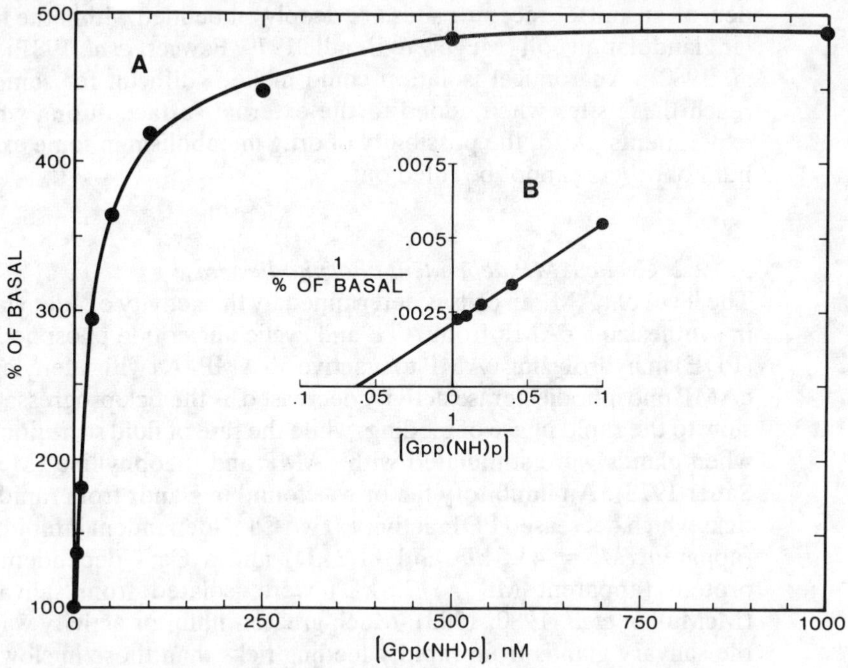

Fig. 3.2 — A. Effects of various concentrations of Gpp (NH)p on adenylate cyclase activity in washed particulates of tick salivary glands. B. Lineweaver–Burke plot of stimulation. Basal adenylate cyclase activity was 57.6 pmoles cAMP/mg protein/min.

Specific DA inhibitor (+)-butaclamol effectively inhibits the ability of DA to stimulate both fluid secretion and adenylate cyclase. Butyrophenone-type drugs such as spiperone and pimozide, both D_2 blockers, did not block but potentiated the fluid stimulatory action of DA at very low (femtomolar) concentrations. These drugs, by themselves, were ineffective stimulants of fluid secretion. Sulpiride blocked the potentiation by spiperone without blocking the effect of DA. The butyrphenone drug, haloperidol, had no effect on DA-stimulated adenylate cyclase (Schmidt *et al.* 1981).

The ergot alkaloids, ergotamine and ergonovine, were potent agonists of fluid secretion. However, the effect of ergonovine was not inhibited by (+)-butaclamol but was inhibited by sulpiride without antagonizing the action of DA. The results suggested a stimulatory pathway independent of cAMP. However, the effects of ergot alkaloids on salivary gland cyclase have not been investigated. Kaufman (1983) has hypothesized at least three separate receptors in the glands: 1) a butaclamol-sensitive receptor, 2) a sulpiride-sensitive ergonovine receptor, and 3) a sulpiride-sensitive butyrophenone receptor.

It is difficult to assess the significance of these results, but a physiological event such as fluid secretion probably consists of many events any of which could conceivably be affected by drugs. Also, synaptic regions and potential neurotransmitter receptor sites are deeply embedded within the multicellular glandular alveoli (Megaw & Beadle 1979, Fawcett *et al.* 1981a, Krolak *et al.* 1982). Anatomical isolation could make it difficult for some drugs to reach these sites when added to the external surface during whole-gland experiments. Also, the possibility of drug metabolism in some experiments but not others cannot be ruled out.

3.2.2.2 *Cyclic AMP dependent phosphodiesterase*

The level of cAMP in cells is determined by the activity of adenylate cyclase in synthesizing cAMP from ATP and cyclic nucleotide phosphodiesterases (PDE) in hydrolyzing cAMP to inactive 5′-AMP. As with adenylate cyclase, cAMP phosphodiesterase activity decreased as the tick progressed from the slow to the rapid phase of feeding, while the rate of fluid secretion increased when glands were stimulated with cAMP and theophylline (McMullen & Sauer 1978). An inhibitory factor was found in glands from rapidly feeding ticks which decreased PDE activity. Two Ca^{2+}-dependent inhibitor proteins (apparent M_r = 43.5 kD and 11.2 kD) and a Ca^{2+}-dependent activator protein (apparent Mr = 17.8 kD) were isolated from salivary glands (McMullen *et al.* 1980, 1983). Much greater inhibitor activity was found in the salivary glands from rapidly feeding ticks than those in slowly feeding ticks. The inhibitor proteins prevent the activation of a cytosolic high K_m cAMP-PDE by the Ca^{2+}-dependent activator. Although not all the criteria have been met, it is possible that the activator protein is calmodulin. Activity of PDE in the presence of an activator reached a maximum at a Ca^{2+} concentration of about 10^{-7}. With increasing Ca^{2+}, PDE activity in the presence of either inhibitor dropped sharply, but activity remained high if the inhibitors were absent. These results demonstrated that a higher level of Ca^{2+} is necessary for inhibition of PDE than for its activation, but the sensitivities of all three protein modulators are within the ranges of Ca^{2+} believed to be present in cell cytoplasm (Rasmussen & Goodman 1977).

A low K_m cAMP-PDE bound to membranes was also found in tick salivary glands (McMullen *et al.* 1983). Its activity was not affected by modulator proteins, and its activity was slightly higher in glands of rapidly feeding ticks. Prestimulation of glands with 1 μM dopamine increased the V_{max} of the membrane-bound cAMP-PDE but had no effect on the kinetic properties of the cytosolic PDE.

Thus, it appears that two different controls may act on cAMP-PDE in the tick salivary glands. The cytosolic (soluble) activity becomes inhibited as a result of an increase in the level of inhibitory proteins (McMullen *et al.* 1980). These proteins, along with the activator (possibly calmodulin), also make PDE activity sensitive to changes in free Ca^{2+} levels and thus may provide another level of control (McMullen *et al.* 1980). The particulate activity is affected by dopamine or other effectors of membrane receptors (Schmidt *et al.* 1981, 1982). The rise in particulate activity may result from increased dopamine stimulation on entering the rapid phase of feeding. Thus, cAMP and possibly fluid secretion may be regulated in two different ways via the PDE activities (McMullen *et al.* 1983). As with adenylate cyclase, it is not known how changes in the nature and/or magnitude of the bloodmeal affect levels or activities of factors affecting PDE activity.

3.2.2.3 *Cyclic AMP and calcium*

Changing activities of adenylate cyclase and cAMP-PDE during tick feeding correlate with an increase in the rate of salivary fluid secretion as the tick progresses from the slow to rapid stage of feeding (Sauer *et al.* 1979). Thus, it is significant that levels of cAMP in the salivary glands of feeding ticks were affected differently by an agonist of adenylate cyclase (dopamine) and an antagonist (theophylline) of cAMP-PDE in glands of slowly and rapidly feeding ticks (Hume *et al.* 1984). Higher levels of cAMP were produced in salivary glands of rapidly feeding ticks than in salivary glands of slowly feeding ticks. After stimulation by dopamine was removed, cAMP decreased faster in glands from slowly feeding ticks than in glands of rapidly feeding ticks. Cyclic AMP in the salivary glands of unfed females did not increase when glands were incubated with 0.1 μM dopamine and 10 mM theophylline, consistent with results showing an absence of dopamine-sensitive adenylate cyclase activity in the salivary glands of unfed females.

Calcium and cAMP are important intracellular components regulating the activities of many cells (Rasmussen 1981). The importance of Ca^{2+} concentration in affecting modulators of salivary gland cytosolic PDE has already been described. It is likely that Ca^{2+} is important in other ways during secretion. For example, when salivary glands were stimulated to secrete fluid (*in vitro*) while immersed in a solution with little or no Ca^{2+}, the rate of secretion was depressed (Needham & Sauer 1979a,b). Interestingly, glands stimulated by catecholamine showed temporary independence of external Ca^{2+} initially, as glands secreted at near normal rates during the first few minutes of the experiment (Fig. 3.3). These results suggested utilization of intracellular stores of Ca^{2+} which were then lost via the secreted saliva or across the plasma membrane into the support medium when Ca^{2+} was missing from it. Results of $^{45}Ca^{2+}$ flux studies and reductions in total gland Ca^{2+} during stimulation support this hypothesis (Table 3.1).

Removal of Ca^{2+} from the support medium prevents cAMP-stimulated secretion (Needham & Sauer 1979a), demonstrating the importance of both in controlling fluid secretion. Thus, neither Ca^{2+} nor cAMP alone is capable of initiating fluid secretion. Only the primary stimulus (dopamine) appears to create the necessary conditions for secretion by inducing formation of cAMP through the activation of an adenylate cyclase and the mobilizing of Ca^{2+} (Needham & Sauer 1979a).

Many of the agonists which induce Ca^{2+} changes in cells stimulate a specific hydrolysis of plasma membrane phosphatidylinositol (PI) (Fain & Berridge 1979a,b). Plasma membrane PI is converted to diacylglycerol (DG) and inositol phosphate. Diacylglycerol can activate protein kinase C (Berridge 1984) or can be phosphorylated to phosphatidic acid (PA). Phosphatidic acid can recombine with inositol to replenish the pool of PI. Diacyglycerol may also serve as a precursor for the release of arachidonic

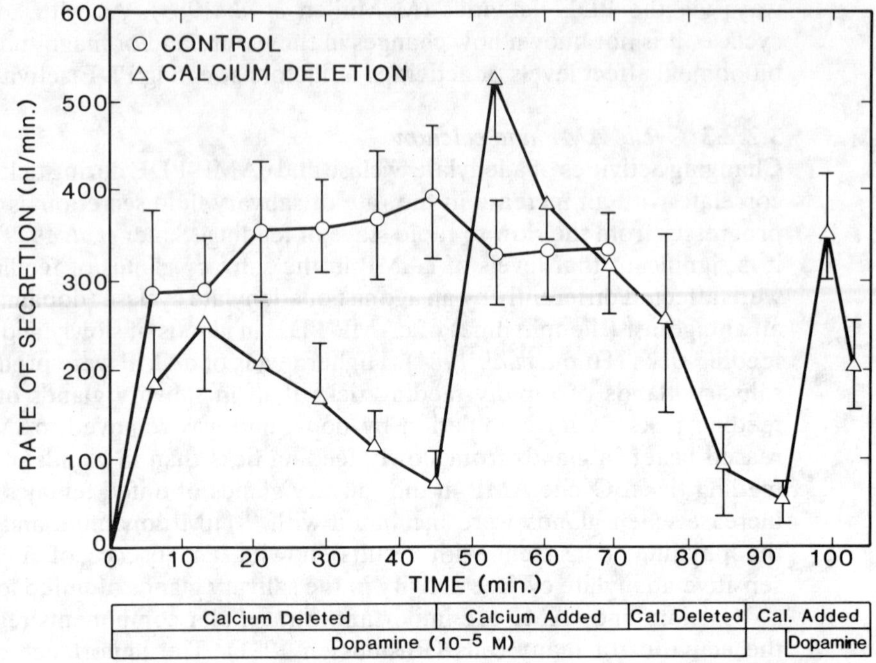

Fig. 3.3 — Effect of removing and adding Ca^{2+} on the rate of fluid secretion by isolated salivary glands of an ixodid tick stimulated by dopamine (△, N=5). Rates of secretion by control glands where Ca^{2+} was present in the support medium throughout are included for purposes of comparison (O,N=7). Verticle lines represent ± S.E. From Needham & Sauer (1979a) with permission.

Table 3.1 — The effect of dopamine on $^{45}Ca^{2+}$ efflux by isolated salivary glands.

5-min incubation			10-min incubation		
C*	D†	Δ%‡	C*	D†	Δ%‡
275	314	+14	242	402	+66
240	387	+61	214	294	+37
181	270	+49	175	230	+31
243	258	+6	190	224	+18
68	109	+60	184	221	+21
$\ddot{x}$		+38±12 (SE)			+34±9 (SE)

*C = CPM/μl bathing medium; control non-stimulated gland.
†D = CPM/μl bathing medium; dopamine (10^{-5}M)-stimulated gland.
‡Δ% = [(D)/(C) × 100] − 100.

From Needham & Sauer (1979a) with permission.

acid that can be converted to prostaglandins (Berridge 1984). Michell (1975) first suggested that PI hydrolysis opened cell surface Ca^{2+} channels, but this hypothesis is now considered untenable (Hawthorne 1982). More recently, Salmon & Honeyman (1980) suggested that the PA formed during the PI hydrolysis functions as an ionophore to gate the entry of Ca^{2+} or effect the release of Ca^{2+} from intracellur stores. Streb *et al.* (1983) and Prentki *et al.* (1984) found that release of Ca^{2+} from a non-mitochondrial intracellular store was stimulated by inositol 1,4,5-triphosphate, a water soluble product of polyphosphoinositide hydrolysis.

Phosphatidylinositol turnover is probably initiated by conversion to polyphosphoinositides (Berridge 1981, 1984). Like PI, polyphosphoinositides can be hydrolyzed to DG and the corresponding inositol phosphates as discussed earlier, but at an even faster rate than PI (Kirk *et al.* 1981). These compounds may also be of great importance because they bind Ca^{2+} strongly (Berridge 1982).

Sarkadi *et al.* (1983) and Enyedi *et al.* (1983) demonstrated that plasma membrane preparations from lymphocytes, platelets, and red blood cells are phosphorylated in the presence of [γ-^{32}P]ATP. Interestingly, the dissociated catalytic subunit of cAMP-dependent protein kinase increased the ^{32}P labelling of proteins and polyphosphoinositides. They suggested that a membrane phosphoprotein causes the phosphorylation of the membrane lipids in these preparations. Demonstration of the ability of cAMP-dependent protein kinase to modulate plasma membrane phospholipid metabolism in other tissues, and possibly to affect the mobilization of Ca^{2+}, would provide strong support for the widely held view that Ca^{2+} and cAMP messenger systems are intimately interrelated (Cohen 1982, Rasmussen 1981). Unfortunately, nothing is yet known about the metabolism of membrane lipids in tick salivary glands during secretion or the possible significance of these events to the mobilization of Ca^{2+} or other phenomena in these cells.

3.2.2.4 *Cyclic AMP-dependent protein kinases and phosphorylation of endogenous proteins*

The only known mode of action of cAMP in eukaryotes is to activate protein kinases that catalyze the phosphorylation of endogenous proteins (Walsh *et al.* 1968, Cohen 1982). Phosphorylated proteins then effect, in some way, the biological response of cells to the hormone or neurotransmitter. Cyclic AMP-dependent protein kinases of most animals are of two types, I and II. Each of the two enzymes is a tetramer consisting of a dimer of regulatory (R) and two catalytic (C) subunits. The enzymes differ in the size of the cAMP binding regulatory subunit and the ability of Type II enzyme to phosphorylate its own regulatory subunit (Erlichman *et al.* 1974). The Type II enzyme expressed in certain tissues can be further resolved by immunological and structural differences in their regulatory subunits into neural specific and non-neural protein kinases in a variety of species (Erlichman *et al.* 1983, Stein *et al.* 1984).

Tick salivary gland cAMP-dependent protein kinases are similar in that they appear to consist of at least two isozymes with receptor cAMP binding proteins and catalytic subunits (Mane *et al.* 1985). This enzyme, like that in other tissues, is Mg^{2+}-dependent and is present in glands at all stages of tick feeding. The receptor subunits have molecular weights near those of the Type I and II subunits in other animals (McSwain *et al.* 1985a). Differences include only minor phosphorylation of the high molecular weight subunit as compared to the low molecular weight subunit. The low molecular weight subunit may consist of two subtypes (Mane *et al* 1985, McSwain *et al.* 1985a). It is not known if these proteins are autophosphorylated or phosphorylated by another kinase(s). Autophosphorylation is possible because phosphorylation is increased after stimulating whole glands with dopamine and broken cells with cAMP. Other unusual characteristics include a 10-fold higher concentration of cAMP required to stimulate the enzyme maximally (Fig. 3.4).

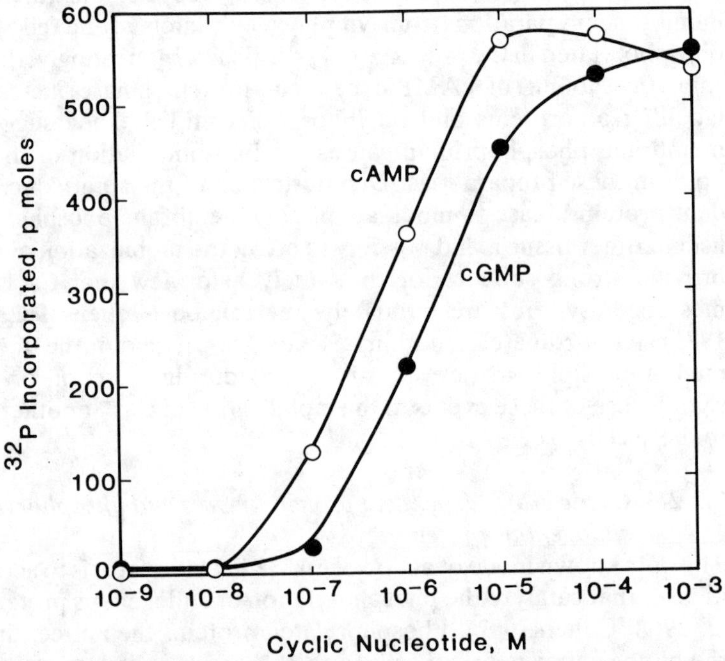

Fig. 3.4 — Effect of cyclic nucleotide concentration on phosphorylation of histone H-1. Varying concentrations of cAMP (○) and cGMP (●) were incubated with 3.2 μg of salivary gland crude homogenate protein (34 000 *x*g supernatant) for 10 min at 37°C.

We have initiated studies to identify endogeneous proteins phosphorylated by these protein kinases during fluid secretion. When isolated glands were incubated with dopamine and cAMP, phosphate incorporation was increased into at least 12 endogenous proteins (McSwain *et al.* 1985a). Eleven were shown to lose phosphate after stimulation with dopamine was terminated with dopaminergic antagonists, thioridazine, and d-butaclamol. Of the 12 proteins which incorporated additional phosphate, those with molecular weights of 62 kD, 47 kD and 45 kD (as determined on 7.5% SDS polyacrylamide gel electrophoresis) were the most prominent. Cyclic AMP stimulated the phosphorylation of at least 14 proteins in broken cell preparations of tick salivary glands, 9 of which were phosphorylated in response to dopamine and cAMP/theophylline in whole-gland experiments. The most prominent phosphoproteins, as in whole-gland experiments, had molecular weights of 62 kD, 47 kD, and 45 kD.

We have investigated which of these phosphoproteins are found in the plasma membranes of tick salivary glands by electrophoresis and autoradiography (McSwain *et al.* 1985b). Three methods were used. In method one, whole salivary glands were preincubated for 1 h in oxygenated TC-199 containing $^{32}P_i$ to label endogenous ATP. Following incubation, glands were stimulated for 5 min with 10^{-5} M dopamine. Following stimulation, glands were homogenized and fractionated. An aliquot of the plasma membrane-rich fraction was subjected to SDS polyacrylamide gel electrophoresis and autoradiography to identify phosphoproteins. In method two, glands were homogenized, fractionated, and an aliquot of each fraction incubated with or without 10 μM cAMP for 5 min and then subjected to electrophoresis and autoradiography as above. In method three, glands were homogenized and incubated with or without 10 μM cAMP. Following stimulation, glands were fractionated and then subjected to electrophoresis and autoradiography as above. In all methods, proteins having molecular weights of 62 kD and 45 kD were heavily labelled with phosphate in response to changes in cAMP in the plasma membrane enriched fraction. Proteins in the plasma membrane enriched fraction having molecular weights of 29, 30, 32, and 37 kD were also substantially phosphorylated by cAMP in the broken cell method.

The function of plasma membrane phosphoproteins as effectors or modulators of fluid secretion are unknown. This question is possibly the most important and formidable challenge in future studies to determine the role of cAMP as an intracellular modulator of fluid secretion in tick salivary glands.

3.2.2.5 *Protein synthesis and secretion*

As described, ixodid salivary glands gain competence to secrete only after the onset of feeding (Kaufman 1976, Sauer *et al.* 1979). Radical cytological changes are correlated with these phenomena. The salivary glands increase greatly in size and protein content after the onset of feeding (McSwain *et al.* 1982). SDS-polyacrylamide gel electrophoresis was used to study changes in individual proteins in the salivary glands of female ixodid ticks during tick feeding (McSwain *et al.* 1982). Changes in adenylate cyclase, and cAMP phosphodiesterase activities in salivary glands during tick feeding have already been described (Schramke *et al* 1984, McMullen *et al.* 1978, 1980, 1983). Increases in salivary gland Na^+, K^+-ATPase activity during tick feeding have also been demonstrated (Kaufman *et al.* 1976). Because of these results, it was not unexpected that several new kinds of proteins were seen in the glands by SDS polyacrylamide gel electrophoresis soon after attachment of the tick to a host (McSwain *et al.* 1982). Other proteins, present in unfed ticks, increase in quantity during the later stages of tick feeding. Mating stimulates additional feeding and an increase in the amounts, but not the number, of proteins. Almost all the protein species that disappear do so after mating (i.e. 5–7 days post-attachment).

We have initiated more detailed studies of protein synthesis in the salivary glands of feeding ticks. For protein synthesis to increase, ample quantities of messenger (mRNA), ribosomal (rRNA), and transfer (tRNA) ribonucleic acid must be synthesized and aggregated into active units of protein synthesis called polysomes.

Uptake of 5′-^{3}H-uridine into RNA of whole glands for 90 min was measured. Very little uptake of isotope was observed in glands from unfed ticks. After 24 h or more of tick feeding, however, glands became much more permeable to 5′-^{3}H-uridine and began to sythesize both nuclear and cytoplasmic RNA (Shelby *et al.* 1986). However, most of the radioactive uridine was apparently incorporated into 4–5 s RNA (i.e. tRNA) and a smaller amount into ribosomal RNA with little evidence of mRNA synthesis. On the other hand, other studies demonstrated considerably more poly(A$^+$)mRNA in glands of ticks fed for 48 h as compared to ticks that had fed for 24 h and unfed ticks, suggesting that the 90 min for the pulse labelling in previous experiments was insufficient time to detect the synthesis of mRNA. This is unusual in that in tissues of other animals, <25 min is usually sufficient to observe incorporation of 5′-^{3}H-uridine into mRNA (Boshes 1970). Reasons for this apparent slow synthesis of mRNA in tick salivary glands under these conditions are unclear, but it may suggest that other media factors or homones which are present in the intact animal are required.

Polysomes in glands from unfed and partially fed ticks were also measured. Ticks fed for increasingly longer periods had increasing amounts of polysomes as well as ribosomal subunits. Glands from unfed ticks had almost undetectable levels of polysomes (Shelby *et al.* 1986).

Thus, from a variety of data it is safe to conclude that protein synthesis occurs in salivary glands after the onset of tick feeding. Important questions center on the control of these changes and how these changes correlate with changes in gland morphology, adenylate cyclase, cAMP-PDE, Na^+, K^+ATPase, and other factors affecting the competence of glands to secrete fluid (Sauer & Essenberg 1984). Arthropod hormones such as ecdysteroids and juvenile hormone can stimulate new gene activity as evidenced by enhanced RNA and protein synthesis (Marmaras & Sekeris 1972, Shaaya & Levenbook 1982, Sridhara *et al.* 1980). These substances or others may be involved in the new gene activity seen in the tick salivary gland. Evidence has accumulated in recent years demonstrating that ecdysteroids and juvenile hormone-like activity are present in ticks and are of functional significance (Chapter 7 and 8; also, Solomon *et al.* 1983). Knowledge of regulatory factors that induce salivary gland differentiation may assist in culturing tick cells and provide host cells capable of being infected. For example, in both *Babesia* and *Theileria* a final cycle of replication is initiated in the salivary glands when the tick takes a bloodmeal, and parasites (sporozoites) infective for the mammalian host are produced (Kurtti *et al.* 1982), Kurtti & Munderloh 1981). Perhaps factors that regulate salivary gland differentiation and its ability to secrete also provide the signal or conditions for these parasites to undergo replication and produce infective sporozoites before transmission to the host.

3.3 SUMMARY AND CONCLUSIONS

The salivary glands of female ixodid ticks are morphologically and physiologically complex, and controversy still exists with respect to the numbers, types, and detailed structure of the cells (Krolak *et al* 1982). Additional complexity is introduced after the onset of feeding where almost all cells in Types II and III undergo marked change (Fig. 3.5). It is likely that trans-

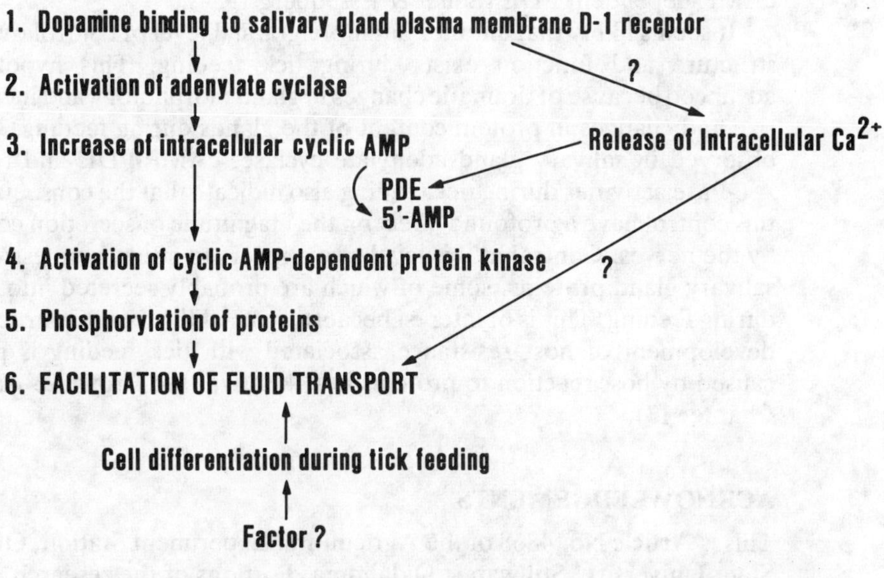

Fig. 3.5 — Schematic representation of the sequence of events by which dopamine (DA) stimulates fluid secretion by ixodid female salivary glands. More cAMP can be produced by stimulating salivary glands of rapidly feeding ticks with DA than slowly feeding ticks because of higher adenylate cyclase and lower cAMP phosphodiesterase activities. It is hypothesized that salivary glands of rapidly feeding ticks have the capacity to secrete more fluid than salivary glands of slowly feeding ticks at least partly because of these differences. The mechanism by which gland stimulation causes release of intracellular Ca^{2+} and how the divalent cation facilitates secretion are unknown.

formed granular cells and expanding ablumenal interstitial cells play key roles in fluid secretion because of a proliferation of membranes and closely associated mitochondria, features common to fluid transporting epithelia. Fawcett *et at.* (1981b) proposed a mechanism for fluid transport. They suggested that transformed granular cells in Type III alveoli produce a primary secretion hypotonic or isotonic to the tick's hemolymph, and that the ablumenal interstitial cells pump Na^+ into the basolateral labyrinth resulting in reabsorption of water from the lumen through 'leaky' septate junctions between cells bordering the lumen. This hypothesis remains to be tested.

The glands are controlled by nerves, and dopamine appears to be the neurotransmitter at the neuro-effector junction which stimulates fluid secretion (Fig. 3.5). Dopamine activates an adenylate cyclase which causes the formation of cAMP which activates protein kinases to phosphorylate proteins that are of importance to secretion. The functions of these phosphoproteins are unknown but some appear to be associated with the plasma membrane. The precise role(s) of Ca^{2+} in secretion is (are) unknown, but one way may be to help regulate cAMP by regulating protein modulators of cAMP-dependent PDE (Sauer & Essenberg 1984).

It seems likely that another but more gradual level of control over gland structure and function exists during tick feeding. This hypothesis is advanced because of dramatic changes in gland morphology and increases in size and changes in protein content of the glands during feeding. Changes observed in salivary gland adenylate cyclase, cAMP-PDE and Na^+, K^+ AT-Pase activities during tick feeding also indicate that the consequences of this control have a profound effect on the magnitude of secretion controlled by the nerves. Control at this level also appears to affect synthesis of other salivary gland proteins, some of which are probably secreted into the host during feeding. This is of interest because much of the tissue destruction and development of host resistance associated with tick feeding is probably caused by host reaction to proteins in tick saliva (see Brown & Askenase, Chapter 13).

ACKNOWLEDGEMENTS

This is Article No. 4868 of the Agricultural Experiment Station, Oklahoma State University, Stillwater, Oklahoma. Portions of the research reported herein were supported by NIH Grant AI-13535 from the National Institutes of Health and NSF Grant No. DCB 8415668 from the National Science Foundation. We thank Linda Hope and Trena Farris for typing the manuscript and Glen R. Needham for critically reviewing the manuscript.

REFERENCES

Balashov, Y. S. (1972) A translation of bloodsucking ticks (Ixodoidea)-vectors of diseases of man and animals. *Misc. Publ. Entomol. Soc. Am.* **8**, 161–376.

Barker, D. M., Ownby, C. L., Krolak, J. M., Claypool, P. L., & Sauer, J. R. (1984) The effects of attachment, feeding and mating on the morphology of the Type I alveolus of salivary glands of the lone star tick, *Amblyomma americanum* (L.). *J. Parasitol.* **70**: 99–113.

Beaumont, C., Piau, J. P., Fischer, S., Delaunay, J., & Schapira, G. (1979) Stimulation of erythroid cells adenylate cyclase by soluble factors. *Biochem. Biophys. Res. Comm.* **91**: 1250–1257.

Berridge, M. J. (1981) Phosphatidylinositol hydrolysis: a multifunctional transducing mechanism. *Molec. Cell Endocrinol.* **24**: 115–140.

Berridge, M. J. (1982) Regulation of cell secretion: The integrated action of cyclic AMP and calcium. In: Kebabian, J. W. & Nathanson, J. (ed.), *Cyclic nucleotides II — Physiology and pharmacology*. New York, Spinger-Verlag pp 227–270.

Berridge, M. J. (1984) Inositol triphosphate and diacylglycerol as second messengers. *Biochem. J.* **220**: 345–360.

Binnington, K. C. & Stone, B. F. (1977) Distribution of catecholamines in the cattle tick *Boophilus microplus*. *Comp. Biochem. Physiol.* **58C**: 21–28.

Boshes, R. A. (1970) *Drosophila* polyribosomes. The characterization of two populations by cell fractionation and isotopic labelling with nucleic acid and protein precursors. *J. Cell Biol.* **46**: 477–490.

Claypool, D. M., Krolak, J. M., Ownby, C. L. & Sauer, J. R. (1986) Effect of attachment, feeding and mating on the structure of the salivary glands of the lone star tick *Amblyomma americanum* (L.): Type II and III alveoli. (In prep.).

Clement-Cormier, Y. C., Parrish, R. G., Petzold, G. L., Kebabian, J. W., & Greengard, P. (1975) Characterization of a dopamine-sensitive adenylate cyclase in the rat caudate nucleus. *J. Neurochem.* **25**: 143–149.

Cohen, P. (1982) The role of protein phosphorylation in neural and hormonal control of cellular activity. *Nature* **296**: 613–620.

Enyedi, A., Farago, A., Sarkadi, B., Szasz, I., & Gardos, G. (1983) Cyclic-AMP-dependent protein kinase stimulates the formation of polyphosphoinositides in the plasma membranes of different blood cells. *FEBS Letters* **161**: 158–162.

Erlichman, J., Rosenfeld, R., & Rosen, O. M. (1974) Phosphorylation of a cyclic adenosine 3′:5′-monophosphate-dependent protein kinase from bovine cardiac muscle. *J. Biol Chem.* **249**: 5000–5003.

Erlichman, J., Bloomgarden, D., Sarkar, D., & Rubin, C. S. (1983) Activation of cyclic AMP-dependent protein kinase isoenzymes: Studies using specific antisera. *Arch. Biochem. Biophys.* **227**: 136–146.

Fain, J. N. & Berridge, M. J. (1979a). Relationships between hormonal activation of phosphatidylinositol hydrolysis, fluid secretion and calcium flux in the blowfly salivary gland. *Biochem. J.* **178**: 45–58.

Fain, J. N. & Berridge, M. J. (1979b) Relationship between phosphatidylinositol synthesis and recovery of 5-hydroxytryptamine-responsive Ca^{2+} flux in blowfly salivary glands. *Biochem. J.* **180**: 655–661.

Fawcett, D. W., Doxsey, S., & Buscher, G. (1981a) Salivary gland of the tick vector (*R. appendiculatus*) of east coast fever. I. Ultrastructure of the type III acinus. *Tissue and Cell* **13**: 209–230.

Fawcett, D. W., Doxsey, S., & Buscher, G. (1981b) Salivary gland of the tick vector (*R. appendiculatus*) of east coast fever. II. Cellular basis for fluid secretion in the type III acinus. *Tissue and Cell* **13**: 231–253.

Filburn, C. R. & Wyatt, G. R. (1976) Adenylate and guanylate cyclases of cecropia silkmoth fat body. *J. Insect. Physiol.* **22**: 1635–1640.

Guenther, P. E., Barker, D. M., & Sauer, J. R. (1980) Sodium, chloride and water balance in feeding Gulf Coast Ticks, *Amblyomma maculatum* (Koch), *Ann. Entomol. Soc. Am.* **73**: 485–488.

Hawthorne, J. N. (1982) Is phosphatidylinositol now out of the calcium gate? *Nature* **295**: 281–282.

Hsu, M. H. & Sauer, J. R. (1975) Ion and water balance in the feeding lone star tick. *Comp. Biochem. Physiol.* **52A**: 269–276.

Hume, M. E., Essenberg, R. C., McNew, R. W., Bantle, J. A. & Sauer, J. R. (1984) Adenosine-3′ ,5′-monophosphate in salivary glands of unfed and feeding female lone star ticks, *Amblyomma americanum* (L.) *Comp. Biochem. Physiol.* **79C**: 47–50.

Kaufman, W. R. (1976) The influence of various factors on fluid secretion by *in vitro* salivary glands of ixodid ticks. *J. Exp. Biol.* **64**: 727–742.

Kaufman, W. R. (1977) The influence of adrenergic agonists and their antagonists on isolated salivary glands of ixodid ticks. *Eur. J. Pharmacol.* **45**: 61–68.

Kaufman, W. R. (1979) Control of salivary fluid secretion in ixodid ticks. In: Rodriguez, J. G. (Ed.) *Recent advances in acarology*, Vol. 1. New York, Academic Press pp. 357–363.

Kaufman, W. R. (1983) The function of tick salivary glands. In: Harris, K. F. (Ed.) *Current topics in vector research, Vol. 1.* New York, Praeger Scientific pp. 215–247.

Kaufman, W. R., Diehl, P. A., & Aeschlimann, A. A. (1976) Na, K-ATPase in the salivary gland of the ixodid tick *Amblyomma americanum* (Koch) and its relation to the process of fluid secretion. *Experientia* **32**: 986–987.

Kebabian, J. W. & Calne, D. B. (1979) Multiple receptors for dopamine. Nature **277**: 93–96.

Kirk, C. J., Creba, J. A., Downes, C. P., & Michel, R. H. (1981) Hormone-stimulated metabolism of inositol lipids and its relationship to hepatic receptor function. *Biochem. Soc. Trans.* **9**: 377–379.

Krolak, J. M., Ownby, C. L., & Sauer, J. R. (1982) Alveolar structure of salivary glands of the lone star tick, *Amblyomma americanum* (L.).: Unfed females. *J. Parasitol.* **68**: 61–82.

Kurtti, T. J. & Munderloh, U. G. (1981) Tick cell culture: characteristics, growth requirements and applications to parsitology. In: Maramorosch,

K. & Mitsuhashi, J. (ed). *Invertebrate cell culture applications.* New York, Academic Press pp. 195–232.

Kurtti, T. J., Munderloh, U. G., & Samish, M. (1982) Effect of medium supplements on tick cells in culture. *J. Parasitol.* **68**: 930–935.

MacNeil, S., Crawford, A., Amirransooli, H., Johnson, S., Pollock, A., Ollis, C., & Tomlinson, S. (1980) Stimulation of hormone-responsive adenylate cyclase activity by a factor present in the cell cytosol. *Biochem. J.* **188**: 393–400.

Mane, S. D., Darville, R. G., Sauer, J. R., & Essenberg, R. C. (1985) Cyclic AMP-dependent protein kinase from the salivary glands of the tick, *Amblyomma americanum*: Partial purification and properties. *Insect Biochem.* **15**: 777–787.

Marmaras, V. J. & Sekeris, C. E. (1972) Pattern and activity of polyribosomes from the integument of blowfly larvae during development. Stimulation of polyribosome formation with β-ecdysone. *Exp. Cell Res.* **75**: 143–153.

McGowen, M. J., Barker, R. W., Homer, J. T., McNew, R. W., & Holscher, K. H. (1981) Success of tick feeding on calves immunized with *Amblyomma americanum* (Acari: Ixodidae) extract. *J. Med. Entomol.* **18**: 328–332.

McMullen, H. L. & Sauer, J. R. (1978) The relationship of phosphodiesterase and cyclic AMP to the process of fluid secretion in the salivary glands of the ixodid tick *Amblyomma americanum. Experientia* **34**: 1030–1031.

McMullen, H. L., Bantle, J. A., Essenberg, R. C., & Sauer, J. R. (1983) Changes in cyclic nucleotide phosphodiesterase activity in the salivary glands of female *Amblyomma americanum* (L.) ticks during feeding. *Insect Biochem* **13**: 585–592.

McMullen, H. L., Sauer, J. R., Bantle, J. A., & Essenberg, R. C. (1980) Regulation of fluid secretion by calcium-dependent modulator proteins of 3′:5′-cyclic AMP phosphodiesterase. *Biochem Biophys. Res. Comm.* **95**: 1555–1562.

McMullen, H. L., Sauer, J. R., & Burton, R. L.(1976) Possible role in uptake of water vapour by ixodid tick salivary glands. *J. Insect. Physiol.* **22**: 1281–1285.

McSwain, J. L., Essenberg, R. C., & Sauer, J. R. (1982) Protein changes in the salivary glands of the female lone star tick, *Amblyomma americanum* during feeding. *J. Parasitol.* **68**: 100–106.

McSwain, J. L., Essenberg, R. C., & Sauer, J. R. (1985a) Cyclic AMP mediated phosphorylation of endogenous proteins in the salivary glands of the lone star tick, *Amblyomma americanum* (L.). *Insect Biochem.* **15**: 789–802.

McSwain, J. L., Sauer, J. R., Schmidt, S. P., Claypool, D. M., & Essenberg, R. C. (1985b) Subcellular location of phosphoproteins in salivary glands of *Amblyomma americanum* (L.) (submitted).

Megaw, M. W. J. (1977) The innervation of the salivary gland of the tick, *Boophilus microplus. Cell Tiss. Res.* **184**: 551–558.

Megaw, M. W. J. & Beadle, D. J. (1979) Structure and function of the salivary glands of the tick, *Boophilus microplus*, Canestrini (Acarina: Ixodidae). *Int. J. Insect Morphol. and Embryol.* **8**: 67–83.

Megaw, M. W. J. & Robertson, H. A. (1974) Dopamine and noradrenaline in the salivary glands and brain of the tick, *Boophilus microplus*: Effect of reserpine. *Experientia* **30**: 1261–1262.

Michel, R. H. (1975) Inositol phospholipids and cell surface receptor function. *Biochim Biophys. Acta.* **415**: 81–147.

Moorhouse, D. E. (1969) The attachment of some ixodid ticks to their natural hosts. *Proc. 2nd Int. Congr. Acarology. 1967.* pp. 319–327.

Needham, G. R. & Sauer, J. R. (1979a) Involvement of calcium and cyclic AMP in controlling ixodid tick salivary fluid secretion. *J. Parasitol.* **65**: 531–542.

Needham, G. R. & Sauer, J. R. (1979b) Dynamics of calcium, catecholamines and cyclic AMP in control of salivary fluid secretion by female ixodid ticks. In: Rodriguez, J. G. (ed.) *Recent advances in acarology*, Vol. I. New York, Academic Press pp. 413–419.

Perkins, J. P. (1973) Adenyl cyclase. In: Greengard, P. & Robison, G. A. (ed.) *Advances in cyclic nucleotide research*, Vol. 3. New York, Raven Press pp. 1–64.

Prentki, M., Biden, T. J., Janjic, P., Irvine. R. F., Berridge, M. J., & Wollheim, C. B. (1984) Rapid mobilization of Ca^{2+} from rat insulinoma microsomes by inositol-1,4,5-triphosphate. *Nature* **309**: 562–564.

Rasmussen, H. (1981) Calcium and cAMP as synarchic messengers. New York. John Wiley, 370 p.

Rasmussen, H. & Goodman, D. B. P. (1977) Relationship between calcium and cyclic nucleotides in cell activation. *Physiol Rev.* **57**: 421–509.

Ross, E. M. & Gilman, A. G. (1980) Biochemical properties of hormone-sensitive adenylate cyclase. *A. Rev. Biochem.* **49**: 533–564.

Rudolph, D. & Knülle, W. (1974) Site and mechanism of water vapour uptake from the atmosphere in ixodid ticks. *Nature* **249**: 84–85.

Rudolph, D. & Knülle, W. (1978) Uptake of water vapour from the air: Process, site and mechanism in ticks. In: Schmidt-Neilson, K., Bolis, L. & Maddrell, S. H. P. (ed.) *Comparative physiology — water, ions and fluid mechanisms.* Cambridge University Press, pp. 97–113.

Salmon, D. M. & Honeyman, T. W. (1980) Proposed mechanisms of cholinergic action in smooth muscle. *Nature* **284**: 344–345.

Sarkadi, B., Enyedi, A., Farago, A., Meszaros, G., Kremmer, T., & Gardos, G. (1983) Cyclic AMP-dependent protein kinase stimulates the formation of polyphosphoinositides in lymphocyte plasma membrane. *FEBS Letters* **152**: 195–198.

Sauer, J. R. (1977) Acarine salivary glands — physiological relationships. *J. Med. Entomol.* **14**: 1–9.

Sauer, J. R. & Essenberg, R. C. (1984) Role of cyclic nucleotides and calcium in controlling tick salivary gland function. *Amer. Zool.* **24**: 217–227.

Sauer, J. R., Needham, G. R., McMullen, H. L., & Morrison, R. D. (1979) Cyclic nucleotides, calcium and salivary fluid secretion in ixodid ticks. In: J. G. Rodriquez (ed.) *Recent advances in acarology*. Vol. 1. New York, Academic Press pp. 365–373.

Schmidt, S. P., Essenberg, R. C., & Sauer, J. R. (1981) Evidence for a D_1 dopamine receptor in the salivary glands of *Amblyomma americanum* (L.) *J. Cyclic Nucleotide res.* **7**: 375–384.

Schmidt, S. P., Essenberg, R. C., & Sauer, J. R. (1982) A dopamine sensitive adenylate cyclase in the salivary glands of *Amblyomma americanum* (L.) *Comp. Biochem. Physiol.* **72C**: 9–14.

Schramke, M. L., McNew, R. W., Schmidt, S. P., Essenberg, R. C., & Sauer, J. R. (1984) Changes in dopamine-sensitive adenylate cyclase activity in salivary glands of female lone star ticks, *Amblyomma americanum* (L.), during feeding. *Insect Biochem.* **14**: 595–600.

Seeman, P. (1981) Brain dopamine receptors. *Pharmacol. Rev.* **32**: 229–313.

Shaaya, E. & Levenbook, L. (1982) 20-hydroxyecdysone and the occurrence of giant hnRNA in early third instar *Calliphora vicina* larvae. *Insect Biochem.* **12**: 663–667.

Shelby, K. S., Bantle, J. A., & Sauer, J. R. (1986) Control of protein synthesis and cell differentiation in the tick salivary gland. (in prep.),

Solomon, K. R., Mango, C. K. A., & Obenchain, F. D. (1983) Endocrine mechanisms in ticks: Effects of insect hormones and their mimics on development and reproduction. In: Obenchain, F. D. & Galun, R. (ed.) *Physiology of ticks*. London, Pergamon Press pp. 399–438.

Sridhara, S., Katulu, K. S., & Gilbert, L. I. (1980) Hormonal control of RNA synthesis in wing epidermis. In: Lock, M. & Smith, D. S. (ed.) *Insect biology in the future*. New York, Academic Press pp. 439–462.

Stein, J. C., Sarkar, D. & Rubin, C. S. (1984) Tryptic peptide mapping studies on the regulatory subunits of Type II protein kinases from cerebral cortex and heart. Evidence for overall structure divergence and differences in the autophosphorylation and cAMP-binding domains. *J. Neurochem.* **42**: 547–553.

Streb, H., Irvine, R. F., Berridge, M. J., & Schulz, I. (1983) Release of Ca^{2+} from a nonmitochondrial intracellular store in pancreatic acinar cells by inositol-1,4,5-triphosphate. *Nature* **306**: 67–69.

Tatchell, R. J. (1967) Salivary secretion in the cattle tick as a means of water elimination. *Nature* **213**: 940–941.

Tatchell, R. J. (1969) The ionic regulatory role of the salivary secretion of the cattle tick, *Boophilus microplus*. *J. Insect. Physiol.* **15**: 1421–1430.

Walsh, D. A., Perkins, J. P., & Krebs, E. G. (1968) An adenosine 3′,5′-monophosphate-dependent protein kinase from rabbit skeletal muscle. *J. Biol. Chem.* **243**: 3763–3765.

Wong, D. L.-P. & Kaufman, W. R. (1981) Potentiation by spiperone and other butyrophenones of fluid secretion by isolated salivary glands of ixodid ticks. *Eur. J. Pharmacol.* **73**: 163–173.

4

The paralyzing toxin and other immunogens of the tick *Ixodes holocyclus* and the role of the salivary gland in their biosyntheses

Bernard F. Stone and Keith C. Binnington

4.1 INTRODUCTION

Each year the Australian paralysis tick, *Ixodes holocyclus,* causes tick paralysis in thousands of domestic pets and livestock along the eastern coast of Australia (Stone *et al.* 1979, B. F. Stone unpublished data, I. Lewis personal communication). Occasional cases of paralysis and many cases of allergic hypersensitivity occur in humans.

Over the past five years, progress has been made in research aimed at the development of a vaccine to immunize animals at risk against tick paralysis and the development of an antitoxin, safer for humans than the canine antiserum currently used (Stone *et al.* 1982b, Wright *et al.* 1983, Stone & Neish 1984, Stone *et al.* in manuscript 1985a). However, it is necessary to know much more of the nature of the paralyzing toxin ('holocyclotoxin' — Stone & Wright 1981) and of the mechanism of holocyclotoxin biosynthesis to facilitate a possible biotechnological approach to the production of a commercial vaccine. It would also be desirable to know more of the induced immune response to salivary immunogens as a prerequisite for the development of a more effective desensitizing agent for human allergies to *I. holocyclus*.

The present report outlines recent advances in our understanding of the toxic and allergic effects on the host, of the biochemistry of the toxin, and, for the first time, as a result of ultrastructural studies made on the salivary glands, defines more precisely the likely sources of holocyclotoxin and the allergens.

4.2 TICK FEEDING AND PARALYSIS

Effects on the host of the female Australian paralysis tick, *Ixodes holocyclus,* vary according to how long the tick has been attached (about 6–8 days is required for it to engorge). Recently attached ticks are able to produce severe allergic reactions in sensitized hosts (Trinca 1964, Banfield 1966, Sutherst & Moorhouse 1971) but do not cause paralysis. On the other hand, ticks which have been attached for 4–5 days are capable of causing a lethal paralysis (Murray & Koch 1969, Goodrich & Murray 1978, Stone *et al.* 1979). The increase in the ability of the female tick to cause paralysis with increased time of attachment is matched by an increase in the toxicity of salivary gland extracts (Stone *et al.* 1979, Binnington & Stone 1981). This temporal relationship has been explored in an attempt to identify the cells responsible for production of the toxin (see section 4.5).

Tick paralysis due to *I. holocyclus* has been known since the 19th century (Hovell 1824 — see Scott 1921, Backhouse 1843 — see Stanbury & Huyck 1945, both cited by Gregson 1973). However, it was not until Ross (1926, 1935) demonstrated toxic effects on mice and dogs injected with extracts of salivary glands that paralysis induced by *I. holocyclus* became associated with the glands of the tick. It was verified only recently and then only indirectly by Stone *et al.* (1983a) that the toxin was secreted into the host with saliva. In this study *I. holocyclus* females which had been prefed on mice for 4 days and then attached to an artificial membrane, repeatedly sucked and salivated with a behavioural pattern similar to that of naturally feeding ticks (Fig. 4.1).

The feeding medium was shown to contain a toxin similar in toxinological characteristics to that present in salivary gland extracts (Stone *et al.* 1983a). Paralyzing fractions separated from the tissue culture medium into which *I. holocyclus* females had secreted toxin were compared with the unused medium by isoelectrofocusing. Four distinct major and numerous minor protein bands were detected in the former sample, but the distinct major bands did not appear in the latter sample. The major bands must have been salivary proteins presumably including holocyclotoxin (Fig. 4.2). The paralyzing component was present in the chromatographic fraction obtained from Sephacryl S200 gel filtration and had similar elution characteristics to the fraction from salivary gland extracts containing holocyclotoxin (Stone *et al.* 1983a and Peak III — Fig. 4.3).

4.3 THE NATURE OF HOLOCYCLOTOXIN

4.3.1 Biochemistry

This toxic fraction separated from salivary gland extracts had been shown (Stone *et al.* 1979, Stone 1979, Stone *et al.* in manuscript 1985) to contain proteinaceous material much of which was acidic (two-directional electrophoresis — Fig. 4.4).

Studies using proteolytic enzymes (Stone 1979, Stone *et al.* in manuscript 1985) also indicated that holocyclotoxin was proteinaceous. Papain caused

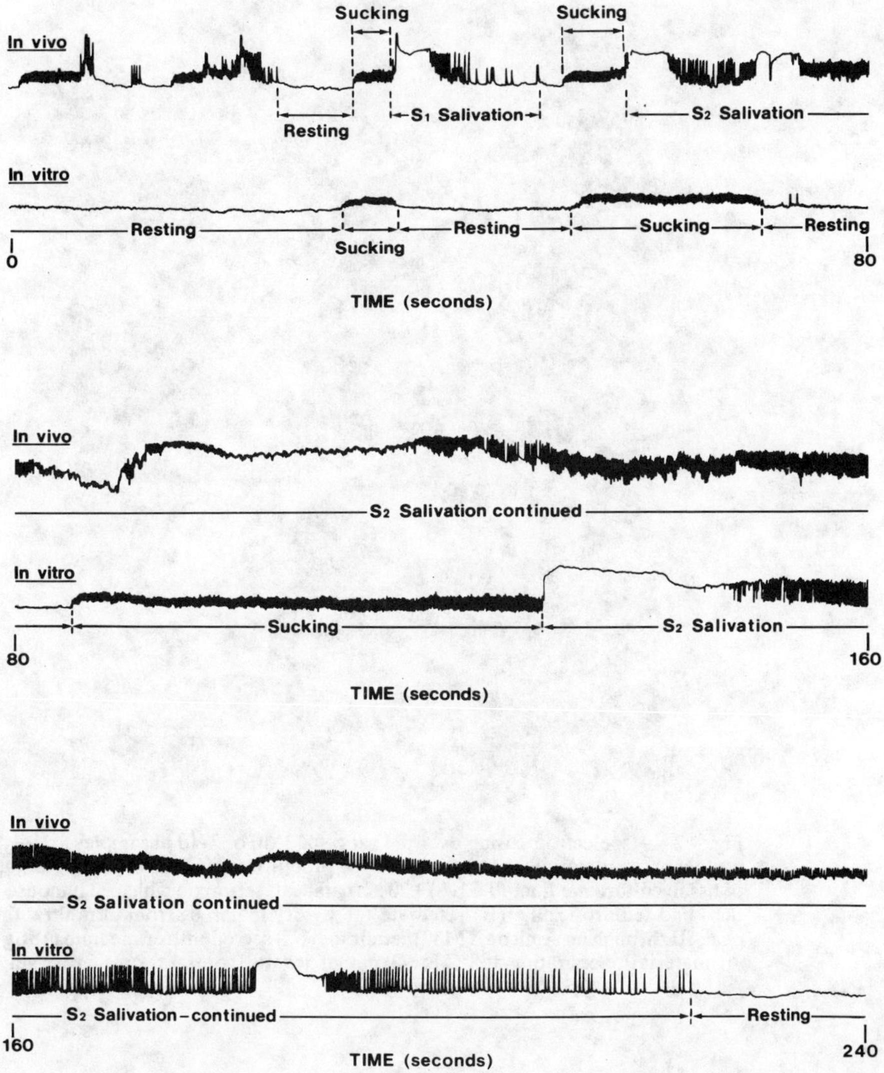

Fig. 4.1 — Similar salivation/feeding patterns of *I. holocyclus* female ticks attached to a mouse (*in vivo*) and to an artificial membrane (*in vitro*). Ticks were included in a circuit consisting of a dorsal electrode contacting the tick, a high-input-impedance electronic control unit, a Washington electro-oscillograph, and an electrode immersed in the feeding medium. The medium consisted of TC 199 (Commonwealth Serum Laboratories, Melbourne) and various additives such as foetal calf serum. Figure reproduced from Stone *et al.* (1983a) with the permission of Pergamon Press Ltd and Australian Society for Parasitology.

a 50% loss of paralyzing activity and hydrolysed 88% of the protein present in the crude extract of salivary glands. However, pronase, which is a commercial mixture of proteolytic enzymes, caused no loss of paralyzing

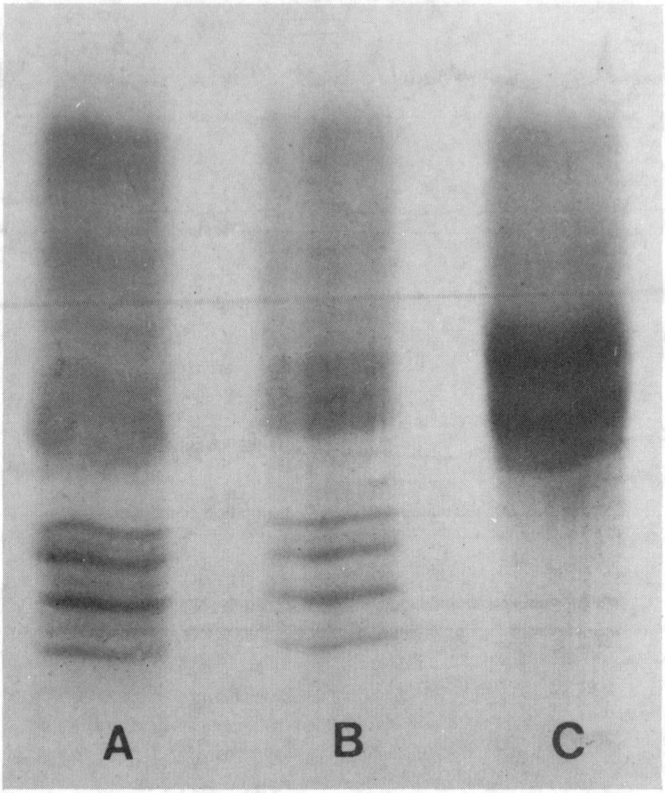

Fig. 4.2 — Isoelectrofocusing over pH range of 3–10 of 7–10 µl samples (1.1–1.8 µl protein or dry material/µl. (A) peak III (8 µg protein) from Ultragel chromatography of tissue culture medium (TC199)+20% foetal calf serum on which 72 h mouse-fed ticks had fed artificially. (B) retentate (18 µg dry material) from ultrafiltration of peak III through an Amicon YM30 membrane. (C) secretion-free medium (18 µg dry material). Separation at 8°C for 40 min at 5 watts (constant) on agarose gel.

activity but digested up to 78% of the total protein in the crude salivary gland extract. Thus holocyclotoxin appears to be resistant to detoxication by pronase. Vosbeck *et al.* (1974) showed that pronase lacks prolidase activity, only digesting peptides until proline residues are reached, e.g. $\alpha + \beta$ chains of human haemoglobin were only slightly digested by pronase. It is not known if holocyclotoxin is resistant for a similar reason.

Histochemical studies on the salivary gland of *I. holocyclus,* using the periodic acid–Schiff (PAS) method (Binnington & Stone 1981) indicated the presence of glycoproteins in the putative toxin cell 'b'. This cell may produce one or more salivary glycoproteins, but it is not yet known if holocyclotoxin is one of these. More work is required to clarify this and other biochemical aspects.

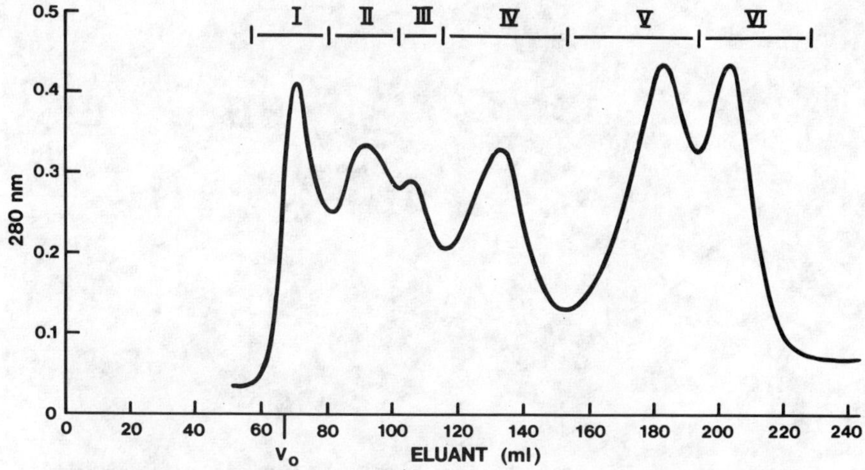

Fig. 4.3 — Gel filtration of 25 mg of protein from crude extract of salivary glands of 5-day fed Australian paralysis ticks *Ixodes holocyclus,* on a Sephacryl S200 column (100 × 2.6 cm). Elution with 0.1 M NH_4HCO_3 at 10 ml/h; 3 ml fractions. Paralyzing effects were largely restricted to peak III although some effects occurred in peaks II and IV.

4.3.2 Mode of action

No recent studies have been carried out, and earlier published research has been largely limited to those of Copper & Spence (1976) who worked with isolated hemidiaphragm preparations from mice with respiratory paralysis induced by the feeding of nymphal ticks. There was a temperature-dependent, reversible paralysis apparently due to inhibition of transmitter release at neuromuscular junctions. The reduced effect at lower temperatures, the relatively slow development of symptoms of paralysis (minimum of 5–7 h; B. F. Stone unpublished data), the lack of evidence of interference with nerve conduction (Cooper & Spence 1976), and the appearance or worsening of symptoms even after ticks are removed, are consistent with the biochemical evidence of a proteinaceous, toxic (possibly also tenacious) molecule. Such a molecule may well have an affinity for ion channels in nerve terminals or interfere in other ways with transmitter release at neuromuscular junctions.

4.4 HOST RESPONSE TO IMMUNOGENS

4.4.1 Holocyclotoxin

The development, decline, and quantitation of immunity to tick paralysis in Australia has been studied by Stone & Wright (1980), Stone *et al.* (1983b), Wright *et al.* (1983) and Stone & Neish (1984). This research has extended the earlier work of Ross (1926, 1935) and of Oxer & Ricardo (1942) who established that dogs may develop immunity as a result of natural or artificial tick infestation. Kaire (1966) gave a preliminary indication that toxic extracts of engorged female *I. holocyclus* may also induce immunity.

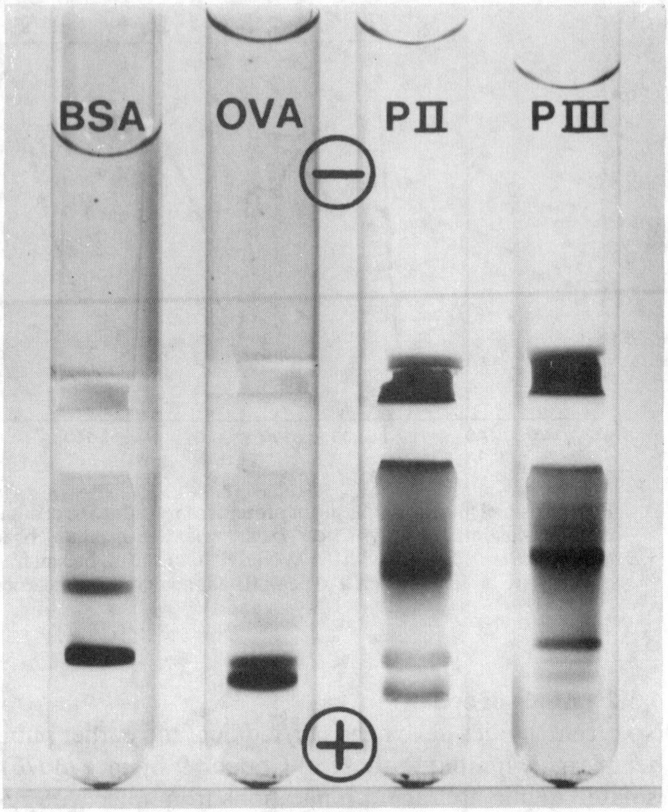

Fig. 4.4 — Two-directional polyacrylamide gel electrophoresis of Peaks II and III from Sephacryl S200 separation compared with reference compounds bovine serum albumen (BSA) and ovalbumen (OVA). Buffers pH 8.5 tris/glycine (0.04 M); running conditions = 3 mA/tube for 3/4 h — method of Broad *et al.* (1979). Samples applied at centres of gels. Under the conditions of the test, many of the proteins in Peaks II and III migrated towards the anode and therefore were negatively charged or acidic.

The kinetics of the immune responses to holocyclotoxin was studied for the first time by Stone & Wright (1980) and Stone *et al.* (1983b). This study was facilitated by the development of a more sensitive and more precise *in vitro/in vivo* neutralization test (Stone *et al.* 1982a) than that originally reported by Kaire (1965). In kinetic studies carried out over a period of 180 weeks, beagle dogs were immunized against tick paralysis by slowly increasing the numbers of female ticks allowed to feed until serum antitoxin titres reached a peak. This was followed by a period of freedom from ticks during which serum antitoxin titres declined. This cycle was repeated three times, with the number of infesting ticks being increased more rapidly in each succeeding cycle. Larger numbers of ticks were required to produce comparable titres in each succeeding cycle (Fig. 4.5), but the reason for this is

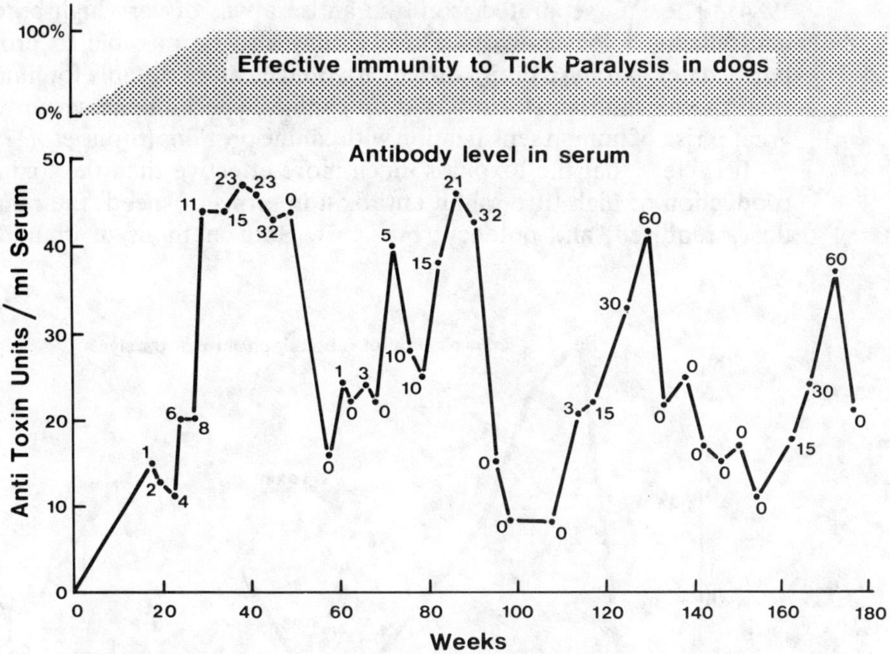

Fig. 4.5 — Antitoxin titres of beagle antiserum 8 days after ticks detached. Blood samples taken for assay as shown by points on the graph. The figures represent the number of ticks applied about 15 days before sampling. The number of antitoxin units/ml are equivalent to the number of standard paralyzing doses 50% neutralized by 1 ml antiserum. Residual toxicity was assessed using 4–5 g mice (Stone *et al.* 1982a). The first 14 weeks represent a priming stage during which a different female tick was fed for 2, 3, 4, 5, or 6 days on consecutive weeks with repetitions and rests of 3–4 weeks between later infestations. Shaded area represents estimated period of immunity to holocyclotoxin. Figure adapted from Stone *et al.* (1983b) and reproduced with the permission of the Australian Veterinary Association.

unknown. Protection from the toxic effects of a potentially lethal challenge of ticks persisted in these beagles for at least 53 weeks and in foxhounds for at least 102 weeks after the final immunizing infestation.

Serum antitoxin titres assayed in the neutralization test (Stone *et al.* 1982a) gave a good indication of effective immunity to tick paralysis during the initial development of hyperimmunity, but were less indicative thereafter. Once maximal levels of circulating antitoxin had been established it seemed that, even if titres had declined to base level, immunity to tick challenge may be present for some months.

Hopes were raised of producing a vaccine against tick paralysis when rabbits were successfully immunized against the effects of holocyclotoxin by injecting them with partially-purified toxin preparations from female salivary glands (Stone *et al.* 1982b). Even more promising was the high degree of immunity obtained when rabbits were injected with glutaraldehyde-detoxified partially-purified holocyclotoxin from salivary glands (Stone & Neish

1984). The IgG separated from the antisera was of very high potency in neutralizing holocyclotoxin. Thus, it may also be possible to produce a purified tick-paralysis antitoxin which would be more suitable for human use than the existing antitoxin based on canine hyperimmune serum with its greater risk of human sensitization with canine proteins (Stone *et al.* 1982b).

It is clear that the toxoid is much more effective than the toxin in the production of high titre rabbit antitoxin in terms of speed, the number of doses required, and potency (Fig. 4.6). Both methods of immunization

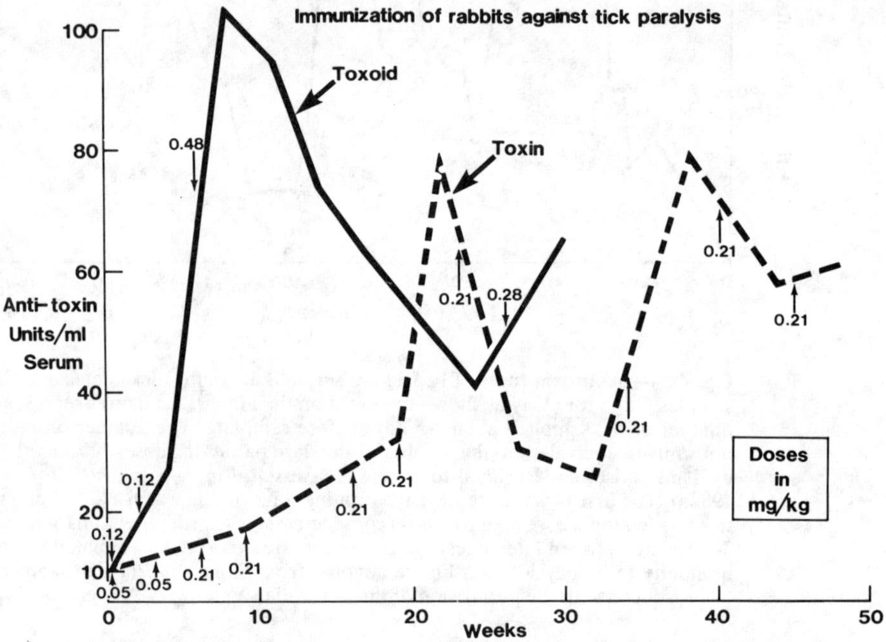

Fig. 4.6 — Comparison of the effectiveness of injection of holocyclotoxin and its derived toxoid in stimulating serum antitoxin titres in rabbits. Blood samples taken 2–3 weeks after injection. Antigens emulsified in Freund's Complete Adjuvant. Doses (mg antigen protein/kg) shown as numbers along curves; timing indicated by arrows. Broken line is in estimated position (no titres). Data based on that of Stone *et al.* (1982b) and Stone & Neish (1984) and reproduced with the permission of the *Australian Journal of Experimental Biology and Medical Science* and the University of Adelaide Press.

carried prospects, in principle, for the vaccination of domestic pets and livestock as evidenced by the immunity of rabbits to challenge by injected holocyclotoxin and by feeding ticks. Therefore the extension of this research to dogs was the next logical step as dogs are *one* of the principal sufferers from tick paralysis in Australia. Crude toxic extracts of salivary glands, injected into beagle dogs produced reasonably high antitoxin titres and immunity to tick paralysis (Wright *et al.* 1983). Fig. 4.7 compares the rate of

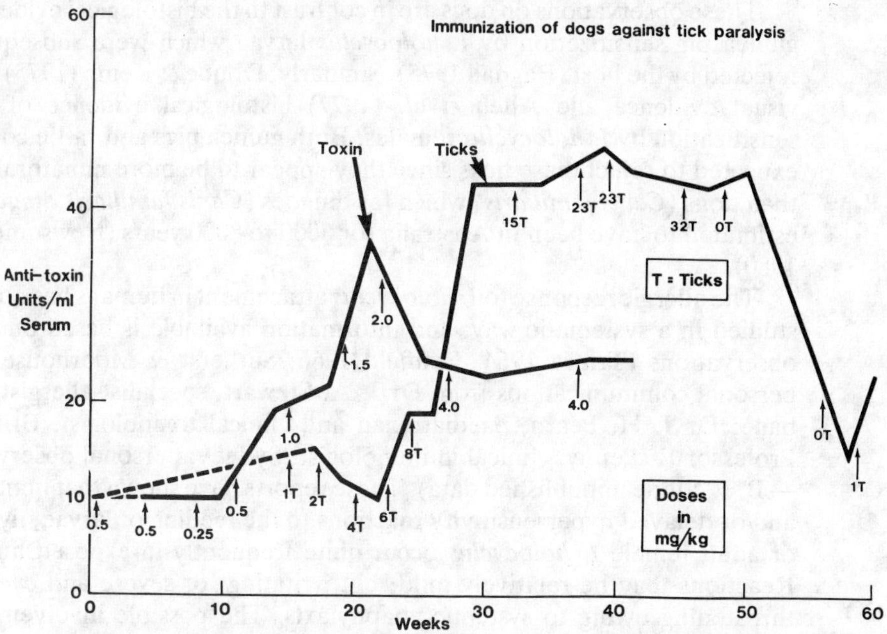

Fig. 4.7 — Comparison of the effectiveness of the feeding of *I. holocyclus* female ticks and of the injection of holocyclotoxin in stimulating serum antitoxin tires in beagle dogs. Blood samples taken 8 days after ticks detached or three weeks after injection. The number of ticks applied shown as 1T, 3T etc. Injected doses expressed as mg antigen protein/kg and indicated by arrows. Additional details as for Figs 4.5 and 4.6. Data based in part on that of Stone & Wright (1980) and Wright *et al.* (1983) and reproduced with the permission of the Australian Veterinary Association.

development of serum antitoxin levels in dogs by this method and by the feeding of ticks. Clearly, injection of toxoid preparations was the most effective practical method of immunization (Stone *et al.* 1986).

These immunological studies support the concept that holocyclotoxin is proteinaceous, and the enhancement of its immunogenicity by treatment with glutaraldehyde, which may produce higher molecular weight antigen proteins, provides further support for this argument.

4.4.2 Allergens

No clear evidence of a tick rejection mechanism which could be attributed to cutaneous hypersensitivity or other allergic reactions was noted during the studies on beagles. However, it is reported (K. C. Curtin, personal communication) that some foxhounds which have been used repeatedly for the commercial production of tick paralysis antiserum by the feeding of 100 or more ticks during each of several infestations each year for 11 years, showed some evidence of serous exudate at the site of attachment of the engorged ticks and some loosening of ticks. This did not appear to interfere unduly with the feeding of the ticks.

These observations on dogs are in contrast to the histological evidence of guinea pig sensitization by *I. holocyclus* larvae which were subsequently rejected by the host (Bagnall 1975). Similarly, Doube & Kemp (1975) found visual evidence and Allen *et al.* (1977) histological evidence of cattle sensitization by *I. holocyclus* females. Both guinea pigs and cattle could be expected to reject these ticks since they appear to be more unnatural hosts than dogs (*Canis familiaris*) which [as dingoes (*Canis familiaris dingo*)] are estimated to have been in Australia for 3000 to 8000 years (Newsome *et al.* 1980).

The allergic response to *I. holocyclus* attachment in humans has not been studied in a systematic way, and information available is based on casual observations (Trinca 1964, Banfield 1966, Sutherst & Moorhouse 1971; personal communications from Dr J. A. Stewart, specialist-allergist, Brisbane; Dr J. H. Pearn, paediatrician and clinical toxinologist, Brisbane; Professor R. Penny, clinical immunologist, Sydney; personal observations — B. F. Stone unpublished data). These reports have shown that immediate and/or delayed hypersensitivity reactions to the feeding of larvae, nymphs, or adult female *I. holocyclus* occur quite frequently in exposed humans. Reactions may be relatively mild, but irritating, or severe and even life-threatening owing to systemic anaphylaxis. The possible involvement of salivary gland secretory material in the reaction is discussed in section 4.5.

4.5 SALIVARY GLANDS

4.5.1 General organization

The morphology of the salivary gland of female *I. holocyclus* is similar to that of other ixodid ticks. Three types of acini are present, an agranular acinus I, and two granular types, II and III (Binnington & Stone 1981). The *I. holocyclus* male differs from other male ticks studied in not feeding on the host and in having only two types of acini, I and IV.

4.5.2 Development of salivary cells during attachment and feeding

Any account of the structure of tick salivary glands must take into consideration the particular demands imposed on the glands by tick feeding. After an attachment phase ixodid ticks grow slowly for some days during which time cuticle growth and the development of a large feeding lesion in the host skin allow rapid engorgement to take place before the tick detaches (Binnington & Kemp 1980). Light microscopic studies on *Boophilus microplus* (Binnington 1978) and *I. holocyclus* (Binnington & Stone 1981) have shown that the glands of unfed females contain stored granular material which is presumably rapidly utilized as the tick establishes itself on the host. Thus in *B. microplus* some cell types, particularly those probably containing attachment cement precursors (Binnington 1978), are replete with secretion granules which are apparently released shortly after attachment. Similarly in *I. holocyclus,* cells of acinus III appear to secrete stored material within a short time of attachment (Binnington & Stone 1981). Indications from light microscopy are that cells in acinus III of both ticks contribute less to the subsequent production of saliva than those of acinus II. In *I. holocyclus,* cell 'a' of acinus II remains rich in granules throughout feeding and cell 'b' greatly increases its apparent secretory activity after about 3 days of attachment (Binnington & Stone 1981). The application of electron microscopy to the salivary glands of *I. holocyclus* shows that the activation of acinus II is more dramatic and sustained than that of acinus III. However, it also provides information which suggests that acinus III continues to secrete a significant quantity of material, at least during the early stages of feeding.

4.5.3 Acinus II

Cell 'a' (terminology of Binnington & Stone 1981) occupies the zone around the duct of acinus II and, in the unfed female, contains granules which stain pink following the PAS reaction for carbohydrates. The cytoplasm which surrounds the granules reacts strongly for non-specific esterases. In the female which has fed for 2 days two distinct zones of secretion granules are seen; the basal half of cell 'a' is occupied by densely staining granules and the apical half by smaller, paler granules. The intensity of esterase staining in the cytoplasm is also less than in the glands from the unfed stage (Binnington & Stone 1981). Electron microscopy shows that in the unfed female the basal half of cell 'a' is occupied by large electron-dense granules which have a narrow, more electron-lucent cortex. The granules of the apical half which are pale following PAS staining (Binnington & Stone 1981) are more electron-lucent than the basal granules (Figs 4.8 & 4.9).† This difference in morphology between basal and apical granules indicates that there is a maturation of granules as they are transported to the lumen. Also present in cell 'a' of the unfed female are many free ribosomes and undilated cisternae of RER (rough endoplasmic reticulum) (Figs. 8, 9). These features and the inconspicuous nature of the Golgi elements all reflect the relatively dormant nature of salivary cell 'a' in non-feeding ticks. After the tick begins to feed there is an increase in the activity of rough endoplasmic reticulum (RER) and Golgi as indicated by an increase in profiles of dilated cisternae of these elements. Associated with the Golgi are large vacuoles containing a fine granular material and, by comparison with findings in exocrine glands of other animals (Doine *et al.* 1984), are probably immature secretory granules (Figs 4.10, 4.11). Cell 'a' continues to become larger as further attachment time elapses, but otherwise there is little change in the morphology of this cell type as the tick changes from being barely toxic at 2 days to when it becomes highly toxic at 4–5 days (Binnington & Stone 1981).

Cell 'b', which occupies the fundus of acinus II, is known from light microscopy to be largely undeveloped in the unfed female but to increase markedly in size after 2 days of attachment. By 3 days, when toxicity first becomes apparent, cell 'b' is rich in PAS positive granules (Binnington &

† Figs 4.8–4.22 are all electron micrographs of the salivary glands of female *I. holocyclus* except for Fig. 4.22 which is from a male *I. holocyclus*.

8

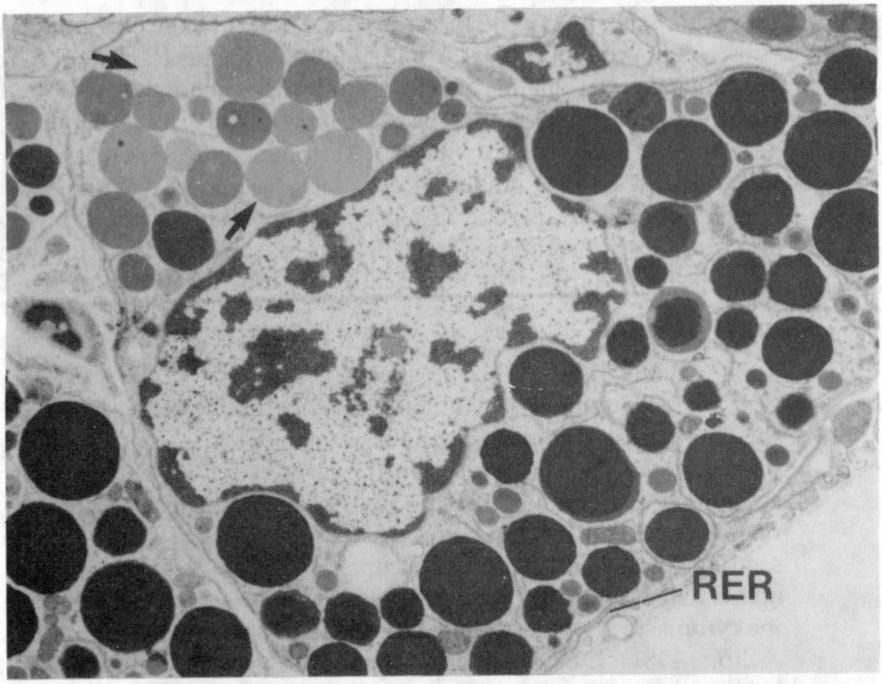

9

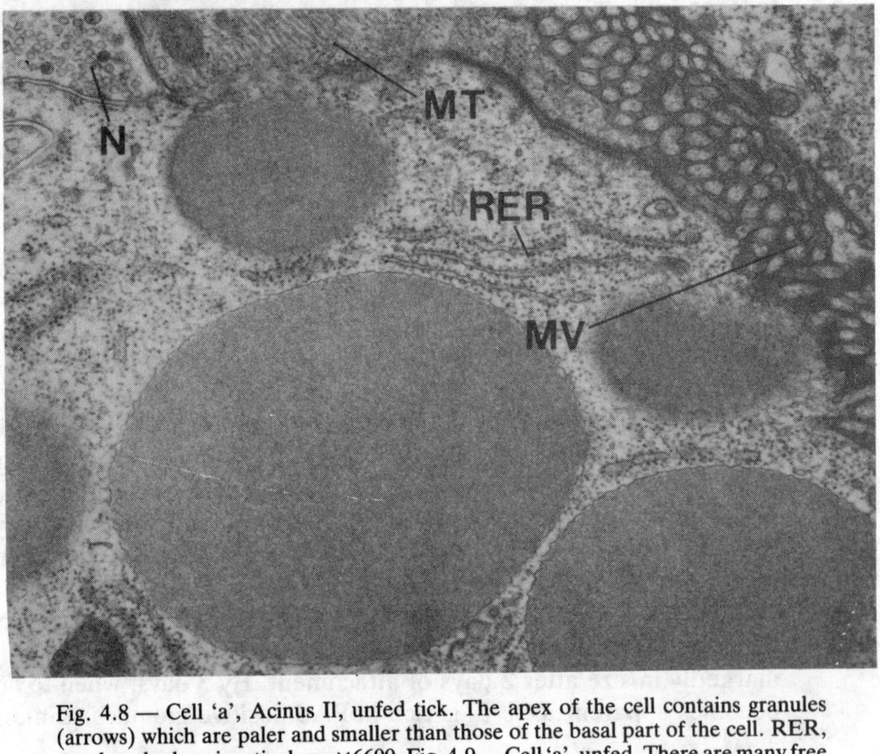

Fig. 4.8 — Cell 'a', Acinus II, unfed tick. The apex of the cell contains granules (arrows) which are paler and smaller than those of the basal part of the cell. RER, rough endoplasmic reticulum. ×6600. Fig. 4.9 — Cell 'a', unfed. There are many free ribosomes, and cisternae of RER are not distended. MT, microtubules of an interstitial adluminal cell; MV, microvilli; N, nerve terminal. ×31800.

10

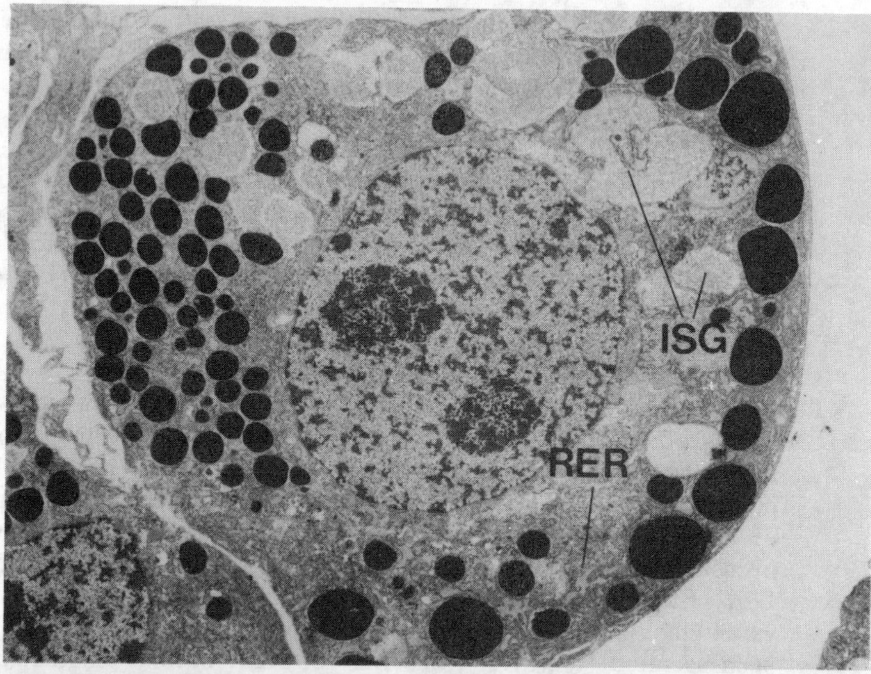

11

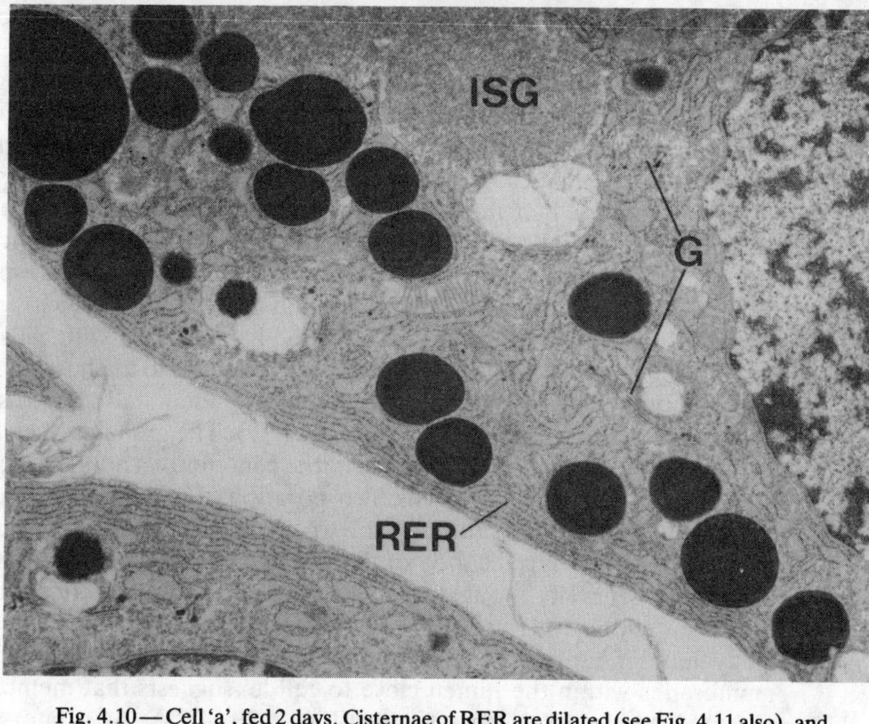

Fig. 4.10 — Cell 'a', fed 2 days. Cisternae of RER are dilated (see Fig. 4.11 also), and there are many immature secretory granules (ISG). ×4000. Fig. 4.11 — Cell 'a', fed 2 days. Higher magnification of cell 'a' fixed during the stage of high synthesizing activity showing dilated RER and Golgi cisternae (G). Note the more electron-lucent cortex of the granules in the basal part of the cell. ISG, immature secretory granules. ×12200.

Stone 1981). Ultrastructural examination generally confirms these light microscopic observations. Compared to later stages, cell 'b' of the unfed tick has scant cytoplasm which contains mainly mitochondria, free ribosomes, a small number of flattened cisternae of RER, and a few granules (Fig. 4.12). In ticks fed for 2 days the cells are larger; they contain dilated cisternae of RER, Golgi bodies are more apparent, and there are a small number of granules at the apex of the cell (Figs 4.13, 4.14). By 4 days these apical granules are the dominant feature of cell 'b', and the cytoplasmic organelles continue to show signs of a high level of synthesizing activity (Fig. 4.15).

4.5.4 Acinus III

Cell 'd', which surrounds the duct of acinus III, was considered from light microscopical observations to actively secrete for a short time only (Binnington & Stone 1981). In the unfed female the cell is packed with PAS positive granules. However, after half a day the cell is smaller with a decrease in the volume occupied by granules. By 3 days the nuclei are pycnotic and surrounded by very little cytoplasm (Binnington & Stone 1981). Electron microscopy indicates that this is probably an oversimplified view of the role of cell 'd' in feeding. After the large granules (Fig. 4.16) have been secreted (Fig. 4.17), cisternae of RER become dilated, and it is possible that this cell type continues to secrete during the period from attachment to at least 2 days following attachment. An interesting feature of cell 'd' is the presence of membranous whorls within the granules (Fig. 4.16). Furthermore the membranous material is secreted into the lumen (Fig. 4.16) from where it must ultimately enter the feeding lesion in the host skin.

Cell 'e', which occupies the fundus of acinus III, also loses stored granules rapidly when the tick begins to feed. From light microscopy, it appeared that the PAS positive granules were replaced by apical clear vacuoles (Binnington & Stone 1981). However, electron microscopy shows that at least some of these vacuoles may be granules which are more electron-lucent than those of unfed ticks (Fig. 4.18). As with cells 'a', 'b', and 'd', cell 'e' appears to be activated by attachment to the host. In recently attached ticks, cisternae of RER are more distended, and various stages of forming secretory granules are seen (Fig. 4.19). The granules of cell 'e' fuse so that during a single exocytotic event, the contents of a number of granules may be released (Fig. 4.20). The incorporation of granule membranes into the luminal plasma membrane, an expected consequence of exocytosis, must increase the surface area of the plasma membrane enormously. This increase is generally thought to be compensated for by retrieval of membrane via endocytotic vesicles, but there is no evidence from the current study that this happens in cell 'e' of *I. holocyclus*. The presence of profiles of membranes within the lumen close to cell 'e' suggests that membranes of some of the granules are secreted (Figs 4.20, 4.21). At a later stage of attachment (4 days) the apex of cell 'e' is occupied by a mixture of granules of moderate electron density and almost empty vesicles (Fig. 4.21).

4.5.5 Salivary glands of male ticks

Light microscopy of the single type of acinus (acinus IV) of male ticks indicated that there was only one cell type (Binnington & Stone 1981). Electron microscopy shows that this conclusion is probably correct. However, it needs to be noted that in some cells the granules are closely packed and have very dense contents, whereas in others there is more space between the granules and the granules have less dense contents (Fig. 4.22). A factor which argues against more than one cell type in the male is that the cells showing these different features appear to have a random spatial arrangement within acinus IV in contrast to the way in which the different cell types of the female occupy constant positions within the acini. The agranular cells of the male, which include those of acinus I and the adluminal and abluminal interstitial cells of acinus IV, resemble those of the female gland.

4.5.6 Cellular sources of holocyclotoxin and allergens

The increase in information provided by electron microscopy on the changes in morphology of acinus III does not challenge the previous conclusion that cell 'b' of acinus II is the most likely candidate as a source of the paralyzing toxin (Figs 4.23 and 4.24 — adapted from Binnington & Stone 1981). At the time of maximum toxicity, cells 'd' and 'e' contain little stored material compared with cells 'a' and 'b'. Furthermore, since allergic responses may occur very rapidly (within 1 h after tick attachment) the stored material in acinus III is still the best candidate as a source of allergenic material (Fig. 4.23).

4.6 SUMMARY

Advances that have been made over the past five years in research aimed at the development of a vaccine against *I. holocyclus* tick paralysis are reviewed. In particular, the nature of the paralyzing toxin (holocyclotoxin), the induced immune response to this and other immunogens (including allergens), and the biosynthetic physiology of the salivary glands is considered.

Studies on tick physiology, biochemistry, immunology, artificial feeding, and toxin mode of action have supported the concept of a proteinaceous toxin secreted via the salivary gland into the host. A toxoid obtained by glutaraldehyde detoxication of a holocyclotoxin preparation was much more immunogenic than holocyclotoxin itself, either injected or from feeding ticks. The toxoid offers prospects for the development of a tick paralysis vaccine. The degree of immunity to tick paralysis may only be partially related to levels of circulating serum antitoxin. Allergic reactions to *I. holocyclus* varied according to the host but could be very serious in humans.

Electron microscopy has added to our understanding of how the morphology of the various cell types of the salivary gland changes after tick attachment. This is particularly so for cells 'd' and 'e' of acinus III which apparently have the ability to secrete stored material before their synthesiz-

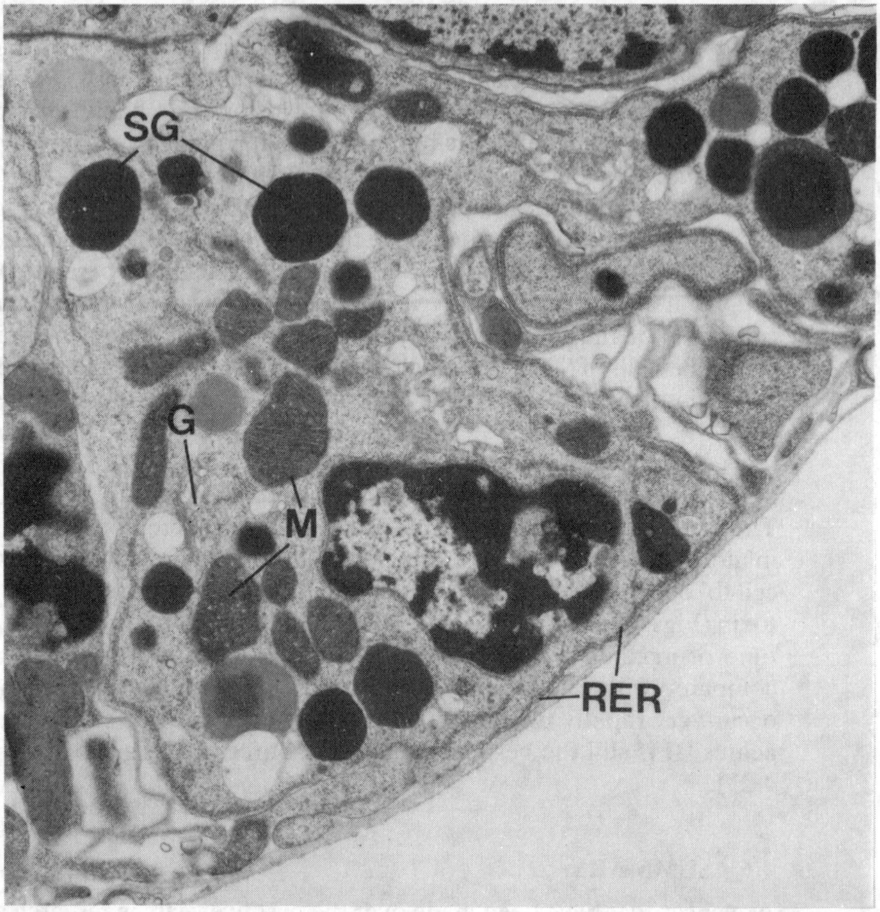

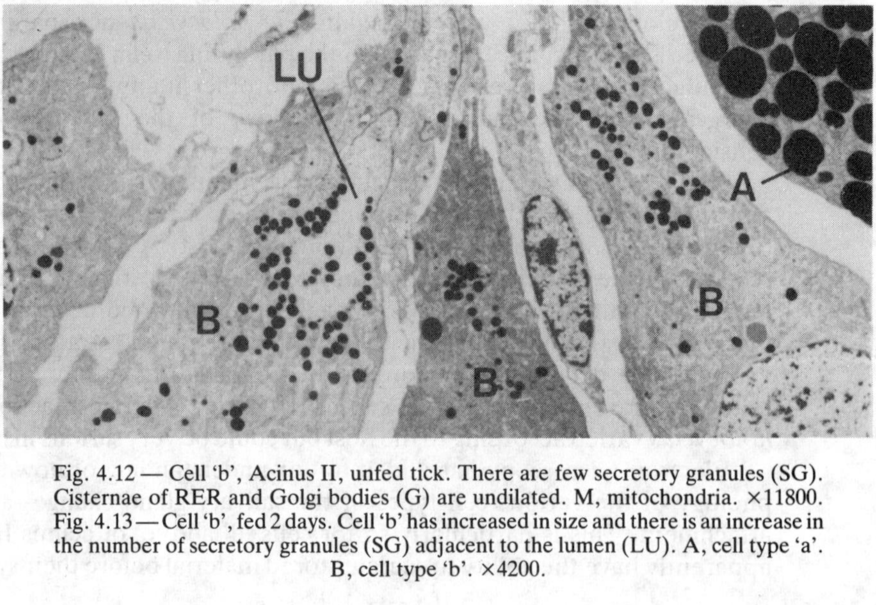

Fig. 4.12 — Cell 'b', Acinus II, unfed tick. There are few secretory granules (SG). Cisternae of RER and Golgi bodies (G) are undilated. M, mitochondria. ×11800.
Fig. 4.13 — Cell 'b', fed 2 days. Cell 'b' has increased in size and there is an increase in the number of secretory granules (SG) adjacent to the lumen (LU). A, cell type 'a'. B, cell type 'b'. ×4200.

14

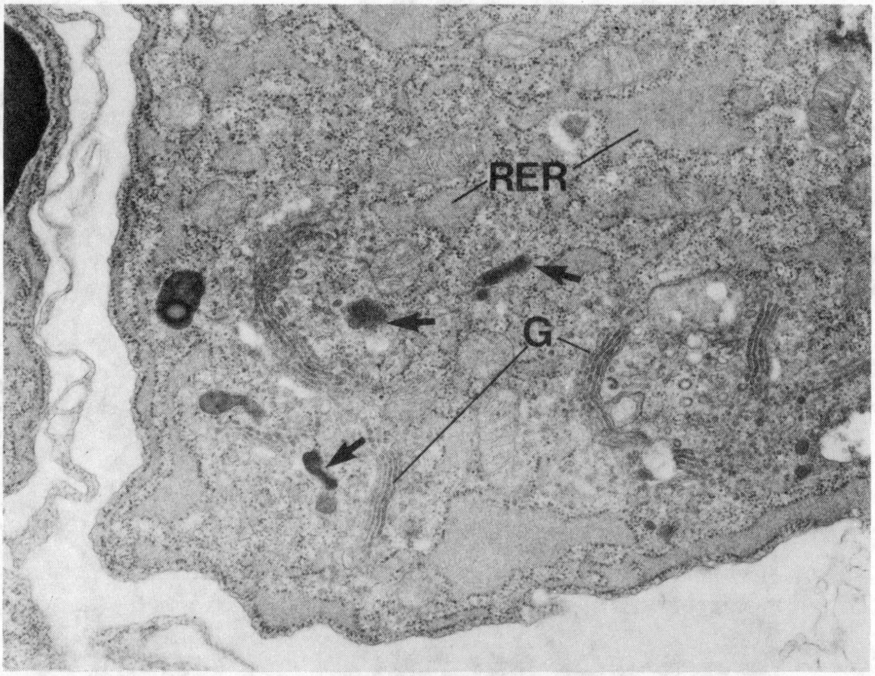

15

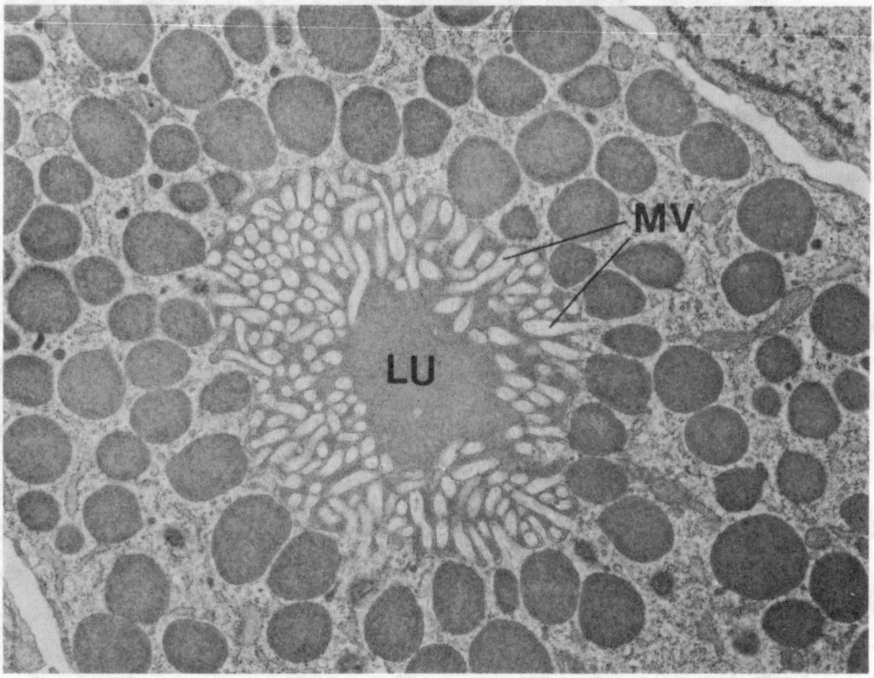

Fig. 4.14 — Cell 'b', fed 2 days. Cisternae of RER are now grossly distended and Golgi bodies (G) have electron-dense secretory material (arrows) associated with them. ×15600. Fig. 4.15 — Cell 'b', fed 4 days. In this highly toxic stage of tick, the apex of the cell contains many secretory granules. LU, lumen. MV, microvilli. ×12800

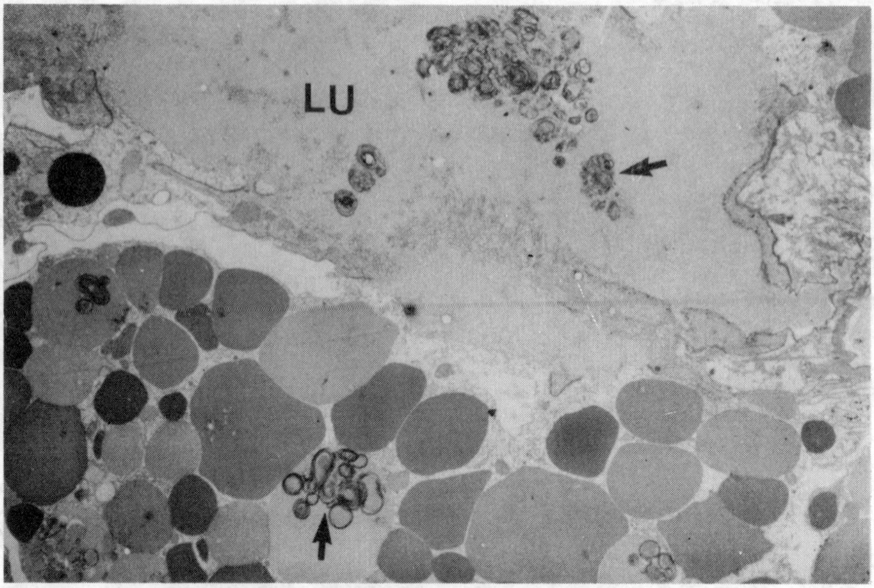

16

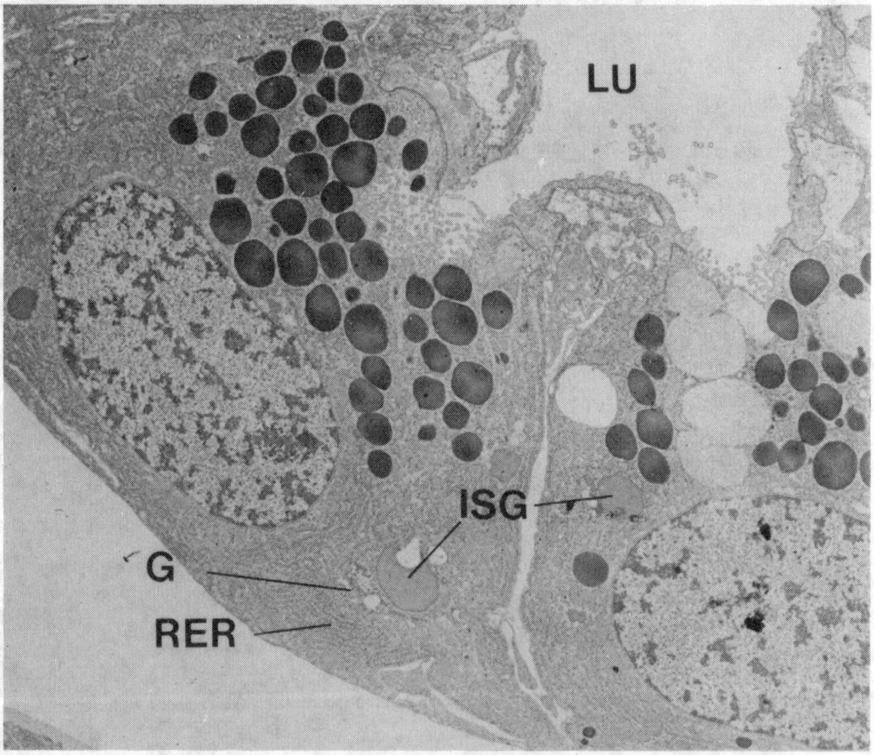

17

Fig. 4.16 — Cell 'd', Acinus III, unfed tick. Very large granules occupy most of the cytoplasm. In some granules there are membranous structures (arrows) which are also seen in the lumen (LU) of the Acinus. ×4500. Fig. 4.17 — Cell 'd', fed 2 days. The large granules seen in the unfed stage have been replaced by smaller, more electron-dense granules. Also present are immature secretory granules (ISG). G, Golgi bodies; RER, rough endoplasmic reticulum. LU, lumen. ×5500.

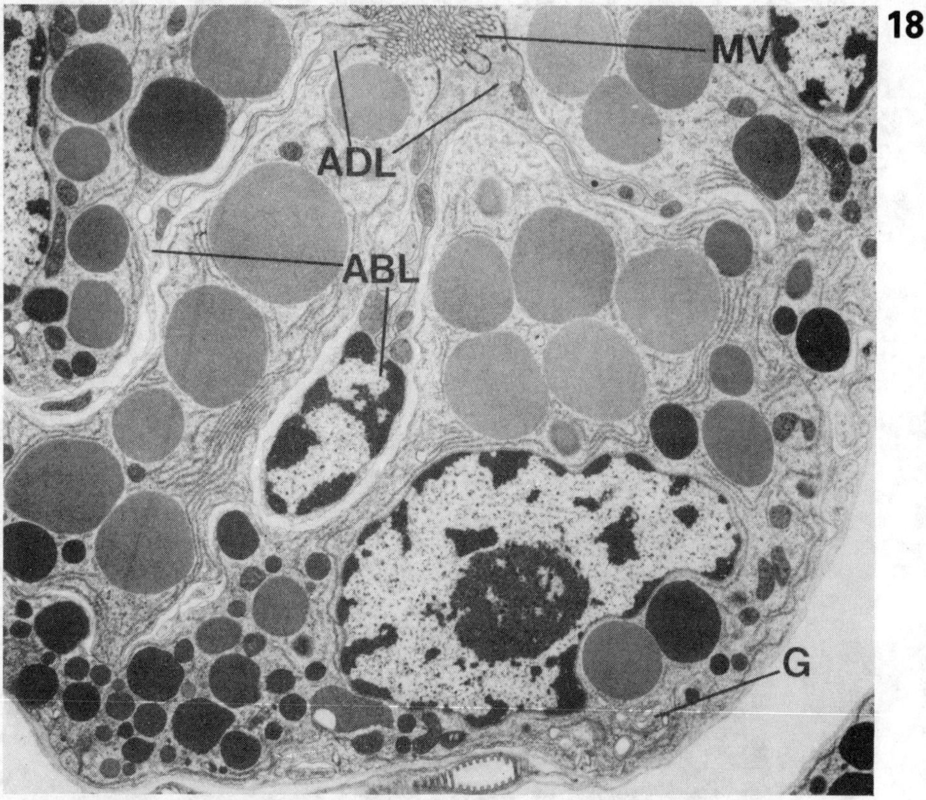

Fig. 4.18 — Cell 'e', Acinus III, unfed tick. The cells are rich in secretion granules, but cisternae of RER and the Golgi bodies (G) contain little secretory material. ABL, Abluminal interstitial cell; ADL, adluminal interstitial cell; MV, microvilli. ×7000.

ing mechanisms are 'switched on' by attachment, to produce granules of a different morphology. It is confirmed that cell 'b' of acinus II contains little secretory material during the time that the salivary gland is non-toxic, but is rich in granules when ticks are able to paralyze the host and toxin can be readily extracted from the glands. Thus cell 'b' remains as the best candidate as the source of toxin.

ACKNOWLEDGEMENTS

The authors wish to thank Drs Ken Bremner, David Kemp, and Ian Wright for valuable comments on the manuscript and Ms Margaret Commins and Mr Alan Neish for technical assistance.

REFERENCES

Allen, J. R., Doube. B. M., & Kemp, D. H. (1977) Histology of bovine skin reactions to *Ixodes holocyclus* Neumann. *Canadian Journal of Comparative Medicine* **41**: 26–35.

19

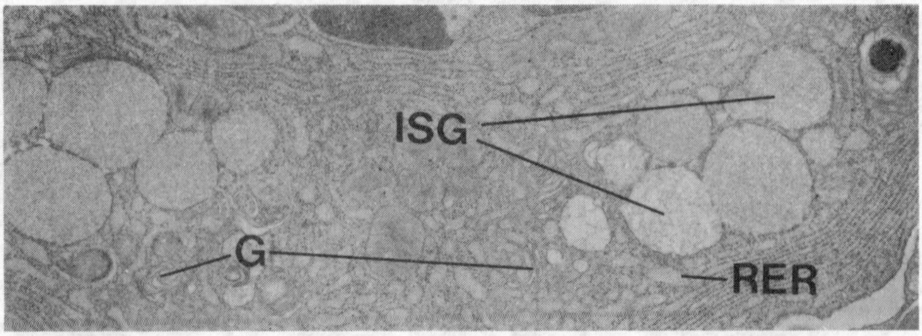

20

Fig. 4.19 — Cell 'e', fed 1 day. During the early stage of feeding, cell 'e' contains distended cisternae of RER and immature secretory granules (ISG). G, Golgi bodies. ×20400. Fig. 4.20 — Cell 'e', fed 1 day. There is continuity between the lumen (LU) and the remnants of fused granules (asterisks), some of which have not yet lost all of their contents (arrow). ×7900.

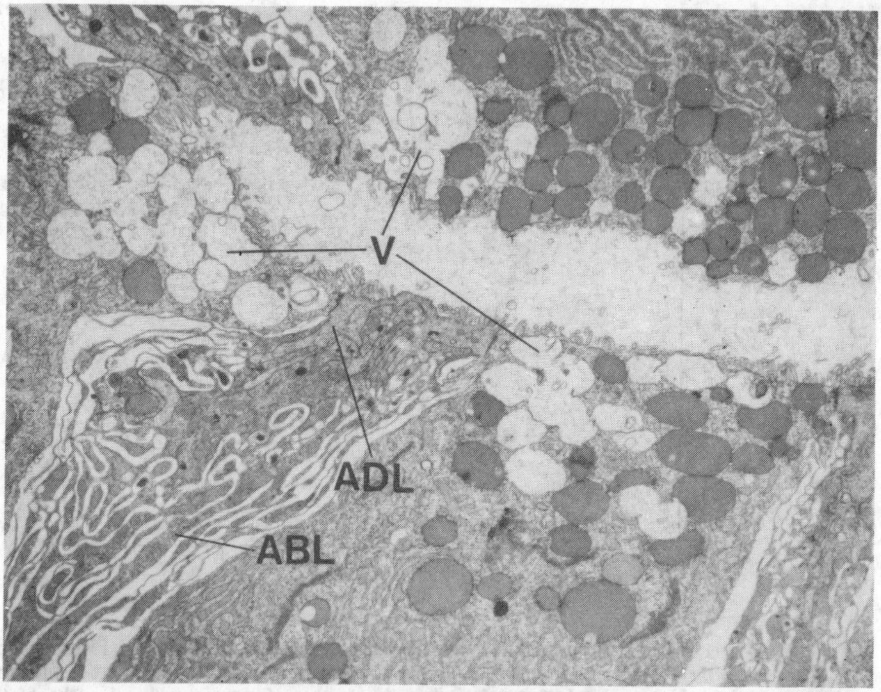

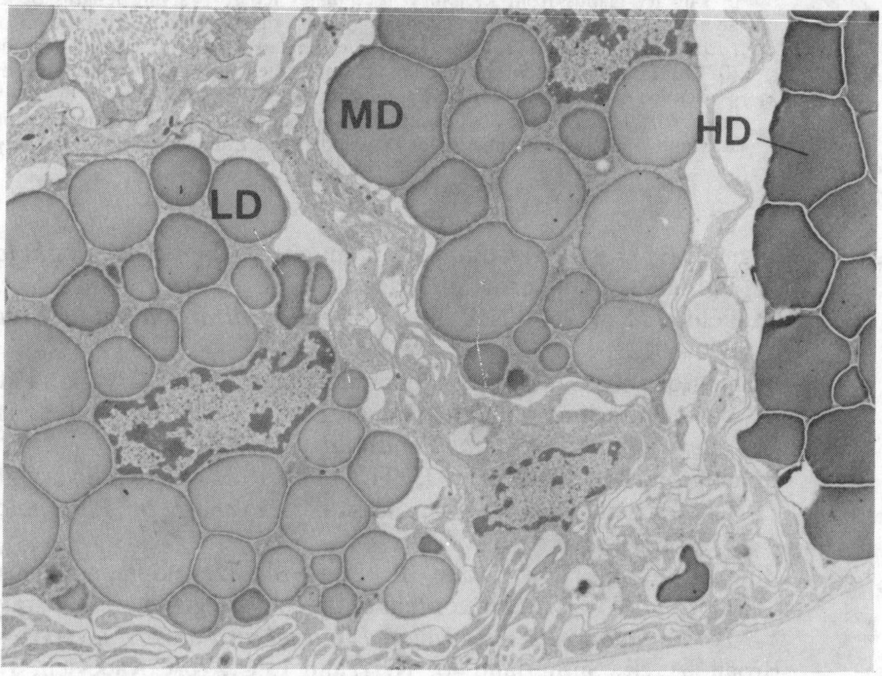

Fig. 4.21 — Cell 'e', fed 2 days. The apex of the cell is occupied by vacuoles (V) and some secretory granules. Cisternae appear less distended than in the 1 day fed female (see Fig. 4.12). ABL, abluminal interstitial cell; ADL, adluminal interstitial cell. ×5100. Fig. 4.22 — The salivary gland of male *I. holocyclus* probably contains one secretory cell type. However, there is variation between the cells in density of secretory granules. The granules may be of low (LD), moderate (MD), or high (HD) density. ×5200.

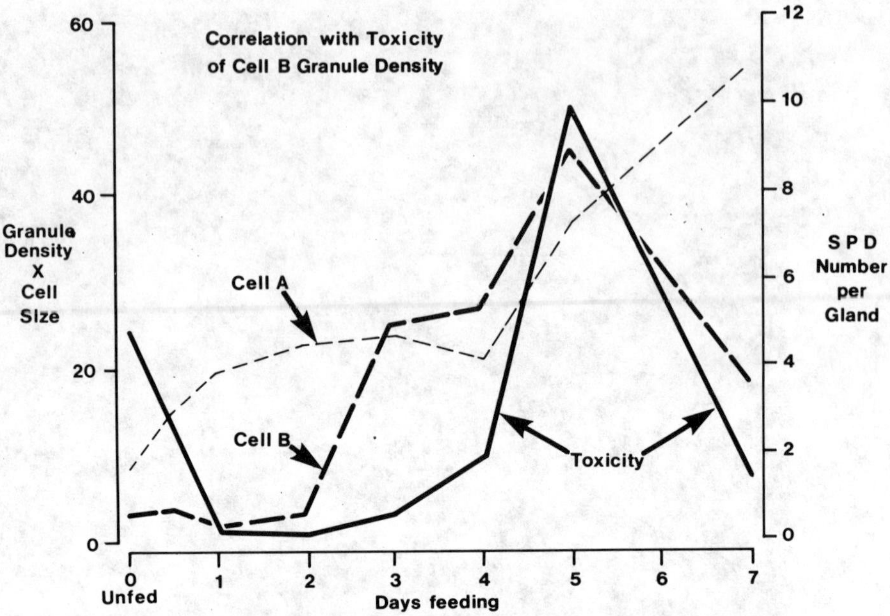

Fig. 4.23 — Relationship between change in the secretory material within salivary cells of types 'a' and 'b' (A and B in figure) of Acinus II, and toxicity of homogenates of salivary glands from *I. holocyclus* females fed for varying lengths of time. Secretory material was calculated as follows: distance from cell base to apex (μm) × distance across cell base (μm) × estimated cytoplasmic density of granules (0–4+). Points represent the mean of at least 3 determinations from each of 3 ticks. Toxicity is expressed as standard paralyzing doses (SPDs) per gland. Figure illustrates correlation with toxicity of cell 'b' granule density — adapted from Binnington & Stone (1981) with permission of Pergamon Press Ltd and Australian Society for Parisitology.

Bagnall, B. G. (1975) Cutaneous immunity to the tick *Ixodes holocyclus*. PhD thesis, University of Sydney.

Banfield, J. F. (1966) Tick bites in man. *Medical Journal of Australia* **2**: 600–601.

Binnington, K. C. (1978) Sequential changes in salivary gland structure during attachment and feeding of the cattle tick, *Boophilus microplus*. *International Journal for Parasitology* **8**: 97–115.

Binnington, K. C. & Kemp, D. H. (1980) Role of tick salivary glands in feeding and disease transmission. *Advances in Parasitology* **18**: 315–339.

Binnington, K. C. & Stone, B. F. (1981) Developmental changes in morphology and toxin content of the salivary gland of the paralysis tick *Ixodes holocyclus*. *International Journal for Parasitology* **11**: 343–351.

Broad, A. J., Sutherland, S. K., Tanner, C., & Covacevich, J. (1979) Electrophoretic, enzyme and preliminary toxicity studies of the venom of the small-scaled snake, *Parademansia microlepidotus* (Serpentes: Elapidae) with additional data on its distribution. *Memoirs of the Quensland Museum* **19**: 319–329.

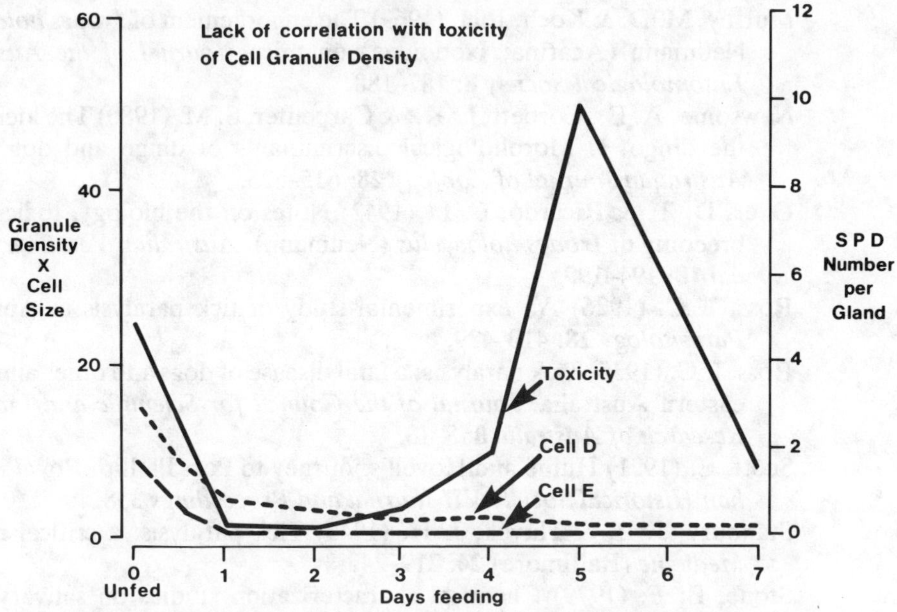

Fig. 4.24 — Relationship between change in the secretory material within salivary cells of types 'd' and 'e' (D and E in figure) of Acinus III, and toxicity of homogenates of salivary glands from *I. holocyclus* females fed for varying lengths of time. Secretory material was calculated as follows: distance from cell base to apex (μm) × distance across cell base (μm) × estimated cytoplasmic density of granules (0–4+). Points represent the mean of at least 3 determinations from each of 3 ticks. Toxicity is expressed as standard paralyzing doses (SPDs) per gland. Figure illustrates lack of correlation of toxicity with cell 'd' and 'e' granule density — adapted from Binnington & Stone (1981) with permission of Pergamon Press Ltd and Australian Society for Parasitology.

Cooper, B. J. & Spence, I. (1976) Temperature-dependent inhibition of evoked acetylcholine release in tick paralysis. *Nature* **263**: 693–695.

Doine, A. I., Oliver, C., & Hand, A. R. (1984) The Golgi apparatus and GERL during postnatal differentiation of rat parotid acinar cells: an electron microscopic cytochemical study. *Journal of Histochemistry and Cytochemistry* **32**: 477–485.

Doube, B. M. & Kemp, D. H. (1975) Paralysis of cattle by *Ixodes holocyclus* Neumann. *Australian Journal of Agricultural Research* **26**: 635–640.

Goodrich, B. S. & Murray, M. D. (1978) Factors influencing the toxicity of salivary gland extracts of *Ixodes holocyclus* Neumann. *International Journal for Parasitology* **3**: 313–320.

Gregson, J. D. (1973) *Tick paralysis: an appraisal of natural and experimental data,* Canada Department of Agricultre Monograph No. 9.

Kaire, G. H. (1965) A method for assay of canine anti-tick serum. *Australian Veterinary Journal* **41**: 239–240.

Kaire, G. H. (1966) Isolation of tick paralysis toxin from *Ixodes holocyclus*. *Toxicon* **4**: 91–97.

Murray, M. D. & Koch, J. H. (1969) The engorgement of *Ixodes holocyclus* Neumann (Acarina: Ixodoidea) on mice. *Journal of the Australian Entomological Society* **8**: 187–188.

Newsome, A. E., Corbett, L. K., & Carpenter, S. M. (1980) The identity of the dingo. I. Morphological discriminants of dingo and dog skulls. *Australian Journal of Zoology* **28**: 615–625.

Oxer, D. T. & Ricardo, C. L. (1942) Notes on the biology, toxicity and breeding of *Ixodes holocyclus* (Neumann). *Australian Veterinary Journal* **18**: 194–199.

Ross, I. C. (1926) An experimental study of tick paralysis in Australia. *Parasitology* **18**: 410–429.

Ross, I. C. (1935) Tick paralysis: a fatal disease of dogs and other animals in eastern Australia. *Journal of the Council for Scientific and Industrial Research of Australia* **8**: 8–13.

Scott, E. (1921) Hume and Hovell's journey to Port Phillip. *Royal Australian Historical Society VII Journal and Proceedings* 358.

Stanbury, J. B. & Huyck, J. H. (1945) Tick paralysis: a critical review. *Medicine* (Baltimore) **24**: 219–242.

Stone, B. F. (1979) Chemical characterization studies on salivary gland toxins of the paralysis tick *Ixodes holocyclus*. In: I. W. Chubb, & Geffen, L. B. (eds.) *Neurotoxins: fundamental and clinical advances*. Adelaide, Adelaide University Union Press, p. 273.

Stone, B. F., Doube, B. M., Binnington, K. C., & Goodger, B. V. (1979) Toxins of the Australian paralysis tick *Ixodes holocyclus*. In: J. G. Rodriguez (ed.), *Recent advances in acarology*, vol. 1. New York, Academic Press, pp. 347–356.

Stone, B. F. & Wright, I. G. (1980) Toxins of *Ixodes holocyclus* and immunity to paralysis. In: Johnston, L. A. Y. & Cooper, M. G. (eds.) *Ticks and tick-borne diseases*. Sydney, Australian Veterinary Association, pp. 75–78.

Stone, B. F. & Wright, I. G. (1981) Tick toxins and protective immunity. In: Whitehead, G. B. & Gibson, J. D. (eds.) *Tick biology and control*. Grahamstown, Tick Research Unit, Rhodes University, pp. 1–5.

Stone, B. F., Cowie, M. R., Kerr, J. D., & Binnington, K. C. (1982a) Improved toxin/antitoxin assays for studies on the Australian paralysis tick *Ixodes holocyclus*. *Australian Journal of Experimental Biology and Medical Science* **60**: 309–318.

Stone, B. F., Neish, A. L., & Wright, I. G. (1982b) Immunization of rabbits to produce high serum titres of neutralizing antibodies and immunity to the paralysing toxin of *Ixodes holocyclus*. *Australian Journal of Experimental Biology and Medical Science* **60**: 351–358.

Stone, B. F., Commins, M. A., & Kemp, D. H. (1983a) Artificial feeding of the Australian paralysis tick, *Ixodes holocyclus* and collection of paralysing toxin. *International Journal for Parasitology* **13**: 447–454.

Stone, B. F., Neish, A. L., & Wright, I. G. (1983b) Tick (*Ixodes holocyclus*) paralysis in the dog — quantitative studies on immunity following artificial infestation with the tick. *Australian Veterinary Journal* **60**: 65–68.

Stone, B. F. & Neish, A. L. (1984) Tick-paralysis toxoid: an effective immunizing agent against the toxin of *Ixodes holocyclus. Australian Journal of Experimental Biology and Medical Science* **62**: 189–191.

Stone, B. F., Neish, A. L., & Gauci, M. (in manuscript 1985) Preliminary characterization of the paralyzing toxin from the Australian paralysis tick *Ixodes holocyclus.*

Stone, B. F., Neish, A. L., Morrison, J. J., & Uren, M. F. (1986) Toxoid stimulation in dogs of high titres of neutralizing antibodies against holocyclotoxin the toxin of the Australian paralysis tick *Ixodes holocyclus. Australian Veterinary Journal* **63:** 125–127

Sutherst, R. W. & Moorhouse, D. E. (1971) *Ixodes holocyclus* larvae and 'scrub-itch' in south east Queensland. *Southeast Asian Journal of Tropical Medicine and Public Health* **2**: 82–83.

Trinca, J. C. (1964) Insect allergy in Australia: results of a five year survey. *The Medical Journal of Australia* **2**: 659–663.

Vosbeck, K. D., Greenberg, B. D., & Awad, W. M. Jr. (1974) The proteolytic enzymes of the K-1 strain of *Streptomyces griseus* obtained from a commercial preparation (Pronase). Specificity and immobilization of aminopeptidase. *Journal of Biological Chemistry* **250**: 3981–3987.

Wright, I. G., Stone, B. F., & Neish, A. L. (1983) Tick (*Ixodes holocyclus*) paralysis in the dog — induction of immunity by injection of toxin. *Australian Veterinary Journal* **60**: 69–70.

5

Water balance by ticks between bloodmeals

Glen R. Needham and **Pete D. Teel**

5.1 INTRODUCTION

It is not unusual for an adult tick to live for several years without feeding. This can place a great burden on the tick's ability to maintain water balance in a desiccating environment off the host. A three-host ixodid tick may spend more than 98% of its life off the host animal (Norval 1977). Most multi-host argasids exceed this in their time off the host (Balashov 1968).

Ticks have evolved numerous capabilities which are responsible for their impressive longevity. Both the Ixodidae (hard ticks) and the Argasidae (soft ticks) actively absorb water vapor from unsaturated air. They have an integument which can dramatically restrict transpirational water loss. The tracheal system is guarded by spiracular valves which open infrequently during inactive periods. Other factors which contribute to their survival during long off-host periods include their slow metabolic rates which help conserve energy stores; and the seeming ability of tick tissues to tolerate relatively significant changes in the hemolymph which bathes them. Some species can tolerate the loss of up to half of their body water.

In many instances these potential vectors are the long-term reservoirs of disease in nature rather than the hosts they feed upon. It is not unusual for their longevity to exceed that of their hosts. This aspect alone makes it imperative that we understand the processes which enable ticks to survive for extended periods. The purpose of this chapter is to draw attention to past studies which should be re-examined in light of new insights provided by recent advances in water balance physiology. By doing so we hope to point out the numerous gaps in our understanding of these sometimes complex and fascinating processes. Much of the literature deals with the ixodids, and our experience is primarily with *Amblyomma* species, thus our report will have an emphasis toward them.

5.2 PASSIVE VAPOR EXCHANGE BETWEEN THE TICK AND ENVIRONMENT

Ticks are relatively small creatures with high internal body water concentrations, surrounded by an environment which is frequently much lower in water concentration. For detached fasting ticks, as well as many other terrestrial arthropods, a net loss (desiccation) usually results owing to this difference in internal and environmental concentrations of water. Water moves inwardly, driven by the vapor pressure of the environmental water, while outward diffusion is driven by the vapor pressure of liquid water in the internal body compartment. Fasting ticks compensate for this deficit by several processes, including catabolic and anabolic water production (Wharton 1985), as well as passive and active water vapor uptake. We discuss active uptake in section 5.3.

Certain fundamental concepts must be understood before one can appreciate vapor exchange dynamics between the fasting tick and its environment. Two and sometimes three or more separate processes of water movement are occurring in the tick at any given time (Fig. 5.1). Passive influx and efflux rates are independent of one another. The critical equilibrium activity (CEA; activity = relative humidity/100) is the activity at which water uptake from the environment equals loss from the body. For this value to be lower than about 0.99 (hemolymph activity) the tick must have the ability to actively absorb water from the air. The lowest vapor activity (or humidity) at which active sorption can occur is therefore the CEA. The separate active and passive absorption rates for *A. americanum* adult females are given in Fig. 5.2. Factors governing water flux between a tick and its microenvironment include: (1) water activity, (2) temperature, (3) permeability of the integument to water, (4) area of the integumental exchange surface, and (5) respiratory exchange when the spiracles of nymphs and adults are open (larvae have no spiracles). All of these affect vapor exchange, but the roles they play are sometimes not appreciated.

The rationale for using activity as the measure of water concentration has been discussed by others (Edney 1977, Wharton & Richards 1978, Arlian & Veselica 1979, Wharton 1985). Ideally, activity of water is its mole fraction, or the ratio of moles of solvent (water) to moles of solute plus moles of solvent (Table 5.1). Except for very dilute solutions of non-electrolytes, this

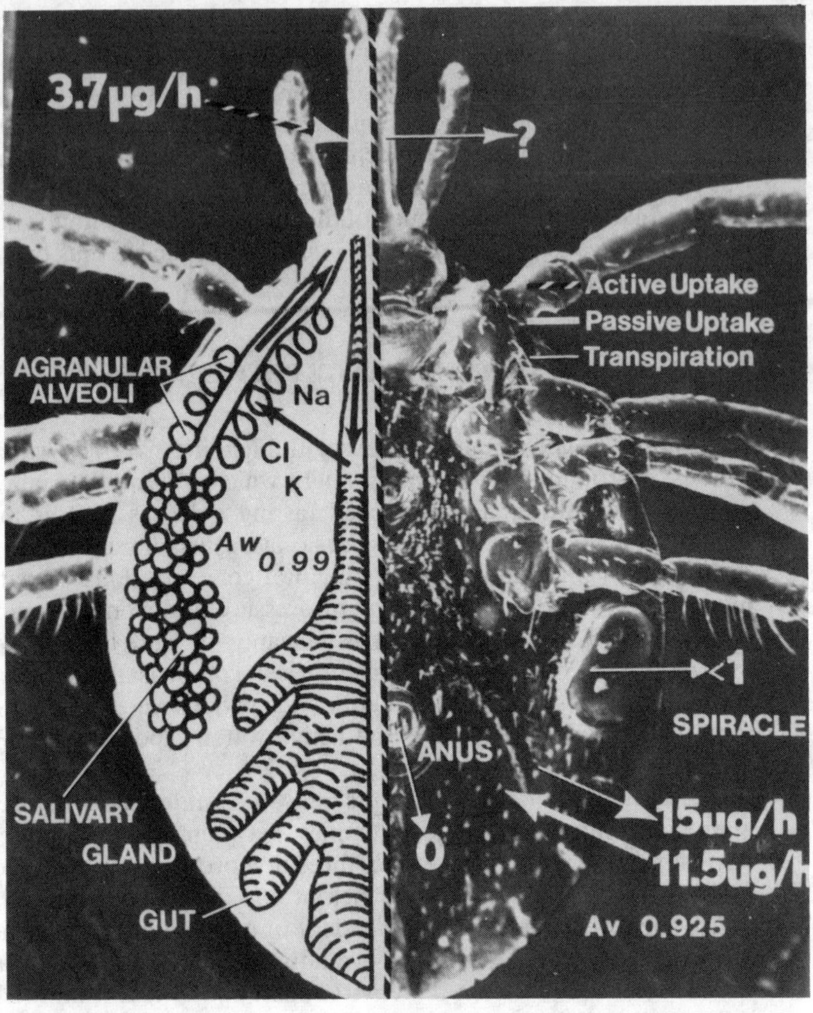

Fig. 5.1 — Modified scanning electron micrograph of an unfed adult ixodid *Amblyomma americanum* showing the primary avenues and rates of water gain and loss in an environmental water vapor activity of 0.925 at 25°C. Rates for active uptake via the mouth and loss from the spiracles are somewhat misleading because they are known to be discontinuous. Possible routes of ion movement during active rehydration through the mouth are shown. Values are from ten ticks which were 6–8 weeks post-fed (see Freda & Needham 1984).

value should be multiplied by an activity coefficient which corrects for any deviation from ideal conditions. When activity is used for water in liquid and vapor phases it is easy to visualize water movement by simple diffusion down activity gradients. For example, it is difficult to comprehend water exchange when the environmental relative humidity is given to define conditions for uptake, and milliosmoles or vapor pressure in atmospheres is given for

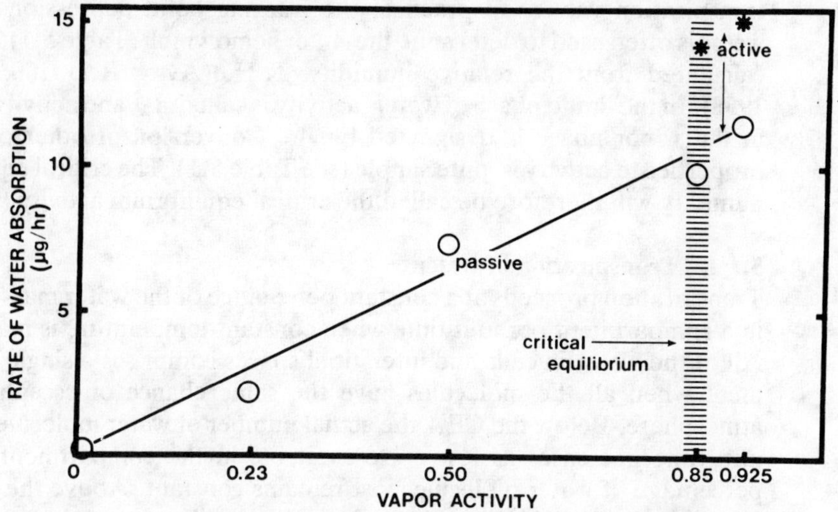

Fig. 5.2 — Passive and active water absorption by unfed *Amblyomma americanum* adult females in five water vapor activities at 25°C. The critical equilibrium value is the activity (relative humidity/100) at which active water uptake can commence, and the rate of uptake (passive and active) equals the rate of loss. Passive rates at 0.85 and 0.925 av were determined by blocking the mouthparts of ticks and subtracting these rates from values obtained without the mouthparts blocked. At least ten ticks were sampled at each activity, and tritiated water was used as a tracer. * indicates active plus passive (Freda & Needham 1984).

Table 5.1 — Interconversion of measurements for colligative properties governing the diffusion of water within the range of critical equilibria and hemolymph water activities of ticks

a_w	Osmoles	Δ°C	π_{25}(atm)
0.9950	0.2789	0.5183	6.2485
0.9925	0.4195	0.7793	9.3985
0.9900	0.5607	1.0418	12.5619
0.9875	0.7026	1.3055	15.7411
0.9850	0.8453	1.5706	18.9359
0.940	3.5431	6.5830	79.3796
0.850	9.7955	18.2001	219.458
0.750	18.5027	34.3780	414.534

a_w=55.508/(55.508+osm), osmoles=(55.508–55.508 a_w)/a_w, Δ°C=1.858[(55.508–55.508 a_w)/a_w], π_{25}=22.404[(55.508–55.508 a_w)/a_w], 55.508=moles water/kg water. Adapted from Arlian & Veselica (1979).

transpiration valves. In practice, the freezing point depression of liquid water is often used to determine the A_w of hemolymph (Table 5.1), and A_v is calculated from the relative humidity, R.H. (Av = R.H./100; Penman 1958). In the liquid phase of water, activity is called A_w, and activity of water in the vapor phase is designated by A_v. Conversion of other colligative properties to activity is quite simple (see Table 5.1). The critical equilibrium humidity will therefore be called the critical equilibrium activity (CEA).

5.2.1 Transpiration of water

Transpiration proceeds at a constant percentage of the water mass available in a compartment per unit time when constant temperature is maintained. All of the water in cells and interstitial spaces comprises a single compartment when all the molecules have the same chance of escaping to the atmosphere. Below the CEA the actual number of water molecules lost per unit time decreases as fewer are available in the compartment, but the percentage of water molecules lost remains constant. Above the CEA the actual amount lost by transpiration increases while water is being absorbed actively and passively for a net gain of water, but the rate (%/hh) remains constant. When equilibrium is achieved (sorption = transpiration) the number of water molecules lost remains unchanged because the percentage of a stable water mass is a constant. Therefore the instantaneous transpiration rate (mg/hh) from a compartment can be estimated by multiplying the rate constant (%/hh) times the water mass in that compartment. The transpiration rate constant for unfed female *Amblyomma americanum* ticks 4–6 weeks post-molt is 0.5%/hh at 26°C (Freda & Needham 1984). Why is this rate relatively constant? If a tick were to lose 50% of its water, and for simplification this came from a single compartment represented by the hemolymph, the concentration of solutes would approximately double. If the original osmotic pressure increased from 0.2789 to 0.5607 osmoles after desiccation (Table 5.1) there would still be very little effect on the water activity, and thus the driving force for transpiration remains relatively stable. Solute concentrations are generally so dilute that even when doubled the A_w decreases only slightly. The surface area does not change, and the permeability of the cuticle does not seem to make acute adjustments in response to ticks being transferred to different water vapor activities (Freda & Needham 1984). The aging process may, however, affect transpiration rates, and this possiblity is explored elsewhere (section 5.4).

5.2.2 Passive water vapor absorption

Passive water absorption rate is proportional to the ambient water vapor activity (Fig. 5.2). The rate of inward diffusion at constant atmospheric pressure and temperature is a function of tick surface area, integumental permeability, and the ambient A_v. Below the CEA, passive sorption can be accounted for by water uptake across the integument. For unfed *A. americanum* females passive sorption accounted for 66–77% of the total at 0.85 and 0.93 A_v, respectively, with active uptake providing the difference (Freda & Needham 1984). At relatively high ambient A_v's passive sorption still accounts for most of the water vapor sorption in *A. americanum*, and this is probably so for other ticks as well.

The fundamental difference between passive uptake and transpiration rates is that the environment offers an infinite source of water for sorption compared to the finite quantity available in the tick for transpiration. At a given A_v below the CEA a virtual equilibrium water mass will be approached. This equilibrium results because as body water content decreases the actual number of water molecules leaving decreases and will ultimately equal the number entering passively. If the water concentration is too low and internal solutes become too concentrated the tick may die. When a tick is maintained below the CEA and water content is plotted versus time, one sees a decrease in the rate of loss (see Hafez *et al.* 1970b). That is, the steepness of the weight loss curve decreases. This has been interpreted by some investigators to mean that water loss is being reduced as a result of lowered cuticular permeability. In fact, the rate remains constant, but as the tick dehydrates there are fewer water molecules available to be lost, giving the appearance of reduced cuticular permeability. If this is plotted as a semi-log function of the water mass versus time (at 0 a_v), one sees a straight line where the slope is the rate constant for transpiration. If the line deviates from linearity, then changes in cuticular permeability, or losses from compartments which have different transpiration rates, may have resulted. Freda & Needham (1984) observed two compartments from which water was lost from *A. americanum* adult females. One may be cuticle surface moisture and the other internal. A fast compartment for house dust mites (*Dermatophagoides*) is due to evaporation of water from the supracoxal gland openings, while the hemolymph is probably the slow compartment (Arlian & Wharton 1974).

A fact which is sometimes not appreciated is that the arthropod integument is symmetric in its permeability to water, that is, passive influx and efflux are restricted to the same degree. What misled some early investigators was their belief that the integument was the site of active uptake as well as passive sorption (Machin 1979). In most instances specific sites (e.g. mouth or anus) are responsible for active uptake. When these areas are blocked, active absorption is inhibited. The earliest studies which demonstrated the symmetric nature of the arthropod integument with respect to water permeability were carried out by Richards *et al.* (1953).

5.2.3 Relationship of critical equilibrium activity and cuticle permeability to humidity requirements

The concept of passive uptake and transpiration occurring simultaneously and independently is fairly straightforward, but when you consider other variables, water balance physiology can quickly become complicated. For example, the CEA corresponds to the lowest or threshold activity at which active water uptake can commence (Knülle & Devine 1972). Some investigators erroneously believe that there is a direct correlation of this lowest threshold with the humidity requirements for a given stage or species. Certainly there are individual examples which could support this hypothesis;

however, when we plotted the estimated CEA values versus transpiration rate constants (%/day) determined in dry air for three argasid and eight ixodid species this correlation was not evident (Fig. 5.3). It seems that the

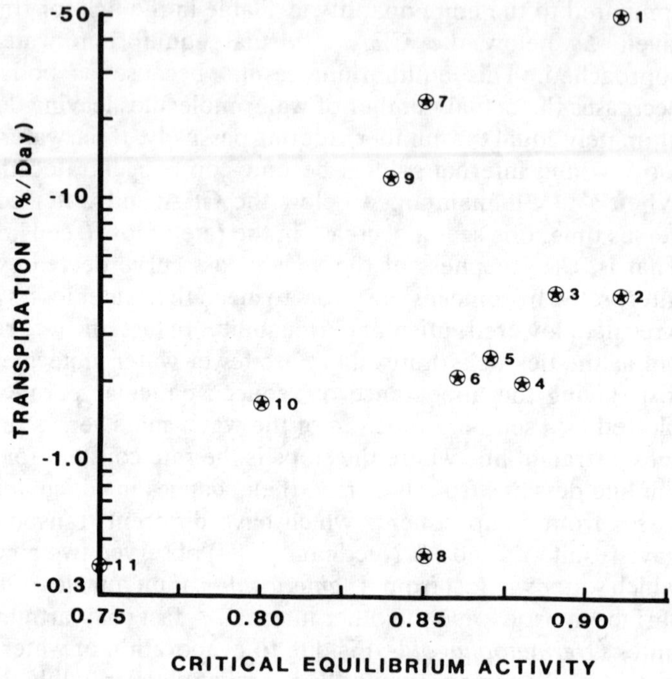

Fig. 5.3 — Transpiration rates for ticks held in dry air versus the estimated critical equilibrium activities for 3 argasid and 8 ixodid species. The number by each data point indicates the species. Data from species 1–6 were from Lees (1946) and were measured at 25°C. 1–*Ixodes ricinus* adults, 2–*Amblyomma cajennense* females, 3–*A. maculatum* females, 4–*Dermacentor andersoni* females, 5–*Rhipicephalus sanquineus* females, 6–*Ornithodorus moubata* nymphs, 7–*D. variabilis* larvae, 25°C (Knülle & Devine 1972), 8–*Argas arboreus* adults, 28°C (Hefnawy *et al.* 1975), 9–*Amblyomma americanum* females, 25°C (Sauer & Hair 1971, Freda & Needham 1984), 10–*Hyalomma dromedarii* adults, 26°C (Balashov & Filippova 1964), 11–*O. savignyi* nymphs, 1st instar, 28°C (Hafez *et al.* 1970b). Note the low degree of correlation, r=0.56.

transpiration rate, which reflects integumental permeability to water, is really more important than the threshold vapor activity at and above which active water uptake can occur. This is not an original concept but should be emphasized for those who are investigating tick distributions, habitat preferences, and habitat modification for tick control. In other words, the CEA range for a species should not be used to describe humidity requirements. Species which have similar CEA values but different integumental permeabilities may be found in microhabitats with very different water vapor activities. Also, caution must be used when calling a tick mesophilic or xerophilic because the microhabitat where it is usually sequestered may be very different from the surrounding macrohabitat. This is particularly true for ticks which live in burrows, nests, and fissures where the A_v may be quite high for a significant percentage of the time.

Environmental moisture stress is much greater in ticks with very permeable integuments, and they have higher rates of water turnover. Therefore, they seek more humid microhabitats to avoid desiccation and spend energy reserves in the process. If the integument is fairly impermeable to water the tick may survive for extended periods in an environment below the CEA, especially if it remains inactive to reduce losses. It is therefore probable that a considerable energy saving should be realized by less permeable ticks because active absorption would be required less frequently. This hypothesis is supported by the longevity of some argasid species which have integuments with low water permeability (Balashov 1968). Even though active sorption requires little energy for terrestrial arthropods (Edney 1977), the more frequent need for active uptake will contribute to the exhaustion of energy reserves.

5.2.4 Respiration and water vapor exchange

A significant avenue for water loss from nymphs and adults can occur during periods of ventilation from the respiratory system through the spiracles (Rudolph & Knülle, 1979). Respiration rates increased dramatically in active *Amblyomma variegatum* females, from 1–2 times/hh when inactive to about 15 times during locomotor activity at 20°C. A dramatic increase in water loss resulted when these ticks were exposed to high CO_2 concentrations (which stimulates the spiracles to remain open). This loss could be prevented by blocking the spiracles, and as expected, larvae (which have no spiracles) were not affected by the CO_2 exposure (cited in Knülle & Rudolph 1982). An active tick, especially during locomotion, will not only consume significant energy reserves but it will also lose a detectable amount of water as measured by continuous weight recordings with an electrobalance. On the other hand, when the tick is inactive the spiracle closing mechanism and slow metabolism offer considerable water and energy savings for the tick.

Wharton (1985) has proposed a process by which arthropods may gain net water by respiration which occurs in an intermittent and infrequent fashion in a sufficiently high humidity. Ticks fit this respiratory pattern when inactive (Rudolph & Knülle 1979). The mechanism might work by allowing a bulk flow of humid air through spiracular valves into the respiratory system. Bulk inward flow of moist air may be possible owing to two key factors. One is the net loss of N_2 across the cuticle driven by a greater concentrtation in the body than in the atmosphere. This would help provide a gradient for greater air flow into the respiratory system. Also, as O_2 is consumed in metabolism and CO_2 converted to bicarbonate in the hemolymph and tissues, more respiratory volume would be made available in the tracheae for the entry of moisture laden air. When the bicarbonate buffering system becomes sufficiently saturated, CO_2 concentrations stimulate the atrial (spiracular) valves to open. The result is that more water would be taken in than would be lost by less frequent expirations containing water. This mechanism would not function when the tick is active and respiring rapidly. Several methods exist for determining the validity of this hypothesis for water vapor uptake in ticks, and it should be tested.

5.2.5 Molting and passive water flux

The effects of development on integumental permeability have been studied for the rabbit tick *Haemaphysalis leporispalustris* (Davis 1974a, b). Permeability decreased gradually over 14 days after nymphal engorgement. Hafez *et al*. (1970a) demonstrated a significant effect of molting on transpiration in fed ammonia killed nymphs of *Hyalomma dromedarii* and *Ornithodoros savignyi*. They found no significant change in cuticlar permeability just prior to ecdysis, even though the endocuticle was mostly digested by molting fluid leaving only the thin exocuticle and epicuticle. One hour after molting the transpiration rate had increased 25 times for the ixodid and almost 15 times for the argasid. By 96 hr cuticular permeability had returned to the values seen prior to molting. Interestingly, both species achieved their cuticle 'impermeability' faster when held at 0.42 a_v than at 0.85, yet the level of permeability three days post-molt was essentially the same.

5.2.6 Temperature and passive water flux

Temperature affects passive sorption and transpiration exponentially (Arlian & Veselica 1979). At low temperatures transpiration and passive sorption rates are greatly reduced. Thus, ticks which are exposed to low temperatures are not as affected (less net loss) by low ambient A_v's because water exhange is slower (Sauer & Hair 1971). Balashov (1968) showed that unfed *Ixodes persulcatus* survived 1.5–2 years at 0–3°C and 0.98 A_v, for 3–4 months at 25°C and the same A_v, but for less than a month at 25°C and 0.80 A_v. These results should be kept in mind when long term maintenance of a laboratory colony is desired.

Finally, we consider critical transition temperature (CTT). This has been defined as the temperature at which waterproofing epicuticular lipids become reoriented to allow a dramatic increase in the transpiration rate from the arthropod (Edney 1977). Questions have been raised about the interpretation of CTT data (Seethaler *et al*. 1979, Toolson 1978, Arlian & Veselica 1982). Before examining these data we should focus our attention on water and re-examine the effect temperature has on its kinetic energy. The most significant component of the temperature-dependent increase in water loss from a tick can be explained by the Boltzmann function of temperature, where the increased energy content of water molecules at higher temperatures is responsible for a greater percentage escaping to the atmosphere from the hemolymph of the tick. Using an Arrhenius equation one can predict the fraction of water molecules with the critical activation energy necessary to escape the surface limiting structure (cuticle) at a given temperature (Williams & Williams 1967). If there is a change in the cuticle permeability then a new activation energy (kcal/mole) is indicated by a change in the slope of an Arrhenius plot (Fig. 5.4a, b). The intersection of the lines that define the two activation energies will be a critical transition temperature (Wharton 1985). Observe Fig. 5.4 to see how transpiration rates for ixodid *Hyalomma dromedarii* and argasid *Ornithodoros savignyi* females held in dry air compare as linear (zero order, transpiration vs *T*, Hafez *et al*. 1970a) and semi-log Arrhenius (first order, transpiration vs absolute temp., 1/*T*) plots of rate constants against temperature or the inverse of the absolute temperature. All points were included in the regression analysis for the semi-log plot. It seems, however, that some points were excluded from the linear plot of Hafez *et al*. (1970a), and this accounts for part of the discrepancy in CTT values determined using the two plots. One should first notice that the slope changes were not as dramatic for the Arrhenius plot. Nevertheless, a CTT was present for the linear and semi-log plots. A critical transition temperature of 57°C was estimated for *O. savignyi* and 51°C for *H. dromedarii* using the Arrhenius plot. If CTTs are above temperatures normally encountered by the tick, then the biological importance of these values seems negligible. An interesting result of plotting these data in this manner is that the slopes for the two species are remarkably similar both above and below their CTTs. This is somewhat misleading because when we compare data between 30 and 50°C, the rate constant increases 1.156%/°C for the ixodid and 0.016%/°C for the argasid. In this biological temperature range the ixodid tick integument was about 72 times more permeable to water than the argasid's.

The main point of this discussion is that the increase in the fraction of water molecules escaping as temperature increases is due to the increase in kinetic energy and not because of a change in cuticle permeability. Above the CTT for each set of experimental animals we see integuments with greater permeabilities and water escaping at different constant rates representing new activation energies. If a slight curve is seen it may be due to a gradual continuous change in the epicuticular permeability as the temperature increases (Toolson *et al*. 1979, Hadley 1981, Arlian & Veselica 1982).

If one is to study these critical transition temperatures, under what conditions should it be done? If tritiated water is not to be used then the study must be completed in dry air to eliminate the active uptake component. The resultant data should be expressed as a percentage of the water escaping per unit time versus the inverse of the absolute temperature. If a change in slope is seen, then the cuticle has been altered and a new CTT is detected. One should consider using freshly killed specimens to avoid problems associated with increased movement and respiration by the tick during the experiment.

5.3 ACTIVE UPTAKE BY TICKS BETWEEN BLOODMEALS

Active water vapor uptake occurs in three-host ixodids, and the mobile stages (those off the host) of one- and two-host ticks which have been examined. The stages which remain on the host, that is nymphs of some two-host and nymphs and adults of some one-host species, may be unable to

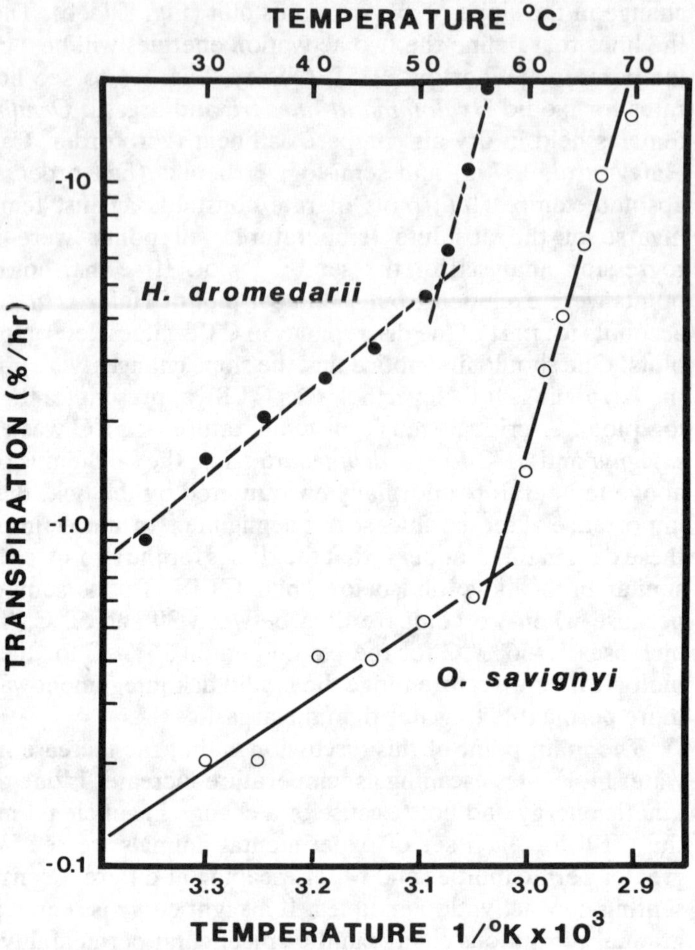

Fig. 5.4a — Arrhenius plot of transpiration rate constants versus the inverse of the absolute temperature for unfed female *Hyalomma dromedarii* and *Ornithodoros savignyi* ticks held in dry air (from Hafez *et al*. 1970a). The slope gives the activation energy for water to escape in that temperature range. The line intersection gives the critical transition temperature. Numbers on the abscissa of the semi-log plot have been reversed for comparison to the linear plot of the same data in Fig. 5.4b. Coefficients of linear correlation, $r>0.99$ for both temperature ranges for *H. dromedarii*; $r>0.95$ (30–55°C), $r>0.99$ (60–70°C) for *O. savignyi*.

actively absorb water from unsaturated air (Knülle & Rudolph 1982). These ticks sometimes live for very short periods when removed from the host, and most of them probably die of desiccation. An exception may be the one-host *Dermacentor albipictus* which as unfed nymphs or adults often live for several weeks to several months off the host in unsaturated air at temperatures ranging from 10–25°C under laboratory conditions (Teel, unpublished observation). Newly molted ticks (Lees 1946, Balashov 1968, Jaworski *et al.*

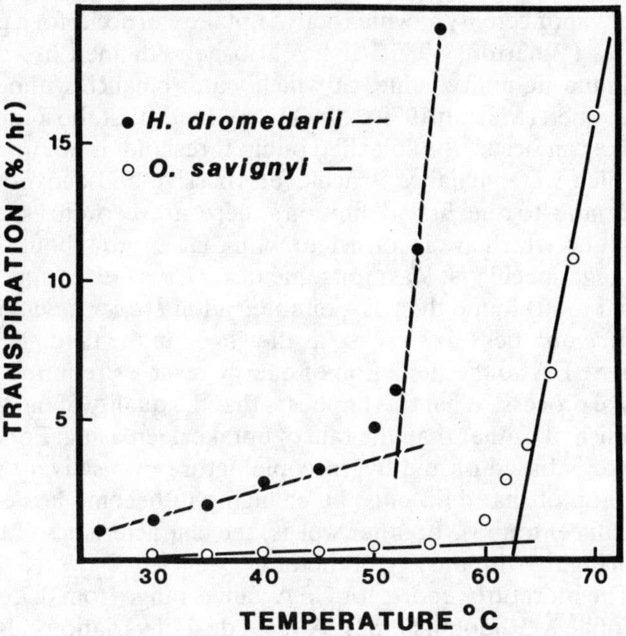

Fig. 5.4b — Linear plot of the transpiration rate versus temperature (redrawn from Hafez *et al.* 1970a).

1984) and recently engorged female ixodids (Sauer & Hair 1971) lack the ability to actively remove water from the air, but engorged or ovipositing females may do so (Barker *et al.* 1984). Argasid ticks likewise actively absorb water from the air (Knülle & Rudolph 1982); in fact any tick with a critical equilibrium activity (CEA) below 0.99 a_v has this ability. Without this ability ticks would either be restricted to habitats with high water vapor activities or be forced to take more frequent bloodmeals to obtain sufficient water. The combination of an active water uptake ability, slow metabolism while inactive, and a relatively impermeable integument greatly extends the life of a tick off the host.

The active vapor uptake mechanism for ticks and other terrestrial arthropods is not well understood. In general, uptake occurs at a localized site (e.g. mouth or anus), it involves a hygroscopic fluid, and water vapor condenses and is carried to a distant site (e.g. gut) for incorporation into the hemolymph (Machin 1979). The site of uptake for many arthropods has only recently been determined (see Machin 1979, Rudolph & Knülle 1982). For excellent reviews on ticks the reader is directed to articles by Rudolph & Knülle (1978), Knülle & Rudolph (1982), and Knülle (1984).

5.3.1 Critical equilibrium activity: a re-evaluation

Before discussing the active uptake mechanism it seems appropriate to re-examine the critical equilibrium activity concept relative to ticks. First, the working definition used by most investigators is that it is the lowest threshold water vapor activity at which active uptake can occur for a particular stage or species (Wharton, 1985). It has also been defined by some to mean the minimum humidity values at which water balance without feeding can be maintained (Machin 1979). Machin contends that the minimum A_v at which uptake can occur, the so-called pump threshold, is not necessarily the point at which water balance is achieved (passive and active uptake = efflux). Returning to the first definition, there are certain aspects of the CEA definition which are inconsistent with that known about ticks. The CEA is not stage specific, at least for some ticks. Lees (1946) and Williams *et al.* (in preparation) found the CEA of unfed adult *Ixodes ricinus* and *Amblyomma americanum* ticks to increase as they aged in the field. They attributed this higher CEA to the depletion of energy reserves required to fuel the active uptake process. What this implies is that the quality of the uptake 'pump' has diminished, rather than the rate of uptake decreasing. For example, if vapor uptake is based on the hygroscopic nature of a salivary solution then the make-up of that fluid must be changing to become less effective (possibly less concentrated). In other words, the characteristics of the secretion may be changing with the age of the tick.

The literature reports for CEA values range from 0.75 to 0.94 according to Knülle & Rudolph (1982). A great deal of variation exists in these values not only between species and stages, but also between ticks of a given stage even when great care is taken to standardize the test population with regard to age and physiological condition. Jaworski *et al.* (1984) found the CEA to range from 0.80 to 0.90 a_v for standardized *A. americanum* females observed in the laboratory. They saw no indication of an increased CEA over the 12-month period, so variation between individuals was responsible. Therefore, to state that the CEA is a certain value for a tick species or even developmental stage is misleading. We suggest that when CEA values are reported, the range should be given with the age, prior hosts, temperature, and the physiological condition as well as it is known. Owing to this variability the importance of determining a specific CEA is somewhat diminished. When maintaining a laboratory colony the best results will probably result from keeping ticks at an A_v above the range of estimated CEA values, yet not so high that problems with fungus or condensation develop. The higher A_v will increase passive uptake, and energy will be conserved because the uptake mechanism will be required to operate less frequently. Energy conservation resulting from less frequent operation of the active uptake process may explain why certain desert species of argasids live so long.

If we focus on the increasing 'pump' threshold A_v or CEA as ticks age, we may lose sight of another important change which may affect water balance. That is, the integument may become less effective as a barrier to water loss. If one uses Machin's (1979) concept of CEA, that is, water balance is not necessarily the point at which the uptake 'pump' begins to function but is the equilibrium activity, then an increasing CEA may be due to increased permeability of the cuticle. It is not too difficult to imagine that lipids which waterproof the cuticle may become less plentiful because of their utilization for energy. As the need arises for replenishing them owing to abrasion or other causes, they will not be replaced as readily. The active uptake mechanism would then have to operate either more frequently or for longer periods, taxing even further the energy reserves. Experiments to determine if changes in the transpiration rates occur with age or with chronic exposure to high or low vapor activities would be very instructive.

Another point is that many of the published CEAs may be lower than the actual value. The reasoning is summarized by Wharton (1985). Mites were shown to attract water vapor as hygroscopic objects, making it seem as though active absorption had occurred. Active water uptake was actually initiated at higher A_v's in the mites studied by Arlian & Veselica (1981).

5.3.2 Active uptake: what is known

Little is known about the factors which initiate active uptake. All we can do at this time is describe the conditions under which it occurs. It seems that no consistent percentage weight loss initiates the process as determined by electrobalance recordings (Rudolph & Knülle 1978). Adult *A. variegatum* fluctuated ±2% of their original weight at 0.93 a_v, with no specific lower value initiating active uptake. Also, transferring a tick from below its CEA to an activity above its CEA does not ensure that it will immediately extract water from the air (Needham, unpublished observation).

We know the 'pump' is activated without a changing A_v because continuous weight measurements demonstrate intermittant active uptake at constant water vapor activities (Rudolph & Knülle 1978). Dried salivary residue in and on the mouthparts may deliquesce to signal an A_v at which the uptake mechanism can function, but these residues may not be present except when the tick is severely dehydrated. We have blocked Haller's organ where humidity receptors are located (see Chapter 16), transferred them to high humidity, and found that some ticks were able to actively absorb water. It was not possible to tell if the wax seals on the distal end of leg I had failed or if the uptake mechanism was stimulated by other processes still unknown. More elaborate experiments involving the blocking or removing of these receptors and exposing the tick to a rehydrating atmosphere might indicate whether they are involved as the first step in initiating active uptake. It is tempting to speculate that receptors sense changes in hemolymph ion concentrations or osmotic pressure to initiate the mechanism. Whatever is involved, it seems that it is more than a simple feedback mechanism from humidity receptors on leg I.

Specifically, what is actually known about the uptake mechanism? The mouth is the primary site of uptake for ixodids (Rudolph & Knülle 1974, 1978, Knülle & Rudolph 1982, McMullen *et al*. 1976) and argasids (Knülle & Rudolph 1982). Since locating the site of uptake investigators have speculated that the salivary glands of the tick were involved. This was based on finding dried salt residues on the mouthparts of severely desiccated ticks, and knowing that these glands were capable of producing a salt solution based on their ultrastructure. The type I agranular acini were identified as the likely production sites of a concentrated salt solution. Balashov (1968) compared them to the nasal salt glands of marine birds which secrete a concentrated solution. The structure and possible function of the acinus are discussed in section 5.3.6 and in Chapter 1.

The hypothesis which is currently used to explain water vapor uptake by ticks involves the secretion of a fluid from the type I acini of the salivary glands at irregular intervals. This fluid is thought to have hygroscopic properties which allow it to become vapor enriched on the mouthparts at or above the CEA. Apparently the fluid is swallowed into the gut where redistribution of water, ions, and other components is carried out (Fig. 5.1).

5.3.3 Active uptake mechanism: a re-evaluation

The saliva from an actively rehydrating tick has not been analyzed, but dried saliva residue on the mouthparts of severely dehydrated *A. variegatum* adults contained sodium, potassium and chlorine in molar ratios of 1.1:1.7:3.8 (Devine 1982). This same dried residue begins to attract water vapor at humidities near the CEA for this tick. This may be significant in that when the tick encounters a water vapor activity in which the uptake mechanism can function, the dried residue liquifies so that it can be swallowed. This would reclaim the residue and might also signal that active vapor uptake is possible. Secretion of the salts in the dehydrated condition may have little to do with the hygroscopic properties of the saliva during rehydration; instead, they may be secreted there for storage excretion to help maintain hemolymph ion concentrations and osmotic pressure. Dopamine has been used to stimulate unfed adult *A. americanum* to salivate, and the resultant fluid was hypertonic to the hemolymph but the osmotic pressure and ion concentrations were dilute (Table 5.2). So dilute that it would be in equilibrium with a vapor activity of about 0.992, far short of the CEA for this tick (approx. 0.80–85) (Needham *et al.*,in preparation). We do not know if this is, or even resembles, natural saliva secreted during rehydration. The ion proportions in the dopamine-stimulated saliva were quite different than was found by Devine (1982), and what is seen for feeding ticks (see Table 5.2; Hsu & Sauer 1975). Ultimately, analysis of the rehydration saliva as it is secreted from the gland will provide the clues to explain the role of this fluid in active vapor uptake.

The rates of active uptake determined by tritiated water flux measurements, weight changes, and respiration rates in unfed female *A. americanum* are shown in Fig. 5.1. At 0.925 a_v it averaged 4.7 μg/h at 26°C. On average, active uptake accounted for 23–34% of the water absorbed under these conditons (Fig. 5.2). Most likely the rate of uptake is faster while the 'pump' is functioning because this value represents an average rate over 24 h, and the 'pump' operates intermittently. The active uptake rate was not significantly faster at 0.925 than at 0.85 a_v. Instead, the passive rate seems to account for most of the change in uptake rate. Knülle & Devine (1972) did not detect a change in the rate of active uptake in *D. variabilis* larvae at 0.85 and 0.925 a_v. On the other hand, others have suggested that active sorption

Table 5.2 — Mean osmotic and ionic concentrations in hemolymph and saliva collected from unfed and feeding female *Amblyomma americanum*

	Osmolality mosm/kg	[Na]	[K] Milliequiv/liter	[Cl]
Unfed tick[1] SALIVA	424±113 *N*=5	252±33 *N*=11	68±21 *N*=7	227±63 *N*=7
Unfed tick[1] HEMOLYMPH	370±72 *N*=5	194±21 *N*=4	34±9 *N*=5	123±14 N=6
Feeding tick[2] SALIVA	366 *N*=31	200 *N*=35	15 *N*=37	150 *N*=26
Feeding tick[2] HEMOLYMPH	360 *N*=38	184 *N*=45	15 *N*=63	135 *N*=32

1 Unfed ticks were injected with dopamine; *N* values represent pooled collections from 8–10 ticks; means ±SD (Greenwald & Needham, in preparation).
2 Feeding ticks were injected with adrenaline (Hsu & Sauer 1975).

rates are proportional to a_v (Wharton & Richards 1978, Arlian & Wharton 1974). The discrepency may be due to quantitative measurements made over each 24 h which tend to conceal rate differences by averaging.

The ability of engorged females to actively remove water from the air seems possible but remains unproven. Certain information may have misled us to think otherwise. Engorged females lose weight prior to oviposition, and this may mask active uptake, and the egg mass weight is less than the weight of the female so no obvious uptake occurs during this time. Also, salivary gland degeneration has been observed by several investigators (Balashov 1968; see Chapter 2), and since these glands are involved in active uptake one might speculate that engorged females do not have an uptake 'pump'. Barker *et al.* (1984) have shown that some reorganization of the type I acini occurs and that they reduce in size during feeding, but in fed *A. americanum* females they revert to an ultrastructural form seen in unfed specimens. Coons & Roshdy (1979) also documented the reduction in size of type I acini during feeding by *D. variabilis, Rhipicephalus sanguineus*, and *A. americanum* females. They observed no reduction in size for *Argas* (*Persicargus*) *arboreus* adults, and the same seems true for *Ornithodorus moubata* adults (Dzhafarov 1965, Rosell-Davis & Coons, personal communication). It is not surprising that the type I acini of argasids remain relatively constant in size, because they would be expected to be involved in osmoregulation between successive bloodmeals and ovipositing. Rather, it seems we should be surprised if ixodids do not take water from the air as engorged females. A simple experiment would entail blocking the mouthparts of engorged females prior to oviposition to see if they lose more water than controls without the mouthparts covered. Certainly, this subject should be examined in more detail in other species of ticks.

5.3.4 Active uptake by drinking

One additional source of active water uptake is by drinking. This phenomenon may or may not occur frequently in nature. If it does, its significance compared to active and passive water vapor uptake is not known. Lees (1946) states that *Ixodes ricinus* does not drink water. Wilkinson (1953) observed slightly desiccated *Boophilus microplus* larvae drinking from methylene blue- and neutral red-containing water droplets. These indicators appeared in the gut diverticula. *Amblyomma americanum* larvae and nymphs have also been observed to insert their hypostomes into water droplets and drink (Needham, unpublished observation). Thus the possibility that certain ticks drink water should not be ignored, especially in microhabitats where condensation occurs frequently and offers a reliable moisture source on nearby substrates.

5.3.5 Active uptake: effect of temperature

There is a temperature threshold below which active uptake fails to occur (5–9°C, Sauer & Hair 1971; Lees 1946). The uptake rate (active and passive) is exponentially faster as temperature is increased, but the amount of water taken up is relatively constant (Rudolph & Knülle 1982). The equilibrium water content is the same at a specific A_v above the CEA whether the temperature is 15 or 25°C. McEnroe (1971) has speculated that the northern limit of distribution for the American dog tick *Dermacentor variabilis* is determined by the inability of the active uptake mechanism to function at 5°C or below. He speculates that this causes the ticks to die of desiccation when temperatures remain below the threshold for extended periods.

We are just beginning to appreciate the variation between ticks with respect to their ability to maintain ion and water balance. Active water uptake is a significant component in the overall water balance equation for ticks, but until the vapor uptake mechanism is better understood, a large gap exists in our knowledge of their off-host physiology.

5.3.6 Salivary gland involvement in osmoregulation and active water uptake in unfed ticks

Argasid and ixodid ticks seem to have similar mechanisms for maintaining water balance off the host, however, on the host their osmoregulatory mechanisms are quite different.

The salivary glands of ixodids are important osmoregulatory organs both on and off the host. While an ixodid is feeding, the paired grape-like structures excrete excess ions and water back into the host. The bloodmeal can be concentrated because space is conserved by water elimination; at the same time osmoregulation is accomplished by removal of salts from the hemolymph (see Sauer 1977, Kaufman & Sauer 1983; and Chapters 1, 3, and 11). As ticks go through desiccation–rehydration cycles off the host the salivary glands may remove salts from the hemolymph to the mouth as a form of storage-excretion. These salts may be sufficiently concentrated to absorb water from unsaturated air for a net moisture gain. Recent reports on salivary gland structure and function have assigned an osmoregulatory role for granular acini during feeding, although an ultrastructural study by Barker *et al.* (1984) has shown agranular acini to undergo morphological changes during feeding in *A. americanum* females. An off-host osmoregulatory role has been shown for agranular type I acini (see below).

The argasids osmoregulate in a very different way from ixodids during feeding. Coxal organs between legs I and II use a filtration-resorption mechanism to maintain ion and water balance during and/or immediately after feeding (Kaufman *et al.* 1981). Certain cells in the argasid salivary glands may secrete ions and water during feeding (Needham, personal observation based on their structure). The agranular acini of argasids, as in ixodids, may function in an off-host osmoregulatory capacity as well. (Compare their many ultrastructural similarities to the ixodid agranular type I acinus).

Initial evidence for salivary gland participation in water balance by ixodids off the host came from the pioneering efforts of Rudolph & Knülle (1974, 1978) and McMullen *et al.* (1976). They suggested that type I agranular acini of the gland secrete a salt containing fluid which is hygroscopic. Early ultrastructural examinations of these acini showed them to have the deep basal infoldings and numerous mitochondria often associated with ion transporting epithelia capable of secreting hyperosmotic fluid (see Table 5.3 for references, Fig. 5.5).

Interestingly, the supracoxal glands of the house dust mite *Dermatophagoides farinae* (Brody *et al.* 1976) have many ultrastructural similarities with the agranular acini. These glands secrete a salt solution which is responsible for net water vapor uptake from unsaturated air (Wharton 1985). Rudolph & Knülle (1978) observed fluid pulsing from salivary glands into the salivarium of desiccated ticks after the dorsal capitulum and chelicerae had been removed. They also found that predesiccated *Hyalomma anatolicum excavatum* adults heavily infected with *Theileria annulata,* which selectively destroy granular acini in this tick, were still able to actively remove water from the air, albeit at a slower rate than uninfected controls. It was rationalized that the intact agranular acini must have supplied the hygroscopic saliva so the tick could rehydrate. Thus most subsequent attention has been focused on the agranular acini as the source of a hygroscopic salt containing fluid.

McMullen *et al.* (1976) demonstrated that significant amounts of ^{36}chloride accumulated on the mouthparts and in the salivary glands of *A. americanum* adults which had been desiccated and rehydrated. Subsequently, Coons & Roshdy (1979) localized chloride deposits within 'vacuoles' in the agranular acini of unfed *Dermacentor variabilis* females, although this technique is questionable as a method for chloride localization. In this same species they localized ouabain-sensitive Na,K-ATPase sites in the basally unfolded membranes thought to be responsible for secreting ions in the hygroscopic saliva. The basal infoldings in *A. americanum* (Needham *et al.*, in preparation) also have ouabain-sensitive Na,K-ATPase sites based on a one-step histochemical localization technique (Mayahara *et al.* 1980). The best direct evidence for the involvement of

Table 5.3 — Descriptions of type I agranular salivary gland acini in ixodid ticks

Authors	Dzhafarov (1965) Balashov (1979)	Kirkland (1971)	Coons & Roshdy (1973) Coons & Roshdy (1979)	Meredith & Kaufman (1973)	Megaw & Beadle (1979)	Krolak *et al.* (1982)	Barker *et al.* (1984)	Walker *et al.* (1985)	Needham & Coons (1984) (see Chapter 1)
Species	*Hyalomma asiaticum*	*Haemaphysalis leporispalustris*	*Dermacentor variabilis*	*Dermacentor andersoni*	*Boophilus microplus*	*Amblyomma americanum*	*Amblyomma americanum*	*Rhipicephalus appendiculatus*	*Amblyomma americanum*
Stage	Adults	Nymphs	Males	Female	Female	Female	Females	Males and females	Female
Feeding state	Unfed (>6 mo. Dzhafarov)	Feeding	Unfed	Feeding (6 days)	Unfed and feeding	Unfed	Unfed-feeding	Unfed-feeding	Unfed (4–6 weeks post-fed)
Acinus diameter	57–64 μm		152 μm unfed male 184 μm unfed female	33± μm	20 μm	50 μm	201 μm		
Cell types	Multicellular pyramidal alveolus-pyramidal cells with peripheral and central areas	Presumed to be unicellular; however, one to several nuclei observed	Central cell (1) 'other cells'	Inner cell (1) Fibrillar cells (2)	Central cell (1) Pyramidal cells (3?) Epithelial cells	Central cell (1) Pyramidal cells (≥6) Peritubular cells (4–5) Constrictor cell (1)	Central cell (1) Pyramidal cells (5–7) Peritubular cells (several) Constrictor cell (1)	Central cell (1) Peripheral cells Neck cells Constrictor cell (1)	Central lamellate cell (1) Peripheral lamellate cells Peritubular cells Constrictor cell (1)
Cytoplasmic vacuoles	Vacuoles seen in the basal and apical regions	Small vesicles were seen in the central region of acinus	Vacuoles in the basal, intermediate, and apical regions of acinus	Vacuoles not seen in cytoplasm	Vacuoles not seen in cytoplasm	Vacuoles in central and pyramidal cells, identified as lipid	Lipid vacuoles numerous in unfed ticks, reduced by 12 h depleted by 72 h feeding	Large lipid droples reduced in size and quantity during feeding	Vacuoles in central and peripheral lamellate cells, identified as lipid
Glycogen	Glycogen was detected	No mention of glycogen	Scattered carbohydrate granules	No mention of glycogen	Extensive glycogen-like material in unfed ticks	Glycogen not mentioned	No mention of glycogen	Glycogen reduced with feeding	Glycogen-like material seen
Notes	Vacuoles appear to be lipid, may be energy source for ion transport (Dzhafarov). It was suggested that the vacuoles are secreted (Balashov)	Type I increase greatly in size at the end of feeding	Central cell communicates with acinar lumen, other cells may also contact lumen. Central cell may collect fluid from adjacent cells; Acinus diameter decreased by 47% during feeding by female, 55% by male after 9 days' feeding	Fibrillar cell may not contact the lumen	Glycogen-like material not seen in Type I of fed ticks. Pyramidal cell apical membrane has no interdigitation. Central cell projections extend into pyramidal cells. Suggest acinus may excrete hygroscopic fluid for vapor uptake	Central cell contacts lumen and may collect and store fluid. Constrictor cell may regulate fluid flow from acinus. Peritubular cells line most of acinus lumen	Autophagic activity in the central and pyramidal cells during feeding; acini decrease in size by 20% at 48 h, return to original size at 72 h, then decrease again 96 h. Lipid vacuoles may be secreted as component of attachment cement	Females have 250 (21%), males 150 (11%) of acini Type I. Type I may secrete hygroscopic salts in unfed, salt excretion during feeding. No dramatic changes from unfed to feeding. Lipid and glycogen may be used as energy sources during questing by unfed ticks	Central cell borders hemolymph (basal processes) and acinus lumen. Peripheral lamellate cell does not contact lumen. Glycogen-like material. Lipid reduced in desiccating–rehydrating ticks

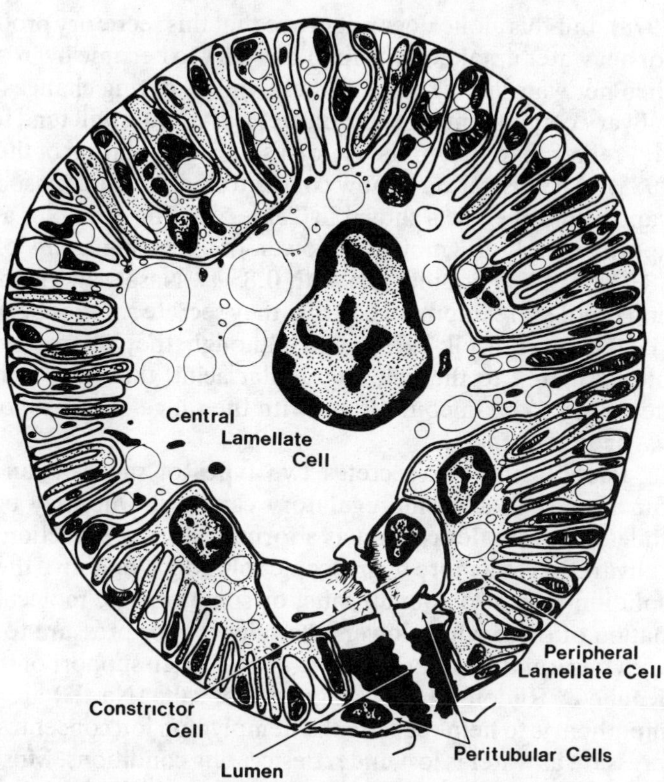

Fig. 5.5 — Diagrammatic representation of an agranular type I salivary gland acinus from an unfed 6–8 week old *Amblyomma americanum* female tick held at 0.85 a_v. Note that the central lamellate cell extends from the hemolymph to the acinus lumen; the peripheral lamellate (pyramidal) cells do not contact the lumen.

agranular acini in the maintenance of water balance and vapor uptake is their dramatic change in mitochondrial conformation and the apparent reduction in number of lipid inclusions when *A. americanum* females are desiccated and rehydrated (Needham & Coons 1984). A discussion of these changes is given below.

Despite this evidence it still remains to be demonstrated that fluid secreted directly from the salivary glands is sufficiently hygroscopic to be responsible for the net uptake of water. Salivary residues from the mouthparts of severely dehydrated adult *A. variegatum* have been analyzed, but the intital concentration of solutes could not be determined because they had dried (Devine 1982). Also, fluid was analyzed after injection of the tick with a stimulant (Kaufman 1976, Rudolph & Knülle 1978; Table 5.2), but the question remains whether this fluid resembles natural 'rehydration' saliva. With either collection method one cannot definitely say that this is the make-up of natural saliva. The dried salivary residue from *A. variagatum* mouthparts deliquesces near the tick's CEA (0.75 a_v, Rudolph & Knülle

1978), but this alone does not prove that this secretory product is responsible for net water uptake. If the fluid must dry to become hygroscopic at the CEA then net water uptake is not likely, unless drying changes the nature of the salivary components. Furthermore, a dry solute will tend to deliquesce at an A_v greater than that of the A_w of a saturated solution of the solute (Wharton 1985). An illustration of how concentrated the fluid must be to attract water vapor from the air is shown in Table 5.1. For example, a solution must be approximately 9.8 molal, or lower the freezing point of water by about 18.2°C to be in equilibrium with 0.85 a_v. Nasal salt glands of marine birds and reptiles are impressive in that they secrete solutions exceeding 1 M NaCl (Peaker & Linzell 1975), and although they share some similarities in ultrastructure to the tick's agranular acini, this value is not close to those necessary to be in equilibrium with the range of CEAs for ticks (0.75–0.94 a_v, or 18.5–3.54 M).

Possibly the tick secretes two types of saliva as an unfed tick, each functioning in an osmoregulatory capacity. One may be secreted during times of severe desiccation as a form of storage excretion, the other during rehydration as a more hygroscopic solution at or above the CEA. This latter solution may also contain salts, or some organic molecule(s), or a combination of these which lowers the saliva vapor pressure to be in equilibrium with unsaturated air at or above the CEA. In support of the first possibility, Knülle & Rudolph (1982) suggest that salts (Na, K) are secreted onto the mouthparts to help regulate the hemolymph ion concentrations and osmotic pressure as water is lost under desiccating conditions. Most of the water used to carry the salts to the mouthparts would be reclaimed when the tick encounters a vapor activity which is sufficiently high to dissolve the dried residue so that it could be swallowed. In this postulated role the saliva does not aid in net water uptake but serves to temporarily extract undesired elements from the hemolymph in times of water stress. Salivary fluid has been seen pulsing on the mouthparts of severely desiccated ixodids (Rudolph & Knülle 1978) but it has also been observed for a rehydrating tick in high humidity as well (Needham, unpublished observation).

Another possibility for saliva involvement in the rehydration process arises from the model suggested by O'Donnell (1982) for the desert cockroach *Arenivega investigata.* Briefly, hairs on paried lateral extensions of the anterior hypopharynx, so-called bladders, are alternately made hydrophilic by the secretion of a dilute fluid from frontal bodies. The water vapor which collects on these hair-covered bladders is swallowed to rehydrate the cockroach. For this to be possible in ticks their mouthparts would have to possess hydrophilic structures accessible to a dilute saliva and still offer the necessary surface area. The only structures in ixodid ticks which might offer this combination of characteristics are the cheliceral sheath denticles (Fig. 5.6). A potential problem with this idea is that although all ticks have cheliceral sheaths, not all have cheliceral sheath denticles upon which moisture could condense. Maybe those ticks which have few or no denticles do not require the additional surface area offered by these structures. Also, we do not know the hydrophilic nature of these structures.

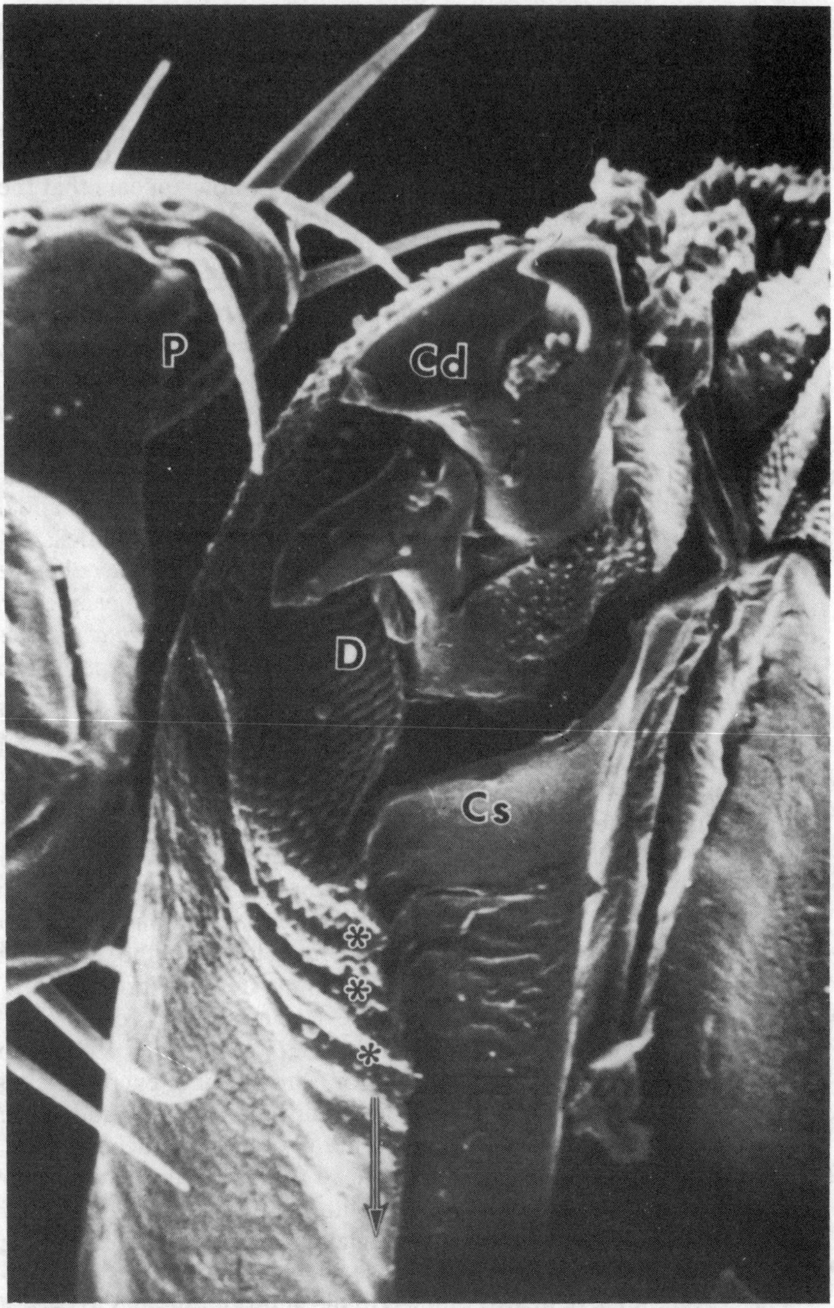

Fig. 5.6 — Scanning electron micrograph of an adult female *Amblyomma americanum* chelicerae and palp. The hypostome has been broken off to reveal the ventral surface of the pliable cheliceral sheaths (Cs) with retrorse denticles (D). Note the lateral trough-like structures in association with the denticles. In high vapor activities (near saturation) moisture will collect on this area (*) and may move toward the mouth by capillary action at the hypostome–cheliceral sheath interface (arrow); Cheliceral digit (Cd); Palp (P).

An additional argument against a structural component of the mouthparts being sufficiently hydrophilic is the evidence that the CEA changes as the tick ages (see section 5.4). In fact, if the CEA does change then this seems to favor a salivary fluid which becomes less hydroscopic as the resouces of the tick are depleted, rather than the cuticle changing in its hydrophilic nature.

Finally, from ultrastructural evidence it now appears that the b-cells of type II acini from rehydrating unfed female *A. americanum* actively secrete granules (Jaworski & Needham, in preparation, Fig. 5.7). This could be the source of hygroscopic organic molecule(s), or a surfactant which allows the salivary fluid to spread out on the mouthparts. This finding suggests that granule-secreting acini may be involved in water vapor uptake, and future investigations should be broadened to include them as well.

5.3.6.1 Type I agranular acini of certain unfed ixodids

Our purpose here is to compare the descriptions of ixodid and argasid agranular acini to highlight important structural similarities and differences in view of their proposed osmoregulatory roles in unfed ticks. We will also discuss the changes observed in these acini after ticks have been exposed to desiccating and rehydrating conditions. The morphology and ultrastructure of the salivary gland of *Rhipicephalus appendiculatus* is described in Chapter 1. Our understanding of the general acinus morphology is shown in Fig. 5.5.

Table 5.3 lists the terminology used by different investigators to describe the agranular acinus of different ixodid species. Discrepancies may have resulted from the following: (1) the physiological condition of the tick (e.g. fed, unfed, desiccated, rehydrated, etc.); (2) the family, genus, or species of tick examined; and (3) differences of interpretation in micrographs.

The following observations on the organization and ultrastructure of the type I agranular acinus in the Ixodidae are based on comparisons of acini from eight species, representing six genera in four of the five subfamilies (see Hoogstraal & Aeschlimann 1982, for systematic groupings). In female *Rhipicephalus appendiculatus,* type I acini account for 21% of the acini, while in males 11% are type I (Table 5.3; Walker *et al.* 1985). If ovipositing females actively remove water vapor from the air or excrete excess ions onto the mouthparts, then having the additional acini to produce more saliva could be advantageous. Males generally remain attached to the host until death and would not necessarily need this added capacity, except for some *Ixodes* males which mate off the host but do not feed. It might be very instructive to examine salivary glands from these non-feeding males to compare the ratio of acini to those which do feed.

Agranular acini are confined to the anterior one-half to two-thirds of the gland, being scattered amongst the granular (types II, III, and IV in males) acini and attached to the main salivary duct. The acini are ensheathed in a thin basal lamina (Figs 5.5, 5.8). Their connection to the duct is by peritubular (or neck) cells which have a cuticular lining. Some disagreement exists as to which cells border the lumen of the acinus and efferent duct (tubule). The observations of Needham *et al.* (in preparation) and those of Barker *et al.*

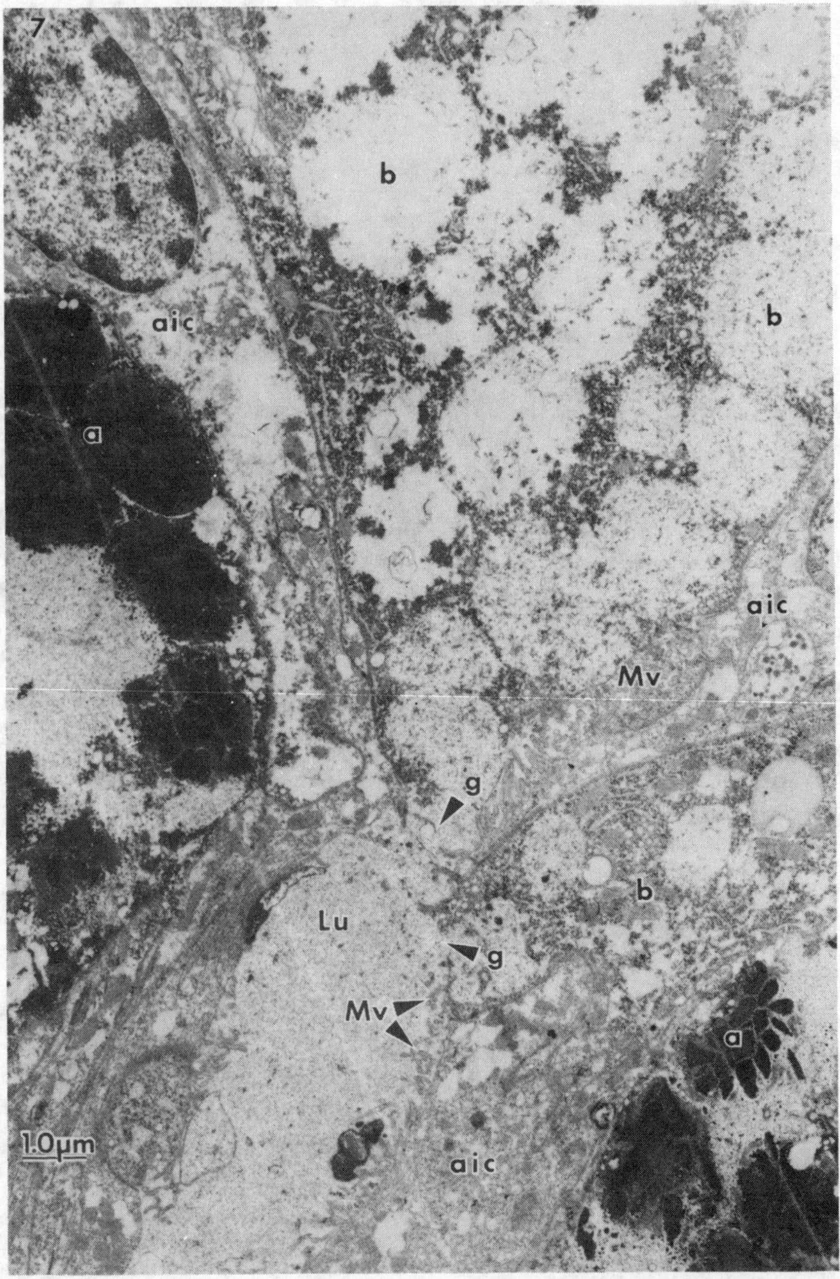

Fig. 5.7 — Granules being secreted from b-cells of the Type II acinus into the lumen. This female *Amblyomma americanum* tick had been in the process of rehydrating based on weight gains in the prior 24 h, and the material being secreted may be involved in the active rehydration process. a = a-cell granule; b = b-cell granule; aic = adlumenal interstitial cell; Mv = Microvilli; Lu = Lumen; g = granule; (Jaworski & Needham, in preparation).

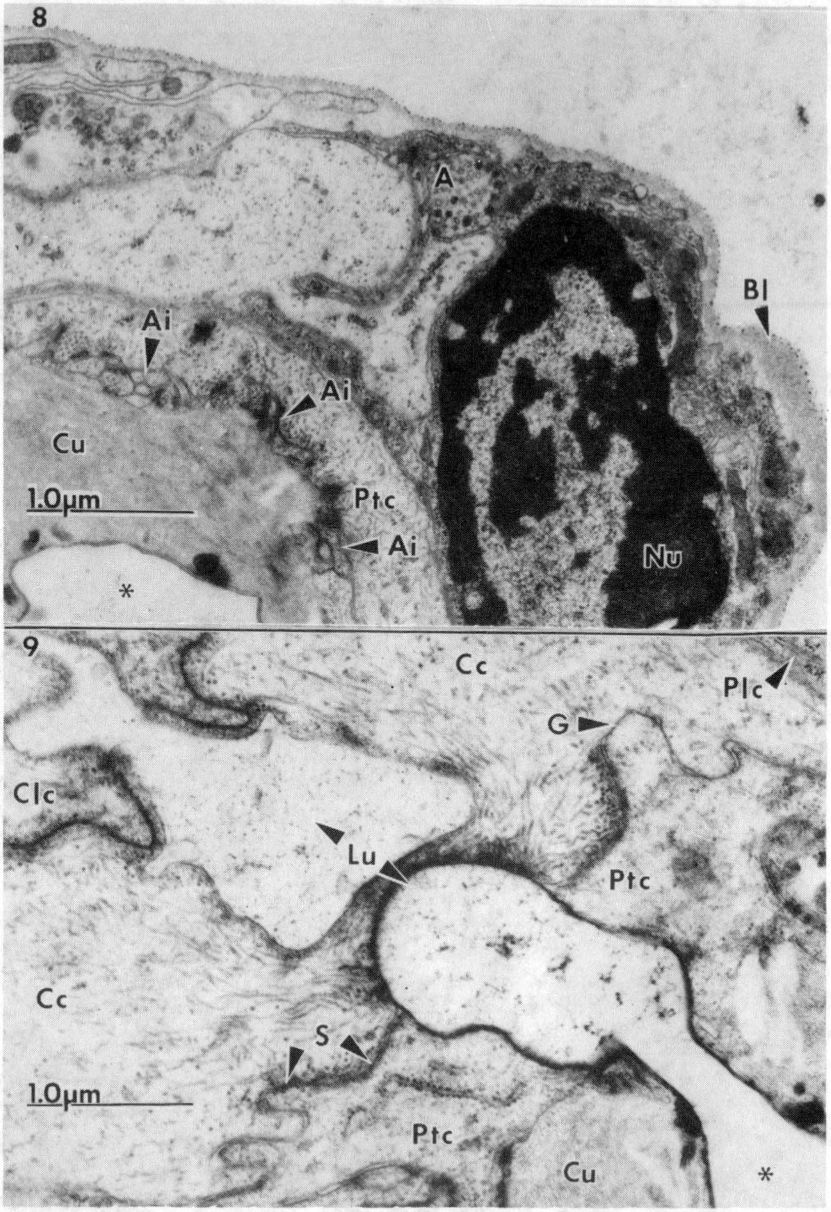

Fig. 5.8 — Micrograph of the type I agranular acinus from a 2-month post-fed female *Amblyomma americanum* female which had been desiccated, rehydrated, and desiccated for 24 h (each treatment). This shows the cuticle-lined efferent duct formed by the peritubular cells (Ptc). Note the modest amount of apical infoldings (Ai) of the peritubular cells and the nerve axon (A). Bl = Basal lamina; Nu = Nucleus; Cu = Cuticle; * = Efferent duct lumen. Fig. 5.9 — Neck region of an agranular type I acinus of an unfed female *A. americanum* exposed to desiccation, rehydration, and desiccation (24 h per treatment). Cells in contact with the lumen: the Central lamellate cell (Clc), Constrictor cell (Cc), and Peritubular cells (Ptc). Notice junctions between the central and peritubular cells. Compare this figure with Fig. 5.14, a similar view of an argasid. S = Septate junction; G = Gap junction; Cu = Cuticle; * = Efferent duct lumen; Lu = Acinus lumen; Plc = Peripheral lamellate cell.

(1984), show that the peritubular cells extending into the acinus are joined to the constrictor cell but do not contact the lumen (Figs. 5.5, 5.9). Therefore, the central lamellate cell is the only cell with an amplified basal surface and associated mitochondria, which is in contact with the acinus lumen. Walker *et al.* (1985), on the other hand, erroneously diagrammed the peripheral lamellate cells to extend between the peritubular and constrictor cells. Their own micrographs can be interpreted to agree with the former arrangement. Subsequently they have revised their interpretation so that we are in agreement on this point (Walker, personal communication). Understanding the arrangement of these cells is critical for formulating a hypothesis for the production of fluid by the acinus.

Both the central and peripheral cells are in contact with the hemolymph side by similar narrow parallel interdigitating basal processes or lamellate sheets (Fig. 5.11). We prefer to call them peripheral and central lamellate cells for this reason (see Table 5.3 for other names). In combination, the basal processes of these two cell types offer a substantial surface area for nutrient and substrate uptake, and fluid transport, especially when compared to their apical surfaces. Understandably, some investigators diagrammed the acinus to have the central cell surrounded by peripheral (pyramidal) cells and not in contact with the basal lamina (Meredith & Kaufman 1973, Megaw & Beadle 1979). They were likely biased by earlier light level micrographs and difficulties in distinguishing cell borders.

Stored energy in the form of glycogen and lipid are in close association with the numerous mitochondria and make the transport of ions the likely function of this acinus. As will be mentioned below, the size, position, and quantity of lipid inclusions depend on the hydration, feeding state, and possibly the age of the tick at the time of dissection. In *A. americanum* the peripheral lamellate cells have a rather straight apical surface in comparison to their highly amplified basal surfaces (Fig. 5.11). Each peripheral cell contacts the central lamellate cell, and some contact the constrictor and peritubular cells (Figs. 5.5, 5.9). One usually sees the nucleus of the peripheral cell near this apical membrane, while the larger central cell nucleus is positioned in the middle of the acinus (Fig. 5.11). Glycogen, lysosomes, and microtubules are commonly observed in the basal processes and apical region of both cell types. Their prominence is also affected by the physiological condition of the tick. For example, in feeding *A. americanum* females, lysosomes are dominant features in the central cell by 24 h, and at 72 h they occur in both peripheral and central lamellate cells (Barker *et al.* 1984).

One of the more interesting features of this acinus is the constrictor cell (Figs. 5.5, 5.9, 5.10), so called by Krolak *et al.* (1982) because of the abundant 'microfilamentous' structures seen in a cell they liken to a myoepithelial cell. They also assign a contractile function to the adlumenal interstitial cell (cap cell) of the type III granular acini. Additional work needs to be done to determine if these structures are microtubles or microfilaments, and if they are capable of contraction.

There is a question of how cells of the type I acinus are joined together.

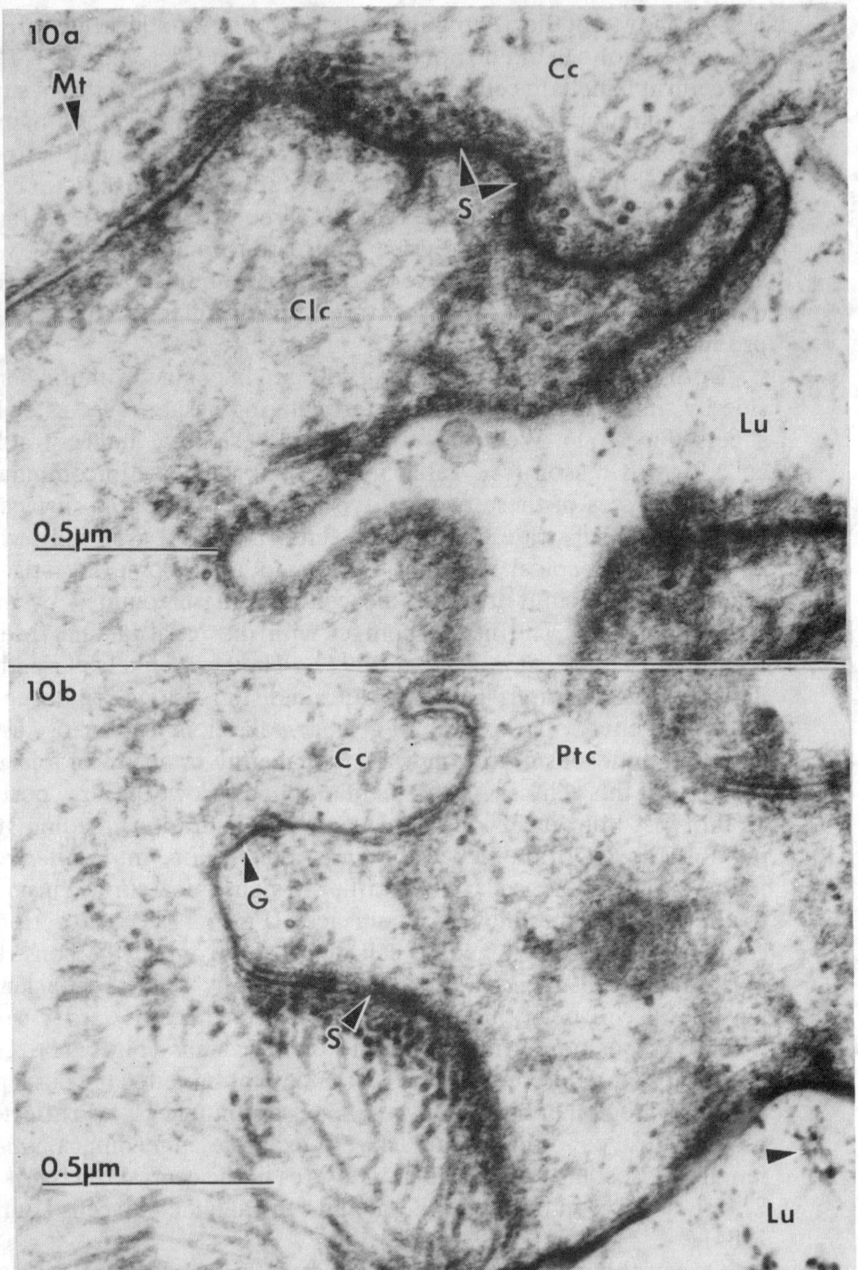

Fig. 5.10a — Higher magnification of Fig. 5.9 to show junctions between the Central lamellate (Clc) and Constrictor cells (Cc). Septa (arrows) are visible in the Septate junction. Microtubules (Mt) are common features of the constrictor cell (see Fig. 5.9 for designations). Fig. 5.10b — Higher magnification of Fig. 5.9 to show Septate (S) and Gap (G) junctions between the Constrictor (Cc) and Peritubular (Ptc) cells adjacent to the lumen of the acinus.

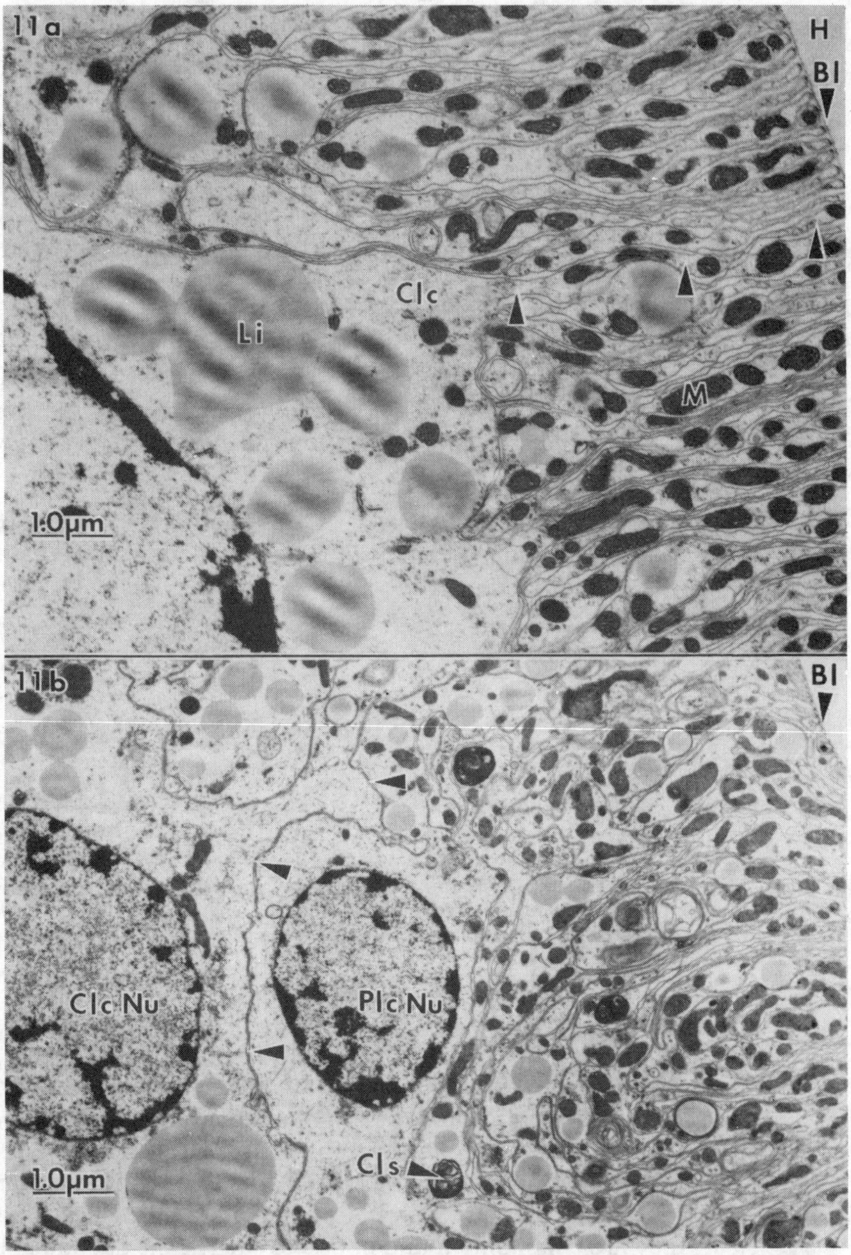

Fig. 5.11a — Basal and central area of an agranular type I acinus from a female *Amblyomma americanum* held at 0.85 a_v for 6–8 weeks. Central lamellate cell extensions project to the hemolymph (arrows). Numerous Lipid inclusions (Li) are present. Mitochondria (M) are in the orthodox configuration which indicates a normal or slow respiratory rate expected in a tick which is not under osmoregulatory stress. Fig. 5.11b — Basal and central area of an agranular Type I acinus physiologically similar to Fig. 5.11a. Notice the relatively straight interface between the Central lamellate (Clc) and Peripheral lamellate (Plc) cells (arrows). Several Cytolysosomes (Cls) which are present in the peripheral lamellate cell are also common in the central cells.

Identification of cell junctions by viewing ultra-thin sections with the electron microscope can allow one to speculate about cell communication, and how 'leaky' or 'tight' the junction may be to ions and small molecules (Lane & Skaer 1980). There is some disagreement about the naming of junctions in the type I acinus, so for camparison we have included Table 5.4.

Table 5.4 — Cell junctions in the type I agranular acini of unfed adult ixodid ticks

Authors	Krolak *et al.* (1982) Barker *et al.* (1984)	Walker *et al.* (1985)	Needham *et al.* (in preparation)
Tick species	*Amblyomma americanum*	*Rhipicephalus appendiculatus*	*Amblyomma americanum*
CELL TYPES			
Constrictor: Central cell[1]	Nexus junction Zonula adherens	Desmosome Septate junction[2]	Desmosome Septate junction Desmosome
Constrictor: Peritubular[1]	—	—	Desmosome Septate junction Gap junction
Constrictor: Peripheral	—	Gap junction	Gap junction
Central: Peripheral	Gap junction	—	Gap junction

[1]The order of these junctions is from the acinus lumen (apical to basal).
[2]Walker (personal communication) agrees that the junction between the constrictor and central cell is a septate junction rather than a zonula adherens.

Figs. 5.9, 5.10 show the junctions in the lumenal area of an unfed *A. americanum* female. The juxtalumenal junctions appear to be desmosomes although they are not clearly seen in these micrographs. Next to them are septate junctions. They join the constrictor cell to both the central and peritubular cells. Septate junctions are characterized by having thin septa at regular intervals, and although their functional significance is not entirely clear, they are commonly found in tissues such as the gut and epidermis of fluid feeders which are subject to rapid dilations (Lane & Skaer 1980). Septate junctions probably aid in intercellular adhesion and they may function as transepithelial permeability barriers to stop or reduce the passive flow of fluids between cells. In some instances these junctions are analogous to the tight functions of vertebrates, with the septa acting to seal the intercellular cleft rather than having the plasma membranes contact as in tight junctions (Lane & Skaer 1980). From these descriptions, the position of circumferential (around the acinus lumen) septate junctions between the constrictor cell and the peritubular and central cells seems plausible. Coons (personal communication) finds the order of junctions next to the lumen of tick tissues he has examined to be: desmosomes, septate junctions, and often gap junctions. This seems to be the case for the type I acinus of two species of ixodids and argasids (Fig. 5.14) examined (Needham *et al.* in preparation).

Krolak *et al.* (1982) used the term 'nexus' to describe the junction between the central and constrictor cells. A nexus junction is a gap junction (Fawcett 1981), and this does not appear to be a gap junction. The intercellular cleft does not narrow as is the case for gap junctions. Also, one would not expect a gap junction at this position in the acinus. They facilitate communication by allowing the diffusion of ions and small molecules from cell to cell by low-resistance pathways (Fawcett 1981). The locations of some gap junctions in *A. americanum* are shown in Figs 5.9, 5.10b, and their cell associations are summarized in Table 5.4. We see these junctions betweeen most cells except the constrictor and central lamellate cell, and few if any could be seen in the basal area of the acinus.

Barker *et al.* (1984) and Walker *et al.* (1985) report that zonulae adherens join the constrictor and central lamellate cells, but they do not agree on location. What these authors call zonulae adherens have thin septa in the intercellular cleft, and seem to be septate junctions. Subsequently, Walker (personal communication) has agreed that the term septate junction rather than zonulae adherens describes the junction between the central and constrictor cells in *Rhipicephalus appendiculatus* type I acini. In addition to their juxtalumenal position, desmosomes are seen between the central and constrictor cell where adhesion is their likely function. Of central importance is the permeability of these junctions to the flow of water, ions, and small molecules. Preliminary studies with ionic lanthanum showed them to be 'leaky', but further experiments are needed to confirm this (Needham *et al.*, in preparation). If these acini produce a highly concentrated salt solution (>10 molal), then understanding the cell junctions in this acinus is crucial to learning how such a fluid is produced.

A commonly visualized feature of the acinus is a filamentous network in the acinus lumen (Figs. 5.9, 5.10). It is not seen in the lumen of the afferent tubule, which is lined with cuticle. Although Coons & Roshdy (1973) suggested it may be an artifact of fixation, the filaments may have some yet undetermined role in the functioning of this acinus. It is a feature consistently observed by most investigators in both ixodids and argasids. These structures may help keep the lumen open or 'wick' fluid away from the apical membrane.

Another common feature of micrographs is a nerve axon in association with the type I acinus (Fig. 5.8). Little is known about the control of salivation by unfed ticks. Pilocarpine has been used to stimulate the secretion of a milky substance which may have been cement (Rudolph & Knülle 1978), and dopamine stimulates secretion of a dilute salt solution (Table 5.2). It seems likely that nerves in intimate association with these cells control their function (Chapters 2 and 3).

Histochemical studies have been completed for several species. The most interesting observation is that females of the one-host tick *Boophilus microplus* have few or no lipid inclusions, and certainly not to the extent we see in other ixodids; instead, the cytoplam contains a great deal of glycogen (Binnington 1978, Megaw & Beadle 1979). The hydration state of these ticks was not reported, so we must qualify our statement about the lack of lipid. It is tempting to speculate that because these ticks may not have the ability that other two- and three-host species have to actively remove water from the air, they do not use lipid reserves. Instead, they may use glycogen as an energy source. Another possibility is that energy used by the mitochondria for fluid transport may come from the hemolymph as it is used, rather than being stored in the peripheral and central lamellate cells. It would be interesting to compare the one-host *Dermacentor albipictus* adult which has the ability to survive for several weeks to several months with *B. microplus* adults, for which the hydration states were known.

5.3.6.2 Changes in agranular and granular acini in an ixodid exposed to desiccating and rehydrating water vapor activities

Ticks naturally undergo periods of desiccation and rehydration while off the host as the ambient water vapor activity fluctuates (Lees 1946). There is a great deal of circumstantial evidence favoring the involvement of the type I acini in vapor uptake; however, more information is required to show what role they and possibly the granular acini play. One indirect approach would be to expose ticks to desiccating and rehydrating conditions, then examine the ultrastructure of the gland for characteristic changes. Needham & Coons (1984) did this and saw some dramatic changes in type I acini ultrastructure after desiccation and rehydration of unfed *A. americanum* females. Ticks were desiccated by exposing them to 0 a_v for 48 h at 26°C. The desiccation–rehydration regime consisted of a 24 h exposure to 0 a_v followed by 24 h at 0.93 a_v. Weight changes determined at 24h intervals showed desiccated specimens to lose 8.7±1.1% (n=6) of their body mass, desiccated–rehydrated (D–R) ticks lost 7.4±0.6% (n=6) in 24 h, and regained 13.7±1.75% 24 h later at 0.93 a_v.

Acini from desiccated ticks were different from controls (ticks which had been maintained at 0.85 a_v) in that lipid inclusions appeared to be more numerous, particularly in the apical regions of the central lamellate cell (Fig. 5.12). It is possible that because the tick is being maintained below its CEA and the uptake mechanism is not operating, lipid accumulates owing to this lack of activity. As we and others have suggested (Lees 1964, Needham & Coons 1984, Walker *et al.* 1985; and Chapter 1), the lipid may serve as an energy reserve for fluid transport or possibly other activities of the tick. On the other hand, this evidence is interesting because during this time we might expect the secretion of some salts onto the mouthparts as a form of storage excretion. Possibly the tick was not sufficiently water stressed for this to have occurred.

Ticks transferred from a desiccating to a rehydrating environment have type I acini which show the greatest changes. Most notably we see a dramatic

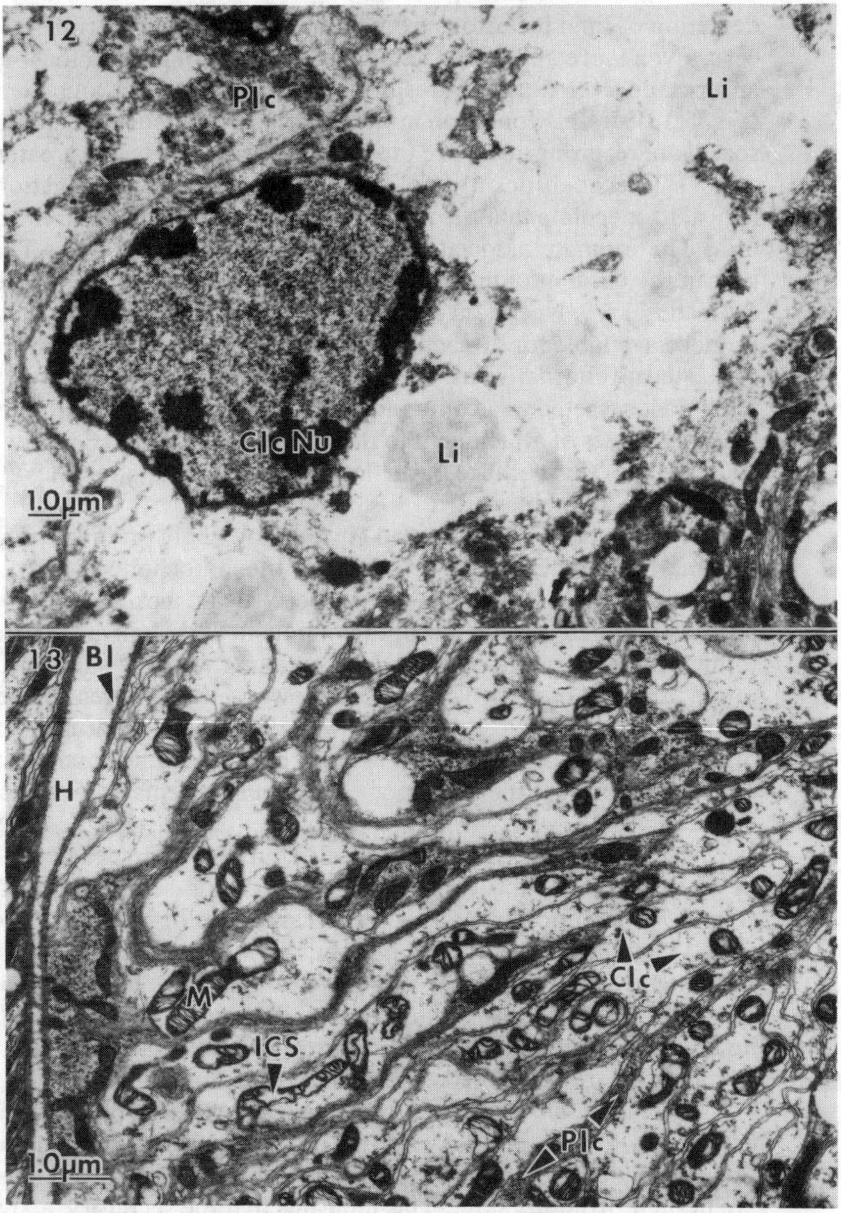

Fig. 5.12 — Central area of an agranular type I acinus from an unfed *Amblyomma americanum* female which had been exposed to dry air at 26°C for 48 h. Numerous lipid inclusions (Li) were consistently seen in both the central and peripheral lamellate cells. Mitochondria appeared to be orthodox in conformation (similar to controls at 0.85 a_v). Fig. 5.13 — Basal area of an agranular type I acinus from an unfed *Amblyomma americanum* female which had been exposed first to dry air at 26°C then transferred to 0.93 a_v for 24 h. Mitochondria (M) in both cell types have transformed to a condensed state which indicates a rapid respiratory rate. Intercristal space (ICS) is obvious in many mitochondria. Much of the lipid is absent and the cytoplasm of the central lamellate cell (Clc) appears to be dilute compared to the peripheral lamellate cell (Plc).

depletion of lipid inclusions from both the central and peripheral lamellate cells. Even more apparent is the transformation of most mitochondria from an orthodox (or normally respiring) conformation, to a condensed state (Fig. 5.13). In this condition mitochondria have a much greater intercristal space, are respiring at a more rapid rate, and are producing greater amounts of ATP (Hackenbrock 1966). At this early stage of investigation it seems logical to speculate that the lipid is a carbohydrate source for the mitochondria. The intimate association of mitochondria with Na,K-ATPase pump sites in the basal infoldings of central and peripheral lamellate cells (Coons & Roshdy 1979, Needham *et al.*, in preparation) would make the ATP they produce available for this active secretory process. Barker *et al.* (1984) have suggested an alternative use for the lipid in feeding ticks. They speculate that the lipid is secreted as a component of the cement which is produced during the early stages of attachment to the host. Fawcett *et al.* (Chapter 1) and we disagree with this because of the lack of evidence for apocrine or holocrine secretion in this acinus.

Alternatively, fatty acids and glycerol are likely products of lipid metabolism, the former being oxidized in the mitochondria for use in the tricarboxylic acid cycle, and the glycerol could be used in glycolysis or possibly secreted as a hygroscopic component of the saliva. We are tempted to hypothesize that glycerol is secreted along with the salts during rehydration, and that these two components lower the vapor pressure of the saliva to make it hygroscopic. Again, we await the testing of 'natural' saliva for its organic and inorganic make-up to confirm or deny this possibility.

In a separate experiment, *A. americanum* females were exposed to desiccating (0.23 a_v, 24 h) and rehydrating (0.93 a_v, 24 h) conditions, and salivary glands were examined ultrastructurally. Granules which appear to be from b-cells of the type II acini were being secreted into the lumen (Fig. 5.7). Secretion from a type III granular cell is noted by Krolak *et al.* (1982) in an unfed *A. americanum*, but no mention is made of its potential significance. These findings suggest to us that the rehydration saliva may contain components from granular as well as agranular acini. In *Rhipicephalus appendiculatus* and *Boophilus microplus* b-cell granules contain polysaccharides and protein (Walker *et al.* 1985, Binnington 1978).

Needham *et al.* (in preparation) have looked at type I acini from ticks exposed to desiccating–rehydrating–desiccating (D–R–D) conditions. The return of lipid inclusions to the basal region can be seen, and mitochondria are in their orthodox state. This indicates that the secretory 'activity' of the acinus has again slowed or stopped because the tick is not in a suitable humidity for active vapor uptake. In the interim, it may replenish energy reserves. We also observed numerous cytolysosomes in the cytoplasm of the central and peripheral lamellate cells of D–R–D ticks. Most likely they are digesting old mitochondria and other cell components in preparation for rehydration.

Also noted were some similarities between type I acini from ticks which had been transferred from 0.85 (control) to 0.93 a_v, to those from desiccated–rehydrated ticks. Rudolph & Knülle (1978) demonstrated that ticks transferred from an activity above the CEA to an even higher one, will take up an additional amount of water and achieve an even higher equilibrium body water mass. This helps confirm the validity of our observations on the transformation of mitochondria and the utilization of lipid in the active uptake process.

A question remaining is, if these acini secrete a concentrated salt solution when the tick is being severly desiccated, then why don't we see a transformation of the acini at that time? Our present answer is that the ticks we examined were not sufficiently water stressed to secrete the concentrated salt solution as storage excretion. A more biochemical approach needs to be taken to answer questions raised by these observations of the tick, and by further electron microscopic examination of glands from one-host ticks and immatures in various states of hydration.

5.3.6.3 Argasid salivary glands: comparison of type I acini to ixodids

The mouth is the site of active water uptake in argasids, as shown by mouthpart blocking experiments (Knülle & Rudolph 1982). The salivary glands of argasids have not received the attention that ixodid glands have; however, a role for them in water vapor uptake and osmoregulation seems very likely. There is little experimental evidence to explain the function of the agranular acini in these ticks, except the ultrastructure of the argasid and ixodid agranular type I acini is similar in many respects (Dzhafarov 1965, Balashov 1968, Roshdy & Coons 1975, Rosell-Davis & Coons, in preparation). Argasid salivary glands have not been examined after desiccating or rehydrating the tick as has been done with an ixodid species. Coons & Roshdy (1981) documented the role for granular type II (or type B) acini during feeding by adult *Argas arboreus*.

Our purpose here is to summarize the apparent ultrastructural similarities and differences between argasid and ixodid salivary glands. It should be noted that some glands were from feeding or recently fed specimens. Thus, these generalisations on cell content should be viewed with some reservation based on the changes seen in ixodid salivary glands during feeding and in different hydration states. Cell organization should not be affected by the physiological condition of the tick. This makes comparisons between families, genera, and species valid. We hope these observations will stimulate further interest in the subject.

There are two acinus types in argasids, rather than three (females) or four (males) as in ixodids. The agranular type I acini of argasids have been called non-granular, Type A, pyramidal alveoli, and 'pyramidenzellen' (Balashov 1968). They are concentrated on the anteriomesal aspect of the main salivary duct, rather than being scattered along the main duct as with ixodids where they sometimes extend two-thirds of its length. Although it is difficult to measure their diameter because of their irregular shape, their size range is comparable to that of ixodids, from 30–40 μm in unfed *Argas arboreus* females (Roshdy 1972) to 45–65 μm in unfed *Ornithodorus moubata* nymphs (Rosell-Davis & Coons, in preparation), and 37.5–50 μm in feeding or fed *O. moubata* adults (Dzhafarov 1965) (compare to Table 5.3

for ixodids). Unlike some ixodids, the type I acini do not seem to decrease in size during feeding. This may indicate a different role for this acinus type off the host or during feeding compared to some ixodid species. As in ixodids, their direct attachment to the main lobular duct is via a short efferent acinar duct which is without a bicuspid valve as is seen in the granular acini of both families. The collapsed efferent acinus duct of *Argas arboreus* is lined by a cuticle layer which is generally thinner than in ixodids (compare Figs. 5.8, 5.14), and the cells which comprise the duct have more complex apical infoldings than for ixodids examined thus far (Fig. 5.10). They may play some resorptive role due to the surface area offered by the apical infoldings near the lumen. If resorption occurs, it is probably a passive process because there are few mitochondria in association with the infoldings to supply energy for active resorption. The concentration of solutes to form a hypertonic saliva may be accomplished here. These cells may be analogous to the peritubular (or neck) cells described in ixodids. Until described otherwise, and for convenience, we will refer to these cells as peritubular lamellate cells. Unlike ixodids, it appears that the argasid peritubular cells have an amplified basal surface similar to peripheral and central lamellate cells (Coons, personal communication). The lumen in the acinus contains a latticework of microtubule-like structures which end where the duct cuticle begins, much like that seen in the ixodids. These structures may help conduct fluid out of the acinus lumen into the efferent acinus duct lumen or may provide the infrastructure to keep the acinus lumen from collapsing.

More observations are required before we can give a detailed description of the argasid type I acinus. However, if Figs. 5.9 and 5.14 are compared one can see several additional similarities between the agranular acini of the two families. In fed adult *A. arboreus* type I acini there is a cell positioned where the ixodid constrictor cell is located. Furthermore, the association of this cell with the central cell is strikingly similar to that seen in *Amblyomma americanum*. There are few organelles in the central acinus compared to the basal region, and microtubules are a dominant feature of the cental and 'constrictor-like' cells. Cell junctions in the lumenal area of *A. arboreus* are similar in arrangement to those seen in *A. americanum*; that is, desmosomes join cells next to the lumen followed by septate and usually gap junctions, respectively (Fig. 5.14). It would be premature at this stage to speculate about the permeability of these junctions, and we can only state that cell adhesion is at least one key function for each type (Lane & Skaer 1980).

The deep basal infoldings with numerous closely associated mitochondria are features common to the peripheral and central lamellate cells of both families. Until a function is discovered for these two cell types in argasids we will employ the names assigned their counterparts in ixodids. Despite the major similarities, an obvious difference between the lamellate cells of these families is that lipid inclusions are not seen in argasid type I cells (unfed *O. moubata* nymphs, Fig. 5.15; unfed and fed adult *A. arboreus* (not shown) — Needham *et al.*, in preparation). Glycogen seems to be the primary stored energy source in argasid type I acini, rather than lipids, as is apparent in most ixodids. Dzhafarov (1965) mentions lipid inclusions in his

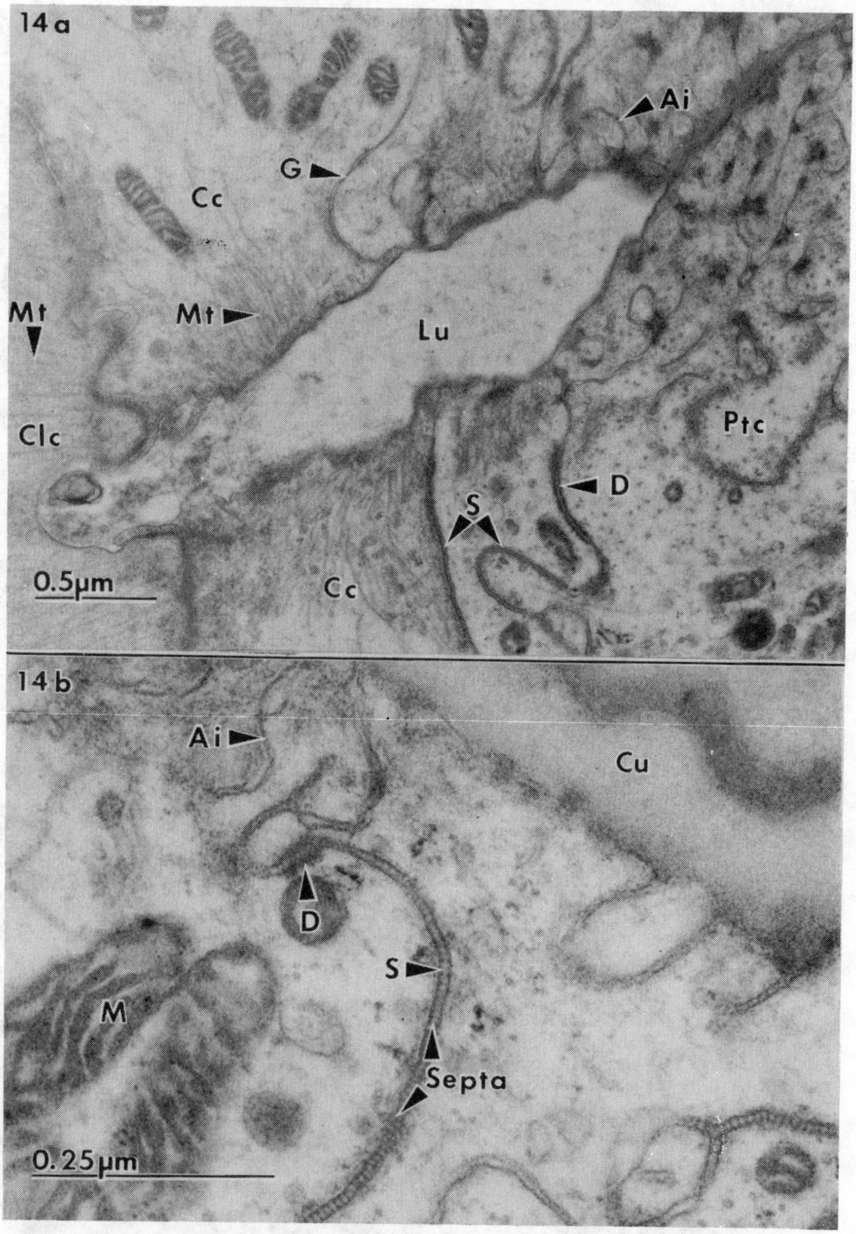

Fig. 5.14a — Lumenal area of an agranular acinus from a fed argasid *Argas arboreus* showing the cell associations and junctional complexes. As in ixodids, the lumen seems to be surrounded by a central lamellate cell (Clc), a cell which may be analogous to the constrictor cell (Cc) of ixodids, and peritubular cells (Ptc). Apical infoldings of the peritubular cells seem to be more extensive than the ixodids (Fig. 5.8), indicating that resorption may occur here. Mt = Microtubules; G = Gap junction; S = Septate junction; D ≡ Desmosome; Lu ≡ Lumen. Fig. 5.14b — Higher magnification of the junctional complex from the same acinus as Fig. 5.14a. Desmosome is followed by septate (S) junctions. Apical infoldings (Ai) of a peritubular cell are present.

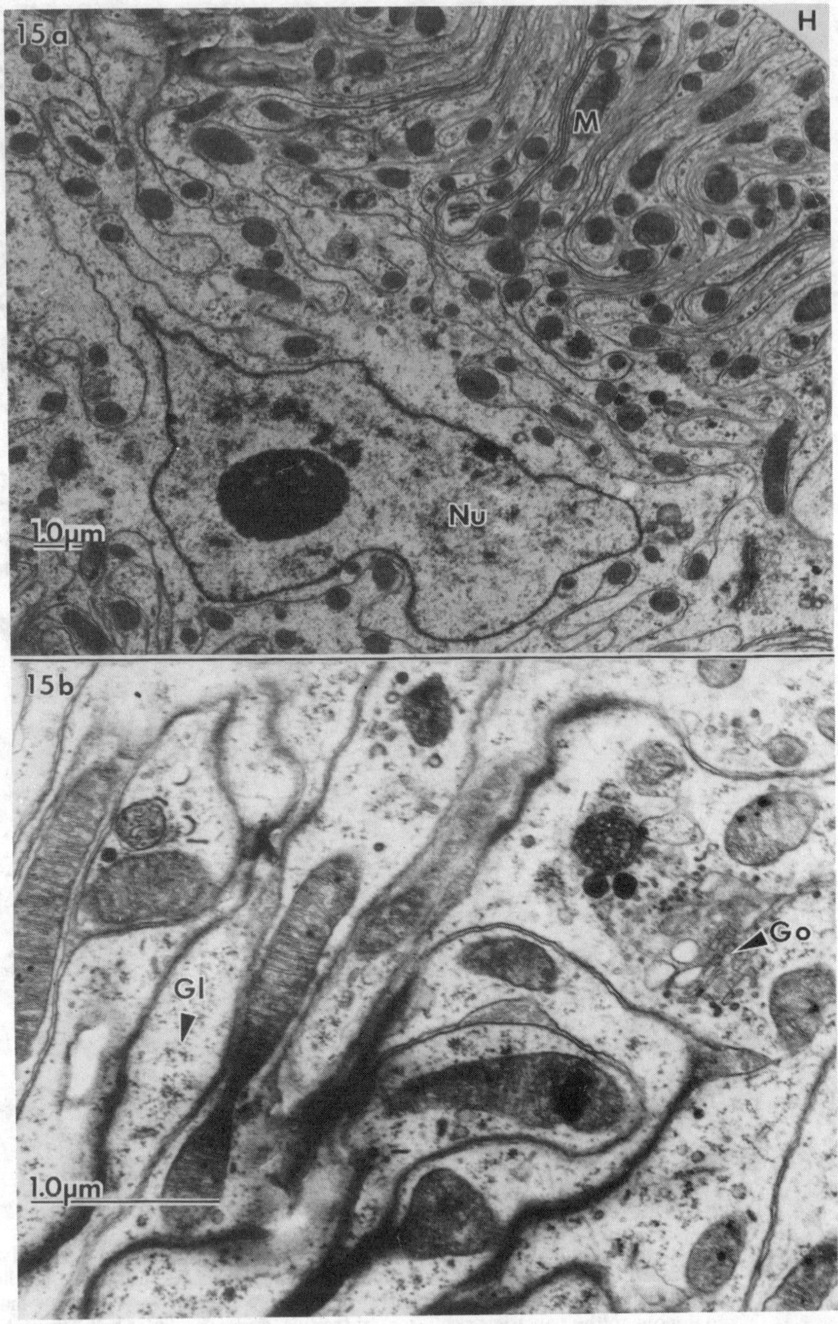

Fig. 5.15a — Basal and central area of an agranular type I acinus from an unfed *Ornithodoros moubata* nymph held at 0.85 a_v. Note the sparsity or absence of lipid inclusions, otherwise this area is very similar to an ixodid type I acinus. Fig. 5.15b — Higher magnification taken from Fig. 5.15a. Note the Golgi (Go) and Glycogen (Gl).

descriptiion of type I acini from feeding and fed adult *O. moubata*, but the inclusions were not as numerous as is seen in ixodids, and their identity was not confirmed histochemically. Therefore, type I acini may not be storage sites for lipid in argasids. This is similar to what is seen in type I acini of the one-host *Boophilus microplus* (Megaw & Beadle 1979, Binnington 1978) where no lipid was observed.

Other features which are consistent with what is seen in ixodids are Golgi bodies (Fig. 5.15b). They are present in association with the lysosomal system, and with microbody formation in order to recycle cellular constituents by autophagic vesicles (Rosell-Davis, personal communication).

The major differences between ixodid and argasid type I acini are the absence of lipid inclusions in argasids and the highly infolded apical surfaces of peritubular-like cells lining the lumen. These cells also have amplified basal surfaces much like the central and peripheral lamellate cells. Ixodids, except for the one-host ixodid female mentioned above, have type I acini with numerous lipid inclusions and less glycogen than argasids. These preliminary observations raise the possibility of differences in secretory products produced by the type I acini of the two families. On the other hand, from structural considerations, they both seem suited for the secretion of salt solutions, and an osmoregulatory role for both seems certain.

In ixodids we cannot rule out the involvement of the granular or type II acini in water vapor uptake. Coons & Roshdy (1981) have described the secretory process for these acini during feeding by *A. arboreus* adults but they did not examine them in different hydration states in unfed specimens.

5.4 OSMOREGULATION BY TICKS OFF THE HOST

Ticks are challenged by fluctuating ambient water vapor activities: the question is, how do they regulate or tolerate the resultant changes in internal body constituents? Also, are there threshold values of water content, ion concentrations, osmotic pressure, hemolymph volume, or other parameters which initiate or terminate water vapor uptake, host seeking behavior, or processes which affect water balance physiology? There must be internal feedback mechanisms which, along with environmental cues, regulate the behavior of ticks. The question of how ticks osmoregulate during desiccation–rehydration cycles off the host has received very little attention. Ticks with relatively impermeable integuments should not experience the rapidly changing concentrations of hemolymph components that a tick with a permeable cuticle might. The 'permeable' tick probably seeks microhabitats with more stable A_v's, or is likely to alternate host questing with more frequent hygropositive responses to compensate for water loss. These ticks may also be more tolerant to acute changes in hemolymph parameters.

Shih *et al.* (1973) exposed unfed female *Amblyomma americanum* ticks to desiccating and rehydrating A_v's and determined the effect on water content, hemolymph volume, osmolarity, and concentrations of Na, K, Mg, and Ca in the hemolymph. Their results are summarized in Fig. 5.16. Perhaps the most significant finding was that during desiccation the hemo-

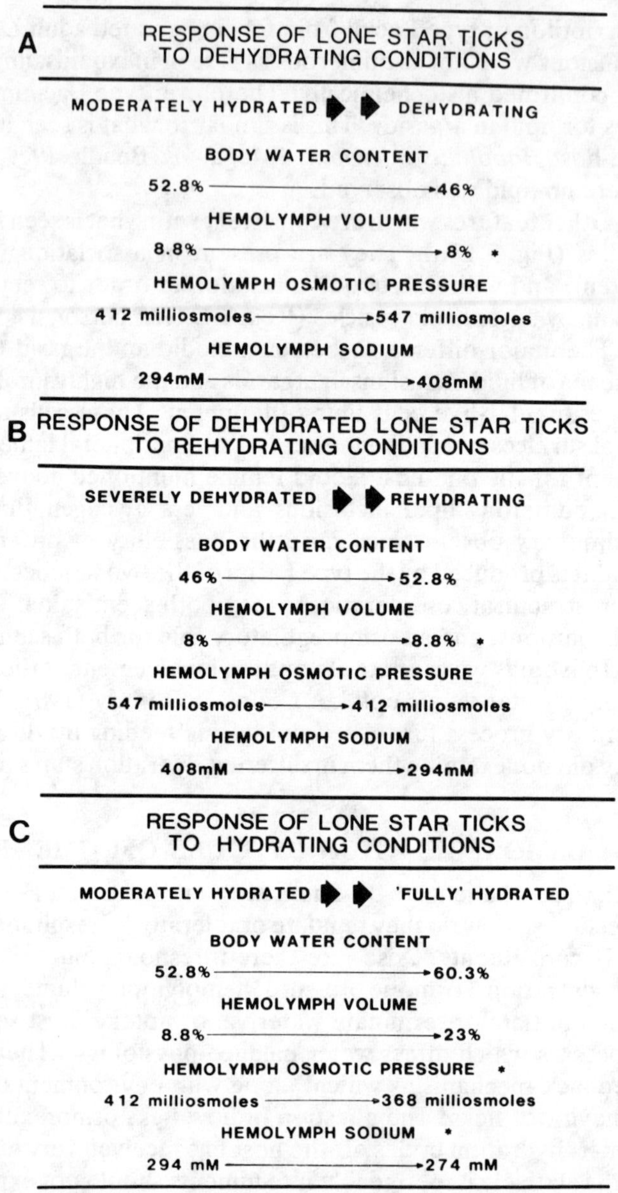

Fig. 5.16 — Effects of desiccation and rehydration on the whole body water content and properties of the hemolymph of unfed adult lone star ticks, *Amblyomma americanum*. Modified from Shih *et al.*, (1973). Note: experimental temperature was room temperature. * indicates no significant change at 0.05 level.

A) Response to dehydrating conditions; after exposure to 0 a_v for 24 h and changing only ±5% body weight ('moderately hydrated') they lost at least 10% of their 24 h pre-desiccation weight after exposure to 0 or 0.42 a_v ('dehydrated').

B) Response to rehydrating conditions; ticks were exposed to 24 h pre-desiccation at 0 a_v and lost at least 10% body weight ('severely dehydrated'), then were transferred to 0.98 a_v ('rehydrating').

C) Response of moderately hydrated ticks to hydrating conditions; after exposure to 0 a_v for 24 h and changing only ±5% body weight ('moderately hydrated'), they were transferred to 0.98 a_v gaining at least 10% body weight ('fully hydrated').

lymph volume remained relatively constant, apparently at the expense of water from elsewhere in the tick (Fig. 5.16A). Sodium and additional solutes accompanied water into the hemolymph from other body compartment(s), but their concentrations were lower than expected and must have been regulated in some way. We speculate that some of the excess solutes which accompanied this water, including Na, Cl, and K, may be secreted by the type I acini onto the mouthparts for temporary storage under desiccating conditions. When the A_v is sufficiently high the dried residue deliquesces so that it can be swallowed and reclaimed. When severely desiccated lone star ticks rehydrate the water is apparently reclaimed first by the tissues which have relinquished it because the body water content increases significantly but the hemolymph volume increases very little (Fig. 5.16B). Sodium seems to play an important role because although the hemolymph volume does not change, the Na concentration decreases significantly. McMullen *et al.* (1976) found that Cl concentrations increased in the mouthparts and salivary glands of rehydrating lone star ticks. Perhaps Na, Cl, and other salivary components are sufficiently concentrated to be hygroscopic and are being used in the active uptake process (section 5.3). When 'moderately hydrated' ticks become even more hydrated ('fully hydrated', Fig. 5.16C) the additional water goes to the hemolymph compartment as shown by the significant increase in its volume. Despite the additional water, hemolymph osmotic pressure and sodium concentrations were maintained at a relatively constant level.

Clearly this tick has the ability to regulate water and ion concentrations within limits. Survival off the host is due in part to the ability of tissues to tolerate changes in hemolymph properties, and because they have some ability to osmoregulate. Ticks compensate by shifting water, ions, and other solutes from one compartment to another. The salivary glands and other tissues (coxal organs and gut) act to maintain hemolymph constituents within tolerable limits so cells can function. Malpighian tubules do not play the primary role in osmoregulation as in hematophagous insects; rather they are responsible for nitrogen excretion toward the end of feeding and after; and the rectal sac cells into which the Malpighian tubules empty may selectively resorb water (Balashov 1968, 1979).

In summary, little is known about how ticks osmoregulate in desiccating and rehydrating conditions except for the study just described (Shih *et al.* 1973). We do not understand how they orchestrate activities associated with osmoregulation during periods off the host. In periods of desiccation do ticks convert free amino acids to dipeptides or polypeptides? This reduces the number of solutes, and in the process liberates a molecule of water for each amino acid added to the chain (Wharton 1985). Other hemolymph components, especially chloride, need to be monitored under desiccating and rehydrating conditions. To our knowledge argasids have not been examined under different hydration states except to look for changes in body water content (Hafez *et al.* 1970a, b, Lees 1946, 1947, Browning 1954, Hefnawy *et al.* 1975). The refinement of sensitive techniques for analyzing numerous elements from a single small sample, and the ability to detect minute concentrations of pharmacologically active molecules, makes this aspect of osmoregulation in ticks an attractive area for future investigations. These studies may tell us if specific threshold concentrations of ions or certain osmotic pressure values affect active vapor uptake or behavioral strategies associated with osmoregulation.

5.5 WATER BALANCE PHYSIOLOGY AND BEHAVIOR

Behavioral and physiological studies of ticks during off-host periods indicate that the two most important survival limitations are energy and water. These two resources must be conserved to enable ticks to live long enough to encounter and feed up on a suitable host. The economic utilization of energy and water are augmented by behavioral strategies often characteristic of the species, its developmental state, and its physiological condition (Lees 1964, 1969).

Before examining these limitations on survival we first consider the general categories of behavior exhibited by ticks off the host. The behavioral strategies of many ixodid ticks which actively seek hosts between bloodmeals fall into relatively low and high levels of activity and thereby impose very different demands on energy and water reserves. One phase is marked by quiescence at the soil–vegetation interface following emergence from the egg or after a molt. This interval may last several days to several months depending on species, stage, and time of year. Individuals surviving more than one season may re-enter this quiescence as a result of seasonal or climatological changes (Belozerov 1982). The other phase is characterized by host-seeking activity which consists of intermittent or single vertical migrations and positioning on substrates to place the tick in an advantageous position for host detection and/or contact. Responsiveness to host stimuli in this phase can change dramatically from aggressive, quick responses to lack of interest (see Waladde & Rice 1982). Considerable variation exists among species and stages within this active phase.

Contrasted with these strategies are ticks which do not seek hosts. That is, they remain sequestered in nests, burrows, or cracks until a potential host comes within their range of detection, at which time they may move toward the animal. This category of behavior describes many of the argasids and some ixodids in the genera *Ixodes, Amblyomma*, and *Aponomma* (Hoogstraal 1978, Petney *et al.* 1983, El-Ziady 1958, Webb 1979). In this kind of situation, the humidity and temperature in their micro-environment may be relatively stable and thus the need to move in response to moisture gradients is not as acute as for other ticks. Many of the argasids are extremely resistant to desiccation, and only long-term deceases in humidity with concomittant losses of significant amounts of water may stimulate movement to a more suitable microhabitat. The energy conserved in association with this more stable micro-environment is often necessary to extend the tick's life span until a host returns 'home' to hibernate, rear young, or seek shelter.

How do energy and water interact to effect the behavior of a tick? The tendency is to think that because the tick season is weeks or months long, individual ixodid ticks actively quest for this length of time, all the while being subjected to the drying powers of the environment and expending valuable energy. While *I. ricinus* were found to survive from 91 to 357 days after making their first ascent on grass, the active period, defined as 'the interval between first and last appearances on vegetation' averaged only 30 days (range 1–123 days, Lees & Milne 1951). Within the active period an average of only 9 days (range 1–54) was spent at the tips of grass with the average single duration at the grass tips being between 2 and 3 days (range 1–24 days). Periods at the tips of grass were interrupted by descents to the near-saturated vegetative mat at the soil surface.

Field (Lees & Milne 1951) and laboratory (Lees 1946) observations of *I. ricinus* posture on grass, responsiveness to host stimuli, and water balance, indicate that the intermittent periods spent in the vegetative mat and the periods of unresponsiveness to stimuli while ticks were at the grass tips were associated with periods devoted to rehydration. The CEA for adult *I. ricinus* was found to be 0.86–0.96 (Lees 1946), and it was speculated that a water loss of 12% or less is required to stimulate this species to seek a more moisture rich microenvironment (Lees & Milne, 1951). As we have already alluded to, the determining factor(s) which drive ticks up and down their questing substrates remain to be established (Rudolph & Knülle 1978).

When considering the comparative length of time spent by ixodids in a questing posture on a substrate, it seems reasonable that if all other factors were equal, species with higher CEAs would descend to a more moisture rich microclimate to prevent desiccation and/or rehydrate on a more frequent basis. This generality can prove to be misleading if the interpretation does not include the consideration of differences in integumental water permeability and microclimate conditions of temperature and relative humidity. Let us take, for example, the two common pest ticks *Amblyomma americanum* and *A. maculatum* which have overlapping general distributions in the southeastern U.S. Within this zone, the habitat association offering greatest longevity for *A. americanum* is defined as oak–hickory woods (Semtner *et al.* 1971) whereas the predominant habitat association for *A. maculatum* is upland prairie (Semtner & Hair 1975). The CEA for adult *A. americanum* is on average lower (reported to be 0.80–0.85, Hair *et al.*, 1975; Sauer & Hair, 1971); while the CEA determinations for *A. maculatum* ranges from 0.88–0.90 a_v (Lees 1946) to 0.92–0.93 (Hair *et al.* 1975). Fig. 5.17 shows the weight loss experienced by females of these two species in an environment of 0.30 a_v and 25°C. Since these test conditions were below the CEA for both species, the weight losses are a reflection of the relative integumental permeability to water. Hair *et al.* (1975) speculated that while *A. maculatum* had the higher CEA, its lower rate of water loss probably enabled this species to successfully inhabit the comparatively more xeric upland prairies.

One might ask if there is any value in associating CEAs and integumental permeability to the interpretation of the questing behavior of *A. americanum* and *A. maculatum*? We constructed two figures from data on *A. americanum* (Fig. 5.18a, Semtner & Hair 1973) and on *A. maculatum*

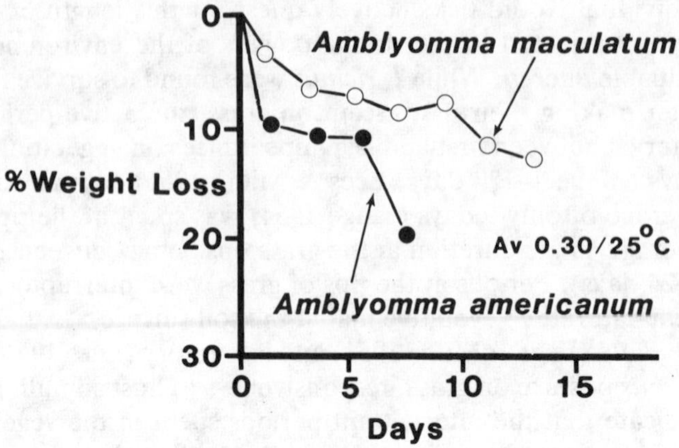

Fig. 5.17 — Weight loss of adult ixodid ticks showing the difference in cuticle permeability when held at an activity below their critical equilibrium activities (redrawn from Hair *et al.* 1975).

(Fig. 5.18b; Fleetwood, unpublished). These field studies used similar procedures to observe tick populations. The behavioral interpretations were made during the normal season of activity for each species. The major difference between them was that *A. maculatum* tended to remain up on the vegetation for extended periods while *A. americanum* exhibited a daily ascent–descent pattern. Observations at 2 h intervals (consisting of height on grass and responsiveness to host stimuli) indicate that *A. maculatum* adults remain on vegetation for days without descending to the duff layer, and that these ticks are most responsive to host stimuli during the hottest part of the day (10:00 to 16:00 h). By monitoring microclimate temperature and relative humidity changes at the soil level and at the questing height on grass, no statistical difference was observed in the saturation deficits during the August–September period. This is when these ticks are most active on the South Texas Plains. Also, during daily periods when the vapor activity was at or above 90, both at ground level and at the questing height, adults on grass were unresponsive to the stimuli of breath and touch. This change in behavior and association to humidity conditions was also observed for the larvae of *A. maculatum* (Fleetwood & Teel 1983). Since there is no major humidity gradient between the questing height and soil level in the prairie environment, it seems logical that a daily descent would be wasteful of energy reserves. Furthermore, their quiescence and unresponsiveness on grass during periods of high humidity may indicate a shift of their priorities from host acquisition (highly responsive/forelegs displayed) to rehydration without vertical migration.

Amblyomma americanum adults, on the other hand, ascend vegetation in early evening when the temperature declines and the relative humidity rises, probably remaining all night, then descend the vegetation to seek refuge in the cooler more humid microclimate of the duff layer during the

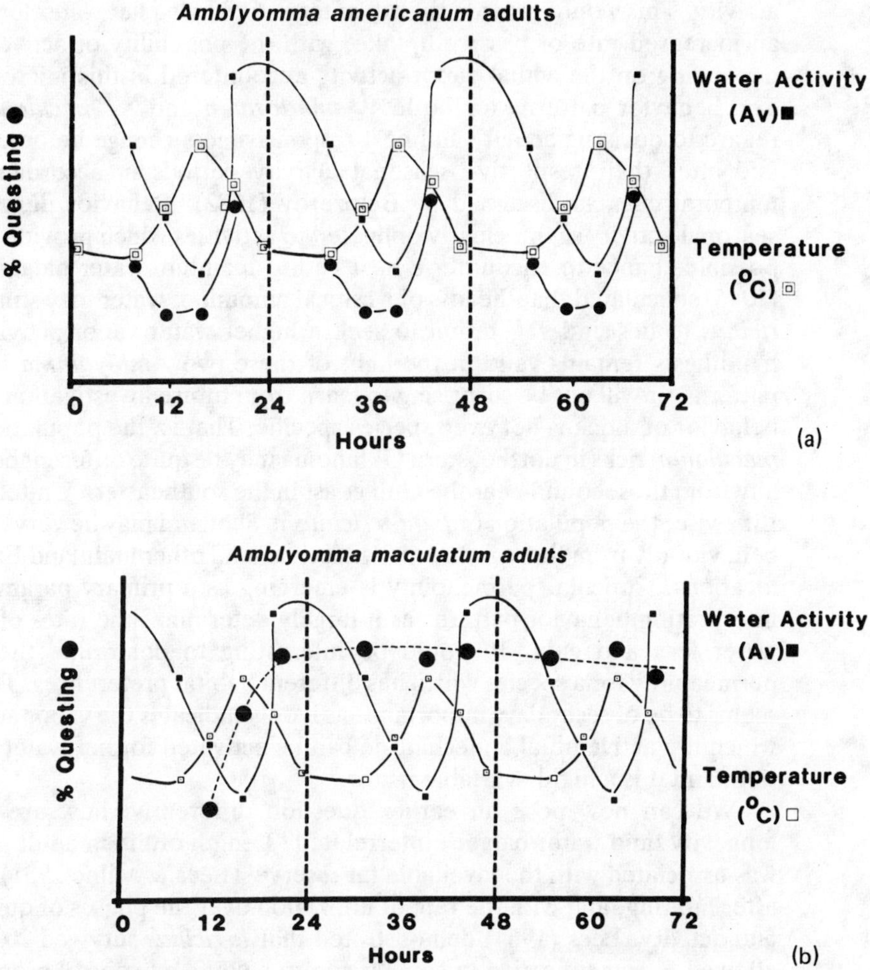

Fig. 5.18a — Questing behaviour of adult lone star ticks in field conditions with changing temperature and water vapor activity (redrawn from Semtner & Hair 1973). Fig. 5.18b — Questing behaviour of adult Gulf Coast ticks in field conditions in the south Texas plains with changing temperature and water vapor activity (from Fleetwood, unpublished). Notice that this species, which is often found in drier microhabitats than lone star ticks, remains on vegetation rather than seeking refuge in the duff layer. This species is more resistant to desiccation and may not need to respond to daily changes in ambient conditions.

heat of the day (Semtner & Hair 1973). Their data plus that of Hoch *et al.*, (1971) indicate that a humidity gradient existed between the duff layer and questing height on vegetation, and that afternoon temperature–humidity conditions exceeded the CEA for this species from early May to late June in the Ozark region of northeastern Oklahoma. The midday return to the duff layer would place *A. americanum* in a microclimate of higher water vapor activity. This would provide the advantage of reduced net water loss due to an increased rate of passive uptake, with the possibility of active uptake depending on the actual vapor activity encountered in this microclimate. The behavior patterns for both *A. americanum* and *A. maculatum* with regard to questing activity and host responsiveness change before, during, and after their respective seasonal activity periods in accordance with temporal cues as discussed by Belozerov (1982). Behavior during adult seasonal activity seems clearly obligated to activities which provide the best possible chance to encounter a host and to maintain water balance. Lees (1946) speculated that the loss of a critical amount of water may stimulate *I. ricinus* to descend vegetation to seek a higher water vapor activity. This hypothesis remains valid in the light of these two *Amblyomma* behavior patterns. It will not be surprising to learn from future investigation that the behavior of a tick is not even species specific. That is, the population of *A. maculatum* ticks in northeastern Oklahoma may be quite different behaviorally from those found near the Gulf coast in the southeastern United States. Likewise, the population of *Ixodes ricinus* in Scotland may be very different behaviorally from those found in Switzerland and other mainland European locations. Cuticular permeability is emerging as a primary parameter for interpreting behavior patterns as it largely determines the rates of passive water loss and gain. It would be interesting to determine the cuticle permeability for a species which has different habitat preferences. The CEA seems to be of secondary importance as it only indicates the vapor activity at which the active uptake mechanism can be activated for net water absorption, and it is a highly variable factor.

We can now pose an earlier question differently: how are energy, longevity, and water balance interrelated? Length of life in adult *I. ricinus* was associated with the 'available fat reserve' (Lees & Milne 1951) present after molting, and with the rate of utilization over the phases of quiescence and activity. Lees (1964) demonstrated that *I. ricinus* survived 203 days if allowed to remain active in comparison to a confined control group at the same temperature and humidity which survived 350 days. The degree of activity was found to affect the CEA of *I. ricinus*. In general, those ticks experiencinge greater activity required that the tick be exposed to a higher humidity to prevent dessication. Field studies with *A. americanum* Williams *et al.*, 1986) support many of Lees' findings with *I ricinus* from England. Adults of this species also require higher humidities to survive as they age, or, in other words, the CEA increases with age under field conditions. On the other hand, laboratory maintained specimens which were not so active, did not seem to have increasing CEA values (Jaworski *et al.* 1984), although the range was considerable (0.80–0.90 a_v). Water content for males and females remained relatively constant during the year-long laboratory and field studies of *A. americanum.* Heme content decreased with age, while lipid decreased only in females which had been fed upon hosts with prior tick feeding.

The evaluation of these behavioral and physiological observations of ixodid ticks suggests that rigorous comparative studies are needed to clarify speculation on how ticks balance their behavioral strategy between hosts while coping with water and energy limitations. Progress has been slow since the stimulating work of Lees, but investigators are taking up the challenge and significant strides should be forthcoming in the future.

As a parting note we offer the following. Lees (1969) poses the question as to whether *Ixodes ricinus* is confined to a humid environment because it has a relatively permeable cuticle, or whether the high permeability has resulted from the relaxation of selection pressure on a species which has for other reasons come to inhabit a humid environment? Comparison of *I. ricinus* from England to those found in Switzerland might provide the clue Lees was seeking.

ACKNOWLEDGEMENTS

The authors thank the following colleagues for their critical comments and assistance on various aspects of the manuscript: Dr L. G. Arlian, Dr L. B. Coons, D. Jaworski, Dr D. E. Johnston, K. Needham, R. Rosell-Davis, Dr J. R. Sauer, N. Sigal, A. Smith, Dr. W. C. Welbourn, and Dr. G. W. Wharton. The assistance of Jack Kocka in the production of figures was greatly appreciated.

REFERENCES

Arlian, L. G. & Veselica M. M.(1979). Water balance in insects and mites. *Comp. Biochem. Physiol.* **64A**: 191–200.

Arlian, L. G. & Veselica M. M. (1981). Reevaluation of the humidity requirements of the house dust mite *Dermatophagoides farinae* (Acari: Pyroglyphidae). *J. Med. Entomol.* **18**: 351–52.

Arlian, L. G. & Veselica M. M. (1982). Relationship between transpiration rate and temperature in the mite *Dermatophagoides farinae. Physiol Zool.* **55**: 344–54.

Arlian, L. G. & Wharton, G. W. (1974) Kinetics of active and passive water exchange between the air and a mite, *Dermatophagoides farinae. J. Insect. Physiol.* **20**: 1063–77.

Balashov, Yu. S. (1968). *Bloodsucking ticks (Ixodoidea) — Vectors of diseases of man and animals.* Nauka, Leningrad. Translation in: *Misc. Publ. Ent. Soc. Am.* **8**: 161–376 (1972).

Balashov, Yu. S. (1979). *An atlas of electron microscopical anatomy of ixodid ticks.* Nauka,. Leningrad. Translation in: Special Publ. Ent. Soc. Am., *An atlas of ixodid tick ultrastructure,* 289 pp. (1983).

Balashov, Yu. S. & Filippova, N. A. (1964). Water balance and critical epicuticular temperature in argasid ticks (Parasitiformes, Argasidae). *Ento. Rev.* **43**: 421–26.

Barker, D. M., Ownby, C. L., Krolak,. J. M., Claypool, P. L. & Sauer, J. R. (1984). The effects of attachment, feeding, and mating on the morphology of the type I alveolus of salivary glands of the lone star tick. *Amblyomma americanum* (L.). *J. Parasitology* **70**: 99–113.

Belozerov, V. N. (1982). Diapause and biological rhythms in ticks. In: *Physiology of ticks,* Obenchain, F. D. & Galun, R. L. (eds.). Pergamon Press Ltd., New York, pp. 469–500.

Binnington, K. C. (1978). Sequential changes in salivary gland structure during attachment and feeding of the cattle tick, *Boophilus microplus*. *Int. J. Parasitology* **8**: 97–115.

Brody,. A. R., McGrath, J. C. & Wharton, G. W. (1976). *Dermatophagoides farinae:* the supracoxal glands. *J. New York Entomol. Soc.* **84**: 34–47.

Browning, T. O. (1954). Water balance in the tick *Ornithodoros moubata* Murray, with particular reference to the influence of carbon dioxide on the uptake and loss of water. *J. Exp. Biol.* **31**: 331–40.

Coons, L. B. & Roshdy, M. A. (1973). Fine structure of the salivary glands of unfed male *Dermacentor variabilis* (Say) (Ixodoidea: Ixodidae). *J. Parasitol.* **59**: 900–12.

Coons, L. B. & Roshdy, M. A. (1979). Functional morphology and cytochemical localization of chloride ions and ouabain sensitive phosphatase activity in the salivary gland transport epithelia of four species of ticks. In: *Recent advances in acarology,* Vol. 1, Rodriguez, J. G. (ed.). Academic Press, New York, pp. 427–34.

Coons, L. B. & Roshdy, M. A. (1981). Ultrastructure of granule secretion in salivary glands of *Argas arboreus* during feeding. *Zeitschrift Parasit.* **65**: 225–34.

Davis, M.-T. B. (1974a). Changes in critical temperature during nymphal and adult development in the rabbit tick, *Haemaphysalis leporispalustris* (Acari: Ixodides: Ixodidae). *J. Exp. Biol.* **60**: 85–94.

Davis, M.-T.B. (1974b). Critical temperature and changes in cuticular lipids in the rabbit tick, *Haemaphysalis leporispalustris*. *J. Insect Physiol.* **20**: 1087–1100.

Devine, T. L. (1982). Neutron activation analysis of saliva from the tick *Amblyomma variegatum*. *Experientia* **38**: 676–77.

Dzhafarov, T. E. (1965). An electron microscopy study of pyramidal alveoli cells of salivary glands of some ticks. *Izd. Akad. Nauk SSSR*, Leningrad, 31–37.

Edney, E. B. (1977). *Water balance in land arthropods: zoophysiology and ecology*, Vol. 9. Springer-Verlag, New York, 282 pp.

El-Ziady, S. (1958). The behavior of *Ornithodoros erraticus* (Lucas 1849), small form (Ixodoidea, Argasidae), towards certain environmental factors. *Ann. Entomol. Soc. Amer.* **51**: 317–36.

Fawcett, D. W. (1981). *The cell.* 2nd ed. W. B. Saunders Co., Philadelphia, 862 pp.

Fleetwood, S. C. & Teel. P. D. (1983). Variation in activity of ageing *Amblyomma maculatum* Koch (Acarina:Ixodidae) larvae in relation to vapor pressure deficits in pasture vegetation complexes. *Protection Ecology* **5**: 343–52.

Freda, T. J. & Needham, G. R. (1984). Water exchange kinetics of the lone star tick *Amblyomma americanum*. In: *Acarology VI*, Vol. 1, Griffiths,D. A. & Bowman, C. E. (eds.). Ellis Horwood Ltd, Chichester, pp. 358–64.

Hackenbrock, C. R. (1966). Ultrastructural bases for metabolically linked mechanical activity in mitochondria. I. Reversible ultrastructural changes with change in metabolic steady state in isolated liver mitochondria. *J. Cell Biol.* **30**: 269–97.

Hadley, N. F. (1981). Cuticular lipids of terrestrial plants and arthropods: a comparison of their structure, composition and waterproofing function. *Biol. Rev.* **56**: 23–47.

Hafez, M., El-Zaidy, S. & Hefnawy, T. (1970a). Biochemical and physiological studies of certain ticks (Ixodoidea). Cuticular permeability of *Hyalomma* (*H.*) *dromedarii* Koch (Ixodidae) and *Ornithodoros savignyi* (Audouin) (Argasidae). *J. Parasitology* **56**: 154–68.

Hafez, M., El-Zaidy, S. & Hefnawy, T. (1970b). Biochemical and physiological studies of certain ticks (Ixodoidea). Uptake of water vapor by the different developmental stages of *Hyalomma* (*H.*) *dromedarii* Koch (Ixodidae) and *Ornithodoros* (O.) *savignyi* (Audouin) (Argasidae). *J. Parasitology* **56**: 354–61.

Hair, J. A., Sauer, J. R. & Durham, K. A. (1975). Water balance and humidity preference in three species of ticks. *J. Med. Entomol.* **12**: 37–46.

Hefnawy, T., Bishara, S. I. & Bassal, T. T. M. (1975). Biochemical and physiological studies of certain ticks (Ixodoidea). Effects of relative humidity and starvation on the water balance and behaviour of adult *Argas* (*Persicargas*) *arboreus* (Argasidae). *Exp. Parasitol.* **38**: 14–19.

Hoch, A. L., Barker, R. W. & Hair, J. A. (1971). Measurement of physical parameters to determine the suitability of modified woodlots as lone star tick habitat. *J. Med. Entomol.* **8**: 725–30.

Hoogstraal, H. (1978). Biology of ticks. In: *Tick-borne diseases and their vectors*, Wilde, J. K. H. (ed.), Edinburgh Univ. Press, pp. 3–14.

Hoogstaal, H. & Aeschlimann, A. (1982). Tick-host specificity. *Bull. Soc. Ent. Suisse.* **55**: 5–32.

Hsu, M. H. & Sauer, J. R. (1975). Ion and water balance in the feeding lone star tick. *Comp. Biochem. Physiol.* **52A**: 269–76.

Jaworski, D., Sauer, J. R., Williams, J. P., McNew, R. W., & Hair, J. A. (1984), Age-related effects on water, lipid, hemoglobin, and critical equilibrium humidity in unfed adult lone star ticks (Acari:Ixodidae). *J. Med. Entomol.* **21**: 100–104.

Kaufman, S., Kaufman, W. R. & Phillips, J. E. (1981). Fluid balance in the argasid tick, *Ornithodorus moubata*, fed on modified blood meals. *J. Exp. Biol.* **93**: 225–42.

Kaufman, W. (1976). Influence of various factors on fluid secretion by *in vitro* salivary glands of ixodid ticks. *Eur. J. Pharmacol.* **45**: 61–68.

Kaufman, W. R. & Sauer, J. R. (1983). Ion and water balance in feeding ticks; mechanisms of tick excretion. In: *Physiology of ticks*, Obenchain, F. D. & Galun, R. L. (eds.), Pergamon Press Ltd, New York, pp. 213–44.

Kirkland, W. L. (1971). Ultrastructural changes in the nymphal salivary glands of the rabbit tick. *Haemaphysalis leporispalustris*, during feeding. *J. Insect Physiol.* **17**: 1933–46.

Knülle, W. (1984). Water vapour uptake in mites and insects: an ecophysiological and evolutionary perspective. In: *Acarology VI*, Griffiths, D. A. & Bowman, C. E. (eds.), Ellis Horwood Ltd., Chichester, pp. 71–82.

Knülle, W. & Devine, T. L. (1972). Evidence for active and passive components of sorption of atmospheric water vapour by larvae of the tick *Dermacentor variabilis*. *J. Insect Physiol.* **18**: 1653–64.

Knülle, W. & Rudolph, D. (1982). Humidity relationships and water balance of ticks. In: *Physiology of ticks,* Obenchain, F. D. & Galun, R. L. (eds.). Pergamon Press Ltd., New York, pp. 43–70.

Krolak, J. M., Ownby, C. L. & Sauer, J. R. (1982). Alveolar structure of salivary glands of the lone star tick, *Amblyomma americanum* (L.): unfed females. *J. Parasitol.* **68**: 61–82.

Lane, N. J. & Skaer, H. IeB. (1980). Intercellular junctions in insect tissues. *Adv. Insect Physiol.* **15**: 35–213.

Lees, A. D. (1946). The water balance in *Ixodes ricinus* L. and certain other species of ticks. *Parasitology* **37**: 1–20.

Lees, A. D. (1947). Transpiration and the structure of the epicuticle in ticks. *J. Exp. Biol.* **23**: 379–410.

Lees, A. D. (1964). The effect of ageing and locomotor activity on the water transport mechanism of ticks. In: *Proc. First Int. Congr. Acarology, Acarologia* hors. ser. **6**: 315–23.

Lees, A. D. (1969). The behavior and physiology of ticks. *Acarologia* **11**: 397–410.

Lees, A. D. & Milne, A. (1951). The seasonal and diurnal activities of individual sheep ticks (*Ixodes ricinus* L.). *Parasitology* **41**: 189–208.

Machin, J. (1979). Atmospheric water absorption in arthropods. *Adv. Insect Physiol.* **14**: 1–48.

Mayahara, H., Fujimoto, K., Ando, T. & Ogawa, K. (1980). A new one-step method for the cytochemical localization of ouabain-sensitive, potassium-dependent p-nitrophenylphosphatase activity. *Histochem.* **67**: 125–38.

McEnroe, W. D. (1971). The limiting equilibrium humidity for *Dermacen-*

tor variabilis Say. and *Dermacentor andersoni* Stiles (Acarina:Ixodidae). *Mass. Agr. Exp. Sta. Res. Bull.* **594**, 15 pp.

McMullen, H. L., Sauer, J. R. & Burton, R. L. (1976). Possible role of uptake of water vapour by ixodid tick salivary glands. *J. Insect. Physiol.* **22**: 1281–85.

Megaw, M. J. W. & Beadle, D. J. (1979). Structure and function of the salivary glands of the tick, *Boophilus microplus* Canestrini (Acarina:Ixodidae). *Int. J. Insect Morph. Embryol.* **8**: 67–83.

Meredith, J. & Kaufman, W. R. (1973). A proposed site of fluid secretion in the salivary gland of the ixodid tick, *Dermacentor andersoni. Parasitology* **67**: 205–17.

Needham, G. R. & Coons, L. B. (1984). Ultrastructural changes in type Ialveoli of the salivary glands from hydrating and desiccating lone star ticks. In: *Acarology VI*, Griffiths, D. A. & Bowman, C. E. (eds.), Ellis Horwood Ltd., pp. 366–73.

Norval, R. A. I. (1977). Studies on the ecology of the tick *Amblyomma hebraeum* Koch in the Eastern Cape province of South Africa. II. Survival and development. *J. Parasitology* **63**: 740–47.

O'Donnell, M. J. (1982). Hydrophilic cuticle — the basis for water vapour absorption by the desert burrowing cockroach, *Arenivaga investigata. J. Exp. Biol.* **99**: 43–60.

Peaker, M. & Linzell, J. L. (1975). *Salt glands in birds and reptiles.* Cambridge Univ. Press, New York, 307 pp.

Petney, T. N., Andrews, R. H. & Bull, C. M. (1983). Movement and host finding by unfed nymphs of two Australian reptile ticks. *Aust. J. Zool.* **31**: 717–21.

Penman, H. L. (1958). *Humidity*. Reinhold, New York. pp 71.

Richards, A. G., Clausen, M. B. & Smith, M. N. (1953). Studies on arthropod cuticle. The asymmetrical penetration of water. *J. Cell Comp. Physiol.* **42**: 395–414.

Roshdy, M. A.. (1972). The subgenus *Persicargas* (Ixodoidea, Argasidae *Argas*) **15**. Histology and histochemistry of the salivary glands of *A*. (*P.*) *persicus* (Oken). *J. Med. Ent.* **9**: 143–48.

Roshdy, M. A. & Coons, L. B. (1975). The subgenus *Persicargas* (Ixodoidea: Argasidae: *Argas*) **23**. Fine structure of the salivary glands of unfed *A*.(*P.*) *arboreus* Kaiser, Hoogstraal, & Kohls. *J. Parasitology* **61**: 743–52.

Rudolph, D. & Knülle, W. (1974). Site and mechanism of water vapour uptake from the atmosphere in ixodid ticks. *Nature* **249**: 84–85.

Rudolph, D. & Knülle, W. (1978). Uptake of water vapour from the air: Process, site and mechanism in ticks. In: *Comparative physiology — water, ions and fluid mechanics*, Schmidt-Nielsen, K., Bolis, L. & Maddrell, S. H. P. (eds.), Cambridge University Press, pp. 97–113.

Rudolph, D. & Knülle, W. (1979). Mechanisms contributing to water balance in non-feeding ticks and their ecological implications. In: *Recent advances in acarology,* Vol. 1, Rodriguez, J. G. (ed.), Academic Press, New York, pp. 375–83.

Rudolph, D. & Knülle, W. (1982). Novel uptake systems for atmospheric water vapor among insects. *J. Exp. Zool.* **222**: 321–33.

Sauer, J. R. (1977). Acarine salivary glands — physiological relationships. *J. Med. Entomol.* **14**: 1–9.

Sauer, J. R. & Hair, J. A. (1971). Water balance in the lone star tick (Acarina: Ixodidae): the effects of relative humidity and temperature on weight changes and total water content. *J. Med. Entomol.* **8**: 479–85.

Seethaler, H., Knülle, W. & Devine, T. L. (1979). Water vapor intake and body water (HOH) clearance in the housemite (*Glycyphagus domesticus*). *Acarologia* **21**: 440–50.

Semtner, P. J. & Hair, J. A. (1973). The ecology and behavior of the lone star tick (Acarina: Ixodidae). IV. The daily and seasonal activity patterns of adults in different habitat types. *J. Med. Entomol.* **10**: 337–44.

Semtner, P. J. & Hair, J. A. (1975). Evaluation of CO_2-baited traps for survey of *Amblyomma maculatum* Koch and *Dermacentor variabilis* Say (Acarina: Ixodidae). *J. Med. Entomol.* **12**: 137–38.

Semtner, P. J., Barker, R. W. & Hair, J. A. (1971). The ecology and behavior of the lone star tick (Acarina: Ixodidae). II. Activity and survival in different ecological habitats. *J. Med. Entomol.* **8**: 719–25.

Shih, C.-Y., Sauer, J. R., Eikenbary, R., Hair, J. A. & Frick, J. H. (1973). The effects of desiccation and rehydration on the lone star tick. *J. Insect. Physiol.* **19**: 505–14.

Toolson, E. C. (1978). Diffusion of water through the arthropod cuticle: thermodynamic consideration of the transition phenomenon. *J. Thermal Biol.* **3**: 69–73.

Toolson, E. C., White, T. R., & Glaunsinger, W. S. (1979). Electron paramagnetic resonance spectroscopy of spin-labelled cuticle of *Centrupoides sculpturatus* (Scorpiones: Buthidae): correlation with thermal effects on cuticular permeability. *J. Insect Physiol.* **25**: 271–75.

Waladde, S. M. & Rice, M. J. (1982). The sensory basis of tick feeding behaviour. In: *Physiology of ticks*, Obenchain, F. D. & Galun R. L. (eds.), Pergamon Press, New York, pp. 71–118.

Walker, A. R., Fletcher, J. D. & Gill,. H. S. (1985). Structural and histochemical changes in the salivary glands of *Rhipicephalus appendiculatus* during feeding. *Intl. J. Parasitol.* **15**: 81–100.

Webb, J. P. (1979). Host-locating behavior of nymphal *Ornithodoros concanensis* (Acarina: Argasidae). *J. Med. Entomol.* **16**: 437–47.

Wharton, G. W. (1985). Water balance of insects. In: *Comprehensive insect physiology, biochemistry and pharmacology*, Kerkut, G. A. & Gilbert,

L. I. (eds.), Pergamon Press, New York, **14**: 565–601.

Wharton, G. W. & Richards, A. G. (1978). Water vapor exchange kinetics in insects and acarines. *Ann. Rev. Entomol.* **23**: 309–28.

Wilkinson, P. R. (1953). Observations on the sensory physiology and behavior of larvae of the cattle tick, *Boophilus microplus* (Can.). *Aust. J. Zool.* **1**: 345–56.

Williams, J. P., Sauer, J. R., McNew, R. W. & Hair, J. A. (1986). Physiological and biochemical changes in unfed lone star ticks, *Amblyomma americanum* (Acari: Ixodidae), with increasing age. *J. Med. Entomol.* (in press).

Williams, V. R. & Williams, H. B. (1967). *Basic physical chemistry for the life sciences.* W. H. Freeman Co., San Francisco, 382 pp.

6

Ultrastructure of the Tick Neuroendocrine system

K. C. Binnington

6.1 INTRODUCTION

The neuroendocrine system of arthropods was originally understood to encompass central neurosecretory cells directing their secretions through axons to a neurohaemal organ from which neurohormones are released into the circulation to act on peripheral organs such as endocrine glands. Since its inception, however, this concept of the neuroendocrine system has undergone modification to accommodate observations which do not fit comfortably into the scheme. Tissues have been found to be innervated directly by 'neurosecretory' axons; a diffuse system of neurohaemal structures additional to the corpus cardiacum — the perisympathetic organs — were described in insects (reviewed by Maddrell 1974); some neurosecretory cells were found to contain adrenergic transmitters; and the concept of neuromodulators was developed (reviewed by Evans 1980). Thus the distinction between motorneurones or interneurones employing relatively non-specific chemical transmitters acting over very small distances, and neurosecretory cells releasing highly specific neurohormones to diffuse through relatively large volumes of haemolymph, has become increasingly blurred. It has also been realised that peptides may play a much greater role than previously suspected in central neurotransmission or neuromodulation in insect nervous systems.

Studies on tick neurosecretion have generally not kept pace with our understanding of the insect system. In this chapter, I will present new and recent data on the ultrastructure of tick neurones, neurohaemal structures, and endocrine glands and discuss these findings in the context of current concepts of insect neurosecretion. Broader aspects of tick neurophysiology will be touched upon as will the structure of tick endocrine glands since these subjects are inextricably associated with neurosecretion.

6.2 'NEUROSECRETORY' AND 'NON-NEUROSECRETORY' CELLS

The application of classical neurosecretory stains to the synganglion of ticks has demonstrated an abundance of cells rendered basophilic by oxidation (Gabe 1955, Ioffe 1964, Dhanda 1967, Eichenberger 1970, Binnington &

Tatchell 1973, Eisen *et al.* 1973, Obenchain 1974, Obenchain & Oliver 1975), which can therefore be considered analogous to the type I or A cells of insects (Nayer 1955, Delphin 1965). Obenchain (1974) found that ticks also have cells of the other three insect types; these are acidiphilic following oxidation (II or B), stain with azan (III or C), or stain variously with basic and acid dye combinations after oxidation, but do not stain with azan (IV or D). In some cases, Obenchain (1974) found it difficult to determine if a cell was 'neurosecretory'. The reasons for this become apparent when the ultrastructure of the tick synganglion is understood.

Electron-dense granules are obvious in profiles of cell bodies in most sections through the synganglion of the cattle tick *Boophilus microplus*. The globuli cells of the olfactory lobes with their scant, granule-free cytoplasm are exceptional (Fig. 6.1). Other cell types have from few (Figs. 6.2, 6.3) to an abundance of granules (Figs. 6.5, 6.8). The cells shown in Figs. 6.2 and 6.3 are quite possibly all of the same type but chance sectioning through a cluster of secretory granules could determine if an individual cell is classified as 'neurosecretory' rather than 'non-neurosecretory'. It is unlikely that all granule-containing neurons of ticks secrete neurohormones into the haemolymph.† Granule-containing cells should not therefore all be regarded as true endocrine cells. Motor and inter-neurones of ticks may employ a wide variety of neurotransmitters including acetylcholine, dopamine, noradrenaline, glutamate, and GABA (reviewed by Binnington & Obenchain 1983) and, since the neuropile contains 'neurosecretory' type dense granules (Fig. 6.6), peptides.

There are no data showing axonal continuity between neurohaemal terminals and perikarya within the central nervous system (which would unequivocally define such neurones as endocrine). However, comparison of the morphology of secretory granules in cell bodies and terminals is of some value in suggesting communication between the two neuronal elements.

6.3 PROBABLE NEUROENDOCRINE CELLS

Two cell types (Figs. 6.4, 6.5) of *Boophilus* contain granules larger than any observed in terminals within the neuropile but of a similar size to ones seen in neurohaemal terminals of the perineurial sheath of the synganglion (see section 6.6). An objection to the use of this criterion in suggesting continuity between such cell bodies and terminals is the observation that, in *Rhodnius prolixus,* the diameter of granules within a particular neurone can vary with stages of the secretory cycle (Morris & Steele 1977). However, large granules are consistently observed in these *Boophilus* neurohaemal terminals and in the two cell types regardless of the developmental stage of the adult female tick. These two cell types can be located and identified in the synganglion with relative ease (Binnington 1981a) making a more detailed sequential study of their morphology during developmental or physiological changes within ticks feasible.

† Figs. 6.1—6.14 are all from *Boophilus microplus* except for Fig. 6.13 which is from *Ixodes holocyclus*.

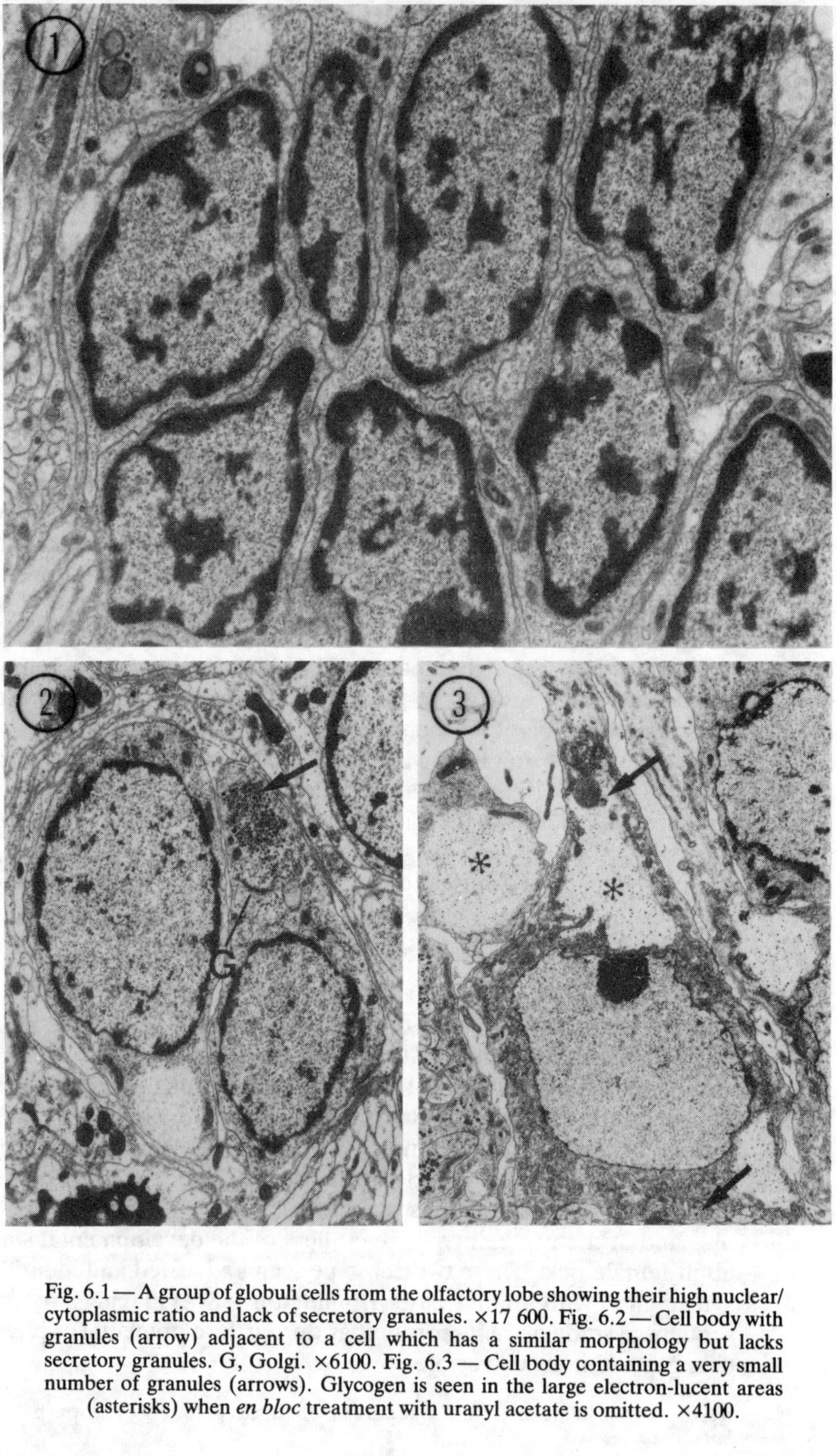

Fig. 6.1 — A group of globuli cells from the olfactory lobe showing their high nuclear/cytoplasmic ratio and lack of secretory granules. ×17 600. Fig. 6.2 — Cell body with granules (arrow) adjacent to a cell which has a similar morphology but lacks secretory granules. G, Golgi. ×6100. Fig. 6.3 — Cell body containing a very small number of granules (arrows). Glycogen is seen in the large electron-lucent areas (asterisks) when *en bloc* treatment with uranyl acetate is omitted. ×4100.

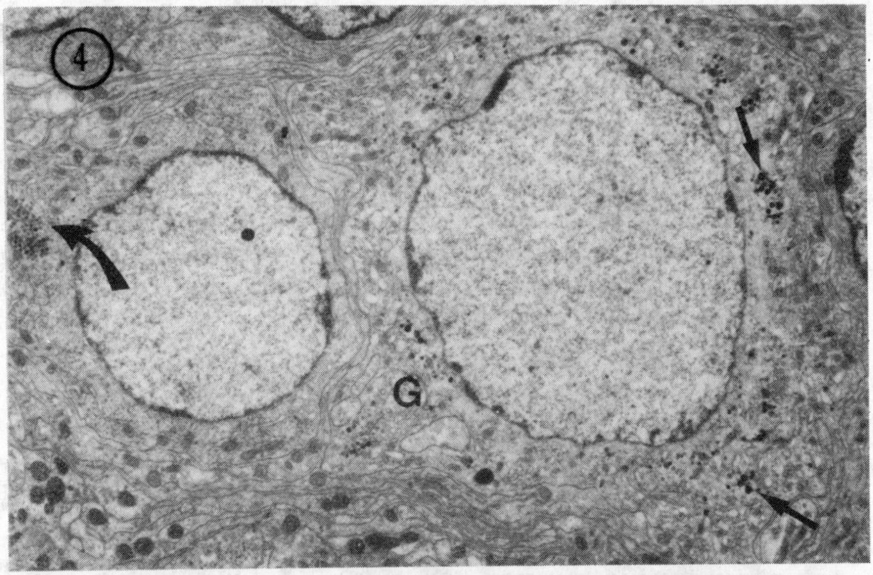

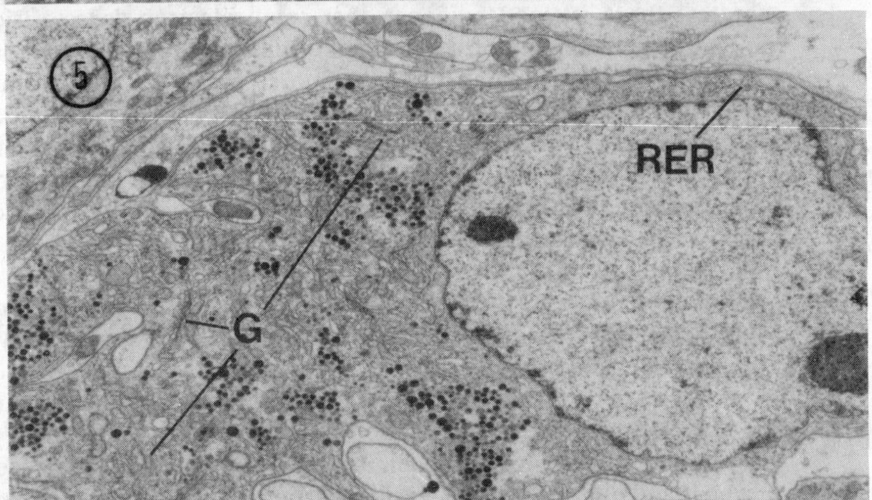

Fig. 6.4 — Cell body of a type seen only in the dorsal cortex surrounding the oesophagus containing large granules (straight arrows). It is adjacent to a cell which contains granules (curved arrow) similar to those in the cell in Fig. 6.2. G, Golgi. ×6200. Fig. 6.5 — Cell body seen only in the dorsal cortex and containing an extensive system of RER and Golgi (G) as well as large numbers of secretory granules (inset). ×7700.

6.4 POSSIBLE NEUROENDOCRINE CELLS

The cell type shown in Fig. 6.2 contains granules of a size which are found in terminals both within the neuropile (Fig. 6.6) and in neurohaemal structures. Without information on the three-dimensional morphology of such neurones it is impossible to define them as neuroendocrine. Fig. 6.7 shows a cell type found in the ventral cortex near the oesophagus and in the posterior of the synganglion adjacent to the opisthosomal ganglia. This cell has a highly convoluted border and is very rich in rough endoplasmic reticulum (RER) and Golgi bodies but contains only a few granules. This morphology suggests a cell whose secretory product is transported away from the cell body immediately after being formed.

6.5 POSSIBLE ADRENERGIC CELLS

The synganglion of *B. microplus* is known to contain a number of cell bodies and areas of neuropile which react positively to the Falck–Hillarp fluorescence reaction for catecholamines (Binnington & Stone 1977). In vertebrates, there is good evidence that small electron-dense vesicles are generally characteristic of adrenergic terminals. Granules of 100 nm diameter are found in the small, intensely fluorescent cells of the rat and guinea pig sympathetic ganglia (reviewed by Eranko 1978). Two cell types of *B. microplus* contain granules of sizes that could be adrenergic. In one type the granules (70–80 nm) are markedly pleomorphic (Fig. 6.8) whereas the other, smaller, cell type, distinguished by a nucleus with a distinctive coarse clumping of chromatin, contains granules of a uniform appearance (Fig. 6.9). Nerve terminals containing small, possibly adrenergic granules are seen in the neuropile (Fig. 6.6), as well as in peripheral organs, providing support for the view that adrenergic neurotransmission probably occurs both centrally and peripherally in ticks (Binnington & Stone 1977).

If we are to distinguish true endocrine from motor and interneurones in the tick central nervous system we must apply techniques which provide information on the immunochemical specificities of neuronal secretory products and on the three-dimensional arrangement of cell bodies and their processes. Further study by electron microscopy and amine histochemistry will also be needed. Only from information of this type will it be possible to provide a sound structural basis for studies on the biochemistry and physiology of the tick nervous system.

6.6 NEUROHAEMAL STRUCTURES

Electron microscopy has clarified our understanding of tick neurohaemal structures. Studies on *B. microplus* (Binnington, 1983) indicate that the neural lamella/perineurial sheath is the major site for the release of neurosecretion (Fig. 6.10), confirming earlier light microscopic observations of neurosecretion associated with the surface of the tick synganglion (Gabe 1955, Binnington & Tatchell 1973, Obenchain & Oliver 1975, Binnington, 1981b). The retrocerebral organ complex and lateral organs, formerly suggested to be neurohaemal, have been shown not to be involved in release of neurosecretion into the haemolymph of *B. microplus* (Binnington 1981c, 1983). Ticks appear to have relatively primitive neurohaemal structures which may be analagous to those of annelids (reviewed by Golding & Whittle 1977) and molluscs (Boer *et al.* 1968, Joose 1976). These animals release neurohormones directly from the sheath of the central nervous system, in contrast to insects and crustaceans in which a peripheral neurohaemal organ such as the corpus cardiacum or sinus gland, is the site of neurohaemal release.

Neurosecretory terminals are also present in the neural lamella of the peripheral nerves of *B. microplus* (Binnington 1983) although there is no concentration of such nerves into areas containing large numbers of terminals like those seen in the peri-sympathetic organs of insects.

The mean size of granules within individual terminals of *B. microplus* range from 60 to 240 nm, with most terminals containing granules 80 to 120 nm in diameter. Five types of terminals can be distinguished on the basis of granule size (Binnington 1983). This suggests that a number of different hormones are released from the sheath of the synganglion, including adrenergic compounds (suggested by the diameter of the smallest sized population of granules).

Tick neurohaemal terminals contain electron-dense 'T-bar' structures and clustered microvesicles which make the terminals indistinguishable, in all respects, except for their contiguity with the haemocoel, from some neuropilar and peripheral nerve terminals (Figs. 6.11, 6.12) (Binnington 1980, 1983). Similar 'T-bars' are present in insects (Finlayson & Osborne 1975, Binnington & Lane 1982) and crustaceans (May & Golding 1983). This appears to be a general phenomenon in arthropods.

6.7 'NEUROSECRETORY' INNERVATION OF PERIPHERAL ORGANS

Electron-dense granules are present in nerve terminals in a variety of peripheral tissues of ticks. These include salivary glands of *B. microplus* and *Ixodes holocyclus* (Fig. 6.13) and other ticks (Coons & Roshdy 1973, Megaw 1977), lateral organs of *B. microplus* (Binnington 1981c), and muscles of *Amblyomma variegatum* (Hart *et al.* 1980). Nerve terminals in the salivary glands contain small or large granules (Fig. 6.13). Terminals containing smaller granules are probably the same as those which react positively to the Falck–Hillarp fluorescence reaction for catecholamines (Binnington & Stone 1977), and the larger granules could contain a peptide. The neurochemistry of the direct innervation of peripheral organs is similar to neurotransmission across synapses, for example the neuromuscular junc-

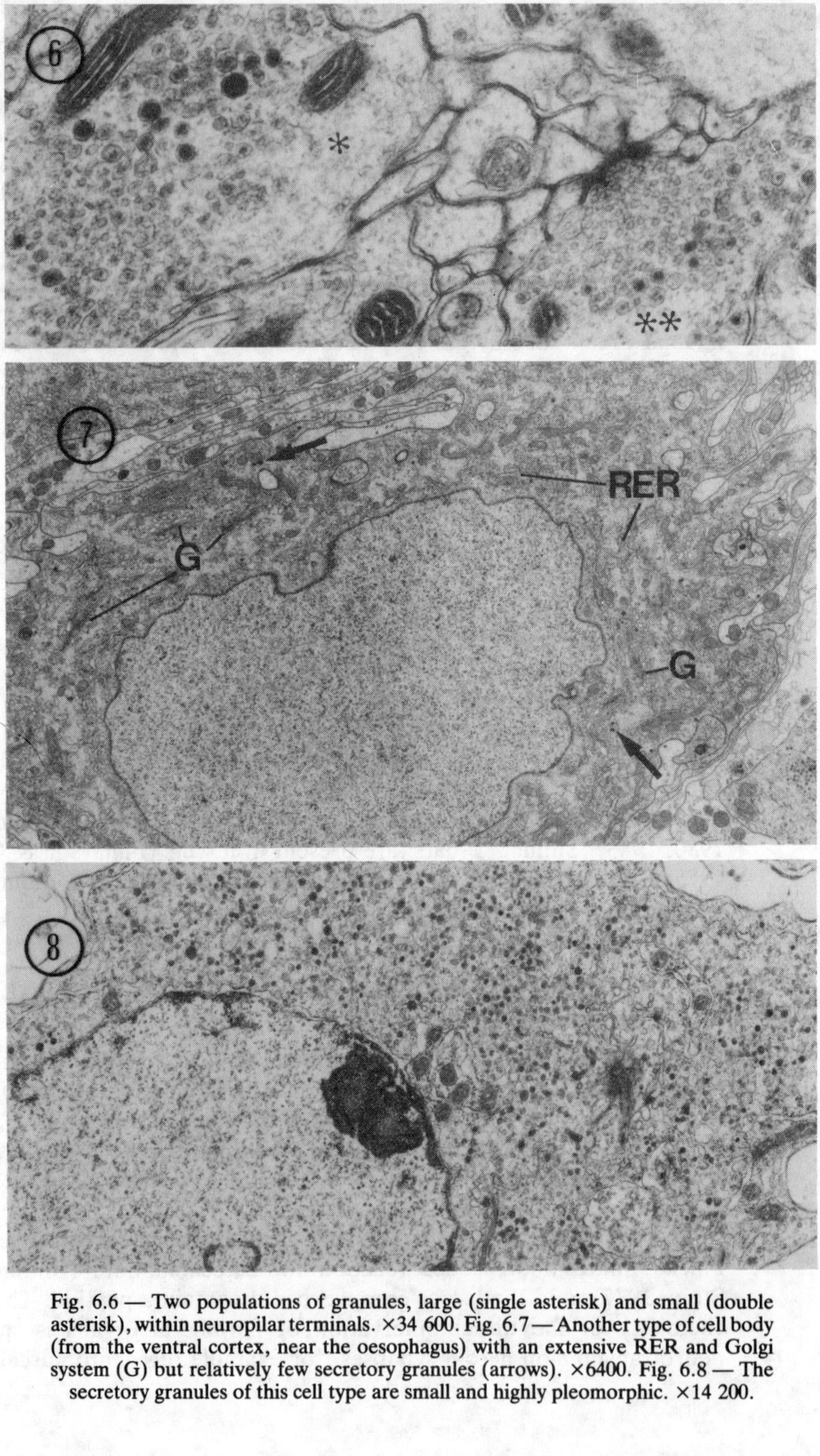

Fig. 6.6 — Two populations of granules, large (single asterisk) and small (double asterisk), within neuropilar terminals. ×34 600. Fig. 6.7 — Another type of cell body (from the ventral cortex, near the oesophagus) with an extensive RER and Golgi system (G) but relatively few secretory granules (arrows). ×6400. Fig. 6.8 — The secretory granules of this cell type are small and highly pleomorphic. ×14 200.

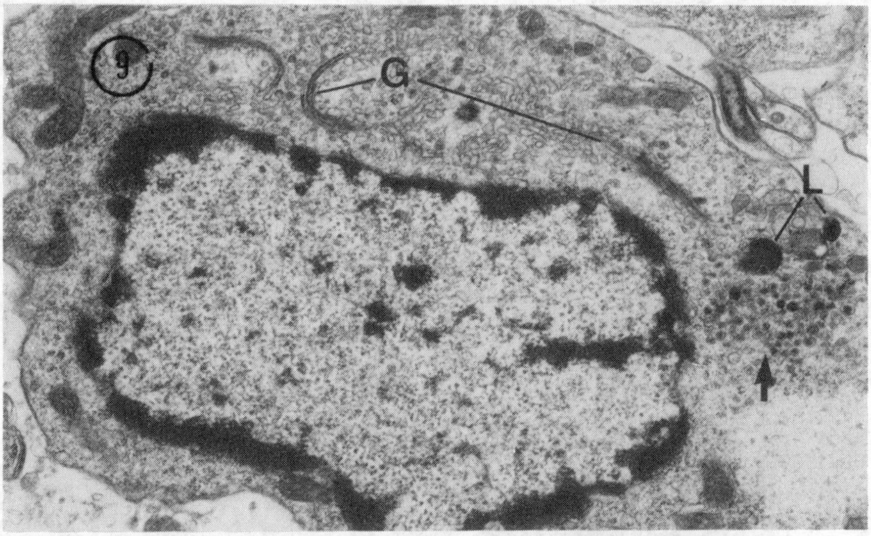

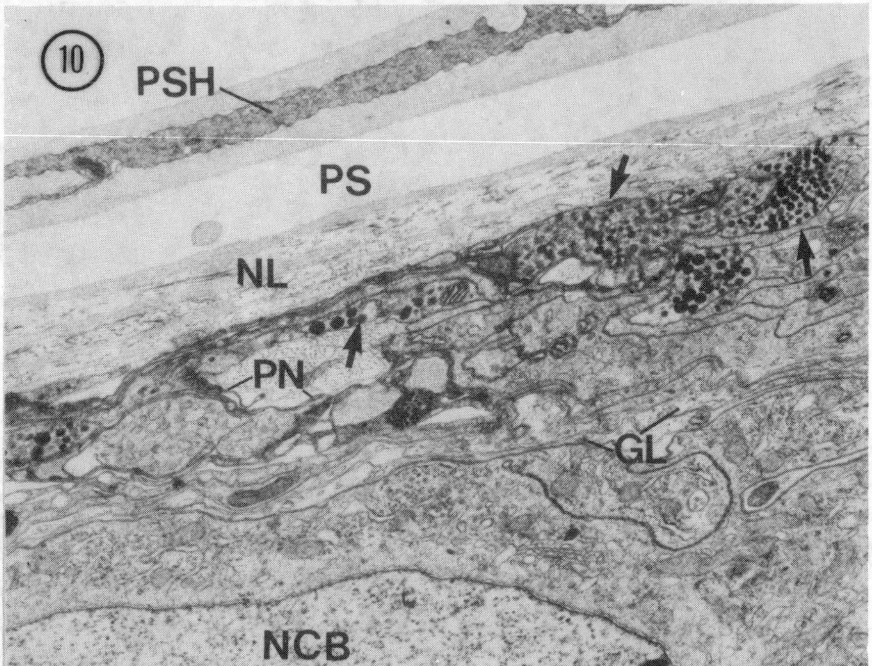

Fig. 6.9 — Cell body with densely staining nucleus and small granules (arrow). G, Golgi; L, lysosomes. ×15 200. Fig. 6.10 — Section through a part of the neural lamella (NL) perineurial (PN) sheath which is rich in neurohaemal terminals (arrows). GI, glial cell processes; NCB, nerve cell body; PS, periganglionic sinus; PSH, periganglionic sheath. ×13 300.

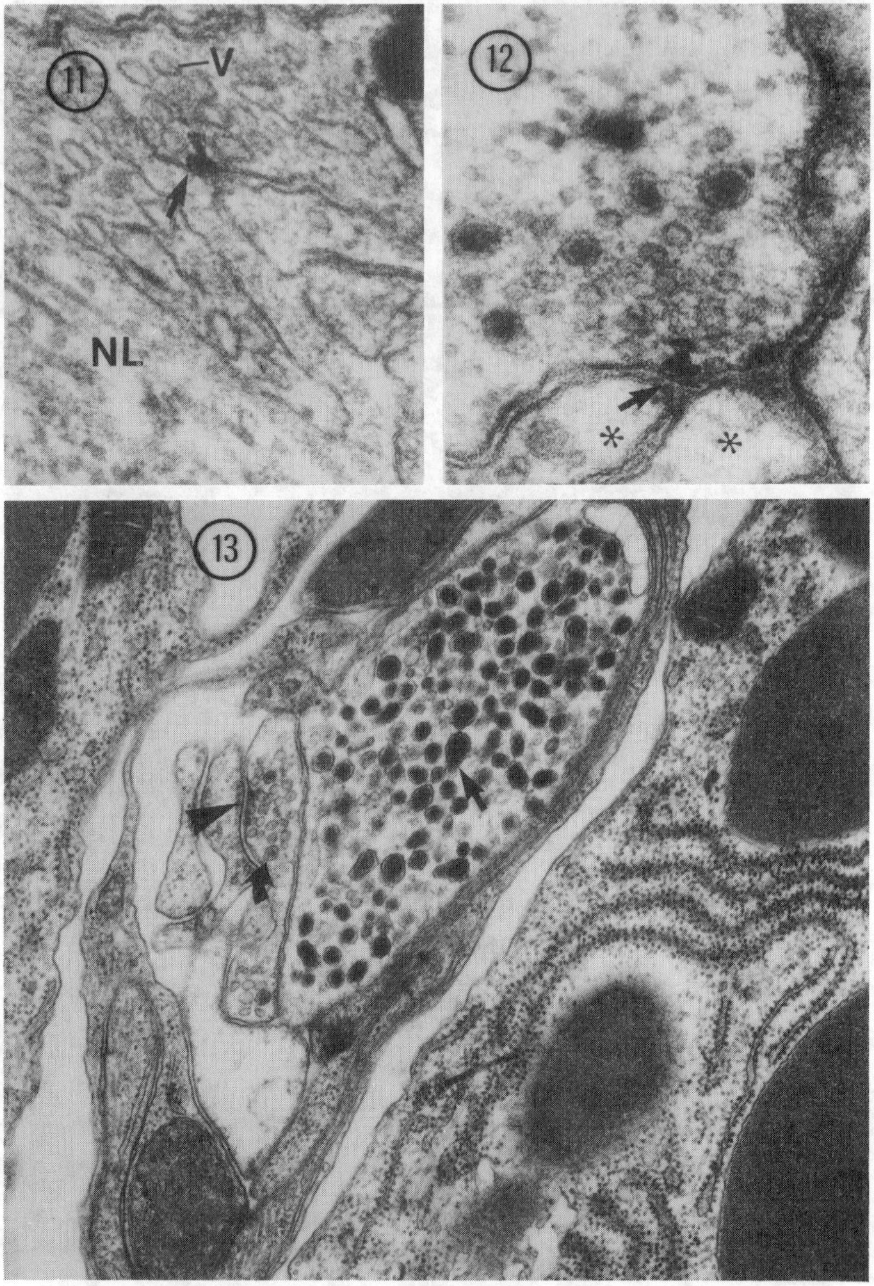

Fig. 6.11 — Section through a tick neurohaemal terminal showing the 'T' configuration. NL, neural lamella; V, vesicles. ×99 000. Fig. 6.12 — A 'T' bar in neuropilar terminal. As in the neurohaemal terminal (Fig. 6.11), clear vesicles surround the dense material, and electron-dense granules occupy a more distal position. Note the narrow synaptic gap (arrow) between the terminal and post-synaptic processes (asterisks). ×98 000. Fig. 6.13 — Two types of nerve terminals are found in the salivary glands of *Ixodes holocyclus*. One contains large electron-dense granules (straight arrow) and the other much smaller electron-dense granules (curved arrow). Once again clear vesicles surround an electron-dense configuration (arrowhead). ×31 200.

tion, and differs from a neuroendocrine control in which neurohormones are transported through the haemolymph.

6.8 ENDOCRINE GLANDS

In *B. microplus,* two structures (lateral organs and the retrocerebral complex) have been described whose morphology is compatible with an endocrine function (Binnington 1981c, 1983). The lateral organs are rich in smooth endoplasmic reticulum (SER) (Fig. 6.14), and more SER forms as female ticks feed and produce eggs (Binnington 1981c). The abundance of

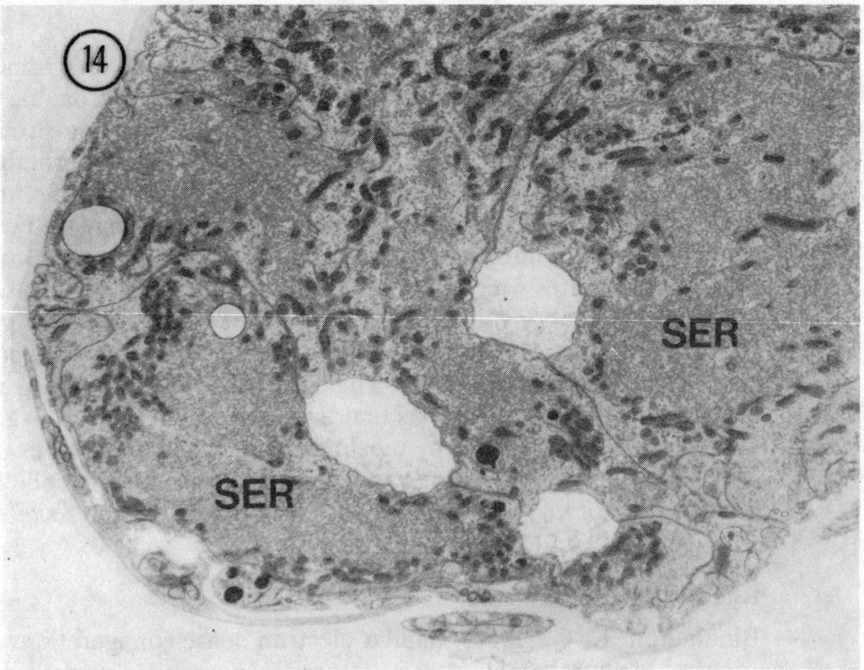

Fig. 6.14 — Lateral organ rich in smooth endoplasmic reticulum (SER). ×73 000.

smooth endoplasmic reticulum suggests that lateral organs secrete a lipidic product. This might be ecdysone or juvenile hormone. The increase in lateral organ activity during oogenesis (Panfilova 1978, 1980) might suggest an association with juvenile hormone secretion, but a better understanding of the role of the lateral organs in tick endocrinology awaits fruther comparative and biochemical studies. The retrocerebral complex (Obenchain & Oliver 1975) of *B. microplus* contains predominantly RER with little or no SER, and it secretes electron-dense granules which could contain peptides (Binnington 1983). However, there is no definitive information on this, or on the times at which the retrocerebral complex is active.

Ecdysone is known to occur in ticks (Delbecque *et al.* 1978, Diehl *et al.* 1982) and to affect their physiology (Kitaoka 1972, Mango *et al.* 1976, Khalil *et al.* 1984), but its source remains unknown. The lateral segmental organs are a possible source although there is no evidence for a relationship between their activity and moulting. The fact that the retrocerebral complex contains little SER means it is unlikely to secrete ecdysone. Another structure demanding detailed examination is the 'peritracheal gland' described by light microscopy in *Argas (P.) arboreus* by Roshdy & Marzouk (1982).

6.9 SUMMARY

The tick synganglion contains a variety of neurone types only some of which can be classified as endocrine or non-endocrine with reasonable certainty. Some endocrine neurones have a highly characteristic fine structure, and studies on them during tick development should be a profitable line of research for increasing the understanding of the roles of tick neurohormones. The use of electron microscopy has greatly increased our understanding of tick neurohaemal structures. Ticks can now be grouped with those invertebrates which have a more 'primitive', diffuse release of neuroendocrine secretions; interestingly, this type of neurohaemal system appears to be present in all invertebrates which have not developed a blood-brain barrier (Binnington 1983). The ultrastructure of lateral organs and the retrocerebral complex suggests that they function as endocrine glands, but this requires confirmation by biochemical and physiological examination. Lateral organs are promising candidates as a source of juvenile hormone, but it is still not known where ecdysone is synthesized and secreted.

REFERENCES

Binnington, K. C. (1980). Similar electron dense configurations at release sites in neurohemal/neurosecretory and neuropilar/conventional terminals in an arthropod. *Eur. J. Cell Biol.* **22:** 187 (abstract).

Binnington, K. C. (1981a). Nervous, neuroendocrine and endocrine systems of ticks. PhD thesis, University of Cambridge.

Binnington, K. C. (1981b). Innervation of coxal muscles, heart and other organs in the cattle tick *Boophilus microplus* Canestrini (Acarina: Ixodidae), *Int. J. Insect Morphol. Embryol.* **10:** 109–119.

Binnington, K. C. (1981c). Ultrastructural evidence for the endocrine nature of the lateral organs of the cattle tick *Boophilus microplus*. *Tissue Cell* **13:** 475–490.

Binnington, K. C. (1983). Ultrastructural identification of neurohaemal sites in a tick: evidence that the dorsal complex may be a true endocrine gland. *Tissue Cell* **15:** 317–327.

Binnington, K. C. & Lane, N. J. (1982). Presence of T-bars, intramembranous particle arrays and exocytotic profiles in neuroendocrine terminals of an insect. *Tissue Cell* **14:** 463–474.

Binnington, K. C. & Obenchain, F. D. (1983). Structure and function of the circulatory, nervous and neuroendocrine systems of ticks. In *Physiology of ticks* (Obenchain, F. D. & Galun, R. eds). Pergamon. pp. 351–398

Binnington, K. C. & Stone, B. F. (1977). Distribution of catecholamines in the nervous system of the cattle tick *Boophilus microplus* Canestrini. *Comp. Biochem. Physiol.* **C 58:** 21–28.

Binnington, K. C. & Tatchell, R. J. (1973). The nervous system and neurosecretory cells of *Boophilus microplus* (Acarina, Ixodidae). *Z. Wiss. Zool. Abt.* **A 185:** 193–206.

Boer, H. H., Douma, E., & Koksma, J. M. A. (1968). Electron microscope study of neurosecretory cells and neurohaemal organs in the pond snail *Lymnaea stagnalis*. *Symp. Zool. Soc. Lond.* **22:** 237–256.

Coons, L. B., & Roshdy, M. A. (1973). Fine structure of the salivary glands of unfed male *Dermacentor variabilis* (Say) (Ixodoidea: Ixodidae). *J. Parasitol.* **59:** 900–912.

Delbecque, J. P., Diehl, P. A., & O'Connor, J. D. (1978). Presence of ecdysone and ecdysterone in the tick *Amblyomma hebraem* Koch. *Experientia* **34:** 1379–1381.

Delphin. F. (1965). The histology and possible functions of neurosecretory cells in the ventral ganglion of *Shistocerca gregaria* Forskal (Orthoptera: Acrididae). *Trans. R. Entomol. Soc. London* **117:** 167–214.

Dhanda, V. (1967). Changes in neurosecretory activity at different stages in the adult *Hyalomma dromedarii* Koch 1944. *Nature* **214:** 508–509.

Diehl, P. A., Morici, M., & Bouvier, J. (1982). Metabolism of molting hormones in nymphs of the tick *Ornithodorus moubata* (Argasidae: Ixodoidea). *Gen. Comp. Endocrinol.* **46:** 78.

Eichenberger, G. (1970). Das Zentralnervensystem von *Ornithodorus moubata* (Murray), Ixodoidea: Argasidae, und seine postembryonale Entwicklung. *Acta Trop.* **27:** 15–53.

Eisen, V., Warburg, M. R., & Galun, R. (1973). Neurosecretory activity as related to feeding and oogenesis in the fowl tick *Argas persicus* (Oken.). *Gen. Comp. Endocrinol.* **21:** 331–340.

Eranko, O. (1978). Small intensely fluorescent (SIF) cells and nervous transmission in sympathetic ganglia. *Ann. Rev. Pharmacol. Toxicol.* **18:** 417–430.

Evans, P. D. (1980). Biogenic amines in the insect nervous system. *Adv. Insect Physiol.* **15:** 317–473.

Finlayson, L. H., & Osborne, M. P. (1975). Secretory activity of neurons and related electrical activity. *Adv. Comp. Physiol. Biochem.* **6:** 165–258.

Gabe, M. (1955). Donnies histologiques sur la neurosecretion chez les Archnides. *Arch. Anat. Microsc. Morphol. Exp.* **44:** 351–383.

Golding, D. W., & Whittle, A. C. (1977). Neurosecretion and related phenomena in annelids. *Int. Rev. Cytol. (Suppl.)* **5:** 189–302.

Hart, R. J., Beadle, D. J., & Botham, R. P. (1980). The ultrastructure of somatic neuromuscular junctions in the tick *Amblyomma variegatum* (Fabr.). *Cell Tissue Res.* **206:** 505–508.

Ioffe, I. D. (1964). Distribution of neurosecretory cells in the central nervous apparatus of *Dermacentor pictus*. *Med. Parazitol. Parazit. Bolezni* **34:** 57–63.

Joose, J. (1976). Endocrinology of molluscs. [In 'Actualites sur les hormones d'Invertebres.] *Colloq. Int. C.N.R.S.* No. 251, 107–123.

Khalil, G. M., Shaarawy, A. A. A., Sonenshine, D. E., & Gad, S. M. (1984). β-ecdysone effects on the camel tick, *Hyalomma dromedarii* (Acari: Ixodidae). *J. Med. Entomol.* **21:** 188–193.

Kitaoka, S. (1972). Effects of ecdysones on ticks, especially on *Ornithodoros moubata* (Acarina: Argasidae). *Abstr. 14th Int. Congr. Entomol.* Canberra, 1972. 272.

Maddrell, S. H. P. (1974), Neurosecretion. In: *Insect neurobiology*. (Treherne, J. E. ed.). North-Holland, Amsterdam 307–357.

Mango, C., Odhiambo, T. R., & Galun, R. (1976) Ecdysone and the super tick. *Nature* **260:** 318—319.

May, B. A., & Golding, D. W. (1983). Aspects of secretory phenomena within the sinus gland of *Carcinus maenas* (L.). An ultrastructural study. *Cell Tissue Res.* **228:** 245–254.

Megaw, M. W. J. (1977). The innervation of the salivary gland of the tick *Boophilus microplus*. *Cell Tissue Res.* **184:** 551–558.

Morris, G. P., & Steel, C. G. M. (1977). Sequence of ultrastructural changes induced by activation in the posterior neurosecretory cells in the brain of *Rhodnius prolixus* with special reference to the role of lysosomes. *Tissue Cell* **9:** 547–561.

Nayer, K. K. (1955). Studies on the neurosecretory system of *Iphita limbata* Stal. I. Distribution and structure of the neurosecretory cells of the nerve ring. *Biol. Bull. Woods Hole, Mass.* **108:** 206–307.

Obenchain, F. D. (1974). Neurosecretory system of the American dog tick *Dermacentor variabilis* (Acari: Ixodidea). I. Diversity of cell types. *J. Morphol.* **142:** 433–446.

Obenchain, F. D., & Oliver, J. H. (1975). Neurosecretory system of the American dog tick *Dermacentor variabilis* (Acari: Ixodidea). II. Distribution of secretory cell types, axonal pathways and putative neurohemal-neuroendocrine associations; comparative histological and anatomical implications. *J. Morphol.* **145:** 269–294.

Paniflova, I. M. (1978). Lateral organs of *Ixodes persulcatus* (Parasitiformes, Ixodidae). *Zool. Zh.* **57:** 190–196.

Panifilova, I. M. (1980). Neuroendocrine system activity changes in female *Ixodes persulcatus* in absence of fertilization (Parasitiformes, Ixodidae). *Zool. Zh.* **59:** 1137–1147.

Roshdy, M. A., & Marzouk, A. S. (1982). The subgenus *Persicargas* (Ixodoidea: Argasidae: Argas). 35. The lateral segmental organs and peritracheal gland in immature and adult *A. (P.) arboreus*. *Z. Parasitenkd.* **66:** 335–343.

7

Chemistry, function and metabolism of tick ecdysteroids

P. A. Diehl, J.-L. Connat, and E. Dotson

7.1 INTRODUCTION

Ecdysteroids have been reported in many invertebrate groups including arthropods (Hetru & Horn 1980), molluscs (Romer 1979), nematodes (Koolman *et al.* 1984, Rees & Mendis, 1984), trematodes (Koolman *et al.* 1984, Nirdé *et al.* 1984), annelids (Porchet *et al.* 1984), and coelenterates (Sturaro *et al.* 1982). They have been detected in archaic arthropods such as xyphosurans (*Limulus*: Jegla & Costlow 1979) and pantopods (*Pycnogonum*: Behrens & Bückmann 1983), in myriapods (*Lithobius*: Joly *et al.* 1979, Leubert *et al.* 1982, *Hanseniella*: Juberthie-Jupeau *et al.* 1979), and in arachnid groups such as spiders (*Pisaura*: Bonaric & De Reggi 1977) and ticks (*Amblyomma*: Delbecque *et al.* 1978a, *Ornithodoros*: Germond *et al.* 1982). Except for arthropods, their physiological role has not yet been elucidated. Also, it is often not known if the ecdysteroids are synthesized *de novo* by the animals or whether they are derived from food.

Ecdysteroids have also been detected in many plants, sometimes even in surprisingly high concentrations (Bergamasco & Horn 1983). They could play a protective role directed against nonadapted phytophagous predators through interference with their hormonal system.

Much work has been devoted to the elucidation of the chemistry, biosynthesis, action, and metabolism of ecdysteroids in insects and crustaceans, as found and reviewed in excellent, recently published books edited by Downer & Laufer (1983), Hoffmann (1980), and Hoffmann & Porchet (1984). In contrast, detailed reports on ecdysteroids in organisms other than insects or crustaceans are still rare. Since the last review on tick hormones by Solomon *et al.* (1982), much new information about ecdysteroids has become available. In this review we will first describe the effects of exogenous hormones on various processes such as diapause, molting, oogenesis, spermatogenesis, salivary gland degeneration, attachment and multiplication of cultured tick cells. Then we will present data on the chemical characterization and quantification of the hormones in various life stages and discuss their possible functions. The last part of the review will be devoted to the biosynthesis and the metabolism of ecdysteroids.

7.2 EFFECTS OF EXOGENOUS ECDYSTEROIDS

Owing to the availability of several ecdysteroids, many observations on the effects of exogenous hormones in ticks have been published. Part of this literature has already been discussed by Solomon *et al.* (1982).

7.2.1 Termination of larval diapause

Wright (1969) was the first person to demonstrate an effect of exogenous ecdysteroids in ticks. He was able to terminate larval diapause in *Dermacentor albipictus* by external application of ecdysone (structure in Fig. 7.1(1)), 20-hydroxyecdysone (Fig. 7.1(2)), or 22,25-dideoxyecdysone (Fig. 7.1(3)). A similar effect of ecdysone or 20-hydroxyecdysone was noted in *Rhipicephalus sanquieneus* by Sannasi & Subramoniam (1972).

7.2.2 Influence on molting

In vitro feeding with porcine blood containing 20-hydroxyecdysone accelerated molting in nymphs of the argasid *Ornithodoros porcinus* (Mango, in Solomon *et al.* 1982). Similar treatment of *O. parkeri* nymphs had no effect on the molting period, but did induce a few second molts without an additional blood meal (Campbell & Oliver 1984). Also we observed no effect on the length of the molting period when fifth stage nymphs of *O. moubata* were fed on porcine blood containing 1 μg/ml of 22,25-dideoxyecdysone, a synthetic ecdysteroid which is highly active for the induction of supermolting of females (see below). As in *O. parkeri*, these nymphs underwent additional molts without taking another blood meal; 20% showed one, 50% two, and 30% three additional molts. The ticks were, however, unable to ecdyse successfully and eventually died imprisoned in their cuticles. In addition, these ticks frequently showed malformations or even an absence of appendages.

In ixodid *Hyalomma dromedarii* nymphs, topical application of 10 or 20 μg/tick of 20-hydroxyecdysone during or after feeding speeded up molting considerably. Applications of 1 or 5 μg had a similar effect, but only when applied to feeding nymphs 3 days after attachment to the host (Khalil *et al.* 1984). On the other hand, injection of 10 to 100 ng of 20-hydroxyecdysone into replete nymphs did not influence molting (Dees *et al.* 1984a). Furthermore, no effect was noted when *Dermacentor variabilis* nymphs were treated topically (1 or 10 μg/tick) or by injecting 10 or 100 ng of 20-hydroxyecdysone into each tick (Dees *et al.* 1984a; McDaniel & Oliver 1978).

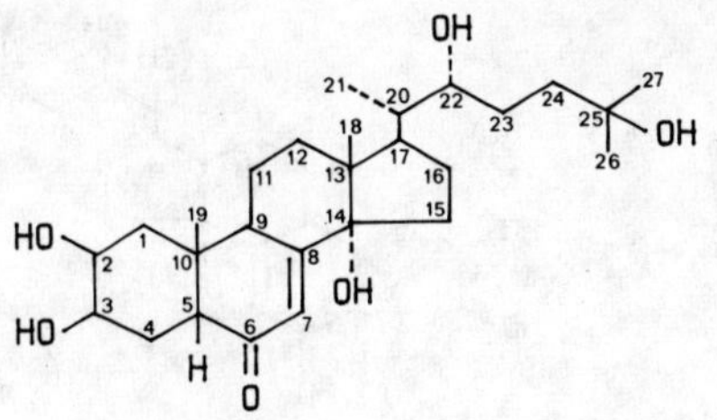

1) ECDYSONE

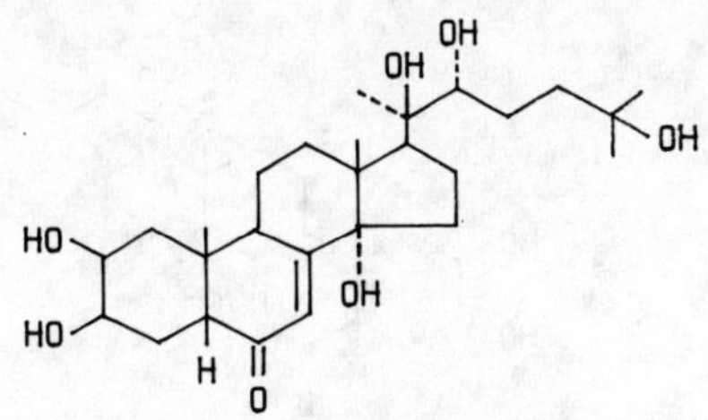

2) 20-HYDROXYECDYSONE

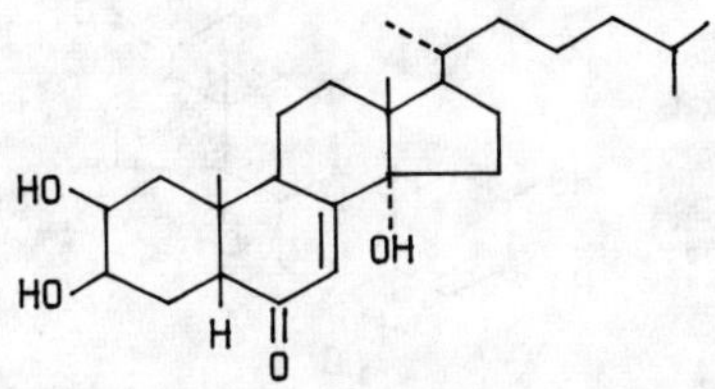

3) 22,25-DIDEOXYECDYSONE

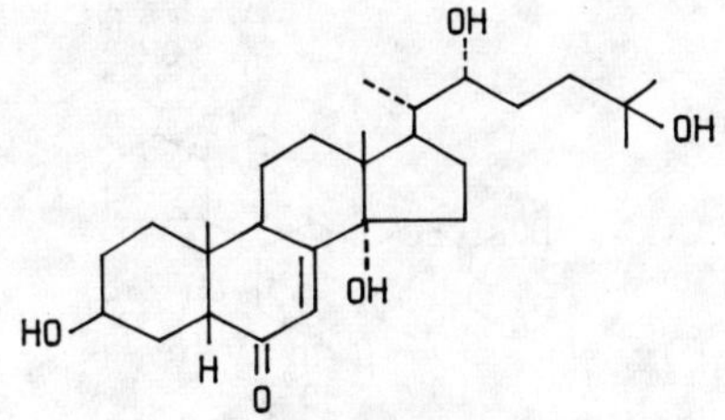

4) 2-DEOXYECDYSONE

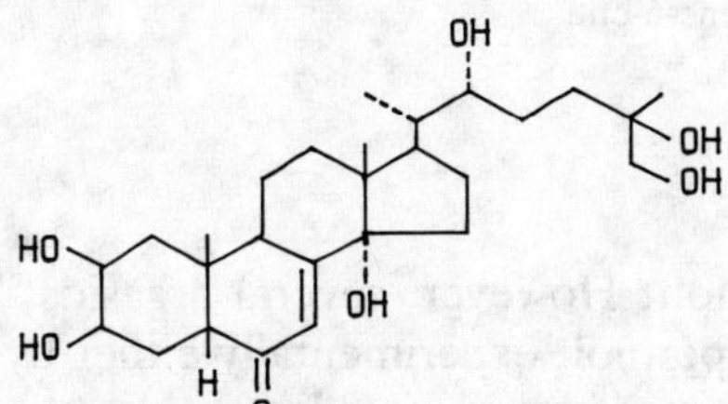

5) 26-HYDROXYECDYSONE

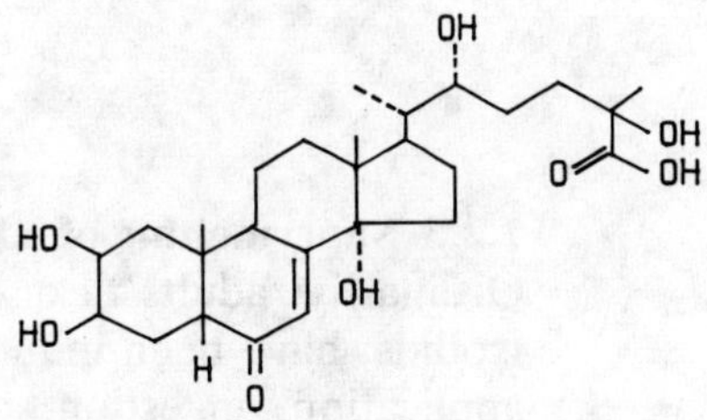

6) ECDYSONOIC ACID

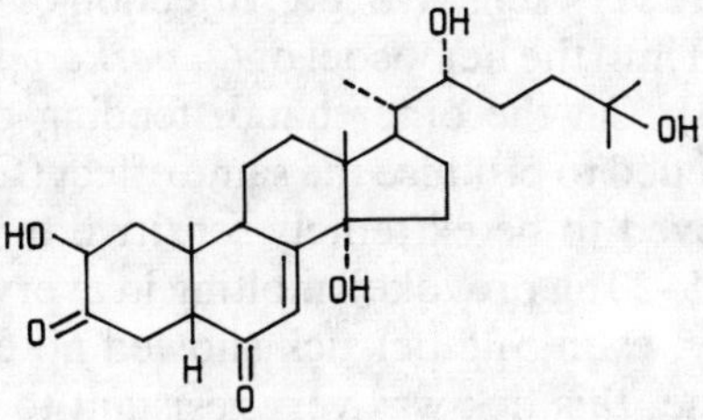

7) 3-DEHYDROECDYSONE

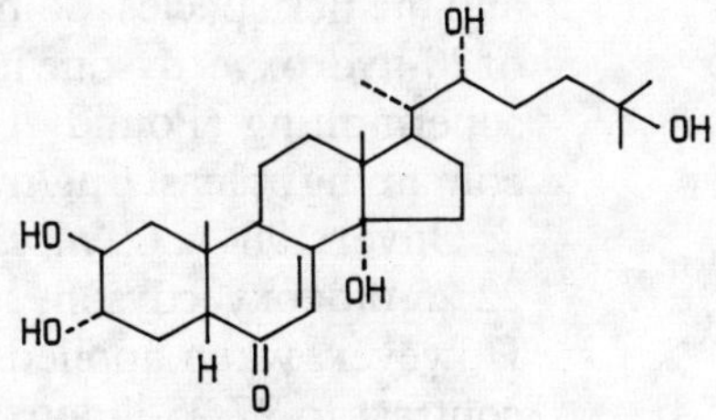

8) 3-EPIECDYSONE

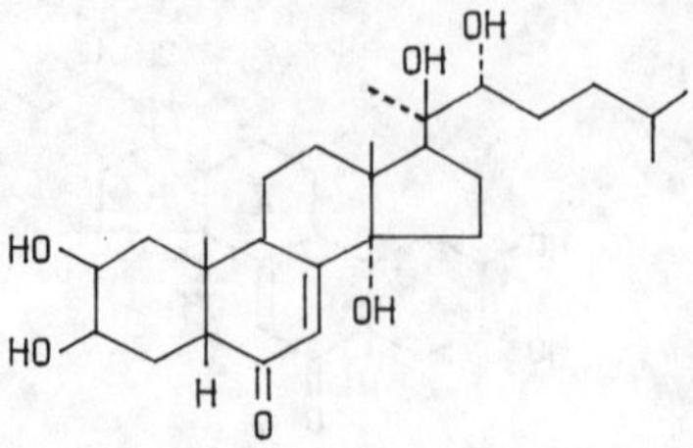

9) PONASTERONE A

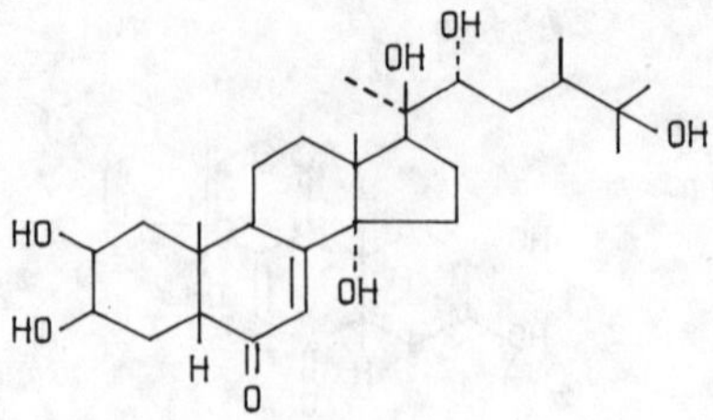

10) MAKISTERONE A

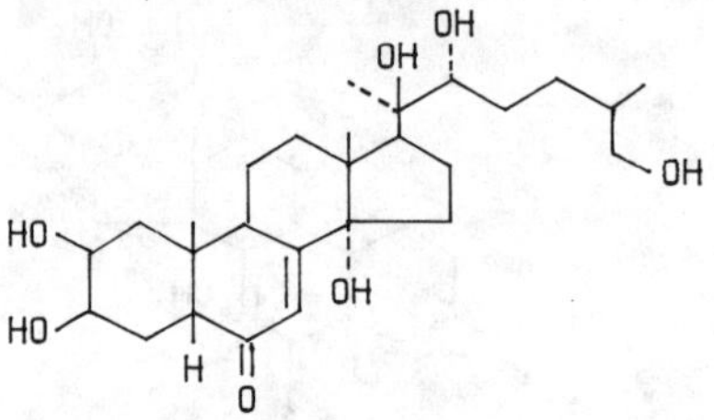

11) INOKOSTERONE

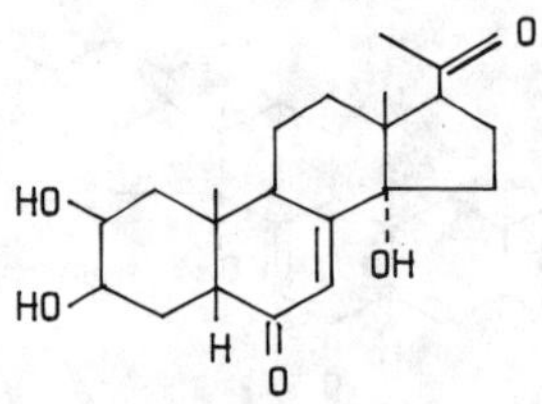

12) POSTSTERONE

Fig. 7.1 — Structural formulae of several ecdysteroids with their trivial names. The systematic name of ecdysone (1) is 2β,3β,14α,22R,25-pentahydroxy-5β-cholest-7-ene-6-one.

7.2.3 Supermolting of adults

Ordinarily, adults ticks do not molt. However, several argasids, but not ixodids, have been induced to supermolt experimentally either by topical application, ingestion, or injection of various ecdysteroids (Ahmed & Bassal 1982, Campbell & Oliver 1984, Connat *et al.* 1983, Kitaoka 1972, Mango *et al.* 1976, Pound *et al.* 1984, Solomon *et al.* 1982).

Great differences of sensitivities in inducing supermolting were noted, depending on the nature of the hormone, the type and timing of application, and the tick species. Some ticks are very sensitive, e.g. injection of only 1 ng of 20-hydroxyecdysone in olive oil into the hemocoel of *O. parkeri* produced supermolting (Pound *et al.* 1984). On the other hand, feeding on blood containing at least 6 μg/ml was needed to produce the same effect (Campbell & Oliver 1984). *O. moubata* proved to be extremely sensitive to ingested 22,25-dideoxyecdysone. About 15–20 ng provoked molting in every female. However, when applied topically, even 5 μg per tick showed no effect. In contrast to 22,25-dideoxyecdysone, this tick was very resistant to ingested ecdysteroids containing a 22-OH group like ecdysone, 20-hydroxyecdysone, ponasterone A (Fig. 7.1(9)), or makisterone A (Fig. 7.1(10)), presumably because of an efficient detoxification system in the midgut (Connat *et al.,* 1986). Doses of at least 10 μg/ml (makisterone A) or 20 μg/ml (ecdysone, 20-hydroxyecdysone) were needed for supermolting (Connat *et al.* 1983). On the other hand, only 50 to 100 ng of ecdysone were necessary when injected into the hemolymph (see Chapter 8). Surprisingly, when compared with the high activity of 22,25-dideoxyecdysone, ingestion of the synthetic 2,22,25-trideoxyecdysone and of 2,22,25-trideoxy-5β-hydroxyecdysone at a dose of 5 μg/ml blood did not produce supermolting (unpublished results). The reason for their ineffectiveness is not yet clear.

In *O. porcinus*, supermolting apparently resulted in perfectly healthy specimens which could successfully feed, supermolt, and lay eggs. It has been speculated that supermolting may occur in nature when the host feeds on plants containing large amounts of phytoecdysteroids (Mango *et al.* 1979). In contrast, supermolting in *O. moubata* frequently resulted in malformations or absence of appendages and failure of ecdysis which rendered subsequent bloodmeals impossible (Connat *et al.* 1983). Ecdysis failures were also reported in *O. parkeri* (Campbell & Oliver 1984).

7.2.4 Effects on oogenesis

In addition to supermolt induction, exogenous ecdysteroids also influence argasid oogenesis (Connat *et al.* 1983, and Chapter 8, Kitaoka 1972, Solomon *et al.* 1982). It appears that molting and vitellogenesis are mutually exclusive processes. For example, in the presence of ingested 22,25-dideoxyecdysone, the ovary of *O. moubata* delayed vitellogenesis until after ecdysis, and fewer eggs reached maturity. After ingestion of high doses of ecdysone, 20-hydroxyecdysone, ponasterone A, or makisterone A, oogenesis was totally suppressed (Connat *et al.* 1983). Injection of small amounts (50 or 100 ng) of ecdysone into the hemocoel during full vitellogenesis on day 8 post-feeding induced egg resorption (see Chapter 8). In contrast to *O. moubata*, the ovary of *O. porcinus* resorbed the already developed eggs during the supermolt cycle (Solomon *et al.* 1982).

Surprisingly, ingested 2,22,25-trideoxyecdysone and 2,22,25-trideoxy-5β-hydroxyecdysone did not inhibit vitellogenesis in *O. moubata*. Doses of 1–5 μg/ml, however, induced egg desiccation, suggesting an influence on the process of egg waxing (see Chapter 8).

Exogenous ecdysteroids can also influence oogenesis in ixodids. For example, Mansingh & Rawlins (1977) observed high mortality and severe inhibition of egg production in *Boophilus microplus* after injection of 120 ng/tick of 20-hydroxyecdysone. Kitaoka (1972), however, did not see any effect when doses of 0.4–2 μg/tick of 20-hydroxyecdysone, ponasterone A, or inokosterone (Fig. 7.1(11)) were injected into the same species.

In *H. dromedarii*, topical application of 20 μg per tick of 20-hydroxyecdysone on day 1 post-attachment caused a drastic reduction of female fecundity and egg hatch (Khalil *et al.* 1984).

7.2.5 Effect on fovea dorsalis development and sex pheromone activity

Injection of 10 ng/tick of 20-hydroxyecdysone into replete *H. dromedarii* nymphs on the day of dropping increased the sex pheromone content in the resulting unfed females. The pheromone gland cells exhibited vesicular disruption and release of secretory droplets in the cytoplasm characteristic of the actively secreting gland. Dosages of 100 ng did not show this effect (Dees *et al.* 1984a).

In contrast, similar experiments with *D. variabilis* did not increase the sex pheromone content, but induced comparable ultrastructural changes in the foveal gland cells. A correlation between ecdysteroids and foveal gland development may therefore exist (Dees *et al.* 1984b).

7.2.6 Effect on ixodid salivary glands

Perfusion or *in vitro* incubation experiments with ecdysone or 20-hydroxyecdysone strongly suggested that these hormones induce degeneration of the salivary glands in replete females of *Amblyomma hebraeum*. They might therefore be identical to the humoral 'tick salivary gland degeneration factor' (Harris & Kaufman 1981). For more details the reader is referred to Chapter 2 (see also section 7.3.4).

7.2.7 Effects on spermatogenesis and male fertility

In *D. variabilis*, injection of 20-hydroxyecdysone into unfed males increased the number of DNA synthesizing germ cells, suggesting an important role of ecdysteroids in spermatogenic induction (unpublished results, see discussion in Dumser & Oliver 1981). Topical application of 1 to 10 μg/tick of 20-hydroxyecdysone to fed nymphs did not influence spermatogenesis (McDaniel & Oliver 1978).

In contrast, topical treatment of *H. dromedarii* nymphs on day 10 post-engorgement with 1,5,10, or 20 μg of 20-hydroxyecdysone reduced male fertility. A similar detrimental effect was observed after treatment of feeding males with a dose of 20 μg per tick (Khalil *et al.* 1984).

7.2.8 Effects on tissue culture cells

Kurtii & Munderloh (1983) described dose-dependent effects of 20-hydroxyecdysone on two cell lines isolated from *Rhipicephalus appendiculatus* and *Anocentor nitens*. The rate of cell attachment to the culture substrate was reduced. Low concentrations (0.2 or 2 nM) stimulated, but higher concentrations (>200 nM), inhibited growth of the *Ripicephalus* cell line. Decrease of growth in the *Anocentor* cell line was less pronounced. Growth suppression was reversible. The hormone also induced cell detachment and vesicle formation.

7.2.9 Mortality

Several authors have observed mortality induced by the application of exogenous ecdysteroids. The lethal doses were generally high and variable, depending on the tick species, the nature of the hormone, the time of treatment, and the method of application.

Ticks appear to be more sensitive to the injection of hormones. For example, 50 ng of ecdysone injected into the hemocoel of *O. moubata* females on day 8 post-feeding induced 80% mortality (unpublished results). 100 ng of 20-hydroxyecdysone in olive oil injected into the hemolymph of *O. parkeri* females caused 40% mortality (Pound *et al.* 1984). 120 ng of 20-hydroxyecdysone induced high mortality in *B. microplus* females (Mansingh

& Rawlins 1977). Kitaoka (1972), however, did not see any lethal effect when doses of 0.4–2 µg/tick of 20-hydroxyecdysone, ponasterone A, or inokosterone were injected into the same species. Similarly, we observed that higher lethal doses were required for the Biarra strain of *B. microplus* (LD_{50} for 20-hydroxyecdysone: about 3–4 µg). 10 µg/tick provoked death in all females within 24 hours. Injection of 5 µg of 22,25-dideoxyecdysone, 2,22,25-trideoxyecdysone, or 2,22,25-trideoxy-5β-hydroxyecdysone were ineffective. In contrast to injection, topical application of 10 µg/tick of 20-hydroxyecdysone in 4 µl acetone was without effect. This clearly demonstrates that the route through the cuticle is generally much less efficient than injection.

A high sensitivity was noted for *Amblyomma variegatum* when the females were allowed to feed through capillaries on TC 199 containing ecdysone, 20-hydroxyecdysone, or 22,25-dideoxyecdysone before being placed on bovine hosts. For example, 1 µg/tick of ingested 20-hydroxyecdysone induced death in two out of three females. 1 or 10 µg of 22,25-dideoxyecdysone had no effect. Topical applications of 10 µg of 22,25-dideoxyecdysone or 50 µg of ecdysone per tick were ineffective. 50 θg of 20-hydroxyecdysone produced smaller replete females (unpublished results).

In conclusion, exogenous ecdysteroids exert several effects in ticks. In particular, the induction of supermolting and acceleration of molting, and the induction of salivary gland degeneration, demonstrate that ecdysteroids might be true endogenous tick hormones.

7.3 CHEMISTRY AND FUNCTIONS OF TICK ECDYSTEROIDS

Rapid progress in our knowledge of arthropod ecdysteroid chemistry, biosynthesis, metabolism, and mode of action has recently become possible through the introduction of several powerful analytical methods such as radioimmunoassay (RIA), high-performance liquid chromatography (HPLC), gas chromatography (GC), mass spectrometry or fragmentography, and NMR spectroscopy. For detailed reviews see Hetru & Hoffman (1983), Lafont & Koolman (1984), and Lafont *et al.* (1980). The application of these methods has allowed substantial progress to be made in our knowledge of tick ecdysteroids. In this section we will discuss the chemistry and the possible functions of these hormones for each of the various life stages.

7.3.1 Ecdysteroids in larvae

At present, very little is known about ecdysteroids in larvae. One of us (E.D.) studied the temporal correlation between ecdysteroid titer and the integument structure during the larval development of *O. moubata* at 28°C. Preliminary results indicated that only small amounts (about 13 pg of 20-hydroxyecdysone equivalents per larva) were present during the first 40 hours of development as measured by the RIA of De Reggi *et al.* (1975). Rising titers were observed during apolysis. High titers of about 70 pg/larva were present between 48–72 hours during epicuticle synthesis. The concentration decreased during procuticle synthesis to basal values (about 13 pg) after 80–88 hours, shortly before ecdysis (which occurred after about 128 hours). The fluctuation in hormone titers in relation to cuticular events are very similar to the ones observed in fifth stage nymphs (see section 7.3.2). The chemical nature of the RIA-positive material has not yet been determined.

Dees *et al.* (1984b) found that RIA titer values in crude extracts rose from about 1.1 ± 0.3 pg in unfed *D. variabilis* larvae to a maximum of 23.5 ± 8.9 pg 20-hydroxyecdysone equivalents per tick on day 2 post-feeding. Three-week old unfed nymphs again had low values of 1.7 ± 1.2 pg/tick (Horn I-1 ecdysone antiserum from J. D. O'Connor). HPLC analysis in reverse phase mode demonstrated a RIA-positive fraction migrating like 20-hydroxyecdysone. Another very polar immunoreactive fraction was also present. Suprisingly, the ecdysteroid concentrations were considerably higher after fractionating extracts with HPLC.

7.3.2 Ecdysteroids in nymphs

Germond *et al.* (1982) performed a detailed study on the temporal correlation between integument structure and ecdysteroid titers in fifth stage nymphs of *O. moubata*. The titers, as measured by the RIA of De Reggi *et al.* (1975), in hemolymph or in whole animals were low during the first 3 days post-feeding. We were not able to confirm the small peak reported in an earlier preliminary note (Germond *et al.* 1980). During this period the hypodermal cells increased considerably in size, deposited a few nymphal procuticle lamellae (day 2–3), and initiated the mitotic period on day 2. The hormone titer began to increase between days 3 and 4, with the mitotic period ending on day 4. Apolysis and formation of the exuvial space took place during day 4–5 concomitant with sharply increasing hormone titers. Highest concentrations of about 500 pg 20-hydroxyecdysone equivalents per µl hemolymph and about 11 ng per tick were measured during deposition of the adult epicuticle between day 5–6. The titer began to drop with the beginning of procuticle deposition and digestion of the nymphal cuticle. It reached low values again shortly before ecdysis on day 9–10 post-feeding. Monitoring of the different HPLC fractions by RIA showed that the majority of the immunoreactive material migrated like ecdysone, and predominantly 20-hydroxyecdysone. This was confirmed by GC of the sylilated compounds. In addition, 20-hydroxyecdysone was also identified by the powerful method of GC coupled to mass fragmentography.

Molting hormone activity was also found in the hemolymph of fifth instar nymphs of *O. porcinus* by the *Musca* test and by RIA (Solomon *et al.* 1982). These authors reported an early low peak during day 3–5 post-feeding and a second higher peak at the time of general apolysis (day 9–11) preceeding ecdysis by several days. However, the correlation of the hormone peaks with epicuticle or procuticle deposition was not determined.

Ecdysteroids were also detected in nymphs of the ixodid *A. hebraeum* (Delbecque *et al.* 1978a; Diehl *et al.* 1982). As in the argasid nymph *O. moubata*, rising hormone titers paralleled apolysis and high concentrations correlated with the beginning of adult cuticle deposition. Procuticle synthesis occurred during falling titers of ecdysteroids. A comparable situation has also been seen to occur in *A. variegatum* (Ellis & Obenchain 1984) and *D. variabilis* (Dees *et al.* 1984b).

Most of the immunoreactive material in *A. hebraeum* nymphs migrated like ecdysone and mainly 20-hydroxyecdysone on HPLC reverse phase columns (Diehl *et al.* 1982). This is in accordance with previous observations which identified these two hormones by the use of GC combined with mass spectroscopy or mass fragmentography (Delbecque *et al.* 1978a).

HPLC studies in *D. variabilis* nymphs demonstrated the presence of 3 RIA-positive fractions. One fraction presumably corresponds to 20-hydroxyecdysone because on reverse phase columns it showed the same retention time as the authentic reference compound; the two other fractions were more polar (Dees *et al.* 1984b).

In conclusion, the molting hormones ecdysone and predominantly 20-hydroxyecdysone are present in immature stages of argasid and ixodid ticks, thus resembling most other arthropods where 20-hydroxyecdysone is thought to be the active molt-controlling hormone. Some bugs like *Dysdercus* or *Oncopeltus*, however, are an exception to this general rule because they use makisterone A (Aldrich *et al.* 1982, Gibson *et al.* 1983). In addition, the tick ecdysteroids shows fluctuating titers which temporally correlate with molting events in a manner resembling insects or crustaceans. These observations and the ability of exogenous hormones to induce shortening of the molting period or provoke supermolting in argasids demonstrate that ecdysteroids probably play an important role in the humoral control of tick molting. However, evidence is still based mainly on the analogy to the insect or crustacean model. Thus, to substantiate the hypothesis further, more conclusive evidence such as the existence of specific hormone receptors or an induction of molt-specific genes is clearly needed.

In addition to their role in molting, the ecdysteroids might also be important in the control of spermatogenesis in developing nymphs (unpublished observations in *D. variabilis*, and see discussion in Dumser & Oliver 1981).

7.3.3 Ecdysteroids in males

Very little work has been done on ecdysteroids in males. In *D. variabilis*, the titer of RIA-positive material in crude extracts decreased from 386 ± 122 pg 20-hydroxyecdysone equivalents per tick at emergence to 38 ± 3 pg 2 weeks later. Males on day 8 post-feeding showed slightly higher values of about 131.6 ± 2.3 pg per male. On the other hand, considerably higher values were obtained after HPLC purification (Dees *et al.* 1984b). Similarly, in unfed *H. dromedarii*, titers in crude extracts decreased from 0.87 ± 0.39 ng 20-hydroxyecdysone equivalents in emerging males to 0.48 ± 0.14 ng per male at day 15 post-emergence. Slightly higher values of 2.03 ± 0.51 ng per tick were recorded from 15-day old males which when ecdysing from nymphs had been injected with 10 ng of 20-hydroxyecdysone (Dees *et al.* 1984a).

The only available information on the nature of male ecdysteroids comes from a preliminary study employing HPLC chromatography on reverse phase columns. It revealed three RIA-positive fractions in *D. variabilis*. One was migrating like 20-hydroxyecdysone; the two others were more polar (Dees *et al.* 1984b).

Dumser & Oliver (1981, see discussion) hypothesized that ecdysteroids might be important for the control of spermatogenesis in males. This was based on unpublished observations where injected 20-hydroxyecdysone provoked a very large increase in DNA synthesizing germ cells in unfed *D. variabilis* males. Furthermore, the presence of ecdysteroids in the hemolymph was detected, although the nature and quantity of the hormones and the detection method were not stated. More work is thus needed to explore the interesting possibility that molting hormones might be involved in the control of tick spermatogenesis.

7.3.4 Ecdysteroids in females

A detailed study on the ecdysteroids present during the single gonotrophic cycle of an ixodid tick was performed with *A. hebraeum* (Connat *et al.* 1985). During the first three days of the bloodmeal, the body weight increased slightly, and the cuticular dry weight remained stable. During this time hypodermal cells enlarged considerably, but did not yet secrete new endocuticle. This first phase was characterized by very low titers of about 12 or less pg 20-hydroxyecdysone equivalents per μl hemolymph as in unfed females. The ecdysteroid content in whole ticks increased slightly between days 2 and 3 from about 0.5 ng to 1 ng per female as determined by RIA (De Reggi *et al.* 1975).

The following 4 days showed a moderate increase of body weight. On the other hand, an important synthesis of endocuticular material took place, thus increasing the cuticle dry weight 4.3-fold. During this period hemolymphatic ecdysteroid levels remained low, but titers in whole animals began to rise quickly after day 5. At present it is impossible to decide if the slightly increased hormone level in whole ticks after day 2 might be responsible for the resumption of endocuticle synthesis. This seems rather unlikely, however, because in most immatures of ticks or other arthropods, procuticle synthesis proceeds generally in the presence of low ecdysteroid concentrations.

The expansion phase during the last 12–24 hours is characterized by the ingestion of a huge amount of blood. This was accompanied by a further increase of the hormone level in whole animals to about 10 ng per tick, while hemolymphatic levels remained low. It is not yet known if the increasing hormone concentration could act as a signal for the induction of rapid feeding.

Within about 4 days after dropping from the host, the salivary glands had degenerated in parallel with vitellogenesis which was initiated after day 3. During this period, ecdysteroid concentrations increased considerably in the hemolymph (Fig. 7.2) and in whole body extracts. This could thus indeed be the humoral signal for salivary gland degeneration as suggested strongly

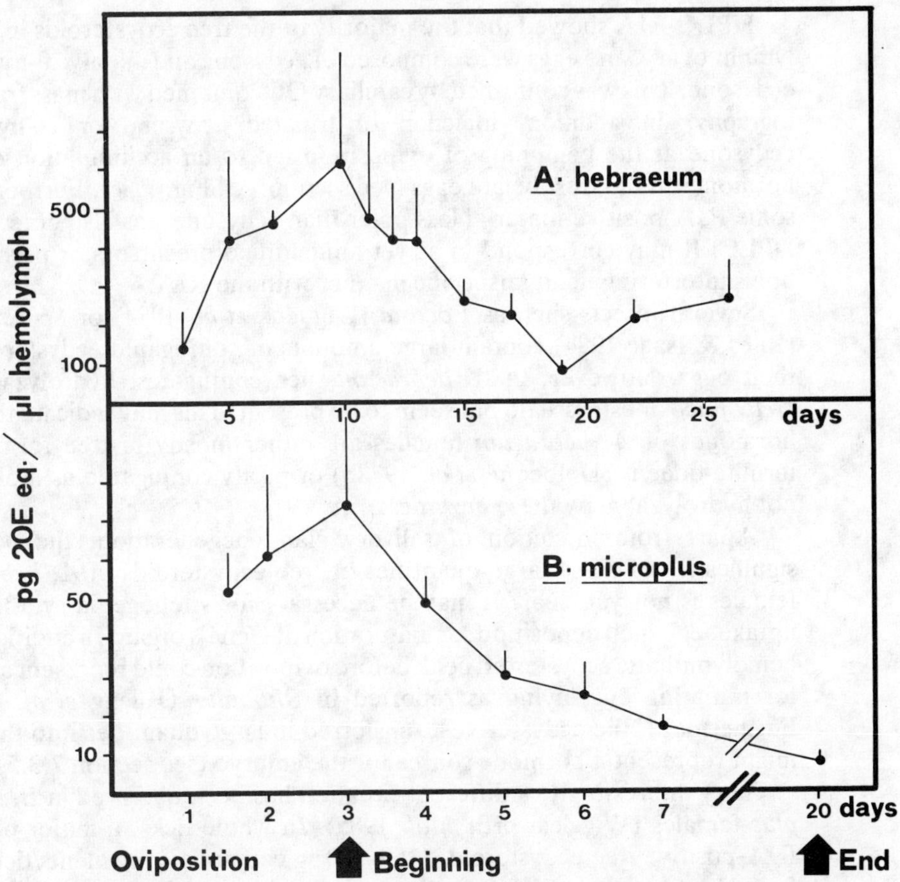

Fig. 7.2 — Hemolymphatic ecdysteroid levels during vitellogenesis and oviposition in females of *Amblyomma hebraeum* and *Boophilus microplus* (Biarra strain). The titers were measured with the RIA of De Reggi *et al.* (1975) at different days post-detachment and are expressed in pg 20-hydroxyecdysone equivalents per µl hemolymph. Each point corresponds to the mean ± S.D. of measurements from 3 females.

by the results of Kaufman's group (see Chapter 2). In addition, rising hormone levels could also play a role in oogenesis as, for example, the induction of vitellogenin synthesis in several mosquitoes or flies (see review by Hagedorn 1983).

During the following period of intense vitellogenesis and oviposition, the ecdysteroid titers in the hemolymph rose to a peak of about 600 pg per µl one day prior to oviposition and decreased thereafter (Fig. 7.2). Hormone levels in whole females continued to increase. About 700 ng of RIA-positive material was found in a single female and its eggs several days after the beginning of oviposition.

The ovary accumulated most of the RIA-positive material. Freshly laid eggs contained about 2744 ± 676 (SD) pg 20-hydroxyecdysone equivalents per mg. It was estimated that a female produced about 3.6 µg of 20-hydroxyecdysone equivalents during the gonotrophic cycle.

HPLC-RIA showed that the majority of the free ecdysteroids in hemolymph, ovary, and eggs were composed of ecdysone and mainly 20-hydroxyecdysone. This was confirmed by capillary GC combined with mass fragmentography. In addition, injection of tritiated ecdysone or 20-hydroxyecdysone at the beginning of oviposition led to an accumulation of free hormones in the freshly laid eggs (see section 7.5, Fig. 7.5c). Furthermore, some RIA-positive material less polar than ecdysone was also detected by HPLC. It may correspond to as yet unidentified precursors or perhaps to lipids interfering in an unspecific manner with the RIA.

Several insects such as *Locusta* (Lagueux *et al.* 1984) or *Schistocerca* (Rees & Isaac 1984) contain large amounts of conjugated ecdysteroids in their eggs. However, in *A. hebraeum* such conjugates, hydrolyzable by *Helix* juice or esterase do not seem to be present. This may indicate that the hormones of *A. hebraeum* females are either mostly in free form as in termite queens (Delbeque *et al.* 1978b) or partly conjugated as molecules not hydrolyzable by these enzymes.

Apart from induction of salivary gland degeneration, the possible significance of these large quantities of free ecdysteroids in *A. hebraeum* females is not yet clear. It may be necessary for vitellogenin synthesis or uptake, egg shell deposition, or maturation of Géné's organ. In addition, the hemolymphatic ecdysteroid peak before oviposition could represent a signal for initiating egg laying as reported in *Rhodnius* (Ruegg *et al.* 1981). Furthermore, the ecdysteroids transferred in large quantities into the eggs might represent a hormone source for the embryo (see section 7.3.5).

Very interestingly, a different situation has been observed in *B. microplus* females (Wigglesworth *et al.* 1985). In whole ticks, a major peak of RIA-positive free ecdysteroids (150–300 ng ecdysone equivalents/tick) was found just before complete engorgement, with very few polar conjugated ecdysteroids being detected. At peak titer, ecdysone was the major hormone, together with small amounts of 20-hydroxyecdysone. This decrease in titer of immunoreactive ecdysteroids before repletion may be accounted for by intense esterification of the hormones with fatty acids. These esters accumulate in the eggs, together with much smaller quantities of free hormones. Similarly, we found, in a preliminary comparative study, an accumulation of apolar hormone conjugates in eggs after injection of tritiated ecdysteroids into females of the Biarra strain of *B. microplus* (see section 7.5, Fig. 7.5e). In addition, we also noticed low titers of RIA-positive material in the hemolymph as compared with the titers found in *A. hebraeum* (Fig. 7.2). As in the latter tick, a peak of about 29.3 ± 7.5 pg ecdysone equivalents per μl (or about 70 pg 20-hydroxyecdysone equivalents per μl) was observed near the beginning of oviposition. However, as reported for whole females or eggs, much of this material was apparently esterified, because treatment with pig liver esterase increased the RIA response 10–15 times. Eggs from the Biarra strain contained lower amounts than *A. hebraeum* when measured in crude methanolic extracts (26 ± 2 pg ecdysone equivalents per mg eggs; or 65 ± 4 pg 20-hydroxyecdysone equivalents per mg). Hydrolysis with esterase increased the hormone concentration to 2–5 ng ecdysone equivalents per mg eggs (or 5–10 ng 20-hydroxyecdysone equivalents per mg) thus confirming the results of Wigglesworth *et al.* (1985).

Therefore, considerable differences between *A. hebraeum* and *B. microplus* females apparently exist. In the former, large quantities of free hormones, mainly 20-hydroxyecdysone, accumulate after repletion in females and eggs with few hydrolyzable hormone conjugates present. In sharp contrast, considerable amounts of free hormones, mainly ecdysone, exist in the latter species prior to complete engorgement which are converted thereafter to esters and transferred to the eggs.

These interesting differences raise many questions. For example, how do ecdysteroids influence salivary gland degeneration in *Boophilus*? It may be that ecdysone is inactive in this respect or that the glands show a delayed response.

Preliminary results by Dees *et al.* (1984b) suggest that *D. variabilis* females resemble *A. hebraeum* with respect to the hormone titer changes. RIA of crude extracts of HPLC–RIA showed a large increase in immunoreactive material near the end of feeding that continues to rise during vitellogenesis. A transient decrease was noticed on day 8 post-attachment. Part of the RIA-positive material appeared to be 20-hydroxyecdysone. In contrast to the other two species, considerable amounts of more polar material were present in two fractions. Very low amounts (68.6 ± 32.9 pg of 20-hydroxyecdysone equivalents per mg) were detected in freshly laid eggs. However, hydrolysis experiments were not performed, and the isocratic HPLC conditions did not elute apolar hormone esters. Therefore, we do not yet know to what extent apolar or polar ecdysteroid conjugates are present in *D. variabilis* females and eggs.

Preliminary experiments using thin layer chromatography combined with RIA, indicate that ecdysone and 20-hydroxyecdysone might be present in *Rhipicephalus appendiculatus* eggs, apparently coupled to a large egg protein (Whitehead *et al.* 1986).

In sharp contrast to immatures or ixodid females, we experienced many unexpected and frustrating difficulties with ecdysteroid analysis in the argasid female *O. moubata* (for a more detailed discussion see Chapter 8). Clearly, RIA-positive material was found in extracts of hemolymph, whole bodies, or eggs. The RIA responses were, however, variable, depending on the treatment before analysis. After simple methanolic extraction, only relatively low values were measured in the hemolymph of vitellogenic females or in eggs (3–7 pg 20-hydroxyecdysone equivalents per mg egg). Part of this material migrated like ecdysone or 20-hydroxyecdysone on reverse phase HPLC columns. However, we have not been able to confirm this by GC combined with mass fragmentography.

In contrast to the confusing situation in normally fed and mated *O. moubata* females, we encountered no difficulties in measuring RIA titer changes in supermolting females. Changes were comparable to the ones in fifth stage nymphs. Very high hemolymph peak titers of about 1800 pg 20-hydroxyecdysone equivalents/μl were temporally correlated with apolysis and deposition of the new epicuticle. Two peaks of RIA-positive material migrated like ecdysone and 20-hydroxyecdysone (see Chapter 8).

In normally fed mated *O. porcinus* females, the hormone titers 'fluctuated rhythmically around mean background levels throughout the period of oogenesis' when assayed by the *Musca* test. Although this mean value was not quantified, it suggests that only a small amount of ecdysteroid material was present. But like *O. moubata*, hemolymph ecdysteroid content increased suddenly on day 10–12 post-feeding in supermolting females, concomitant with termination of vitellogenesis and induction of egg resorption (Solomon *et al.* 1982).

Thus, in contrast to some ixodid females, it is not known if ecdysteroids are present in argasid females and eggs. If they exist, they must be structurally different from classical hormones or masked in such a manner that they avoid induction of a molting cycle and egg resorption. It will be worth investigating whether molting hormones are used as humoral signals in fed, virgin females where oocyte resorption occurs naturally without any signs of supermolting.

7.3.5 Ecdysteroids in embryos

Clearly, freshly laid eggs of two well studied ixodid species contain maternal ecdysteroids, either mostly in free form (*A. hebraeum*) or in bound form as fatty acid conjugates (*B. microplus*). Very small amounts (68.6 ± 32.9 pg 20-hydroxyecdysone equivalents /mg of immunoreactive material) have been found in *D. variabilis* eggs (Dees *et al.* 1984b). They may, however, contain higher levels than suggested by RIA if the hormones are present as conjugates. This has not been tested. TLC–RIA studies revealed unspecified quantities of possibly free ecdysone and 20-hydroxyecdysone in *R. appendiculatus* eggs (Whitehead *et al.* 1986).

The fate of these maternal ecdysteroids during ixodid embryogenesis has not been elucidated in detail. Dees *et al.* (1984b) noted an increase of RIA-positive material in crude extracts from 3 different embryonic stages in *D. variabilis*. Such an increase, however, was not noted after HPLC–RIA. HPLC revealed the presence of 2 RIA-positive fractions, one migrating like 20-hydroxyecdysone and the other like more polar compounds. Wigglesworth *et al.* (1985), noted the occurrence of distinct hormone peaks during *B. microplus* embryogenesis. The free hormones might be enzymatically hydrolyzed from the conjugates which could therefore serve as a hormone storage form. This would be comparable to *Locusta* embryos where the free hormones are liberated from the maternal conjugates and are used for the control of serosal and embryonic cuticle deposition (Lagueux *et al.* 1979).

Freshly laid eggs of *O. moubata* apparently contain very low amounts of RIA-positive material; it is not clear if this response is due to endogenous ecdysteroids (see section 7.3.4). Preliminary investigations suggest titer variations during embryogenesis. However, owing to the presence of very little immunoreactive material and to a rather poor synchronization of embryos, the studies are not yet complete. In contrast to *B. microplus*, *O. moubata* embryos apparently do not use fatty acid hormone esters as a potential hormone source, as was speculated (Connat *et al.* 1984); at least not when the developing oocytes were experimentally loaded with such conjugates after injection of tritiated ecdysone or 20-hydroxyecdysone into the hemocoel of vitellogenic females (see section 7.5.2).

Therefore it appears that much more work is required to elucidate the nature, origin, titer changes, and the probable functions of ecdysteroids during tick embryogenesis. One of the possible functions of these hormones might be to control the deposition of extra-embryonic layers or embryonic cuticles as in *Locusta* (Lagueux *et al.* 1979); formation of such structures has been observed in *O. moubata* embryos (Vogel 1975).

7.4 BIOSYNTHESIS OF ECDYSTEROIDS

It is now generally agreed that in most immature insects prothoracic glands or homologous structures synthesize ecdysone from cholesterol or other phytosterols under the influence of a stimulatory prothoracicotropic hormone (for a review see Hoffmann & Hetru 1983). However, in several cases e.g. *Tenebrio* pupae (Delachambre *et al.* 1984), ecdysteroids can also be produced outside the prothoracic glands, possibly in a generalized tissue such as the hypodermis itself, or in oenocytes. In addition, the follicular cells in some females are capable of ecdysone biosynthesis.

In contrast to insects, ecdysone synthesis in crustaceans occurs in Y-organs under the control of an inhibitory hormone from the X-organ/sinus gland complex. Other sources are not precluded (see Spindler *et al.* 1980). In opilionids belonging to the class Arachnid, oenocytes in the walking legs produce ecdysone and 20-hydroxyecdysone (Romer & Gnatzy 1981).

In ticks, nothing is known about the biosynthesis of ecdysteroids, its possible site(s), or its control. Cytological studies suggested several possible endocrine tissues such as the lateral organs, the retrocerebral complex or the peritracheal glands in addition to the 'brain' with its numerous neurosecretory cells (see Chapter 6). However, experiments such as gland ablation, and demonstration beyond doubt of the endocrine nature of the tissue, are still lacking.

Cox (1960) was the first to attempt to demonstrate the humoral control of tick molting. In 1 out of 3 series, 2 fed females and 1 male of the argasid *Ornithodoros turicata* supermolted after implantation of 3–4 'brains' from donor fed 3rd instar nymphs. In addition, by the use of ligatures, molting in some nymphs could be restricted to the anterior part containing the 'brain'. Cox thus concluded that the presumptive hormone was produced in the anterior part by the 'brain'. However, this remains merely as a suggestion because many nymphs molted in both parts even when ligated very early. In addition, it was not determined if the non-molting posterior parts were still viable and responsive to molting hormones. Thus, although the work by Cox suggests that the 'brain' is important for molting control, it does not yield information on the structure of the supposed 'brain' hormone(s) which might be of neurosecretory nature.

Ellis & Obenchain (1984) tried to study the origin of ecdysteroids in *A. variegatum* nymphs by measuring the amounts of RIA-positive material released into the culture medium by several tissues. Low levels were observed in media from midgut, heart, pericardial tissue, and central nervous system cultures. Titers were higher and variable in medium from fat body associated with tracheal trunk cultures. This evidence suggested that some cell type associated with this explant (e.g. the nephrocytes) could be responsible for ecdysteroid production. This remains speculative, however, because the amounts of RIA-positive material present in the different tissues before or after incubation have not been measured. The results might therefore simply reflect a release, but not biosynthesis, by these tissues. In addition, other tissues, e.g. the hypodermis, were not tested, although in the light of results with *Tenebrio* (Delachambre *et al.* 1984) and *Locusta* nymphs (Porcheron *et al.* 1984) this tissue appears now as a possible site of ecdysone biosynthesis.

In *A. hebraeum* females, the RIA-positive material appeared much later in the ovary than in the rest of the animal (Connat *et al.* 1985). In addition, the vitellogenic ovary avidly took up injected tritiated hormones (see section 7.5). Thus, in contrast to several insects, it appears that the source of maternal ecdysteroids might reside outside the ovary. However, as only total immunoreactivity was measured, we cannot exclude a secretion of low- or non-immunoreactive precursors by the ovary and their subsequent transformation to ecdysone and 20-hydroxyecdysone outside the ovary.

The biosynthetic pathway, presumably from dietary cholesterol up to ecdysone, remains to be elucidated. We noticed, however, in several studies in *O. moubata* nymphs or females, that the possible precursors 22,25-dideoxyecdysone, 2-deoxyecdysone (Fig. 7.1(4)) or ponasterone A did not yield ecdysone or 20-hydroxyecdysone to any appreciable extent, although this has been reported in some insects. Thus, it is possible that these ticks do not utilize these precursors. Alternatively, it may be that these compounds when injected or ingested do not gain access to the ecdysteroid producing tissues.

In short, the site(s), the pathway, and the control of ecdysteroid biosynthesis in ticks are unknown.

7.5 METABOLISM OF ECDYSTEROIDS

After the introduction of powerful new analytical techniques, our knowledge of ecdysteroid metabolism in insects and to a lesser extent in crustaceans increased greatly (for a recent review see Lafont & Koolman 1984). The conversion of ecdysone into 20-hydroxyecdysone by an ecdysone 20-monoxygenase is a reaction occurring in many tissues, but especially in the fat body and the digestive tract. Ecdysone and 20-hydroxyecdysone are further metabolized via several pathways. (1) They are converted into the corresponding 26-carboxyecdysteroids (ecdysonoic acid (Fig. 7.1(6) and 20-hydroxyecdysonoic acid) via the 26-hydroxylated metabolites (26-hydroxyecdysone (Fig. 7.1(5)) and 20,26-dihydroxyecdysone). (2) Many ecdysteroids become conjugated and form either more polar compounds (e.g. phosphates, acetophosphates, nucleotides, sulfates) or less polar compounds (acetates). (3) Injected ecdysone or 20-hydroxyecdysone can lead in some insects to the formation of 3-dehydroecdysone (Fig. 7.1(7)) or 3-dehydro-20-hydroxyecdysone. (4) Several ecdysteroids can undergo epimerization at the C3 position, leading to formation of 3-epiecdysone (Fig. 7.1(8)) or 3-epi-20-hydroxyecdysone. (5) Side chain cleavage between C_{20} and C_{22} might lead to poststerone (Fig. 7.1(12)), although this appears to be a minor reaction. Most of these metabolites are inactivation products which are excreted by the animal. Others, such as ecdysteroid-22-phosphates and/or nucleotides, are stored in eggs where they represent a hormone source for the embryo.

Compared with insects or crustaceans, our research on ecdysteroid metabolism in ticks is only beginning. Nevertheless, studies have already led to the identification of ecdysonoic acids and to the discovery of a class of apolar conjugates which is new for arthropods. We shall now review our current knowledge of hormone metabolism for each of the life stages.

7.5.1 Metabolism in nymphs

In fifth stage nymphs of *O. moubata*, injection of tritiated ecdysone into the hemocoel at various times during the post-feeding stage demonstrated the tick's ability of efficiently transforming this hormone to 20-hydroxyecdysone (Bouvier *et al.* 1982, and see Fig. 7.3a,b). The ticks metabolized 20-hydroxyecdysone further presumably to 20,26-dihydroxyecdysone and then to polar compounds of yet unknown chemical nature (Fig. 7.3a–c). These are ionizable, but do not correspond to any of the known polar compounds from insects such as ecdysonoic acids or conjugates hydrolyzable by *Helix* enzymes (e.g. phosphates). This polar pathway was especially active when endogenous ecdysteroid titers were decreasing, leading to an accumulation of polar products in the digestive tract. It is not operative with ingested ecdysteroids and also not in females (see section 5.2). The pathway can thus be hypothesized to act as an important inactivation mechanism for endogenous nymphal hormones.

In a second, but apolar, pathway, injected ecdysone, 20-hydroxyecdysone, or the probable 20,26-dihydroxyecdysone (after HPLC purification) were rapidly converted into apolar metabolites (Bouvier *et al.* 1982; AP2 in Fig. 7.3a) These metabolites were always prominent, constituting from about 30 to 75% of the injected label after 24 hours of metabolism.

Such apolar metabolites were even more rapidly formed after ingestion of ecdysone or 20-hydroxyecdysone (Diehl *et al.* 1985; and see Fig. 7.3d,e). In some experiments, up to 95% of the ingested hormone was already converted to apolar compounds after 24 hours, even with quantities of up to 35 μg/ml blood. Recent studies with ingested 20-hydroxyecdysone

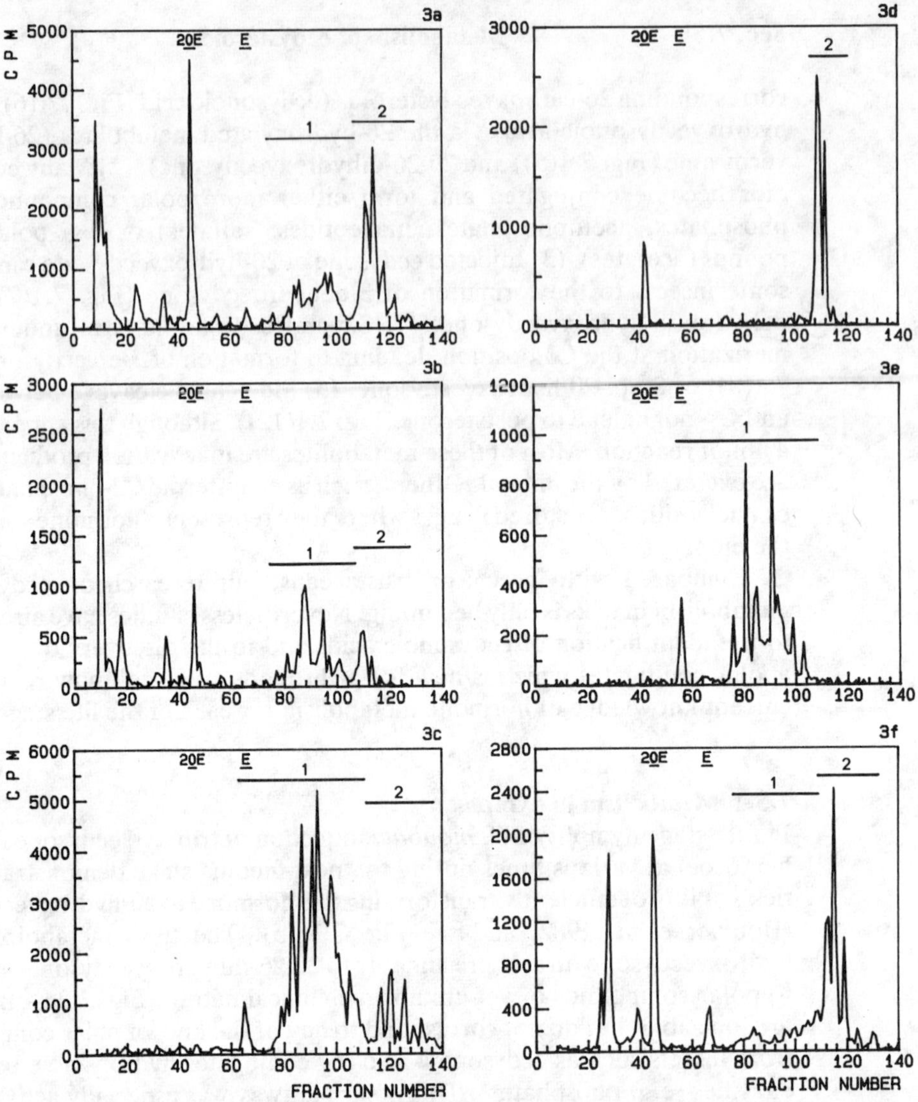

Fig. 7.3 — Metabolism of ecdysteroids in nymphs of *Ornithodoros moubata* and *Amblyomma hebraeum*. Tritiated ecdysteroids and their metabolites were extracted with methanol and subjected to HPLC on a Perkin–Elmer Series 3 chromatograph fitted with a reverse phase column RP-18 (Hibar Lichrosorb RP-18, 7 µm; column dimensions : 4×250 mm). Solvent system: linear gradient of methanol — tris/perchloric acid buffer (20 mM, pH 7.5) at 0.8 ml/min from 30–45% in 10 min, 45% for 15 min, 45–100% for 20 min, followed by 100% methanol for 20 min. Fractions were collected every 30 seconds and their radioactivity (counts per min: CPM) monitored by liquid scintillation counting in a Kontron MR 300 counter. Samples were dissolved in Riatron scintillation cocktail (Kontron; 1:1.75 v/v). The retention times of the internal cold standards ecdysone (labeled E) and 20-hydroxyecdysone (labeled 20E) were monitored with a spectrophotometer operated at 242 nm. For more technical details see Connat *et al.* (1984), or Diehl *et al.* (1985). 3a–c: metabolite pattern in whole animals after injection of ^{3}H-ecdysone (100 000 cpm/animal; sp. activity 50 Ci/mM) into the hemocoel of fifth stage nymphs of *O. moubata*. 3a: injection on day 8 post-feeding, extraction 24 hours later. 3b: ditto, but extraction after 48 hours. 3c: injection 24 hours before feeding, extraction on day 7 post-feeding. Ecdysone was metabolized to 20-hydroxyecdysone and to several polar metabolites (3a,b) and to apolar conjugates AP2 (labeled 2) which are slowly transformed into apolar conjugates AP1 (labeled 1). 3d–e: HPLC radiochromatograms of extracts of fifth stage nymphs of *O. moubata*, 6 hours (3d) and 13 days (3e) after ingestion of 20-hydroxyecdysone (10 µg cold hormone + 100 000 cpm of tritrated hormone/ml pig blood). The ingested 20-hydroxyecdysone was rapidly converted to apolar conjugates AP2 (3d) and later slowly to slightly more polar conjugates AP1 (3e).
3f: metabolite pattern, 6 hours after injection of 100 000 cpm of ^{3}H-ecdysone into *A. hebraeum* nymphs on day 2 post-dropping. The hormone was converted to 20-hydroxyecdysone and to polar compounds (acids ?). Furthermore, apolar conjugates AP2 were rapidly formed, together with a small amount of AP1.

revealed that these apolar compounds (AP2) are conjugates of 20-hydroxyecdysone esterified at C_{22} with the common long-chain fatty acids $C_{16:0}$, $C_{18:0}$, $C_{18:1}$ or $C_{18:2}$ (Fig. 7.4; Diehl *et al.* 1985). This represents a new class

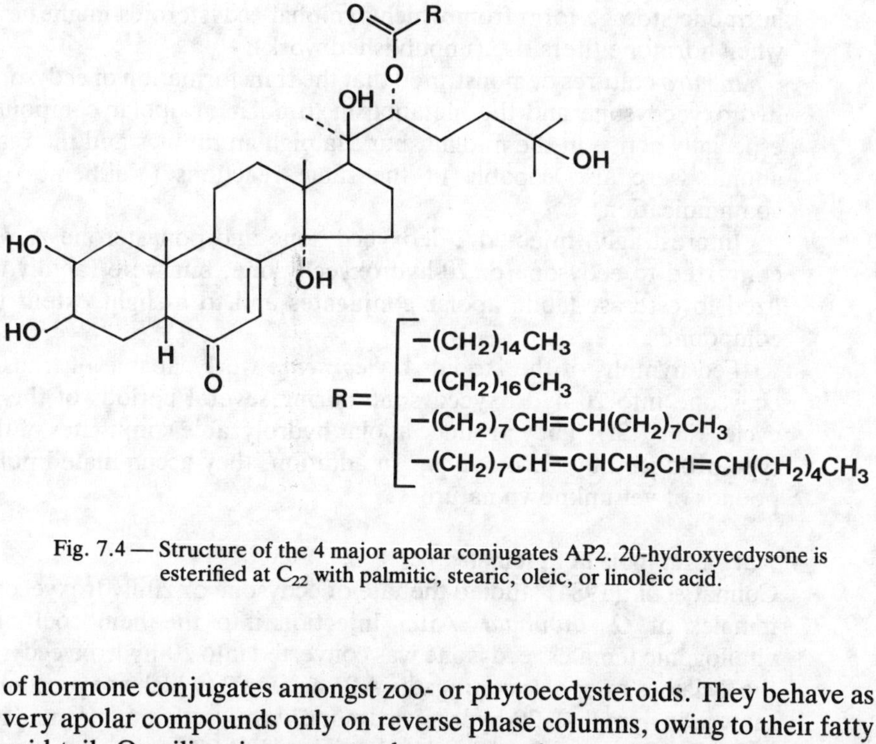

Fig. 7.4 — Structure of the 4 major apolar conjugates AP2. 20-hydroxyecdysone is esterified at C_{22} with palmitic, stearic, oleic, or linoleic acid.

of hormone conjugates amongst zoo- or phytoecdysteroids. They behave as very apolar compounds only on reverse phase columns, owing to their fatty acid tail. On silica, in contrast, they are much more polar than suspected from their behaviour in the reverse phase mode.

Conjugates AP2 were not the final products of the apolar pathway. As they gradually disappeared, other apolar conjugates (AP1) were formed which are slightly more polar (Fig. 7.3a–c,e). We do not know the chemical nature of AP1. These ionizable compounds retain their ester nature because they can be readily hydrolyzed by pig liver esterase to yield the original hormones. AP2 may be transformed into AP1 by a modification of the fatty acid tail through, for example, oxidation. Other possibilities include transesterification or an additional esterification.

A preliminary analysis of the metabolite distribution in various organs demonstrated that most of the ingested hormone was retained in the digestive tract where it was transformed by the midgut cells to AP1 via AP2. AP1 accumulated in the midgut lumen. However, during the first few hours after feeding, a small quantity of free hormone crossed the midgut epithelium and circulated in the hemolymph. Later, only AP2 and some AP1 were found in the hemolymph. The free hormone was apparently quickly removed from circulation by peripheral tissues which tranformed it into AP2 and finally into AP1.

The physiological significance of the apolar pathway is not clear. We postulate that conjugation might inactivate exogenous ecdysteroids which could be present in the blood of herbivorous hosts on which the ticks had fed (see also 7.2). In addition, it could also inactivate endogenous ecdysteroids. Several studies showed that these conjugates apparently do not serve as a hormone storage form from which nymphal ecdysteroids might be derived when hormone titers rise (unpublished work).

In vitro cultures demonstrated that the transformation of ecdysone to 20-hydroxyecdysone and the metabolism to polar or apolar compounds was especially active in the midgut, but Malpighian tubules and the rest of the animal were also capable of the same reactions (Vuillième, personal communication).

Interestingly, injected 2-deoxyecdysone and ponasterone A were not converted to ecdysone or 20-hydroxyecdysone, but were rapidly metabolized to esterase-labile apolar conjugates and to a slight extent to polar compounds.

Fed nymphs of the ixodid *A. hebraeum* were capable of transforming ecdysone into 20-hydroxyecdysone during several periods of the molting cycle (Fig. 7.3f). They produce apolar hydrolyzable conjugates of the same type produced by *O. moubata*. In addition, they accumulated polar compounds of yet unknown nature.

7.5.2 Metabolism in females

Connat *et al.* (1984) studied the fate of ecdysone or 20-hydroxyecdysone in females of *O. moubata*. After injection into the hemocoel of mated vitellogenic females, ecdysone was converted into 20-hydroxyecdysone and the 2 classes of apolar conjugates AP1 and AP2 (which are presumably the same as in nymphs). 20-hydroxyecdysone was directly transformed into AP1 and AP2. Polar products were not produced. AP2 conjugates accumulated principally in the ovaries and eggs, and accounted for about 25% of the total injected hormone by the end of vitellogenesis (Fig. 7.5a). Metabolism in fed virgin females was comparable, but ovaries incorporated less AP2. Cultured ovaries were also capable of hydroxylating ecdysone to 20-hydroxyecdysone and of producing AP2. It was hypothesized that the apolar conjugates accumulating in the eggs could represent a storage form of hormone for the embryo. New studies showed, however, that this might not be the case (see section 7.5.3).

Similarly, injected 2-deoxyecdysone and ponasterone A were converted to apolar products and AP2 accumulated in the eggs. As in nymphs, no ecdysone or 20-hydroxyecdysone were formed from these potential precursors.

Because of the great sensitivity of *O. moubata* females to supermolting after ingesting 22,25-dideoxyecdysone, we investigated the metabolism of several ingested ecdysteroids (Connat *et al.* 1986). As in the nymphs, ecdysone, 20-hydroxyecdysone, 2-deoxyecdysone, or ponasterone A were all rapidly converted to apolar conjugates. Most of the hormone was stored in the midgut mainly in the form of AP1 and some as AP2. In contrast

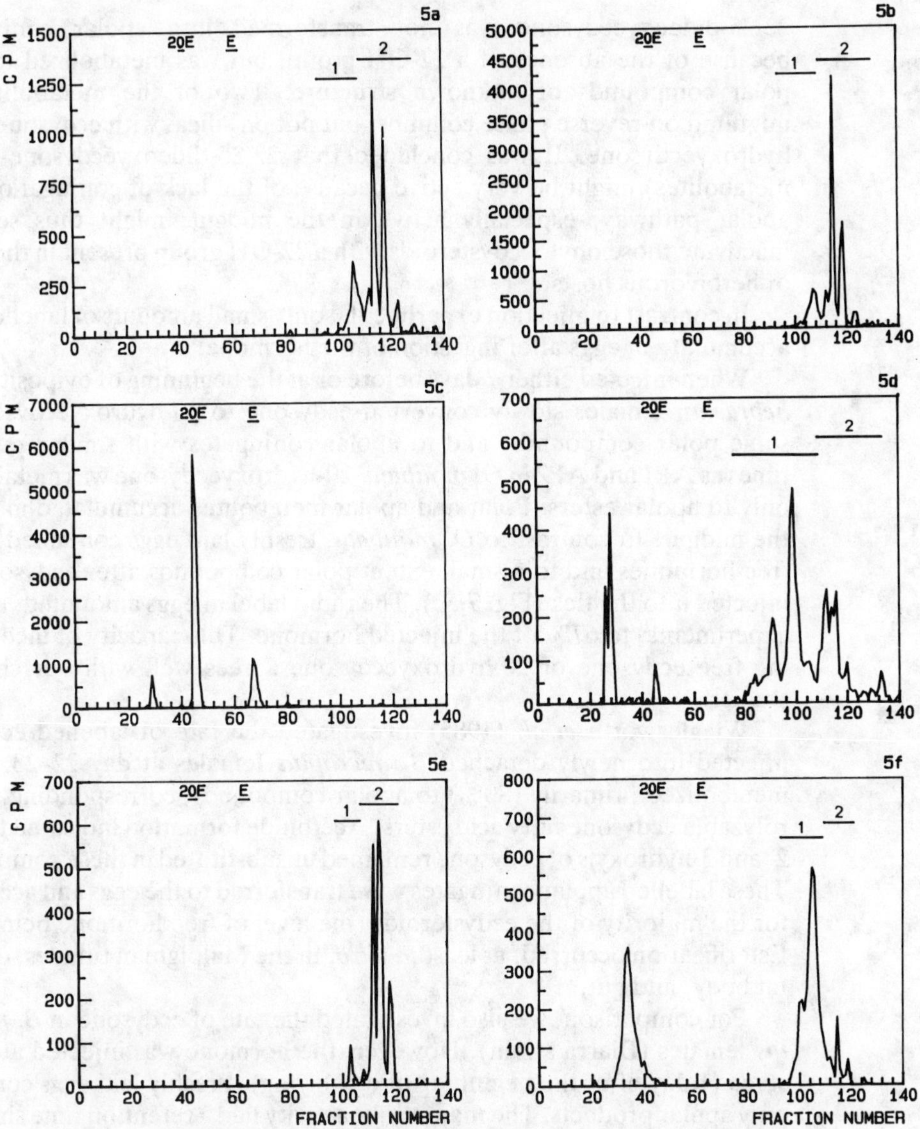

Fig. 7.5 — Ecdysteroid metabolites in freshly laid eggs after injection of ^{3}H-ecdysone into vitellogenic females and their subsequent fate during embryogenesis. For techniques and symbols see Fig. 7.3.

5a: mainly AP2 and small amounts of AP1 accumulated in freshly laid eggs after injection of ^{3}H-ecdysone into vitellogenic females of *O. moubata.* At the end of embryogenesis (day 9 at 28° C), the pattern had hardly changed (5b).

5c: mostly 20-hydroxyecdysone together with small amounts of ecdysone and of polar metabolites (acids ?) were present in freshly laid eggs of *A. hebraeum* which were oviposited 3 days after injection of ^{3}H-ecdysone into females at the beginning of oviposition. The free hormones were converted during embryogenesis to apolar conjugates AP1 and AP2 (5d, embryos shortly before hatching).

5e: only apolar conjugates, mainly AP2, accumulated in eggs of *B. microplus* (Biarra strain), oviposited 3 days after injection of ^{3}H-ecdysone into females at the beginning of egg laying. At the end of embryogenesis, the conjugates AP2 were nearly completely transformed into conjugates AP1 and to polar compounds (presumably ecdysonoic acid and 20-hydroxyecdysonoic acid, 5f).

22,25-dideoxyecdysone was not transformed into apolar conjugates, because of the absence of a 22-OH group, but was metabolized to more polar compounds of unknown structure. Two of the metabolites co-migrated on reverse phase columns, but not on silica, with ecdysone or 20-hydroxyecdysone. It was concluded that 22,25-dideoxyecdysone (or its metabolites) might be very active because of the lack of conjugation. The apolar pathway, especially active in the midgut, might thus serve to inactivate those phytoecdysteroids with a 22-OH group present in the blood of herbivorous hosts.

In contrast to injection experiments, only small amounts of labelled AP2 accumulate in eggs after ingestion of the hormone.

When injected either 7 days before or at the beginning of oviposition, *A. hebraeum* females slowly converted ecdysone to 20-hydroxyecdysone, to some polar compounds, and to apolar conjugates with similar retention times as AP1 and AP2 in *O. moubata*. 20-hydroxyecdysone was metabolized only to apolar esters. Polar and apolar metabolites accumulated mainly in the midgut. In contrast to *O. moubata*, freshly laid eggs contained mainly free hormones and to a small extent polar compounds after ecdysone was injected into the tick (Fig. 7.5c). The radio label in eggs amounted, in some experiments, to 67% of the injected hormone. This capacity of incorporating free ecdysone or 20-hydroxyecdysone agrees well with our chemical studies.

Wigglesworth *et al.* (1985) investigated the fate of labelled ecdysone injected into newly detached *B. microplus* females at day 22–23. It was metabolized primarily (80%) to apolar compounds, corresponding to hydrolyzable ecdysone fatty acid esters. Acetonide formation indicated that the 2- and 3-hydroxyls of ecdysone remained unsubstituted in these compounds. These labelled apolar conjugates were transferred to the eggs and accounted for the majority of the ecdysteroids (the level of free hormone being low). Esterification occurred, at least *in vitro*, in the Malpighian tubules, ovaries, fat body, and gut.

For comparison, we also investigated the fate of ecdysone in *B. microplus* females (Biarra strain). However, the hormone was injected at a later stage (1 day after the beginning of egg-laying). Freshly laid eggs contained only apolar products. The major radioactivity had a retention time similar to AP2 in *O. moubata* nymphs and females. Some small amounts corresponded to AP1 (Fig. 7.5e), and hydrolysis with pig liver esterase liberated ecdysone. About 90% of the injected hormone passed into the eggs, with the remainder being present mainly as AP1 (75%) in the midgut. Our data support the results of Wigglesworth *et al.* (1985), showing a strong tendency of *B. microplus* eggs to accumulate apolar hormone esters.

7.5.3 Metabolism in embryos

Preliminary studies in *A. hebraeum* embryos indicated that labelled free hormones initially present, were converted to apolar products (Fig. 7.5d). This conjugation began in embryos older than 16 days. Production of polar metabolites during embryogenesis apparently did not occur. From our data, however, we can not as yet state unequivocally if the maternal free ecdysteroids are used by the embryo.

Maternal ecdysteroid esters in *B. microplus* embryos are suspected as being inactive hormone storage forms from which free hormones can be enzymically hydrolyzed at certain times during embryogenesis. Other sources are not, however, precluded (Wigglesworth *et al.* 1985). Recent studies showed that the developing embryo metabolized hormones to 26-carboxyecdysteroids (Rees *et al.* 1984). This indicates that the pathway leading to acids is not restricted to insects or crustaceans (see Lafont & Koolman 1984), but is also operative in *B. microplus* embryos.

Our comparative study with the Biarra strain of *B. microplus* appears to support the conclusions of Rees and his co-workers. During embryogenesis, the labelled products corresponding to AP2 initially present were partly converted to more polar metabolites similar to AP1 and to an appreciable extent to other polar metabolites which may be ecdysonoic acids (Fig. 7.5f). Hydrolysis of the apolar fraction produced mostly polar products indicating that, after formation, these metabolites were partly re-esterified.

In contrast to *B. microplus*, studies of *O. moubata* embryogenesis indicated that labelled apolar AP2 conjugates, which accumulated after injection of hormones into vitellogenic females, were not metabolized to any appreciable extent, not even during larval development (Fig. 7.5b). No free labelled ecdysone, 20-hydroxyecdysone, or polar products were detected. This is contrary to the previous hypothesis that apolar conjugates, at least those of experimental origin, may be used as a source of hormone for the developing embryo of this species (Connat *et al.* 1984).

In vitro experiments with *O. moubata* demonstrated that a freshly laid egg and an embryo at any stage during its development could efficiently conjugate hormones to form AP2. On the other hand, the capacity to hydroxylate ecdysone to 20-hydroxyecdysone was manifested only after 24–48 hours of development (at 28°C). As in the *in vivo* condition no formation of polar metabolites was observed.

7.6 CONCLUSIONS

In the preceding sections we reviewed our current knowledge of tick ecdysteroids based on published and unpublished observations. Clearly, ticks possess ecdysteroids (ecdysone, 20-hydroxyecdysone) which are important in controlling processes such as molting or ixodid salivary gland degeneration. In addition, these hormones are strongly suspected of being involved in the hormonal control of other events such as diapause, spermatogenesis, oogenesis, and embryogenesis, but additional work is needed to substantiate these observations. We have also progressed in our understanding of ecdysteroid metabolism. Results indicate that polar metabolites (in part ecdysonoic acid and 20-hydroxyecdysonoic acid) and a new class of apolar conjugates, i.e. ecdysteroid esters with long-chain fatty acids, can be formed.

On the other hand, several important research areas remain totally unexplored. We do not know the biosynthetic pathway, its localization, or its probable control by other hormones such as neurosecretions. In addition, we have no information on the hormone action at the gene level or on possible interactions with other postulated hormones such as juvenile hormone-like substances which appear to be important for the control of oogenesis.

We expect that the results presented here will be of interest not only for tick endocrinology, but also for comparative arthropod endocrinology. In addition, we think that the studies on basic tick endocrinology might also be useful for applied biology. Indeed, resistance to new classical acaricides can develop rapidly. Thus, interference with hormone biosynthesis, action, or metabolism could provide a new approach to tick control. For example, the synthesis of hormone analogues inducing precocious ixodid salivary gland degeneration or the disturbance of molting and oogenesis would be highly desirable. In view of the effects of exogenous ecdysteroids, this does not appear too unrealistic.

ACKNOWLEDGEMENTS

We are grateful for the generous financial support by the Swiss National Science Foundation and by the Emil Barrel Stiftung. We thank Miss M. Zweilin and Mr M. Morici for skilful technical assistance. CIBA–GEIGY (Drs W. Häusermann, R. Immler, Mr H. Bouvard) kindly provided us with ixodid ticks. We are very grateful for many helpful discussions, technical assistance, and the supply of various ecdysteroids or RIA ingredients, to Drs M. A. Delaage, J. P. Delbecque, M. De Reggi, J. P. Dubacq, H. Gfeller, J. P. Girault, C. Hetru, M. Hirn, J. Hoffmann, D. Horn, J. Koolman, F. Lachaise, R. Lafont, J. D. O'Connor, E. Ohnishi, H. Rees, U. P. Schlunegger, and M. J. Thompson.

REFERENCES

Ahmed, S. H. & Bassal, T. T. M. (1982) Induction of adult tick molting by ecdysteroids, *J. Egyptian Soc. Parasitol* **12**:383–387.

Aldrich, J. R., Kelly, T. J., & Woods, C. W. (1982) Larval moulting hormone of trichophoran Hemiptera-Heteroptera:Makisterone A, not 20-hydroxy-ecdysone, *J. Insect Physiol* **28**:857–861.

Behrens, W., & Bückmann, D. (1983) Ecdysteroids in the pycnogonid *Pycnogonum litorale* (Ström) (Arthropoda, Pantopoda), *Gen. Comp. Endocrinol* **51**:8–14.

Bergamasco, R., & Horn, D. H. S. (1983) Distribution and role of insect hormones in plants. In *Endocrinology of insects* (Downer, R. G. H. & Laufer, H., eds), pp. 627–654. Alan R. Liss, Inc, New York.

Bonaric, J. C. & De Reggi M., (1977) Changes in ecdysone levels in the spider *Pisaura mirabilis* nymphs (Araneae,Pisauridae), *Experientia* **33**:1664–1665.

Bouvier, J., Diehl, P. A., & Morici, M. (1982) Ecdysone metabolism in the tick *Ornithodoros moubata* (Argasidae, Ixodoidea), *Rev. Suisse Zool* **89**:967–976.

Campbell, J. D. & Oliver, J. H., (1984) Membrane feeding and developmental effects of ingested β-ecdysone on *Ornithodoros parkeri* (Acari:Argasidae). In *Acarology VI*, Vol. 1 (Griffiths, D. A. & Bowman C. E., eds), pp. 393–399. Ellis Horwood Ltd, Chichester.

Connat, J. L., Diehl, P. A. & Morici, M. (1984) Metabolism of ecdysteroids during the vitellogenesis of the tick *Ornithodoros moubata* (Ixodoidea,Argasidae): Accumulation of apolar metabolites in the eggs, *Gen. Comp.Endocrinol* **56**: 100–110.

Connat, J. L., Diehl, P. A., Gfeller, H., & Morici, M. (1985) Ecdysteroids in females and eggs of the ixodid tick *Amblyomma hebraeum, Int. J. Invert. Reprod. Develop.* **8:** 103–116.

Connat, J. L., Diehl, P. A., Dumont, N., Carminati, S., & Thompson, M. J. (1983) Effects of exogenous ecdysteroids on the female tick *Ornithodoros moubata*: Induction of supermolting and influence on oogenesis, *Z. ang. Ent* **96**: 520–530.

Connat, J. L., Diehl, P. A., & Thompson, M. J. (1986) Possible inactivation of ingested ecdysteroids by conjugation with long-chain fatty acids in the femal tick *Ornithodorus moubata* (Acarini:Argasidae), *Arch. Insect Biochem. Physiol.* (in press).

Cox, B. L. (1960) Hormonal involvement in the molting process in the soft tick, *Ornithodoros turicata* Dugès. PhD thesis, University of Oklahoma (USA).

Dees, W. H., Sonenshine, D. E. & Breidling, E. (1984a) Ecdysteroids in *Hyalomma dromedarii* and *Dermacentor variabilis* and their effects on sex pheromone activity. In *Acarology VI*, Vol. 1 (Griffiths, D. A., & Bowman, C. E., eds), pp. 405–413. Ellis Horwood Ltd, Chichester.

Dees, W. H., Sonenshine, D. E. & Breidling, E. (1984b) Ecdysteroids in the American dog tick, *Dermacentor variabilis* Acari: Ixodidae, during different periods of tick development, *J. Med. Entomol* **21**:514–523.

Delachambre, J., Besson, M. T., Quennedey, A., & Delbecque, J. P. (1984) Relationships between hormones and epidermal cell cycles during the metamorphosis of *Tenebrio molitor*. In *Biosynthesis, metabolism and mode of action of invertebrate hormones* (J. A. Hoffmann & M. Porchet, eds), pp. 245–254. Springer–Verlag, Berlin, Heidelberg.

Delbecque, J. P., Diehl, P. A., & O'Connor, J. D. (1978a) Presence of ecdysone and ecdysterone in the tick *Amblyomma hebraeum* Koch, *Experientia* **34**:1379–1381.

Delbecque, J. P., Lanzrein, B., Bordereau, C., Imboden, C., Hirn, M., O'Connor, J. D., Noirot, C., & Lüscher, M. (1978b) Ecdysone and ecdysterone in physogastric termite queens and eggs of *Macrotermes bellicosus* and *Macrotermes subhyalinus, Gen. Comp. Endocrinol* **36**:40–47.

De Reggi, M. L., Hirn, M., & Delaage, M. A. (1975) Radioimmunoassay of ecdysone: An application to *Drosophila* larvae and pupae, *Biochem. Biophys. Res. Commun* **66**:1307–1315.

Diehl, P. A., Germond, J. E., & Morici, M. (1982) Correlations between ecdysteroid titers and integument structure in nymphs of the tick *Amblyomma hebraeum* Koch (Acarina:Ixodidae), *Rev. Suisse Zool* **89**:859–868.

Diehl, P. A., Connat, J. L., Girault, J. P., & Lafont, R. (1985) A new class of apolar ecdysteroid conjugates:esters of 20-hydroxyecdysone with long-chain fatty acids in ticks, *Int. J. Invert. Reprod. Develop.* **8:** 1–13.

Downer, R. G. H. & Laufer, H. (1983) *Endocrinology of insects*, 707 pp., Alan R. Liss, Inc., New York.

Dumser, J. B. & Oliver, J. H. (1981) Kinetics of spermatogenesis, cell-cycle analysis, and testis development in nymphs of the tick *Dermacentor variabilis, J. Insect Physiol.* **37**:743–753.

Ellis, B. J. & Obenchain, F. D. (1984) *In vivo* and *in vitro* production of ecdysteroids by nymphal *Amblyomma variegatum* ticks. In *Acarology VI*, Vol. 1 (Griffiths, D. A. & Bowman, C. E.), pp. 400–404. Ellis Horwood Ltd, Chichester.

Germond, J. E., Morici, M., & Diehl, P. A. (1980) Correlations between haemolymph ecdysteroids titer, multiplication of hypodermal cells and deposition of cuticle during the last larval instar of the tick *Ornithodoros moubata, Experientia* **36**:695–696.

Gibson, J. M., Majumder, M. S. I., Mendis, A. H. W., & Rees, H. H. (1983) Absence of phytosterol dealkylation and identification of the major ecdysteroid as makisterone A in *Dysdercus fasciatus, Arch. Insect Biochem. Physiol.* **1**:105–120.

Hagedorn, H. H. (1983) The role of ecdysteroids in the adult insect. In *Endocrinology of insects* (Downer, R. G. H. & Laufer, H., eds), pp. 271–304. Alan R. Liss, Inc, New York.

Harris, R. A. & Kaufman, W. R. (1981) Hormonal control of salivary gland degeneration in the ixodid tick *Amblyomma hebraeum, J. Insect Physiol* **27**:241–248.

Hetru, C. & Horn, D. H. S. (1980) Phytoecdysteroids and Zooecdysteroids. In *Progress in ecdysone research* (Hoffmann, J. A., ed.), pp. 247–280. Elsevier/North-Holland, Amsterdam.

Hoffmann, J. A. (1980) *Progress in ecdysone research*, 491 pp., Elsevier/North-Holland, Amsterdam.

Hoffmann, J. A. & Hetru, C. (1983) Ecdysone. In *Endocrinology of insects* (Downer, R. G. H. & Laufer, H. eds), pp. 65–88. Alan R. Liss, Inc., New York.

Hoffmann, J. A. & Porchet, M. (1984) *Biosynthesis, metabolism and mode of action of invertebrate hormones*, 519 pp., Springer Verlag, Berlin, Heidelberg.

Jegla, T. C. & Costlow, J. D. (1979) Ecdysteroids in *Limulus* larvae, *Experientia* **35**:554–555.

Joly, R., Porcheron, P., & Dray, F. (1979) Etude des variations du taux des ecdystéroides au cours du cycle de mue dans l'hémolymphe de *Lithobius forficatus* L. (Myriapode Chilopode) par dosage radio-immunologique, *C. R. Acad. Sci. Paris* **288**:243–246.

Juberthie-Jupeau, L., Strambi, A., De Reggi, M., & Juberthie, C. (1979) The presence of ecdysteroids and the variation of their level during the first adult stage of the myriapod *Hanseniella ivorensis* Juberthie-Jupeau and Kehe (Symphyla), *Experientia* **35**:1406–1408.

Khalil, G. M., Shaarawy, A. A. A., Sonenshine, D. E., & Gad, S. M. (1984) β-ecdysone effects on the camel tick, *Hyalomma dromedarii* (Acari:Ixodidae), *J. Med. Entomol.* **21**:188–193.

Kitaoka, S. (1972) Effect of ecdysone on ticks, especially on *Ornithodoros moubata* (Acarina-Argasidae), *Proc. 14th Int. Congress of Entomology, Australia*, p. 272.

Koolman, J., Walter, J., & Zahner, H. (1984) Ecdysteroids in helminths. In *Biosynthesis, metabolism and mode of action of invertebrate hormones* (Hoffmann, J. A., & Porchet, M. eds), pp. 323–330. Springer–Verlag, Berlin, Heidelberg.

Kurtti, T. J. & Munderloh, U. G. (1983) The effects of 20-Hydroxyecdysone and Juvenile Hormone III on tick cells, *J. Parasitol* **69**:1072–1078.

Lafont, R. & Koolman, J. (1984) Ecdysone metabolism. In *Biosynthesis, metabolism and mode of action of invertebrate hormones* (Hoffmann, J. A. & Porchet, M. eds), pp. 196–226. Springer–Verlag, Berlin, Heidelberg.

Lafont, R., Sommé-Martin, G., Mauchamp, B., Maume, B. & Delbecque, J. P. (1980) Analysis of ecdysteroids by high-performance liquid chromatography and coupled gas-liquid chromatography-mass spectrometry. In *Progress in ecdysone research* (Hoffmann, J. A., ed.), pp. 45–68. Elsevier/North-Holland, Amsterdam.

Lagueux, M., Hetru, C., Goltzené, F., Kappler, C., & Hoffmann, J. A. (1979) Ecdysone titre and metabolism in relation to cuticulogenesis in embryos of *Locusta migratoria, J. Insect Physiol* **25**:709–723.

Lagueux, M., Hoffmann, J. A., Goltzené, F., Kappler, C., Tsoupras, G., Hetru, C., & Luu, B. (1984) Ecdysteroids in ovaries and embryos of *Locusta migratoria.* In *Biosynthesis, metabolism and mode of action of invertebrate hormones* (Hoffmann, J. A. & Porchet, M. eds), pp. 168–180. Springer–Verlag, Berlin, Heidelberg.

Leubert, F., Eibisch, H., Kroschwitz, H., & Scheffel, H. (1982) Ecysteroid-Biosynthese durch das Ovar von *Lithobius forficatus* (L.) (Chilopoda), *Zool. Jb. Physiol* **86**:456–476.

Mango, C., Odhiambo, T. R., & Galun, R. (1976) Ecdysone and the super tick, *Nature* **260**:318–319.

Mansingh, A. & Rawlins, S. C. (1977) Antigonadotropic action of insect hormone analogues on the cattle tick *Boophilus microplus, Naturwissenschaften* **64**:41.

McDaniel, R. S. & Oliver, J. H. (1978). Effects of two juvenile hormone analogs and β-ecdysone on nymphal development, spermatogenesis, and embryogenesis in *Dermacentor variabilis* (Say) (Acari:Ixodidae), *J. Parasitol* **64**:571–573.

Nirdé, P., Torpier, G., Capron, A., Delaage, M., & De Reggi, M. L. (1984) Ecdysteroids in schistosomes and host-parasite relationship. In *Biosyn-*

thesis, metabolism and mode of action of invertebrate hormones (Hoffman, J. A. & Porchet, M. eds), pp. 331–337. Springer-Verlag, Berlin, Heidelberg.

Porcheron, P., Caruelle, J. P., Baehr, J. C., & Cassier, P. (1984) Ecdysteroids and integuments in locusts. In *Biosynthesis, metabolism and mode of action of invertebrate hormones* (Hoffmann, J. A. & Porchet, M. eds), pp. 234–244. Springer-Verlag, Berlin, Heidelberg.

Porchet, M., Gaillet, N., Sauber, F., Charlet, & Hoffmann, J. A. (1984) Ecdysteroids in annelids. In *Biosynthesis, metabolism and mode of action of invertebrate hormones* (Hoffmann, J. A. & Porchet, M. eds), pp. 346–348. Springer-Verlag, Berlin, Heidelberg.

Pound, J. M., Oliver, J. H. & Andrews, R. H. (1984) Introduction of apolysis and cuticle formation if female *Ornithodoros parkeri* (Acari:Argasidae) by hemocoelic injections of beta-ecdysone, *J. Med. Entomol* **21**: 612–614.

Rees, H. H., & Isaac, R. E. (1984) Biosynthesis of ovarian ecdysteroid phosphates and their metabolic fate during embryogenesis in *Schistocerca gregaria.* In *Biosynthesis, metabolism and mode of action of invertebrate hormones* (Hoffmann, J. A. & Porchet, M.), pp. 181–195. Springer-Verlag, Berlin, Heidelberg.

Rees, H. H. & Mendis, A. H. W. (1984) The occurrence and possible physiological significance of ecdysteroids during nematode and cestode development. In *Biosynthesis, metabolism and mode of action of invertebrate hormones* (Hoffmann, J. A. & Porchet, M. eds), pp. 338–345. Springer-Verlag, Berlin, Heidelberg.

Romer, F. (1979) Ecdysteroids in snails, *Naturwissenschaften* **66**:471.

Romer, F. & Gnatzy, W. (1981) Arachnid oenocytes:Ecdysone synthesis in the legs of harvestmen (Opilionidae), *Cell Tissue Res.* **216**:449–453.

Ruegg, R. P., Kriger, F. L., Davey, K. G., & Steel, C. G. H. (1981) Ovarian ecdysone elicits release of a myotropic ovulation hormone in *Rhodnius* (Insecta:Hemiptera), *Int. J. Invert. Reprod.* **3**:357–361.

Sannasi, A., & Subramoniam, T. (1972) Hormonal rupture of larval diapause in the tick *Rhipicephalus sanguineus* (Lat.), *Experientia* **28**:666–667.

Solomon, K. R., Mango, C. K. A., & Obenchain, F. D. (1982) Endocrine mechanisms in ticks: effects of insect hormones and their mimics on development and reproduction. In *Physiology of ticks* (Obenchain, F. D. & Galun, R. eds), pp. 399–438. Pergamon Press, Oxford, New York, Toronto, Sidney, Paris, Frankfurt.

Spindler, K. D., Keller, R., & O'Connor, J. D. (1980) The role of ecdysteroids in the crustacean molting cycle. In *Progress in ecdysone research* (Hoffmann, J. A. ed.), pp. 247–280. Elsevier/North-Holland, Amsterdam.

Sturaro, A., Guerriero, A., De Clauser, R., & Pietra, F. (1982) A new, unexpected marine source of a molting hormone. Isolation of ecdysterone in large amounts from the zoanthid *Gerardia savaglia, Experientia* **38**:1184–1185.

Vogel, B. E. (1975) Morphologische Untersuchungen über Integument und Häutung sowie Entwicklung des vierten Beinpaares bei *Ornithodoros moubata* (Murray) 1877 (Acarina, Ixodoidea, Argasidae). PhD thesis, University of Basle (Switzerland).

Whitehead, D. L., Osir, E. W., Obenchain, F. D., & Thomas, L. S. (1986) Evidence for the presence of ecdysteroids and preliminary characterization of their carrier protein in the eggs of the brown ear tick *Rhipicephalus appendiculatus* (Neumann), *Insect Biochem..* **16:** 121–133.

Wigglesworth, K. P., Lewis, D., & Rees, H. H. (1985) Ecdysteroid titre and metabolism to novel apolar derivatives in adult female *Boophilus microplus* (Ixodidae), *Arch. Insect Biochem. Physiol* **2**:39–54.

Wright, J. E. (1969) Hormonal termination of larval diapause in *Dermacentor albipictus, Science* **163**:390–391.

NOTE ADDED IN PROOF

Ovaries from *Amblomma hebraeum* females massively converted *in vitro* ecdysone to 3-epiecdysone presumably through the intermediate formation of 3-dehydroecdysone. Only small amounts of 20-hydroxyecdysone were produced. Epimerization occurred also to some extent in carcasses, but not in gut (Connat, J. L., Lafont, R., & Diehl, P. A., *Mol. Cell Endocrinol.*, in press).

8

Some aspects of the control of the gonotrophic cycle in the tick *Ornithodoros moubata* (Ixodoidea, Argasidae)

Jean-Louis Connat, Jacques Ducommun, Peter A. Diehl, and **André Aeschlimann**

8.1 INTRODUCTION

Ticks are obligate hematophagous ectoparasites which can transmit many diseases, and they have a high reproductive potential. They are of considerable economic importance and thus reproduction of numerous species has been studied extensively (see reviews of Diehl *et al.* 1982, Oliver 1974, Oliver Chapter 10 of the present book). However, in contrast with insects, very little is known about the hormonal physiology controlling oocyte maturation and oviposition. This fact is due in part to the difficulty in experimenting with this material. For instance, because of the 'sac-like-structure' of ticks, selective ablations of glandular tissues are very difficult and simpler experimental operations such as ligations induce high mortality. The anatomy and cytology of the glandular tissues associated with the brain have been investigated very recently, but in only a few species (see review of Binnington. Chapter 6).

The nature of the circulating hormones, the site of vitellogenin synthesis, and the exact process of ovary development have not yet been described in most species of adult ticks. However, data obtained in our laboratory during the past 20 years with the argasid tick *Ornithodoros moubata* (Murray 1977, *sensu* Walton 1962) permit us to describe various interactions and succession of interdependent steps for the completion of a gonotrophic cycle in this species. We will briefly review these studies, and together with unpublished results and some data from other authors, we will provide a synthetic scheme for the possible control of the gonotrophic cycle in *O. moubata.* Of course, caution must be exercised in generalizing mechanisms described for one species to other species. Similar or different mechanisms controlling the gonotrophic cycle in other species are reviewed in Chapter 10 by Oliver.

8.2 THE GONOTROPHIC CYCLE OF *ORNITHODOROS MOUBATA*

As in other ticks, engorgement and mating are the usual prerequisites for a complete gonotrophic cycle in *O. moubata,* including vitellogenesis and oviposition. Descriptive aspects of oogenesis and cytological changes in the oocytes during vitellogenesis have been studied (Aeschlimann & Hecker 1969, 1970, Diehl 1970) and summarized diagrammatically in Diehl *et al.* (1982). In ixodids (with the exception of a few parthenogenetic species) mating is a requirement for completion of the blood meal (Aeschlimann & Grandjean 1973a, Graf 1974). In contrast, mating is not necessary for engorgement in argasids. This characteristic has allowed us to study separately the influence of the blood meal and mating on successful completion of the gonotrophic cycle in an argasid species.

8.2.1 Importance of the bloodmeal

O. moubata, like other tick species, is an obligate hematophagous ectoparasite. Each immature stage (except the larvae) requires a bloodmeal to complete a molting cycle (Hoogstraal 1956). Autogeny, however, does take place in a very low percentage of mated unfed females, and even two successive cycles have been observed (Aeschlimann & Grandjean 1973a).

In mated females, the size of the bloodmeal determines the number of eggs laid. Experiments using 82 fully engorged and 55 partially fed females demonstrated high linear correlations (correlation coefficient > 0.945) between the weight of the egg-batch, the number of eggs laid, and the weight of the bloodmeal. Such high correlations have also been observed with the weight of the engorged females after excretion of their coxal fluid, one day after feeding (Table 8.1). Later, for simplification, the correlation between the number of eggs laid and the weight of engorged females was used. Five partially fed females (experimentally detached) did not lay eggs. The percentage of the unfed weight represented by the bloodmeal was 52% (Table 8.2). Ingestion of a quantity of blood representing 65% of the unfed weight was sufficient to induce the oviposition of a reduced number of eggs.

In virgin females, as in mated females, feeding initiates the beginning of digestion. Vitellogenins are synthesized and found circulating in the hemo-

Table 8.1 — Linear correlations† between the Bloodmeal weight and the egg-yield (in mg or in number (No) of eggs) in ovipositing *O. moubata* females.

Equations	Correlation coefficients
Egg yield (mg) = 0.502 × Bloodmeal weight − 0.946	0.950
No. of eggs = 0.725 × Bloodmeal weight + 2.089	0.948
Egg yield (mg) = 0.460 × Engorged female weight − 17.34	0.948
No. of eggs = 0.663 × Engorged female weight − 21.422	0.945

†Correlations were established either with the estimated weight of the bloodmeal (difference between the weights of the engorged female 24 hrs after feeding and its unfed weight) or directly with the engorged female weight 24 hrs after feeding.

Table 8.2 — Number of eggs laid by mated female *O. moubata* in relation to the bloodmeal weight expressed as percentage of the unfed weight of the females.

Number of eggs	Number of females	Bloodmeal weight expressed as % of unfed weight† $\bar{x}$	SD
0	5	52.0	32.7
$0 < N \leqslant 14$	7	65.0	15.2
$15 \leqslant N \leqslant 24$	11	97.4	34.8
$25 \leqslant N \leqslant 34$	10	121.9	82.6

†$\bar{x}$, SD: mean and standard deviation of the different estimated percentages respectively.

lymph. Spectrophotometric measurements at 400 nm (maximum absorbance of the dark hemoproteins which constitute vitellogenins) were not different in mated and virgin females at the same time after feeding. The concentration of the protein H8 increases in virgin females during the first 2 weeks post-feeding as in mated females, but, in contrast to mated females, its concentration does not decrease. However, the levels of proteins H6 and H7 in virgin females require several weeks to reach that attained within one week in mated females (Diehl 1969, 1970). These vitellogenins are incorporated into the oocytes. However, the oocytes rarely reach the final stage, and the partially to fully developed eggs are finally resorbed. This 'abortive vitellogenesis' takes about 50 days. One hundred days after feeding, the ovaries of fed virgin females look like those of unfed females (Aeschlimann 1968, Germond & Aeschlimann 1977). Nevertheless, a few fed virgin females are able to oviposit a few non-viable eggs after a prolonged preoviposition period (Aeschlimann 1968). After fifteen years of research in the area we note that these females lay as many eggs as mated females of the same breeding colony (Fig. 8.1A). Generally, they represent a small percentage of the total female population, but surprisingly, this percentage of virgin females ovipositing is very high in some nutrition batches (Fig. 8.1B). The hormonal quality of pig blood on which the ticks fed could be a reason for such variation. Vertebrate steroids play a role in the synchronization of reproduction of certain fleas (Rothschild & Ford 1973). In addition, vertebrate steroids have been recently reported in insects (De Clerck *et al.* 1983, 1984). Thus, an interesting question centres on whether vertebrate hormones can sometimes induce oviposition in virgin female ticks.

To summarize, the bloodmeal constitutes a stimulus for the onset of vitellogenesis, but it is generally not enough to ensure completion of the gonotrophic cycle. In addition, the ingestion of a minimum quantity of blood (corresponding to approximately 60% of the weight of the unfed females) is necessary to induce oviposition in mated females.

8.2.2 Importance of mating

The dependency of oviposition upon when copulation takes place (Fig. 8.2) indicates the importance of mating for completion of a successful gonotrophic cycle. Mating alone, without a bloodmeal, except in certain rare cases of autogeny (Aeschlimann & Grandjean 1973a), does not induce oviposition.

If the female is mated before or soon after the bloodmeal, the isolated engorged female completes digestion of the bloodmeal and completes normal vitellogenesis. Vitellogenins are synthesized, secreted, and circulated in the hemolymph and are incorporated in the oocytes (Diehl 1969, 1970, Chinzei 1983, Chinzei *et al.* 1983). At 27 °C the preoviposition period varies between 10 and 20 days, with the maximum number of females ovipositing at 11–12 days. Disturbance of the female delays but does not reduce the number of eggs laid. Overcrowding also delays, and sometimes inhibits or reduces the number of eggs laid by the female. In *Argas arboreus* similar disturbances inhibit oviposition *via* a pheromone (Khalil 1984).

If a virgin female engorges and is not mated, the 'abortive vitellogenesis' described above takes place. Such virgin females, possessing nutrient reserves in the midgut lumen, may be mated after several months. The

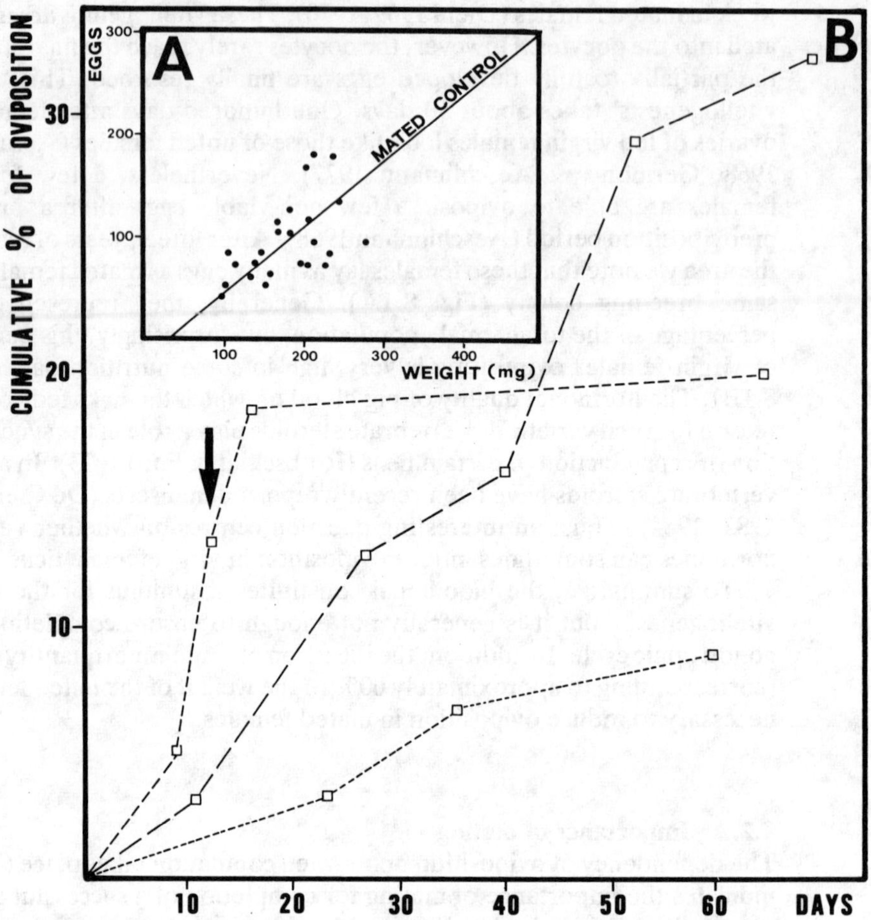

Fig. 8.1A — Occurrence of oviposition in fed virgin female *O. moubata*. The numbers of eggs laid by the ovipositing virgin females of our colony plotted in relation to the weight of the engorged females after excretion of their coxal fluid.

The numbers of eggs laid by these females are very close to those laid by mated females (regression line drawn with 137 mated females). Fig. 8.1B — Occurrence of oviposition in fed virgin female *O. moubata*. Cumulative percentage of ovipositing fed virgin females in 3 different nutrition batches. (The females were fed on defibrinated pig blood through a parafilm membrane). 60 females were used with blood A (----), 31 with blood B (-----), and 35 with blood C (— — — —).

The arrow indicates the usual preoviposition period for fed mated females. Note that virgin females generally oviposit with a very prolonged preoviposition period.

delayed mating triggers digestive activity of the midgut cells, and the mobilization of food reserves stored in the midgut lumen (Aeschlimann 1968, Grandjean 1983). This resumption of digestion is followed by vitellogenesis leading to oviposition of viable eggs. However, the number of eggs laid is lower than if mating coincided with the bloodmeal. Only 14% of the weight of the female tick is converted into eggs instead of about 30% if mating coincides with feeding (Grandjean 1983). Also the 'preoviposition' period (the time between mating and beginning of oviposition) is shorter (6–14 days instead of 10–20 days) (Aeschlimann 1968, Germond & Aeschlimann 1977).

O. moubata females can perform several gonotrophic cycles (up to 8), each cycle needing a bloodmeal. We have noted, however, that only one mating is necessary for the successive gonotrophic cycles, but in this case, the preoviposition periods become successively prolonged and egg viability decreases. Nevertheless, the number of matings or the number of endospermatophores does not influence the number of eggs laid (Aeschlimann & Grandjean 1973b).

8.3 CHARACTERIZATION OF THE STIMULI PRODUCED DURING MATING

8.3.1 Mechanical stimulation

We investigated the role of mechanical stimulation of the female genital ducts by introducing small metal beads into the vagina and the uterus of engorged virgin females more than 3 months after feeding (see details in Ducommun 1984). Upon dissection, vitellogenic oocytes were observed in these females. In addition, the dry weight of the ovaries increased within 7 days after treatment (Table 8.3). The weight of the ovaries of treated females was significantly higher ($P \leqslant 0.05$) than those of untreated females, but not significantly different ($P \leqslant 0.05$) from those of mated females. A high proportion of 'bead' treated females (73%) laid eggs within a slightly prolonged 'preoviposition' period (20–25 days). This percentage was not significantly different from those of mated controls of the same batch ($P =$ 43%, exact probability test of Fisher) (Table 8.4). However, the egg yield expressed as the number of eggs laid/weight of the female, was significantly lower in treated females (0.42 ± 0.19 instead of 0.59 ± 0.10).

Thus, these results clearly demonstrate that mechanical stimulation alone is enough to trigger vitellogenesis *and* oviposition in virgin females in which the ovary was in a resting stage after the 'abortive vitellogenesis'.

8.3.2 Chemical stimulation

Mature spermiophores, taken from the uteri of mated females, induce egg-laying or vitellogenesis when injected (either intact or as homogenate) into the hemocoel of engorged virgin females (Aeschlimann 1968). In addition, if the mature spermiophores are incubated for 20 h in a medium, injection of the washed spermiophores induced egg-laying in the virgins, indicating that the chemical stimulus was not in the seminal fluid but in the spermiophore. Furthermore, injection of homogenates of the accessory glands, in contrast to observations in *O. parkeri* by Oliver *et al.* 1984, did not induce vitellogenesis or oviposition, indicating further the inactivity of seminal fluid in *O. moubata*. Thus, in order to determine where this chemical stimulus origi-

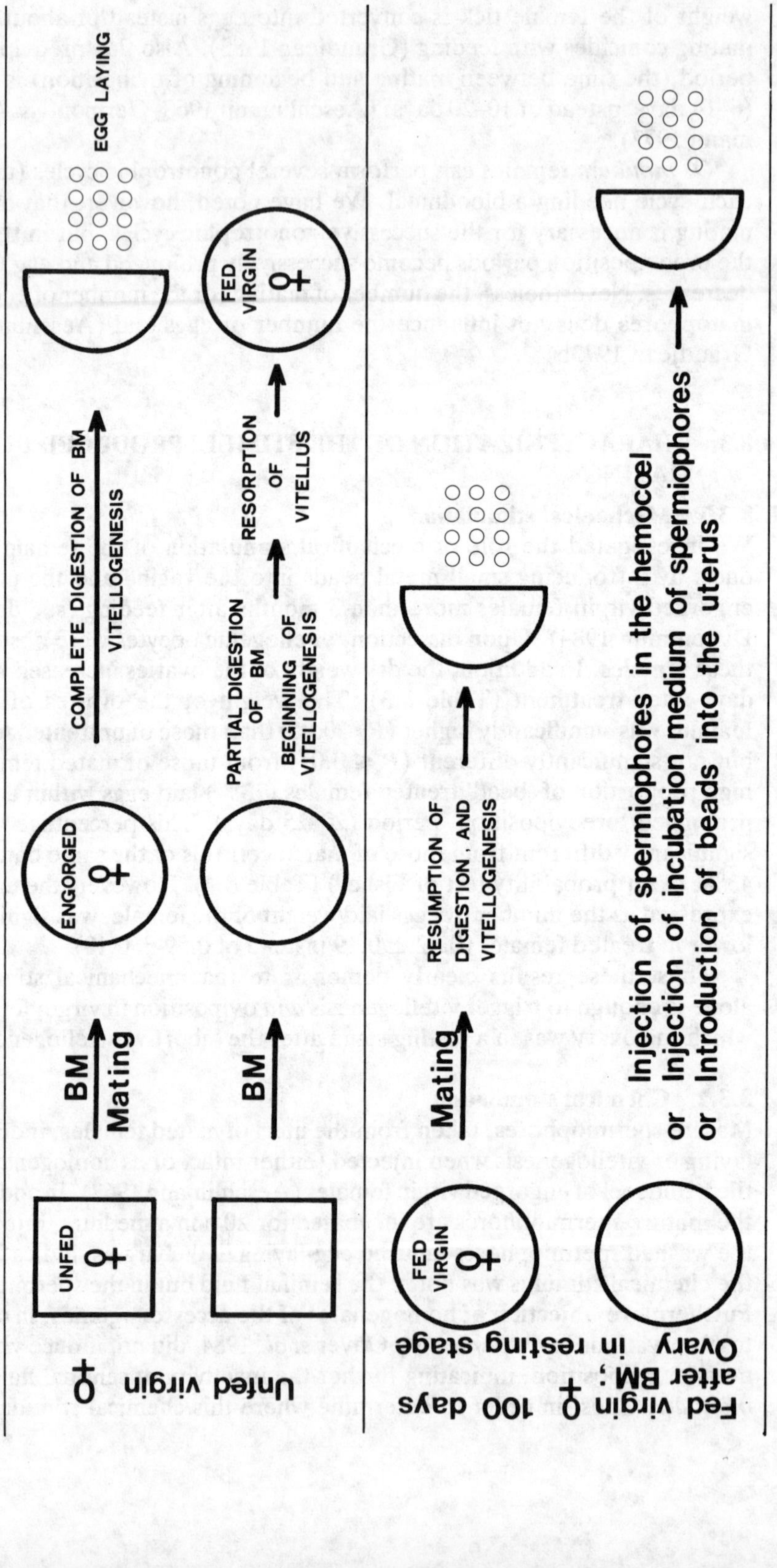

Fig. 8.2 — Diagrammatic summary of the possible inducers of vitellogenesis and/or oviposition in female *O. moubata* (BM = bloodmeal).

Table 8.3 — The role of mechanical stimulation of mating in *O. moubata*: The dry weight of ovaries of virgin females 100 days after feeding was estimated 7 days after implantation of small metal beads into the uterus.

	Number of females	Mean weight of a dry ovary ± SD (μg)
Mated control	5	1500 ± 1360
Virgin control	10	270 ± 170
Beads in the uterus of virgins	10	1060 ± 680

Table 8.4 — The role of mechanical stimulation of mating in *O. moubata*: Number of ovipositing virgin females 100 days after feeding in which beads were introduced into the uterus. The females oviposited approximately 25 days after treatment.

	Number of females	Number of ovipositing females
Mated control	8	7 (88%)
Virgin control	15	0 (0%)
Beads in the uterus of virgins	15	11 (73%)

nates, we investigated the effects of injecting homogenates of testes. Again, in contrast with *O. parkeri* (Oliver *et al.* 1984), vitellogenesis was not observed. Only the injection of homogenates of seminal vesicles distended with immature spermiophores was effective; 71% of the females injected with crude extracts and 50% injected with the extract diluted 5 times produced vitellogenic ovaries. Thus, the chemical responsible for inducing oogenesis was already present in prospermium. Aeschlimann (1968) noted that the medium in which mature spermiophores were incubated was also effective in inducing vitellogenesis when injected into fed virgin females. Recently, Sahli *et al.* (1985) demonstrated, *in vitro,* that the incubation medium of immature, non-devaginated spermiophores did not induce vitellogenesis or oviposition. The medium becomes active only after the spermiophores are almost fully devaginated and the contents of the apical cisternae are liberated. The activity of the *in vitro* incubation medium decreases rapidly (after about 12 h of incubation), but the incubation medium of spermiophores removed from females 48 h after mating still shows high activity. This suggests that, *in vivo,* the chemical stimulus begins to be released by the spermiophore after devagination, that is, soon after mating, and continues to be released for several days thereafter.

Preliminary experiments (Germond & Aeschlimann 1977) showed that the compound inducing oogenesis was heat sensitive; it was eluted in the void volume when passed through a Sephadex G-100 column, suggesting a high molecular weight. Results of Sahli *et al.* (1985) indicated that it could be a protein. The molecular weight of this compound has been estimated after chromatographing an active spermiophore incubation medium on a Bio Gel A5M column and injecting aliquots of eluted fractions into virgin females 3 months after feeding (Fig. 8.3). This bioassay showed a peak of activity corresponding to an apparent molecular weight of 1 600 000 to 1 900 000 daltons.

Analysis of the denatured protein from different incubation media of spermiophores by SDS polyacrylamide electrophoresis demonstrated the presence of two proteins of high molecular weight (between 100 000 and 200 000 daltons) in the bioassay positive supernatants (Sahli *et al.* 1985). The differences between the molecular weights found by gel filtration and electrophoresis may be due to the formation of a proteinaceous complex between different subunits. The fact that treating with urea inactivates the compound and that the activity is recovered only after slow but not after rapid dilution of urea (Sahli personal communication), suggests that the active compound consists of several subunits that can be denatured and then reconstituted. If the active substance is not itself a protein, it must be linked to a protein when active. However, because of the difficulties of working with proteins, investigations on the molecular weight of this compound and the possible existence of subunits perhaps should be carefully repeated.

Finally, Germond & Aeschlimann (1977) demonstrated that this chemical stimulus was not species specific, since injections of spermiophores of *O. tholozani* or *O. tartakowsky* induce oviposition in virgin females of *O. moubata.*

8.4 POSSIBLE PATHWAYS VIA WHICH THE ACTIVE SUBSTANCE ACTS

Although we do not yet know the exact nature of the chemical compound inducing vitellogenesis and oviposition in *O. moubata,* we are interested in knowing how it exerts its effect. We note that this compound is produced by an individual and transferred to another individual of the same species (in the present case, from a male to a female). In addition, it seems to act at very low concentrations. Therefore, it corresponds to what is usually called a *pheromone.* This pheromone is normally introduced into the vagina during mating, and we hypothesize that receptors for this product exist in this organ or in the genital ducts. However, it also exerts its effects when introduced into the hemocoel, suggesting the presence of other receptors. Two possibi-

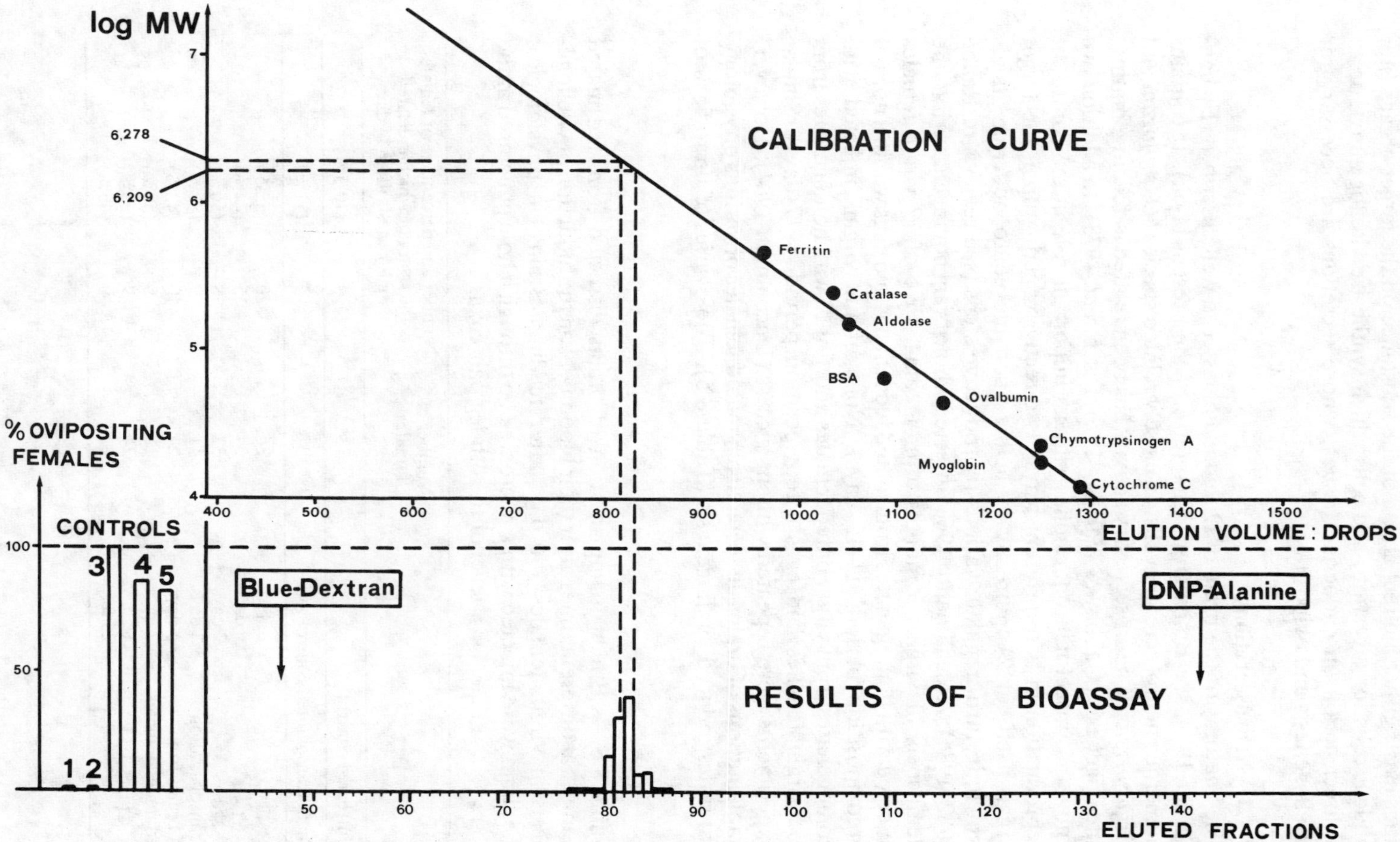

Fig. 8.3 — Estimation of the molecular weight (MW) of the vitellogenesis inducing factor present in the supernatant of incubated spermiophores. The supernatant was separated into fractions of 10 drops (600 μl) on a Bio-Gel A5M column, (length 60 cm, ø 15 mm). A bioassay was performed by injecting aliquots of these fractions into the hemocoel of 11 to 18 virgin females 100 days after feeding. Percentage of ovipositing females are recorded. Controls were completed by injecting the elution solvent (NaCl 2%) into fed virgin females (1) and fed mated females (3). Similar controls were done with Arachnid Ringer, the solution in which the sample was deposited onto the column (2: virgin females; 4: mated females). Control no. 5 corresponded to the injection of an aliquot of the supernatant before chromatography in fed virgin females.

lities exist: either a direct action of the substance on the target organs (i.e. midgut cells for digestion, and cells which synthesize vitellogenins), or an indirect action on endocrine tissues, whose secretions are necessary for vitellogenesis and oviposition.

8.4.1 Neurosecretions

The anatomy of the brain of *O. moubata* has been well described (Eichenberger 1970, see Binnington, Chapter 6 of the present book). It contains numerous neurosecretory cells. In closely related species (*O. tholozani* and *Argas persicus*) neurosecretory activity has been related to feeding, mating, and oogenesis (Gabbay & Warburg 1976, Eisen *et al.* 1973). In addition, we know that insect neurohormones play an important role in the control of reproduction in two different ways: 1) directly by acting upon the target organ itself, or 2) indirectly by acting upon the neuroendocrine glands which, in turn, regulate various reproductive steps (see reviews of Raabe 1982, 1984). Aeschlimann (1968) injected whole brain homogenates into the hemocoels of virgins 3 months after feeding. Five of 12 virgin females injected with homogenates from mated fed females laid eggs within a normal preoviposition period (11–15 days). None of the virgins injected with homogenates from virgin unfed females and 1 of 12 with homogenates from fed virgin females laid eggs (preoviposition period = 17 days). Similar experiments were repeated without success (Ducommun 1984). However, implantations of brains dissected from mated females of various physiological states induced vitellogenesis (Table 8.5). The ovaries of up to 56% of

Table 8.5 — Effect of implantation of brains dissected from fed mated females at different times after their bloodmeal on the induction of vitellogenesis in virgin females 100 days after feeding. Brains, in physiological medium, were injected with a microsyringe through a hole punctured at the level of the 4th leg pair coxa.

		Number of females	Number and percentage of females, the ovary of which indicated vitellogenesis after 14 days	
Untreated control		38	4	(10.5%)
Sham transplanted control		37	4	(11%)
Transplantation of brains from fed mated females	unfed	36	20	(55.5%)
	0 days	34	15	(44%)
	2 days	32	18	(56%)
	3 days	11	9	(82%)
	4 days	24	10	(42%)
	7 days	10	5	(50%)
	12 days	24	9	(38%)
	25 days	22	10	(45%)

transplanted females contained large brown oocytes. Thus, transplantation has a stimulating effect on vitellogenesis but not on oviposition. Unfortunately, however, this effect does not seem to be specific. In fact, implantation of brains from either virgin females of corresponding ages, or males, induced vitellogenesis in about the same proportions (40 to 52% of females oviposited). Other experiments on brain implantation in which the retrocerebral complex has been carefully preserved did not significantly modify the results ($P \leqslant 0.05$). This complex is believed to have a structure very close to that of the corpora allata — corpora cardiaca complex of insects. Abortive vitellogenesis was induced in 44 to 67% of the virgin females (fed 3 months before) with transplanted brains from either virgin females, mated females, or males.

8.4.2 Production of juvenile hormone-like substances

Mating may also act as a stimulus for the production of hormones related to juvenile hormones (JHs) that play an important role in the control of vitellogenesis in insects (see review of Engelmann 1984).

It is possible to induce egg-laying in virgin females (either immediately after feeding or 3 months after feeding) by topical application of JH analogues or JH stereoisomers (Connat *et al.* 1983b). In recently fed virgin females, these compounds are able to maintain the vitellogenic process and cause a high percentage of ovipositing females (up to 70% instead of less than 13% in virgin females). The length of the preoviposition period is inversely related to the dose applied, but the number of eggs laid — though nonviable — is always as high as that expected for mated females of the same weight. Surprisingly, 10 μg of farnesyl methyl ether per female induced hyperstimulation of vitellogenesis leading to a greater egg yield than in mated females.

Topical applications of the same compounds to virgins with a resting stage ovary 3 months after feeding were less potent. A high dose (500 μg) of a 1:1 ratio of JH1:JH3 induced only 58% of the females to oviposit. However, in this case the action of the hormone induces both a resumption of digestion and vitellogenesis. Injecting the hormone into the ticks yielded higher percentages of ovipositing females. Fifteen to 60 μg of a JH1:JH3 mixture (1:1) induced oviposition in about 60% of the females. The preoviposition period was also longer than in the control females. Injections of the natural isomers of insect juvenile hormone were much less effective than stereoisomeric mixtures or JH analogues.

Our results, with large doses of JHs or JH analogues do not necessarily show a direct effect, but suggest, as did the results of Pound & Oliver (1979) with *O. parkeri*, that a hormone structurally related to insect JH may exist in these animals. We have thus attempted in collaboration with Dr B. Mauchamp, CNRA, Lab. of Phytopharmacy, Versailles, France to analyze samples with gas chromatography coupled to mass fragmentometry after extraction and purification. The technique is similar to that used in analyzing insect JHs (Bergot *et al.*, 1981), but without derivative formation since we used a chemical ionization for detection (see Mauchamp *et al.* 1984).

Analysis of 3 different samples of 300 μl haemolymph each from 3 different periods during vitellogenesis were performed. Several products exhibited retention times close to those of the insect juvenile hormone JH3 on gas chromatography. However, analysis of the spectra, corresponding to these products, showed that they did not correspond to insect JHs. At present, it is not yet possible to say if these products correspond to JH-like substances or to lipidic compounds without hormonal properties. A bioassay should be used to answer this question.

Thus, the products detected with the retention time of insect JHs and having some identical fragments are not the insect JHs themselves. This emphasizes the importance of doing precise analysis with reliable methods, to determine the exact nature of possible JH-like hormones in ticks.

8.4.3 Possible role of ecdysteroids

Another hormonal factor which is at present being studied is the role of ecdysteroids during vitellogenesis. In insects, these hormones have various roles in different species. For example, in certain diptera they induce the fat body to synthesize vitellogenins, in *Rhodnius* they trigger oviposition *via* a neurohormone, and in cockroaches oocyte development ceases during a 'pregnancy' stage because ecdysteroids inhibit JH production (see review of Hagedorn 1983). When *O. moubata* was fed with blood containing ecdysteroids, the gonotrophic cycle was inhibited during the induction of a supermolting cycle (Connat *et al.* 1983a). When ecdysone, 20-OH-ecdysone, makisterone A, or ponasterone A, are ingested, quite large doses (more than 10 μg/ml blood) are required for induction of the supermolt, probably owing to an efficient mechanism for detoxifying ingested ecdysteroids by the conjugation of fatty acids to the 22-OH position (Diehl *et al.* 1985, Connat *et al.* 1986). In contrast, ingestion of minute quantities (20–30 ng) of the synthetic 22,25-dideoxyecdysone induced supermolting in all the females and delayed the onset of vitellogenesis (Connat *et al.* 1983a). However, doses of 5 μg or less of ecdysone or 20-OH-ecdysone per ml blood, or doses of 22,25-dideoxyecdysone which did not induce supermolting influenced neither the preoviposition period nor the number of eggs laid.

Suprisingly, 2 compounds closely related to 22,25-dideoxyecdysone: 2,22,25-trideoxyecdysone or 2,22,25-trideoxy,5βOH-ecdysone were unable to induce a supermolt, even when ingested at 5 μg/ml blood (which is the lethal dose for 22,25-dideoxyecdysone). Females treated in this manner oviposit a normal quantity of eggs, but a high proportion of the eggs dessicate, suggesting a possible action of these compounds on the function of Géné's organ. Topical application of 5 μg of these ecdysteroids in 5 μl methanol did not induce the same effects.

When a supermolt cycle is induced by 22,25-dideoxyecdysone, vitellogenesis is completely inhibited. Egg laying occurs 10 days after ecdysis, corresponding to the usual preoviposition period, suggesting that vitellogenesis starts shortly after ecdysis. Nevertheless, the number of eggs laid per supermolting female was lower than in control mated females. The higher the dose of 22,25-dideoxyecdysone ingested, the lower the number of eggs laid (Connat *et al.* 1983a). We investigated, with the use of radioimmunoassay (RIA) (De Reggi *et al.* 1975), the hemolymph ecdysteroid level during the supermolting cycle of these females (Fig. 8.4), and we correlated

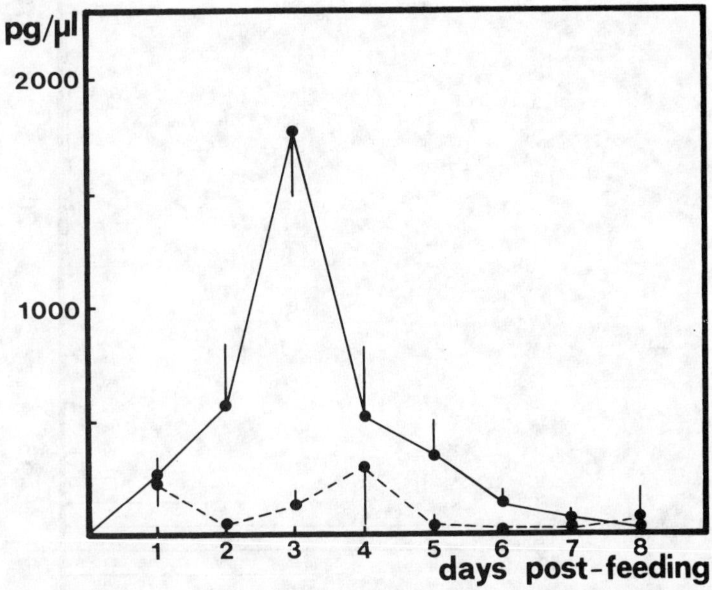

Fig. 8.4 — Hemolymph ecdysteroid levels expressed as pg 20-hydroxyecdysone equivalent/µl hemolymph assayed by RIA (antiserum provided by Drs De Reggi, Hirn and Delaage) in female *O. moubata* either during a supermolting cycle induced by ingestion of 0.04 ng 22,25-dideoxyecdysone per ml blood (●—●) or during vitellogenic cycle (●- - -●). 10 to 30 µl hemolymph were withdrawn, put into 500 µl methanol, vortexed, sonicated and centrifuged. The supernatant was evaporated to dryness and then replaced by 175 µl citrate buffer and 175 µl of 3H-antigen (ecdysone) solution in order to perform the assay in duplicate. Each point corresponds to the mean value obtained with 3 or 4 females. Vertical bars correspond to ±SD.

those results with cytological observations of the integument. Observations of semi-thin sections demonstrated a sequence of cuticular events similar to that observed in fifth stage nymphs (Germond *et al.* 1982). Ecdysteroid levels increased during the first 3 days and reached a peak of 1788 pg 20-OH-ecdysone equivalent per µl hemolymph at the beginning of epicuticle synthesis. It is surprising that this peak, induced by 22,25-dideoxyecdysone ingestion, is about 3 times higher than that in fifth stage nymphs. Then, as in fifth stage nymphs, the ecdysteroid levels gradually decrease during procuticle and cement deposition. RIA coupled with HPLC analysis of the 3rd day hemolymph showed 2 peaks of immunoreactive material comigrating with 20-OH-ecdysone and ecdysone (Fig. 8.5). Surprisingly, however, metabolic

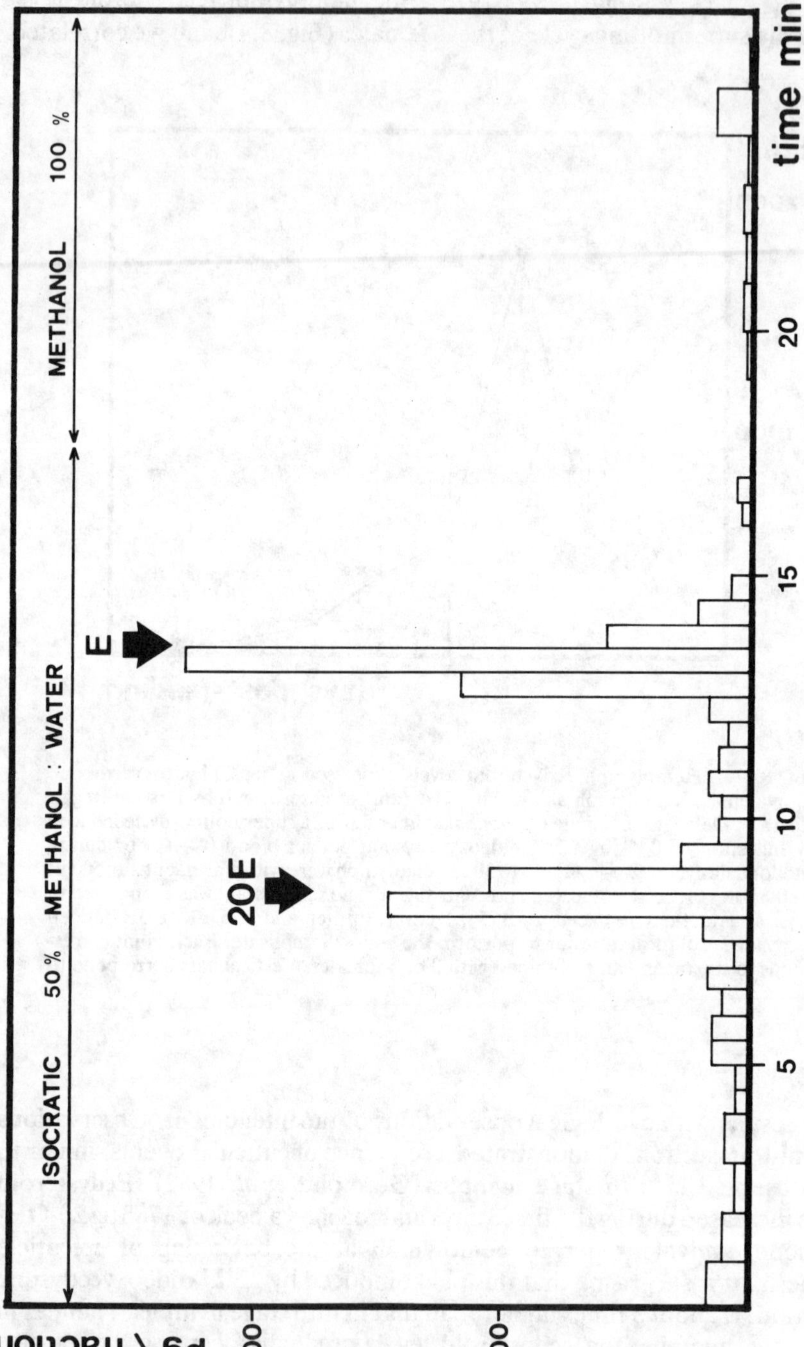

Fig. 8.5 — HPLC coupled RIA analysis of hemolymph from supermolting females *O. moubata* 3 days after the ingestion of 0.04 μg 22,25-dideoxyecdysone per ml blood. 20E and E indicate the retention times of 20-hydroxyecdysone and ecdysone respectively, under the same chromatographic conditions. Results are expressed in pg ecdysone equivalent per fractions. Merck RP-18 reverse phase column was used with a flow of 0.8 ml/min. Fractions were collected every min. or every 30 sec.

studies with ingested ^{14}C-22,25-dideoxyecdysone indicated that this latter compound is not the direct precursor of ecdysone and 20-OH-ecdysone (Connat *et al.* 1986). We can speculate that this ingested hormone may act on tissues which synthesize the molting hormones. Nevertheless, our results indicate that vitellogenesis does not occur when high ecdysteroid levels are present. We have compared, by the same method, the hemolymph ecdysteroid levels in females performing a normal vitellogenic cycle (Fig. 8.4). When they were assayed by RIA after a methanolic extraction, the levels never reached values as high as those obtained during the supermolting cycle. The levels were generally low (less than 100 pg 20-hydroxyecdysone eq./μl). However, on days 1 and 4 higher levels were obtained (301 pg/μl on day 4). Nevertheless, we have noted discrepancies in hemolymph ecdysteroid levels obtained by different extraction or separation procedures. For example, assays of the 3 different fractions (25% methanol, 60% methanol, 100% methanol) obtained after the elution of hemolymph extracts with a C18-SEP-PAK® cartridge (Waters) demonstrated that 2 peaks of immunoreactive material of about 700 pg 20-OH-ecdysone eq./μl hemolymph each were present on days 3 and 9 after feeding. This is different from methanolic extracts where only one smaller peak was obtained on day 4. The poor synchronization of the onset of vitellogeneses in female *O. moubata* could partially account for this difference. New experiments using more females and estimating the stage of vitellogenesis by observing the ovary instead of the time elapsed after the bloodmeal would give more accurate results. Another example of difficulties met with female *O. moubata* was the apparent greater quantity of immunoreactive material (up to 40-fold increase) found when assaying with the use of Silica TLC. These differences could be due to hydrolysis of unstable conjugates which cross-react poorly with our RIA or to unspecific lipid interference (Rash *et al.* 1980). No such artefacts have been observed with nymphal hemolymph or with hemolymph from female ixodids e.g. *Amblyomma hebraeum* or *Boophilus microplus* (see Diehl *et al.* Chapter 7 of the present book). Thus, we do not as yet have a precise idea of the real nature or even the quantities of ecdysteroids present in females during vitellogenesis. Similar difficulties seem to occur in *O. porcinus* where a discrepancy between data obtained by RIA and the *Musca* bioassay were noted (Solomon *et al.* 1982). However, in this closely related species, the hemolymphatic ecdysteroid titers were determined by the *Musca* bioassay during the post-engorgment period of oogenesis and oviposition and compared to the pattern observed in supermolting females where vitellogenesis is stopped (Mango 1979). 'In normally fed and mated females the ecdysterone-equivalent titers fluctuate rhythmically around mean background levels throughout the period of oogenesis'. This suggests that no peak of activity was found during this period with the bioassay. In contrast, 'hemolymph ecdysteroid titers of supermolting females rose abruptly on days 10–12 post-feeding and these increased titers were correlated with the termination of vitellogenesis and the onset of oocyte resorption' (Solomon *et al.* 1982). The authors' conclusions agree with our observations on *O. moubata* i.e. 'Peak titers of ecdysterone activity in supermolting *O. porcinus* females appear to be incompatible with the maintenance of vitellogenesis and oocyte maturation'.

To confirm the hypothesis that high ecdysteroid titers are incompatible with vitellogenesis, we investigated the effect of injecting ecdysteroids into the hemocoel of mated females. Doses up to 10 ng of ecdysone or 20-OH-ecdysone that were injected 2, 4, 6, or 8 days after feeding were totally ineffective, the preoviposition period and egg-yield were absolutely normal. Injection of 50 ng of ecdysone injected 8 days after feeding inhibited oviposition in 8 of 10 females. Later, four died imprisoned in a second synthesized cuticle. Upon dissection, all of the ovaries of these females showed signs of vitellus resorption. However, the 2 females which laid eggs displayed a slightly delayed preoviposition period but a normal egg-yield, indicating an 'all or none' effect. Doses of 100 ng injected at 2, 4, 6, or 8 days after feeding induced supermolting in all the females. However, the animals were unable to ecdyse and died imprisoned in a second cuticle. The injection of 250 ng caused most females to die before synthesizing a second cuticle, and doses of 500 ng rapidly induced mortality. We have reinvestigated the effects of injecting 100 ng of ecdysone into mated females 8 days after feeding to observe the ovocytes cytologically. At the moment of injection, the females had an ovary with fully developed ovocytes. Ovaries were dissected in physiological medium each day for 4 days and then thin-sectioned for observation. After one day, the injection triggered the appearance of numerous lysosomes in the cytoplasm of the oocytes. Two days after injection, the dissected ovaries displayed signs of vitellus resorption identical to that seen during an abortive vitellogenic cycle. The oocytes showed a disorganized cytoplasm with lysed mitochondria, and numerous lysosomes (Fig. 8.6). The egg-shell was also lysed. It was thus possible to induce an abortive vitellogenesis by injection of ecdysone in mated females progressing through a normal gonotrophic cycle. Some preliminary results with RIA indicate that, in contrast to fed mated females, much higher levels of ecdysteroids could be found in the hemolymph of virgin females going through an abortive vitellogenic cycle. Thus, secretion of high levels of ecdysteroids may be the natural way by which the females induce the resorption of vitellus. However, the reason why these females, in contrast to the females in our experiments, do not supermolt remains to be determined. Circulation of these ecdysteroids bound to a protein could be an element of the response.

8.5 CONCLUSIONS

In this last part, we want to present our working hypothesis integrating the above results (Fig. 8.7). We hypothesize that at least two categories of hormones play a role in the control of the gonotrophic cycle of *O. moubata:* Juvenile hormone-like substances and ecdysteroids. The bloodmeal may act on the brain *via*, for example, stretch receptors, to induce the production of a small burst of JH-like substances. This hormone could be responsible for the initiation of digestion and vitellogenesis. However, in the absence of

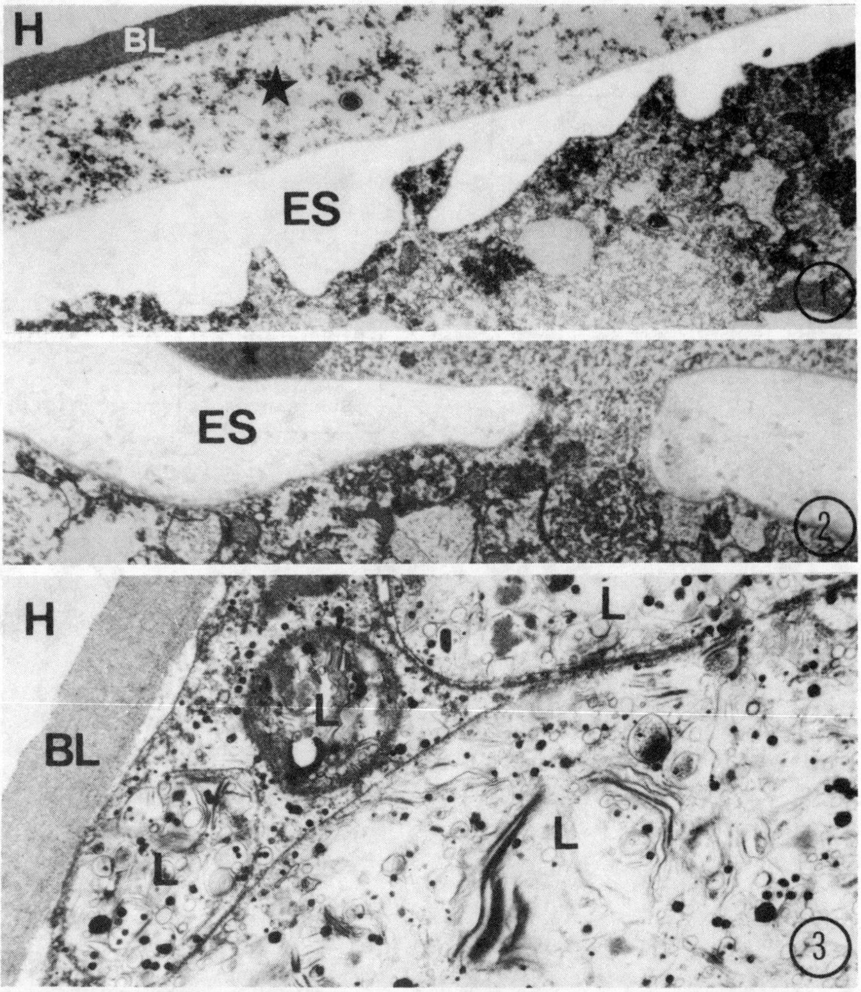

Fig. 8.6 — Cytology of the oocytes in which resorption of vitellus is induced after injection of 100 ng of 20-OH-ecdysone into mated females 8 days after feeding. Observation of resorbing oocytes of fed virgin females showed a similar pattern. ① The inner side of the eggshell (ES) is being lysed. H: Hemocoel, BL: Basal lamina, ★: artifactual space. (× 7900) ② Parts of the eggshell (ES) are completely lysed and oocyte cytoplasm is leaking out. (× 7900) ③ and ④ Advanced disorganisation of the oocyte cytoplasm, with numerous lysosomes (L), partially lysed mitochondria (m) and altered rough endoplasmic reticulum (?) (RER). H: Hemocoel, BL: Basal lamina (× 16100).

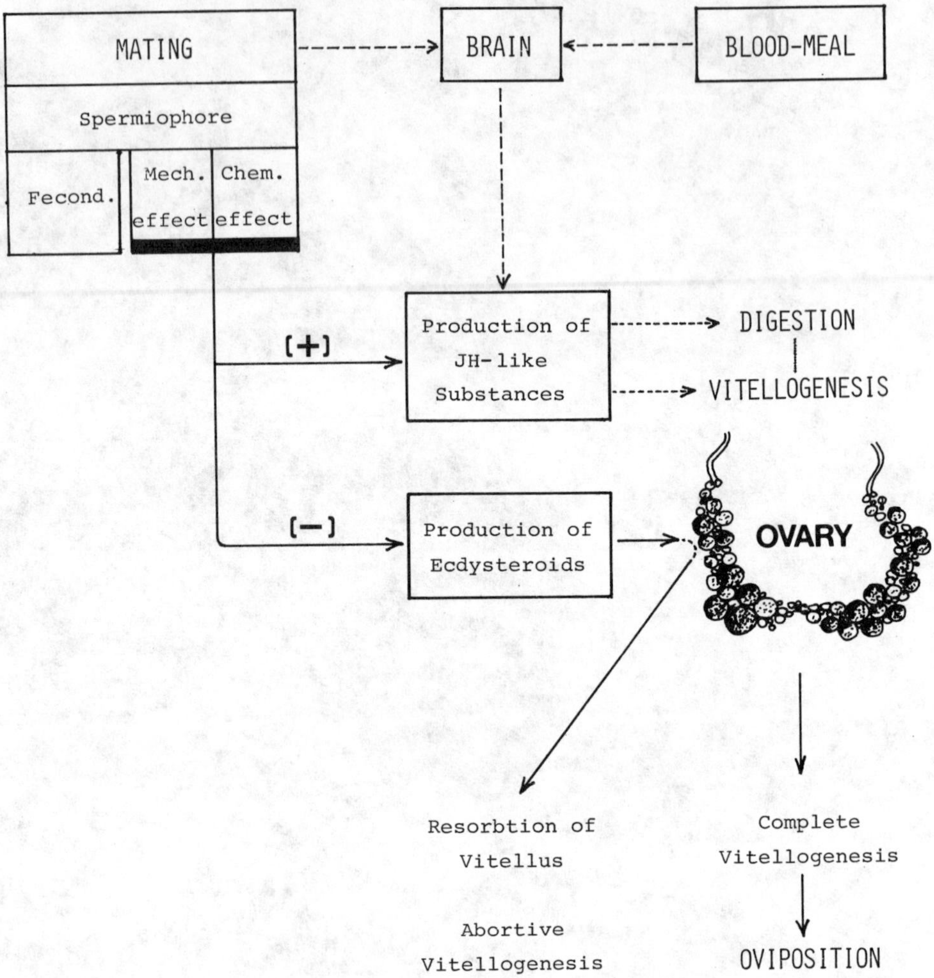

Fig. 8.7 — Scheme of our working hypothesis concerning the possible control of the gonotrophic cycle in *Ornithodoros moubata*.

mating, the secretion of this hormone ceases, and ecdysteroids, the production of which is thought to be inhibited by the presence of high levels of juvenoids, may be secreted. An antagonistic effect of JHs on ecdysteroid production has already been suggested in *O. porcinus* by experiments where topical application of 10 μg JH3 reversed the supermolting cycle induced by 20-OH-ecdysone (Obenchain & Mango 1980). The ecdysteroids could act at the level of the ovary to induce the resorption of the vitellus. Thus, the typical 'abortive vitellogenesis' of virgin females occurs. Mating probably acts as a stimulus, allowing a continuous secretion of JH-like substances. If mating occurs at approximately the same time as engorgement, it maintains the high levels of juvenoids necessary for the completion of digestion and vitellogenesis initiated by a small burst of JH-like substances triggered by the bloodmeal itself. In the case of the delayed mating of engorged virgin females, the production of JH-like substances permits the restoration of digestion and vitellogenesis.

Our results demonstrate that the mating acts through at least two different stimuli (i) a mechanical stimulation by stretching of the vagina. This stimulus could act on the brain to induce JH-like secretion; (ii) a chemical stimulation corresponding to a proteinaceous factor released by the sperm cell after devagination in the female genital ducts. This compound, pheromonal in nature, may also act directly on the brain to induce production of the vitellogenic hormone.

In fact, these two stimuli ensure that mating provides, in females, the onset of a complete gonotrophic cycle and the oviposition of viable eggs. We must note however, that since only a single mating is necessary for several gonotrophic cycles, the mating information is 'memorized' by the female.

This system would provide *O. moubata* with a very good adaptive mechanism to meet its biological and environmental requirements. Hosts are generally not continuously available, and these ticks are opportunistic feeders. Females are able to feed without mating; this property facilitates the opportunity for both sexes to feed independently. In virgin females, the bloodmeal is stored in the midgut until a male is encountered and copulation takes place. This ensures utilization of nutrients only when production of a fertilized viable egg-batch is possible.

ACKNOWLEDGEMENTS

We would like to thank Miss Ellen Dotson for correcting the English of the manuscript and Mrs Denise Schneiter for typing it. The works cited in this text were supported in part by grants of the Swiss National Foundation for Scientific Research.

REFERENCES

Aeschlimann, A., (1968) La ponte chez *Ornithodoros moubata,* Murray (Ixodoidea, Argasidae), *Rev. Suisse Zool.* **75:** 1033–1039.

Aeschlimann, A. & Grandjean, O., (1973a) Influence of natural and "artificial" mating on feeding, digestion and oviposition in ticks (Ixodoidea), *Fol. Parasitol. (Praha)* **20:** 67–74.

Aeschlimann, A. & Grandjean, O., (1973b) Observation on fecundity in *Ornithodorus moubata.* Relationships between mating and oviposition, *Acarologia* **15** (2): 206–217.

Aeschlimann, A. & Hecker, H., (1969) Vitellogenèse et formation cuticulaire chez l'oeuf d'*Ornithodoros moubata.* Etude au microscope électronique., *Acarologia* **11** (2): 180–192.

Aeschlimann, A. & Hecker, H., (1970) Présence de membranes annelées dans les ovocytes d'*Ornithodoros moubata,* Murray (Ixodoidea, Argasidae). Etude au microscope électronique. *Acarologia* **11:** 180–192.

Bergot, B. J., Ratcliff, M., & Schooley, D. A., (1981) A method for quantitative determination of the four known Juvenile Hormones in insect tissue using gas chromatography — mass spectroscopy, *J. Chromatogr.* **204:** 231–244.

Chinzei, Y., (1983) Quantitative changes of vitellogenin and vitellin in adult female ticks, *Ornithodoros moubata,* during vitellogenesis, *Mie Medical Journal* **23** (3): 117–127.

Chinzei, Y., Chino, H., & Takahashi, K., (1983) Purification and properties of vitellogenin and vitellin from a tick, *Ornithodoros moubata, J. Comp. Physiol.* **152:** 13–21.

Connat, J. L., Diehl, P. A., Dumont, N., Carminati, S., & Thompson, M. J., (1983a) Effects of exogenous ecdysteroids on the female tick *Ornithodoros moubata:* Induction of supermolting and influence on oogenesis, *Z. ang. Ent.* **96** (5): 520–530.

Connat, J. L., Ducommun, J., & Diehl, P. A., (1983b) Juvenile hormone-like substances can induce vitellogenesis in the tick *Ornithodoros moubata* (Acarina: Argasidae), *Int. J. of Invertebrate Reprod.* **6** (5, 6): 285–294.

Connat, J. L., Diehl, P. A., & Thompson, M. J., (1986) Inactivation of ingested ecdysteroids by conjugation with fatty acids in 22-OH position in the female tick *Ornithodoros moubata* (Acarina, Argasidae), *Arch. Insect Biochem.* (in press).

De Clerck, D., Eechaute, W., Leusen, I., Diederik, H., & De Loof, A., (1983) Identification of testosterone and progesterone in hemolymph of larvae of the fleshfly *Sarcophaga bullata, Gen. Comp. Endocrinol.* **52** (3): 368–678.

De Clerck, D., Diederik, H., & De Loof, A., (1984) Identification by capillary gas chromatography — mass spectrometry of eleven non-ecdysteroid steroids in the hemolymph of larvae of *Sarcophaga bullata, Insect Biochem.* **14** (2): 199–208.

De Reggi, M. L., Hirn, M. H. & Delaage, M. A., (1975) Radio-immuno-assay of ecdysone. An application to *Drosophila* larvae and pupae, *Biochem. Biophys. Res. Commun.* **66:** 1307–1315.

Diehl, P. A., (1969) Hämolymphenproteine und Vitellogenese bei *Ornithodorus moubata,* Murray (Ixodoidea, Argasidae), *Bull. Soc. ent. suisse* **42:** 117–125.

Diehl, P. A., (1970) Zur Oogenese bei *Ornithodorus moubata* Murray (Ixodoidea: Argasidae) unter besonderer Berücksichtigung der Vitellogenese, *Acta Tropica* **27** (4): 302–355.

Diehl, P. A., Aeschlimann, A., & Obenschain, F. D., (1982) Tick Reproduction: Oogenesis and Oviposition, In *Physiology of ticks* (Obenchain F. D. & Galun R. Eds) Pergamon Press, p. 277–350.

Diehl, P. A., Connat, J. L., Girault, J. P., & Lafont, R., (1985) A new class of apolar ecdysteroid conjugates: esters of 20-hydroxyecdysone with long-chain fatty acids in ticks, *Int. J. Invertebrate Reprod. Develop.*, **8**: 1–13 .

Ducommun, J., (1984) Contribution à la connaissance de la reproduction chez la tique *Ornithodoros moubata,* Murray, 1877; *sensu* Walton, 1962 (Ixodoidea: Argasidae), Thesis Univ. Neuchâtel 138 pp.

Eichenberger, G., (1970) Das Zentralnervensystem von *Ornithodorus moubata* Murray und seine postembryonale Entwicklung, *Acta Trop.* **27:** 15–53.

Eisen, A., Warburg, M. R., and Galun, R., (1973) Neurosecretory activity and oogenesis in the fowl-tick *Argas persicus* (Oken), *Gen. Comp. Endocrinol.* **21:** 331–340.

Engelmann, F., (1984) Regulation of vitellogenesis in insects: The pleiotropic role of juvenile hormones. In *Biosynthesis, metabolism and mode of action of invertebrate hormones* (Hoffmann J. & Porchet M. eds.) Springer-Verlag Berlin Heidelberg p. 444–453.

Gabbay, S., & Warburg, M. R., (1976) Neurosecretory activity as related to feeding, mating and oogenesis in the female tick, *Ornithodoros tholozani., J. Insect Physiol.* **22:** 1291–1301.

Germond, J. E. & Aeschlimann, A., (1977) Influence of copulation on vitellogenesis and egg-laying in *Ornithodoros moubata* (Ixodoidea: Argasidae), In *Advances in invertebrate reproduction* (Adiyodi K. G. & Adiyodi R. G. Eds.) p. 308–318.

Germond, J. E., Diehl, P. A., & Morici, M., (1982) Correlation between integument structure and ecdysteroid titers in fifth-stage nymphs of the tick, *Ornithodoros moubata* (Murray, 1977; *sensu* Walton, 1962), *Gen. Comp. Endocrinol.* **46:** 255–266.

Graf, J. F., (1974) Ecologie et éthologie d'*Ixodes ricinus* L. en Suisse (Ixodoidea, Ixodidae). III. Copulation, nutrition et ponte, *Acarologia* **16** (4): 636–642.

Grandjean, O., (1983) Blood digestion in *Ornithodorus moubata* Murray *sensu stricto* Walton females (Ixodoidea: Argasidae). Modifications of midgut cells related to the digestive cycle and to the triggering action of mating, *Ann. Parasitol. Hum. Comp.* **58:** 493–514.

Hagedorn, H. H., (1983) The Role of Ecdysteroids in the Adult Insect, In *Endocrinology of insects* (Downer R. G. H. & Laufer H. Eds.), Alan R. Liss, Inc., New York p. 271–304.

Hoogstraal, H., (1956) African Ixodoidea. *I. Ticks of the Sudan.*, Dep. Navy, Bur. Med. Surg., Washington, 1101 pp.

Khalil, G. M., (1984) Fecundity-reducing pheromone in *Argas (Persicargas) arboreus* (Ixodoidea: Argasidae), *Parasitology* **88:** 395–402.

Mango, C. K. A., (1979) The effects of beta-ecdysone and ponasterone A on the soft tick *Ornithodoros porcinus porcinus.* Walton, 1962., PhD thesis Univ. Nairobi.

Mauchamp, B., Zander, M., & Wolff, R., (1984) The qualitative and quantitative determination of juvenile hormones by mass spectrometry. In *Biosynthesis, metabolism and mode of action of invertebrate hormones,* (Hoffmann J. A. & Porchet M. Eds.) Springer-Verlag p. 363–372.

Obenchain, F. D. & Mango, C. K. A., (1980) Effects of exogenous ecdysteroids and juvenile hormones on female reproductive development in *Ornithodoros p. porcinus, Am. Zool.* **20** (4): 936.

Oliver, J. H. Jr., (1974) Symposium on reproduction of arthropods of medical and veterinary importance. IV. Reproduction in ticks (Ixodoidea)., *J. Med. Ent.* **11:** 26–34.

Oliver, J. H., Pound, J. M., & Andrews, R. H., (1984) Induction of egg maturation and oviposition in the tick *Ornithodoros parkeri* (Acari: Argasidae), *J. Parasit.* **70** (3): 337–342.

Pound, J. M. & Oliver, J. H., (1979) Juvenile Hormone: evidence of its role in the reproduction of ticks, *Science* **206:** 355–357.

Raabe, M., (1982) Contrôle neuroendocrine de la reproduction des insectes, *J. Physiol.* **78:** 553–557.

Raabe, M., (1984) Insect neurohormones and female reproduction. In *Advances in invertebrate reproduction* (Engels W. *et al.* Eds.) Elsevier p. 163–176.

Rash, J. M., Jerkunica, I., & Sgoutas, D. S., (1980) Lipid interference in steroid radioimmunoassay, *Clin. Chem.* **26** (1): 84–88.

Rothschild, M. & Ford, B., (1973) Factors influencing the breeding of the rabbit flea *(Spilopsyllus cuniculi):* A spring-time accelerator and a kairomone in nestling rabbit urine, *J. Zool. Lond.* **700:** 87–137.

Sahli, R., Germond, J. E., & Diehl, P. A., (1985) Spermateleosis and secretory activity of the sperm cell of the tick *Ornithodoros moubata. Exp. Parasitol.* **60:** 383–395

Solomon, K. R., Mango, C. K. A., & Obenchain, F. D., (1982) Endocrine Mechanisms in Ticks: Effects of Insect Hormones and their Mimics on Development and Reproduction. In *Physiology of ticks* (Obenchain F. D. & Galun R. Eds.) Pergamon Press p. 399–438.

9

Observations on the mating behaviour of ticks

B. Feldman-Muhsam

9.1 INTRODUCTION

In arthropods there is a wide variety of ways and means to achieve the meeting of the gametes. In some insects, sperm is transferred from the male to the female without any meeting of the couple. In others, the couple copulate and the sperm is ejaculated by the male directly into the female's bursa copulatrix. Many intermediate possibilities between these extremes are also encountered. Sperm transfer via a spermatophore is one of the possible means. Also, transfer of the spermatophore from the male to the female may be performed with or without meeting of the sexes. Whenever a meeting of the sexes is required, olfactory, auditory, visual, tactile, and other senses are employed to locate, recognize, and identify the partner of the right species and sex.

In ticks, the sexes meet, sperm is transferred from the male to the female by means of a spermatophore, and an act of copulation takes place. Argasid and some ixodid ticks release assembly pheromones, which facilitate meetings of sexual partners (Leahy *et al.* 1975, Sonenshine *et al.* 1982). In addition, the female releases various compounds which act as sex pheromones that regulate mating behaviour of receptive males. However, these compounds have been found only in a number of ixodid ticks (Wood *et al.* 1975) and in one species of *Ornithodoros* (Schlein & Gunders 1981). Furthermore some of these sex pheromones are not species specific. Another sex pheromone, which appears to be associated with the female genital area of two species of *Hyalomma,* guides the male to the female gonopore (Khalil *et al.* 1983). A pheromone was shown, in two species of *Dermacentor,* to be percepted by the male while probing the female genital aperture (Sonenshine 1984). In the following I will summarize the main features of copulation and sperm transfer as observed in the laboratory, and supplement this with some observations on the conditions under which males and females are ready to copulate.

It should be remembered that in nature argasid ticks and some species of *Ixodes* (Prostriata) copulate off the host whereas the Metastriata copulate on the host. Therefore, the copulatory behaviour of argasids in the laboratory follows the same course as in nature and is easy to observe. Copulation of Metastriata ticks takes place on the host after the male and the female have been attached to the host as a couple, venter to venter, and fed together for some time. To observe copulation of Metastriata ticks *in vitro*, males and females must be fed for some time on separate hosts. If they are removed from their hosts, and placed together in a Petri dish when they are ready, details of copulation can be observed under the dissecting microscope. The copulatory behaviour of ixodids follows the same pattern as argasids, with some differences.

There are still many gaps in our knowledge of the process of copulation and in our understanding of the factors involved. This chapter is not intended to be a complete survey of the present knowledge in the field. Emphasis will be given to behavioural patterns and comparisons between the three main groups of ticks: the family Argasidae, and the two subfamilies of the Ixodidae: Ixodinae (=Prostriata) and Amblyomminae (=Metastriata).

9.2 COPULATION AND SPERM TRANSFER

9.2.1 Precopulatory behaviour

A sexually mature male tick is attracted by a sexually mature female when he perceives a pheromone, released by the female, with receptors located on the 4th article of his palps (Feldman-Muhsam & Borut 1971, Leahy *et al.* 1975). He then climbs first on the back of the female, crawls around the posterior end of her body to the ventral side until his capitulum reaches her genital aperture and their venters face each other. At this stage the male folds his legs with the female, and the couple remain immobile.

There are some differences between the groups of ticks with regard to the folding of their legs. In *Ornithodoros* the male usually folds his first and second pairs of legs with the corresponding pairs of legs of the females. With each leg of the first two pairs of legs the male grasps the female's leg from its anterior aspect, bending his leg backwards over the dorsal aspect of the female's leg. The third and fourth pairs of legs of the male do not grasp the legs of the female, but he rests them on the female's venter, the third and fouth pairs behind the female's third and fourth pairs, respectively (Fig. 9.1). In *Hyalomma,* the male usually folds the first 3 pairs of his legs with the corresponding pairs of the female, each pair of the male's legs behind the corresponding pair of the female, and the fourth pair remaining unfolded (Fig. 9.2). However, if a male happens to be in a slightly more backward position, he may fold his first pair of legs with the second pair of the female's legs etc. Sometimes the legs on the left side follow one of these patterns and

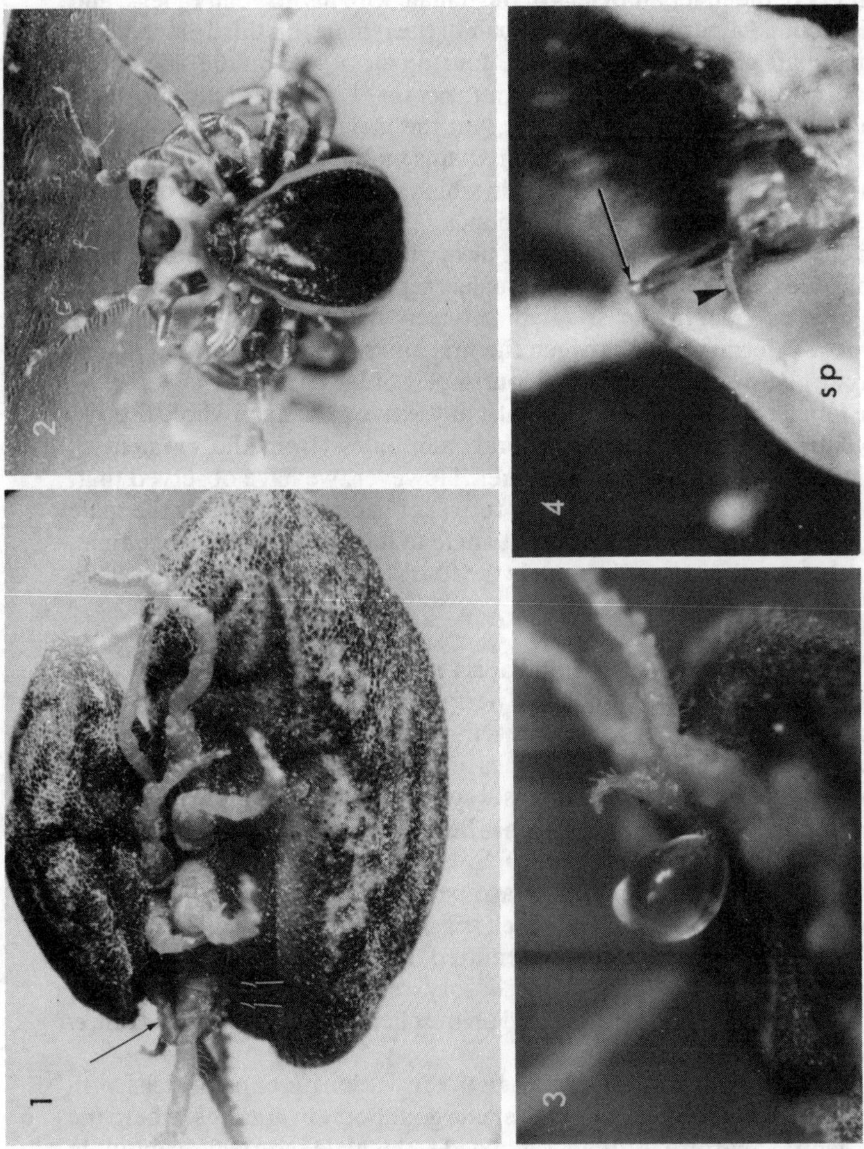

Fig. 9.1 — *Ornithodoros savignyi in copula*, in inverted position. Note folding of legs. Leg 1 of male (↑); tarsus of leg 1 of male (↑↑) × 8. Fig. 9.2 — *Hyalomma excavatum in copula*. Note folding of legs. × 7. Fig. 9.3 — First step in spermatophore formation. Note the empty transparent ectospermatophore on the male's gonopore. (*Ornithodoros savignyi*). × 11. Fig. 9.4 — The male of *Ornithodoros savignyi* holding the spermatophore (sp). The cheliceral digits (↑) are on the tip of the spermatophore, and the palps (▲) on the neck. × 32.

the legs on the right side the other. Each of the male's legs always grasps the female's leg from its posterior aspect and bends forward over the dorsal aspect of the female's leg. In *Ixodes uriae* White, 1852 the male holds the trochanters of the first pair of legs of the female with his first pair of legs, and rests his second pair around the third pair of the female, his third pair around the fourth pair of the female, and the fourth pair of legs on the side of the female (Eveleigh & Threlfall 1975). In *I. ricinus* (Linnaeus) Latreille, 1804 the male and female do not fold legs, but the male grasps the body of the female with his legs while the female remains mobile (Graf 1978). *I. ricinus* is apparently the only species of ticks in which the female has been observed not to be immobilized during copulation.

Differences between the groups of ticks with regard to folding of legs are probably due to the position of the female's genital opening. In argasids, the female genital opening is situated between the first pair of legs, in *Hyalomma* (Metrastriata) between the first and second pair, and in *Ixodes* (Prostriata) between the third and fourth pair of legs.

The female apparently does not take any active part in the whole process of copulation and remains, in general, motionless from the moment of folding of legs until the male leaves her. However, we have observed that, during copulation, the ixodid female turns her capitulum towards the male so that her basis capituli forms a right angle to her body. At the same time, she protrudes and retracts alternately the fourth segment of her palps, while the setae quiver.

9.2.2 Probing of the female genital aperture

At this stage, after pairing, the male performs a very peculiar and remarkable act which consists of probing the female gonopore with its mouthparts. The male of one species may sometimes probe the gonopore of a female of a closely related species, but in such cases copulation is not completed, i.e. no spermatophore will be formed. This has been demonstrated in *Dermacentor variabilis* Say, 1821, and *D. andersoni* Stiles, 1908 (Sonenshine *et al.* 1982). The male probes the female genital aperture apparently to perceive a specific aphrodisiac pheromone, which triggers the formation of the spermatophore, if the pheromone is emitted by a conspecific partner. In *Dermacentor variabilis,* the chemosensory receptors of the aphrodisiac pheromone are assumed to be located on the cheliceral digits (Sonenshine *et al.* 1984).

There are significant differences between various groups of ticks with regard to the manner of probing the female gonopore. In argasids, where the female genital aperture is about 1 mm wide, the male inserts his capitulum (mouthparts and palps) into the female genital aperture. He withdraws and re-inserts it several times before proceeding to the next step in copulation.

In the Prostriata, the male inserts only the mouthparts (chelicerae and hypostome) while the palps remain outside the gonopore and splayed outwards, closely adhering to the female's ventral integument (Arthur 1962, Moorhouse 1966). In this group, the female genital aperture is narrower (0.1–0.2 mm) than in the Argasidae, and more soft and expandable. We have observed that in *Ixodes* the female genital aperture is considerably enlarged after copulation.

In the Metastriata, the female genital aperture is as narrow as in the Prostriata but is flanked by strongly chitinized flaps which prevent its widening, so that after copulation the size of the aperture remains unchanged. In *Hyalomma rhipicephaloides* Neumann, 1911, *Hyalomma excavatum* Koch, 1844, *Rhipicephalus bursa* Canestrini & Fanzago, 1877, and *Amblyomma cayennense* (Fabricius, 1787), the tips of all of the mouthparts of the male touch the gonopore and seem not to be inserted into it, while the palps reach the ventral integument on both sides near the gonopore. However, when a male is slightly and gently pulled away from the female in the course of copulation, the chelicerae of the male are extended from their shafts and seem to be inserted into the female gonopore (Feldman-Muhsam & Borut 1971).

During *in vitro* observations of copulating ixodids, the males may remain for one or two hours motionless in the copulatory position, i.e. with their mouthparts on the female gonopore, before forming the spermatophore. Sometimes the male does not produce a spermatophore even after three hours of observation. In *I. holocyclus* Neumann, 1899 copulation lasts one to three hours (Moorhouse 1966) and according to Graf (1978) copulation in *I. ricinus* may last up to seven days.

9.2.3 Formation of spermatophore

In all ticks, the spermatophore is formed external of the male gonopore after the male has probed the female genital aperture. The male first extrudes a droplet containing mucopolysaccharides, and then a droplet containing protein that is pushed into the first droplet. Thus a two-component transparent drop is formed outside the male genital aperture that remains attached to the tick (Fig. 9.3). Next a mass of sperm in PAS positive spermal fluid is ejaculated into the center of the drop which expands like a balloon and turns opaque. The size of the spermatophore depends on the amount of sperm. Thus, in cases of juvenile sterility (see below) where no sperm is injected, the spermatophore is small. Less than a second later sperm symbionts are injected into the drop (Feldman-Muhsam 1974). Last to be ejected from the male is a special structure, the endospermatophore, (Feldman-Muhsam 1967a), which closes the 'balloon'. All the components are rapidly ejected one after another with precise and perfect timing (Feldman-Muhsam & Borut 1978). The whole process of spermatophore formation lasts ca. 30 seconds.

9.2.4 Structure of spermatophore

The spermatophore of ticks is a pear-shaped body composed of a three-layered external shell, the ectospermatophore, which encloses all the other components, i.e. spermal fluid, sperm, large protein granules, symbionts, and the endospermatophore. The sperm mass fills most of the ectospermatophore. The large granules are embedded around the entire inner layer of the external shell, and symbionts are located around the endospermatophore

(Feldman-Muhsam 1969). The endospermatophore is connected to the rim of the ectospermatophore. This general structure is common to both the Ixodidae and the Argasidae; however, significant differences between the two families exist with regard to the morphology of the endospermatophore. Whereas in ixodids the endospermatophore is a complex structure consisting of a chain of connected organelles (Feldman-Muhsam & Borut 1983), in argasids it is bilobed with each lobe containing a proteinaceous sac. These sacs have been shown to contain a high concentration of spermine phosphate. This is the only reported occurrence of spermine in the Arthropoda (Feldman-Muhsam *et al.* 1980).

9.2.5 Transfer of spermatophore

When the formation of the spermatophore is complete it is still located on the male genital aperture. Since the male has no copulatory organ, the spermatophore is transferred to the female genital aperture by means of a very unusual procedure. In all ticks, the male's mouthparts perform an essential role in this process. However there are differences in the method of transfer between ixodids and argasids. Argasids use their mouthparts to transfer the spermatophore the whole distance from male to female, while ixodids use them only when the spermatophore is near the female genital aperture. Recall that in the Ixodidae the capitulum is situated at the anterior edge of the body of the tick, whereas in the Argasidae it is on the ventral side. Therefore the argasid male can reach its gonopore with its mouthparts and catch the spermatophore while still attached to the gonopore.

Thirty seconds after the spermatophore is completed, the argasid male begins secreting saliva and then grasps the neck of the spermatophore with its palps and applies the cheliceral digits to the tip of the spermatophore (Fig. 9.4). Thus, the male completes the transfer by implanting the tip of the spermatophore into the female genital aperture. The male continues secreting saliva which spreads over the spermatophore, keeping it moist and preventing it from adhering to either the male's or female's cuticle (Feldman-Muhsam *et al.* 1970).

When the ixodid male probes the female genital aperture he bends his capitulum perpendicular to his body, but is unable to reach his own gonopore to catch the spermatophore. Indeed, the ixodid male has never been observed to use his mouthparts to catch the spermatophore from his gonopore. He does not remove his mouthparts from the moment he begins probing the female gonopore until the spermatophore is near the female genital aperture. It can be observed under the dissecting microscope that just before the male begins to form the spermatophore, he bends his body by strongly retracting the area around the gonopore and protruding the posterior part of his body towards the female. In *Hyalomma* and *Rhipicephalus* he presses his anal plates against the body of the female. Thus a space is formed between the venters of the male and the female into which the spermatophore is extruded step by step. After the spermatophore is completed, the male appears to push the spermatophore towards the female gonopore by releasing the posterior end of the retracted area of his venter and applying it to the venter of the female. When the tip of the spermatophore is near the female gonopore, the male withdraws his chelicerae from the female genital aperture as he simultaneously stretches the fourth article of his palps out. He immediately catches the tip of the spermatophore with the digits of his chelicerae for implanting it into the female genital aperture while supporting the spermatophore with the fourth article of its palps (Feldman-Muhsam & Borut 1971, Fig. 3). At this moment the spermatophore is attached to the chelicerae by strands of filamentous material probably derived from dried saliva.

9.2.6 Sperm transfer

The spermatophore itself never penetrates into the female genital tract. However, immediately after the spermatophore is placed on the female, the endospermatophore evaginates into the female genital tract and the whole mass of sperm is pushed into the everted endospermatophore. This process occurs without any intervention by the male. Indeed, if a fully formed spermatophore is removed from the couple, while still on the male genital aperture, the process of evagination occurs as if it were on the female genital aperture if the spermatophore is placed into a liquid medium, such as the male's saliva or an appropriate artificial solution. If the female is removed from the male, while the spermatophore is on the male's mouthparts, the endospermatophore evaginates into the air, (Fig. 9.5).

In the Argasidae, this process starts with the simultaneous elongation of the neck of the ectospermatophore and the straightening of the neck of the endospermatophore (Fig. 9.8). The proteinaceous sacs are the first to leave the ectospermatophore. They are pushed gradually into the tip of the neck of the endospermatophore by pressure created in the ectospermatophore. The sheath of the endospermatophore is then pushed out by the same pressure in the ectospermatophore while turned inside out (Fig. 9.7). Finally the whole mass of sperm and the adlerocysts flow into the everted endospermatophore as one or two capsules (Fig. 9.9). The sperm is stored in these capsules inside the female genital tract as long as is necessary. The empty shell of the ectospermatophore collapses and later drops off along with the long neck of the endospermatophore. The whole process of transfer of the content of the ectospermatophore into the female genital tract is produced by pressure from CO_2 created in the ectospermatophore (Fig. 9.9) (Feldman-Muhsam *et al.* 1973).

In most species of the genus *Ornithodoros* e.g. *O. tholozani* (Laboulbene & Megnin), 1882, *O. moubata* (Murray), 1877, *O. savignyi* (Audouin) 1827, and *O. parkeri* each of the lobes of the endospermatophore forms a capsule (Fig. 9.11). The two capsules remain connected to the elongated neck of the spermatophore, which eventually breaks near the bifurcation leading to the capsules. Although the capsules are still connected, they generally proceed to one lobe each of the spermatophorotheka (=uterus).

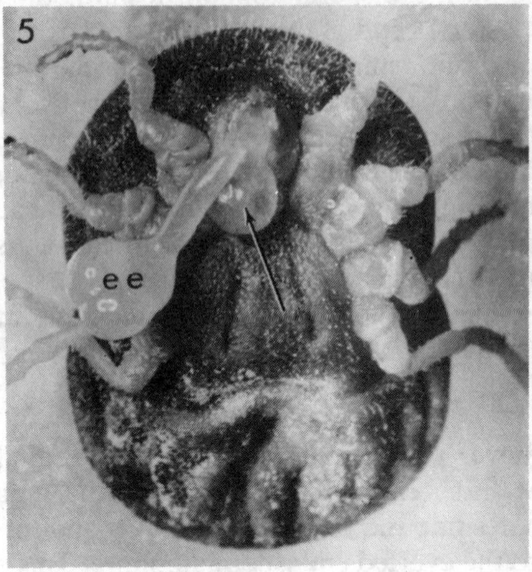

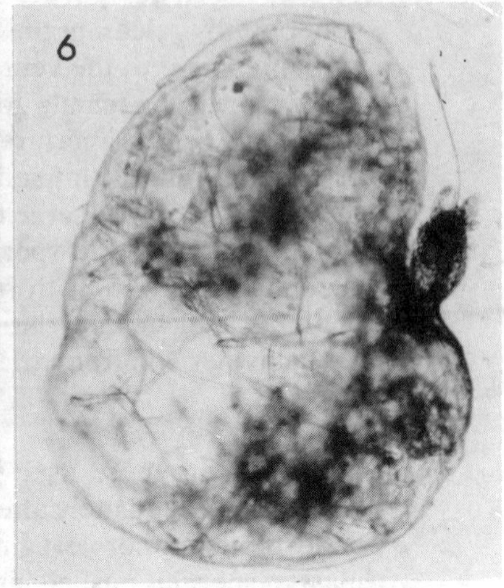

Fig. 9.5 — Spermatophore, completely evaginated on the male's mouthparts. The female had been removed before the male could deposit the spermatophore on the female genital aperture (*Ornithodoros savignyi*). Ectospermatophore (↑); everted endospermatophore (ee). × 9. Fig. 9.6 — A sperm capsule of *Ornithodoros moubata* dissected out of the spermatophorotheka. Note the few spermiophores still in the capsule, and one in course of leaving through the tube. × 255.

The length of time during which they remain connected to one another depends on subsequent copulations and the pressure created by the arrival of additional capsules. In an experiment to test the viability of sperm, starved females of *O. savignyi* were dissected up to one year after copulation; the 2 capsules were often still connected and the spermiophores were still moving in the capsules. When the connecting tube breaks, a short tubelike appendix remains attached to each capsule through which the sperm eventually leave the capsule (Fig. 9.6).

In other species of *Ornithodoros,* the two lobes of the endospermatophore do not form two separate capsules, and the everted endospermatophore assumes a variety of shapes (Feldman-Muhsam 1969). In *Argas persicus* (Oken), 1818, only one capsule is formed, whereas in *Argas reflexus* (Fabricius) Latreille, 1796 the everted endospermatophore becomes two capsules, as in most species of *Ornithodoros*.

In the Ixodidae, the transfer of ectospermatophore contents into the female genital tract begins as in the Argasidae, with the elongation of the neck of the ectospermatophore. Then the organelles of the endospermatophore burst from the ectospermatophore into the female genital tract forming a chain which remains attached to the edge of the ectospermatophore at both ends. Next, the sheath of the endospermatophore evaginates, forming a small bleb which when filled with sperm, becomes the sperm capsule (Feldman-Muhsam and Borut 1983, 1984). In contrast to the capsule in argasids which has only one tube-like appendix, the ixodid capsule has

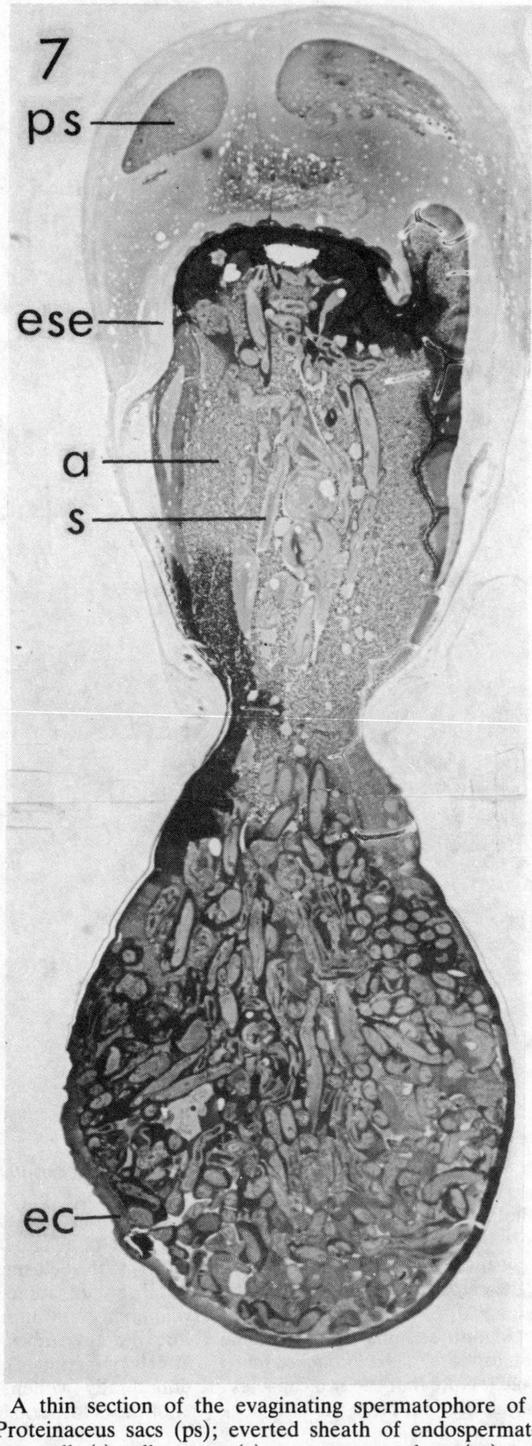

Fig. 9.7 — A thin section of the evaginating spermatophore of *Ornithodoros tholozani*. Proteinaceus sacs (ps); everted sheath of endospermatophore (ese); sperm cells (s); adlerocysts (a); ectospermatophore (ec). × 580.

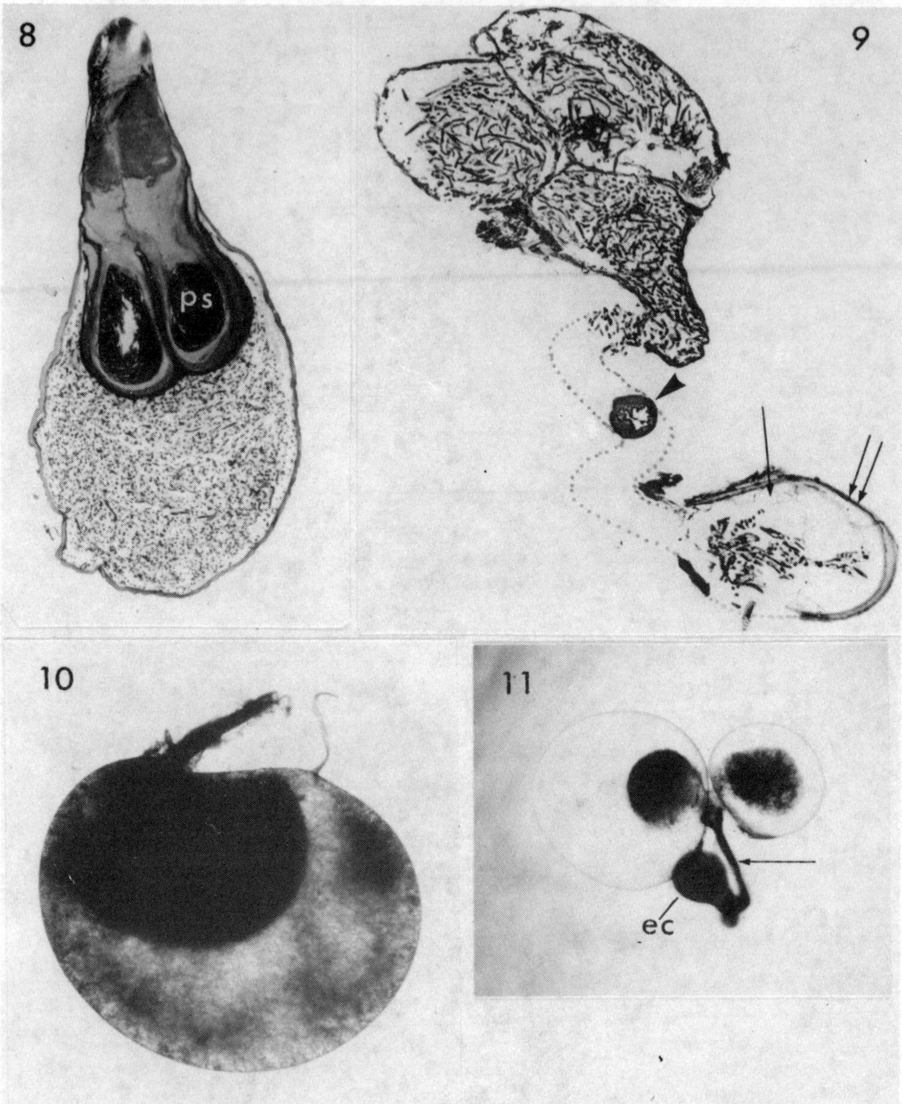

Fig. 9.8 — Histological section of a spermatophore of *Ornithodoros savignyi* at the beginning of evagination *in vitro*. The proteinaceus sacs (ps) are still at the bottom of the bilobed endospermatophore. × 32. Fig. 9.9 — Histological section of a spermatophore of *Ornithodoros savignyi* at the last stage of evagination, the two capsules not yet separated. Note the bubbles (↑) of CO_2 in the ectospermatophore, the neck (▲), the ectospermatophore (↑ ↑) which still contains some sperm at its center. × 52. Fig. 9.10 — A sperm capsule of *Hyalomma excavatum* dissected out of the receptaculum seminis of the female. Note the two tubes. × 55. Fig. 9.11 — Spermatophore of *Ornithodoros savignyi* after evagination of endospermatophore in 5% NaCl. Note that the two capsules are unnaturally swollen, after being left 2 days in the solution. Note the long neck (↑) of the ectospermatophore (ec). × 7.

two such appendices (Fig. 9.10). In both families, the same tube serves as entrance and exit for sperm.

It has been claimed that in *Ixodes* the sperm leaves the capsule only after it ruptures (Arthur 1962). Since we observed, in various species of argasids and ixodids, that the sperm leaves the capsule through the same tube through which it entered without rupturing the capsule, we believe that the same occurs in *Ixodes*.

These intricate processes ensure that there is neither loss of sperm nor possible contamination. Indeed, in all ticks the whole mass of sperm remains in a closed system, from the moment of being ejaculated by the male into the ectospermatophore until it leaves the endospermatophore inside the female genital tract where it proceeds to the oviducts.

9.3 READINESS TO COPULATE

9.3.1 First copulation of males

In the Argasidae, very young unfed males are attracted to and engage in copulatory activities with females, but often do not achieve insemination. Sometimes no spermatophore is formed, in other cases an apparently normal spermatophore is formed and properly deposited on the female genital aperture but the endospermatophore does not evaginate into the female genital tract (=unsuccessful mating) (Feldman-Muhsam 1979). In some cases the endospermatophore evaginates normally into the female genital tract and seems to contain all components except sperm (=juvenile sterility) (Feldman-Muhsam & Havivi 1967).

One or two day old, unfed *O. tholozani,* males were observed to be attracted to females, engaged in precopulatory activity, and remained in the copulatory position, i.e. with the mouthparts on the female gonopore, for many hours without producing a spermatophore. All spermatophores produced by these unfed males during the first two weeks after ecdysis were sterile, although the males copulated up to five times. Fed males of *O. tholozani* often exhibit juvenile sterility at their first copulation, sometimes as long as 14 days after their first bloodmeal.

In *O. savignyi,* where both males and females are sexually very active and vigorous, sterile spermatophores are produced by starved males, but only on the first two days after ecdysis. If fed on the day of ecdysis, the spermatophore contains sperm as early as the following day. Juvenile sterility was also observed in *O. erraticus* (Lucas), 1849, *O. coriaceus* Koch, 1844, and *O. lahorensis* Neumann, 1908, but not in *O. moubata.*

In studies of reproduction in *O. tholozani,* 70% of first copulations of starved males were unsuccessful matings. This percentage decreased to about 18% by the seventh copulation, but increased again thereafter. Unsuccessful matings may occur with fed males of all ages but they are much rarer than if the male is unfed (Feldman-Muhsam 1979).

The youngest age at which the argasid male is able to produce a spermatophore varies widely from species to species. In *O. savignyi,* some unfed males were observed to copulate a few hours after ecdysis, while in *O. tholozani* copulation was possible on the following day. Other species within the genus *Ornithodoros,* e.g. *O. coriaceus* start copulating, whether fed or unfed, much later.

It seems that in all *Ixodes* species the unfed male can fertilize the female (Arthur 1962, Balashov 1968).

Until recently is was believed that all Metastriata ticks, males as well as females, were required to feed for some time before copulation. In *Amblyomma hebraeum* Koch, 1844 the time required to feed before becoming sexually mature varies from 4–6 days (Norval 1974). However, unfed males and females of *Amblyomma triguttatum triguttatum* Koch, 1844 are able to copulate, with viable sperm transferred from the male to the female (Guglielmone & Moorhouse 1983).

9.3.2 First copulation of females

In general, an unfed argasid female attracts a male, and is ready to copulate. In nine species of *Ornithodoros: O. moubata, O. coriaceus, O. savignyi, O. canestrinii* (Birula), 1895, *O. arenicolous* Hoogstraal, 1953, *O. delanoei* Roubaud & ColasBelcour, 1931, *O. tholozani, O. tartakovskyi* Olenev, 1931, and *O. parkeri* Cooley, 1936, females were observed to copulate before feeding (Feldman-Muhsam 1973). It should be stressed however, that not all unfed females attract males. This applies in particular to relatively small females, which probably do not release the required pheromone.

In *O. moubata,* the unfed female may attract the male on the day of her ecdysis. In a series of observations, 29 unfed females of *O. moubata,* less than 24 h old, were put together with 5 males each. All these females attracted males and copulated with them within 24 hours, almost all of them several times and in most cases with different partners.

In *O. tholozani,* males are attracted to unfed females, but only after the females are 3–9 days old. In a series of experiments with *O. tholozani,* 8 out of 27 males refused to copulate with unfed females, although they were kept together as long as 9–43 days and copulated immediately when the unfed female was replaced by a fed female. The other 19 males copulated 44 times with unfed females (Feldman-Muhsam 1967b). These observations seem to be inconsistent with Schlein and Gunders' (1981) claim, that the female 'sex pheromone appears by the fourth day after a bloodmeal'. This inconsistency may be due to the fact that Schlein and Gunders studied *O. erraticus,* which is not among the 9 above-mentioned species of the genus *Ornithodoros.* It is also possible that Schlein and Gunders did not find pheromone in the coxal fluid, before and after the first days of feeding, because it is released otherwise.

The readiness of unfed females of *Ixodes ricinus* and *I. persulcatus* Schulze, 1930 to copulate was observed by Balashov (1968) in nature as well as in the laboratory.

The unique observation of an unfed *Amblyomma triguttatum triguttatum* being able to copulate relates, as mentioned above, to females as well as to males (Guglielmone & Moorhouse 1983).

9.3.3 Repeated copulations of males

Argasid males, whether fed or unfed, may copulate repeatedly. Single starved males of *O. savignyi* kept permanently with a single female, were observed to copulate an average of almost 20 times (maximum 33) during their life span which averaged 78 days (Table 9.1). Although the mean

Table 9.1 — Copulatory behaviour of starved and fed males of *Ornithodoros savignyi*.

	Length of life (days)	Active period (days)	Number of copulations	Interval between copulations (days)	Post-copulatory period (days)
Starved males	78.1 ± 11.4	50.9 ± 7.6	19.6 ± 2.7	2.58 ± 0.3	27.2 ± 7.8
Fed males	88.8 ± 4.8	79.5 ± 4.6	32.3 ± 3.5	2.66 ± 0.4	9.0 ± 2.6

Mean ± standard error.

interval between copulations was 2.5 days, some males copulated daily for several consecutive days and a few twice during the same day. In *O. tholozani,* which is a less sexually active species, a starved male may copulate more than once during one day when isolated from females for a long time.

Fed *O. savignyi* males do not copulate at shorter intervals than unfed males but produce many more sperm cells per copulation. Fed males live slightly longer than unfed males, but their sexually active period is much longer and the post-copulatory period is much shorter; thus they copulate more often than unfed males (Table 9.1).

Balashov (1968) states that males of *O. tholozani* and *O. lahorensis* can withstand long starvation and are able to fertilize females four years after the last meal. We observed that a once-fed male *O. tholozani* lived 275 days and copulated 35 times. His last seven copulations were unsuccessful matings, and in the last five the spermatophores did not contain sperm. Thus, in addition to juvenile sterility, senile sterility exists. Since unsuccessful matings, juvenile sterility, and senile sterility occur with *Ornithodoros* in the laboratory, it appears, that at certain stages of the tick's life, there is sometimes a lack of coordination between factors that regulate the various phases of copulation. For example, a male may perform what appears to be a successful copulation but fails to ejaculate sperm cells into the spermatophore or component(s) responsible for forming CO_2 which then create(s) the pressure needed for the evagination process.

The ixodid male can also copulate more than once, but as far as we know no observations on repeated copulations of ixodid males have been published. However, as is well known, the ixodid female needs only one insemination, completes engorgement after insemination, and descends from the host. Thus, the male has no further opportunity for copulating with the same female. In the laboratory, on the other hand, males have often been observed to remain on the host for weeks or months after all engorged females descended. The male, no doubt, remains there, waiting for new mates to arrive.

9.3.4 Repeated copulations of females

The pattern of multiple copulations of females differs considerably between argasids and ixodids. In general, the female feeds, then copulates, and finally oviposits. The ixodid female usually performs this cycle once in her life while the argasid female performs it several times, and copulation may precede feeding. As is well known, the ixodid female, with the exception of members of the genus *Ixodes,* feeds for a time before copulation takes place and then generally copulates only once. However we found 3 sperm capsules in the receptaculum seminis of a female *Rhipicephalus sanguineus* Latreille, 1806 collected from a dog. In 48 *Hyalomma* females kept with several males, 42% contained one sperm capsule; all the others had more than one, and one female had 5, indicating that she had copulated 5 times.

In most species of argasids, unfed females copulate many times; e.g. a starved female of *O. moubata* which lived for 43 days, and was kept constantly with 5 engorged males, was observed to copulate 33 times. After feeding and ovipositing, the pattern of behaviour of the argasid females resembles that of the unfed females. Generally, female *O. tholozani* require a new meal and a new copulation, before an additional oviposition is possible. However, about 20% of the females oviposit a second batch of eggs after an additional meal without copulation taking place. Conversely, *O. savignyi* females may oviposit up to seven batches of eggs after a single copulation with an additional meal before each oviposition.

We have observed repeatedly that females of *O. tholoazini* copulate during the course of oviposition, and have found mature eggs and capsules full of sperm at the same time in the spermatophorotheka. Thus argasid females are attractive to males before and after engorging, before and after ovipositing, and even during oviposition.

9.4 CONCLUSIONS

Although the present description of the copulatory behaviour of ticks is incomplete, it illustrates both the uniformity and the diversity which characterize the processes involved. Indeed, all the major facets of copulation and sperm transfer are common to all ticks: the venter to venter position of male and female, the probing of the female genital aperture by the mouthparts of the male, the building of the spermatophore outside the male genital aperture, the use of the mouthparts of the male in the transfer of the spermatophore, the evacuation of the sperm in a way which prevents any risk of loss or contamination of the sperm, etc. On the other hand, there are significant differences in many details between the three groups of ticks: the readiness to copulate, the number of times males and females copulate, the morphology of the endospermatophore, etc.

The uniformity presents arguments favouring the assumption of a common phylogenetic origin of the three groups of ticks. The study of the diversity may prove useful for establishing taxonomic relations, not to mention it rendering the subject attractive and rewarding.

REFERENCES

Arthur, D. R. (1962) *Ticks and Disease.* Oxford, Pergamon Press. pp. 445.

Balashov Yu. S. (1968) *Blood sucking ticks (Ixodoidea) — vectors of diseases of man and animals.* Leningrad, Nauka Publishers. (English translation: *Misc. Publ. Ent. Soc. Amer.* 1972).

Eveleigh, E. S. & Threlfall, W. (1975) The biology of *Ixodes (Ceratixodes) uriae* White, 1852 in Newfoundland. *Acarologia* **16** 621–635.

Feldman-Muhsam, B. (1967a) Spermatophore formation and sperm transfer in *Ornithodoros* ticks. *Science* **156** 1252–1253.

Feldman-Muhsam, B. (1967b) Fecundity and viability of sperm cells of *Ornithodoros* ticks. *Wiadomosci Parasitol.* **13** 533–542.

Feldman-Muhsam, B. (1969) The components of the spermatophore, their formation and their biological functions in argasid ticks. In: Evans G. O. (ed.) *Proceedings, Second International Congress of Acarology* 1967. Budapest, Hungarian Academy of Sciences, p. 357–362.

Feldman-Muhsam, B. (1973) Autogeny in soft ticks of the genus *Ornithodoros* (Acari: Argasidae). *J. Parasitol.* **59** 536–539.

Feldman-Muhsam, B. (1974) Extracellular symbionts of tick sperm. In: Arvy, L. (ed.) *Recherches Biologiques Contemporaines.* Nancy, Vagner p. 47–59.

Feldman-Muhsam, B. (1979) Copulatory behaviour and fecundity of male *Ornithodoros* ticks. In: Rodriguez, J. G. (ed.) *Recent Advances in Acaroloty* Vol. 2 London, Academic Press p. 159–166.

Feldman-Muhsam, B. & Borut, S. (1971) Copulation in ixodid ticks. *J. Parasitol.* **57** 630–634.

Feldman-Muhsam, B. & Borut, S. (1978) Further observations on spermatophore formation in argasid ticks. *J. Insect Physiol.* **24** 693–697.

Feldman-Muhsam, B. & Borut, S. (1983) On the spermatophore of ixodid ticks. *J. Insect Physiol.* **29** 449–457.

Feldman-Muhsam, B. & Borut, S. (1984) Differences in the structure of the spermatophore between argasid and ixodid ticks. In: Griffiths, D. A. & Bowman, C. E. (eds) *Acarology VI,* Vol. 1, Chichester, Ellis Horwood p. 504–511.

Feldman-Muhsam, B. & Havivi, Y. (1967) Juvenile sterility in male ticks of *Ornithodoros tholozani. Nature* **213** 422–423.

Feldman-Muhsam, B., Borut, S. & Saliternik-Givant, S. (1970) Salivary secretion of the male tick during copulation. *J. Insect Physiol.* **16** 1945–1949.

Feldman-Muhsam, B., Borut, S., Saliternik–Givant, S., & Eden, C. (1973) On the evacuation of sperm from the spermatophore of the tick *Ornithodoros savignyi*. *J. Insect Physiol.* **19** 951–962.

Feldman-Muhsam, B., Bachrach, U., & Ben-Joseph, M. (1980) On the presence of spermine in the spermatophore of a soft tick. *J. Insect Physiol.* **25** 407–413.

Graf, J.-F. (1978) Copulation, nutrition et ponte chez *Ixodes ricinus* L., (Ixodoidea: Ixodidae) — 2e partie. *Bull. Soc. Entomol. Suisse* **51** 241–253.

Guglielmone, A. A. & Moorhouse, D. E. (1983) Copulation and successful insemination by unfed *Amblyomma triguttatum triguttatum* Koch *J. Parasitol.* **69** 786–787.

Khalil, G. M., Sonenshine, D. E., Sallam, O. A., & Homsher, P. J. (1983) Mating regulation and reproductive isolation in the ticks *Hyalomma dromedarii* and *H. anatolicum excavatum* (Acari: Ixodoidea: Ixodidae). *J. Med. Ent.* **20** 136–145.

Leahy, (Sister) M. G., Karuhize, G., Mango, C., & Galun, R. (1975) An assembly pheromone and its perception in the tick *Ornithodoros moubata* (Murray) (Acari: Argasidae). *J. Med. Ent.* **12** 284–287.

Moorhouse, D. E. (1966) Observations on copulation in *Ixodes holocyclus* Neumann and the feeding of the male. *J. Med. Ent.* **3** 168–171.

Norval, R. A. I. (1974) The life cycle of *Amblyomma hebraeum* Koch, 1844 (Acarina: Ixodidae). *J. Ent. Soc. Sth. Afr.* **37** 357–367.

Schlein, Y. & Gunders, A. E. (1981) Pheromone of *Ornithodoros* spp. (Argasidae) in the coxal fluid of female ticks. *Parasitology* **82** 467–471.

Sonenshine, D. E. (1984) Pheromones of acari and their potential use in control strategies. In: Griffiths, D. A. & Bowman, C. E. (eds.) *Acarology VI.* Vol. 1. Chichester, Ellis Horwood, p. 100–108.

Sonenshine, D. E., Homsher, P. J., Carson, K. A., & Wang, V. D. (1984) Evidence of the role of the cheliceral digits in the perception of genital sex pheromones during mating in the American dog tick, *Dermacentor variabilis* (Acari: Ixodidae). *J. Med. Ent.* **21** 296–306.

Sonenshine, D. E., Khalil, G. M., & Homsher, P. J. (1982) *Dermacentor variabilis* and *Dermancentor andersoni:* genital sex pheromones. *Exper. Parasitol.* **54** 317–330.

Wood, W. F., Leahy, (Sister) M. G., Galun, R., Prestwich, G. D., & Meinwald, J. (1975) Phenols as pheromones of ixodid ticks: a general phenomenon. *J. Chem. Ecol.* **1** 501–509.

10

Induction of oogenesis and oviposition in ticks

James H. Oliver, Jr.

10.1 INTRODUCTION

Reproduction in animals may be regulated by several events depending upon the particular evolutionary state and ecological niche of the species under consideration. Invertebrates display a much greater variation regarding reproductive stimuli and synchronization than do vertebrate animals, as might be expected from the larger number of invertebrate species and varied ecological niches which they occupy. Environmental cues usually related to seasonality or lunar cycles often act as primers from which subsequent stimuli are derived. The relationship between the primers and subsequent stimuli might not be immediately apparent. For example, spawning in certain marine invertebrates appears to be dependent upon phytoplankton blooms in the spring (Himmelman 1981). Another interesting relationship between primary and subsequent cues for reproduction occurs among some marine insects. In the intertidal chironomid midge, *Clunio,* reproduction is based on the interaction between exogenous time cues and endogenous timing mechanisms. The lunar period affects tides (that provide mechanical stimuli via wave action and temperature cues) which together with *Clunio*'s circadian timing system controls the physiological processes of emergence. Timing of emergence guarantees a strict correlation of reproduction with a favourable local intertidal situation that recurs only every 15 days around a distinct time of day. Tides probably act as a selecting influence on the circadian mechanism during the evolution of the various geographic races (Neumann 1981).

Reproduction in ticks is controlled by several factors. Food often acts as a reproductive stimulus in addition to providing the nutritive materials needed as nourishment for developing embryos. Mating may also act as a stimulus for egg development and oviposition. Both feeding and mating are necessary for successful reproduction to occur in most species of ticks. Exceptions to this generalization involve only a few parthenogenetic species (Oliver 1971, 1983) and some argasids that demonstrate autogenic development (Diehl *et al.* 1982, Oliver 1974, Pound *et al.* 1984). In this chapter I intend to ignore these fascinating exceptions and restrict my remarks to the general condition observed in most tick species in which both feeding and mating are necessary prerequisites to the induction of oogenesis and oviposition. Furthermore, I will attempt to partition various components of feeding and mating and analyze critical aspects of these parts. The analysis will rely on both published and unpublished data.

10.2 GONOTROPHIC CYCLES

Ixodids (hard ticks) undergo a single gonotrophic cycle, while argasids (soft ticks) undergo several such cycles. Species of these two families also differ in that ixodid females must mate prior to completion of the bloodmeal, whereas argasids feed to repletion without need of mating. The usual pattern is that metastriate ixodid females attach to hosts and feed to a certain point at which they remain until they mate. Upon mating, a rapid engorgement phase is entered and is finished usually within 24 to 48 hours. Some prostriate ixodids (*Ixodes*) display a similar pattern, but others may mate prior to host attainment. Argasids may mate before or after feeding, but must mate for oviposition to occur. Female ixodids produce one large mass of eggs and die, whereas argasids oviposit several batches of eggs usually separated by time and subsequent bloodmeals. Interestingly, although argasids must mate for initial oviposition to occur, they do not require (but will accept) repeated mating for subsequent gonotrophic cycles.

10.3 EFFECTS OF FEEDING

Vitellogenesis (yolk formation) in ova (eggs) does not proceed to completion unless feeding is achieved. Successful ovum development may be classified into 5 stages (Balashov 1972) including previtellogenic (stages 1 and 2), vitellogenic (stage 3), vitellogenic-pre-ovulate (stage 4), and ovulate (stage 5). Although mating is necessary for normal egg development and oviposition by fed females, in some cases a few individual virgin ixodids and argasids occassionally oviposit. If a critical minimum amount of blood is imbibed, certain females will oviposit a few abortive eggs. The primary regulatory role of mating and the subsequent roles of other factors in controlling vitellogenesis, other aspects of ovum development, and oviposition will be considered in detail later in this chapter, but a few comments are now appropriate relative to vitellogenesis.

Vitellogenin (Vg), a group of proteins, are synthesized exogenously to the ovary and later are incorporated into it. Vg becomes vitellin (Vn), the major yolk protein in the oocyte. Vg is produced primarily by the liver in vertebrates (Deeley & Goldberger 1979) and by the fat body in insects

(Hagedorn & Kunkel 1979). The process of Vg synthesis, subsequent secretion into the blood/hemolymph, and final uptake into the oocyte where they become Vn, is controlled by estrogen in vertebrates (Tata 1978), by juvenile hormone in most insects, and by ecdysone in some dipteran insects (Hagedorn & Kunkel 1979). Other evidence suggests that egg production in ticks is under hormonal control (Shanbaky & Khalil 1975) and that this gonadotropic hormone is a juvenile hormone-like compound (Pound & Oliver 1979, Connat *et al.* 1983). Mites also regulate egg development via a juvenile hormone-like compound (Hänel 1983, Oliver *et al.* 1985).

In ticks some yolk proteins are probably synthesized by oocytes, but most are derived from the fat body (Diehl *et al.* 1982). These data are supported by work showing that Vg and Vn are present in female hemolymph and eggs, respectively (Chinzei *et al.* 1983). Moreover, a variety of methods using *in vitro* tissue culture show that the fat body is the major (perhaps only) site of Vg synthesis in *O. moubata* (Chinzei & Yano 1985). Other evidence derived from the ixodid *Rhipicephalus sanguineus,* however, suggests that both midgut and fat body synthesize Vg (Coons *et al.* 1982).

10.4 EFFECTS OF MATING

Mating is necessary for most ixodids and argasids before they will complete oogenesis and oviposition (see reviews by Diehl *et al.* 1982, Oliver 1974). Attempts by investigators to partition the various components of mating in ticks to determine which are critical variables in the induction of oogenesis and/or oviposition have met with varying degrees of success. At least three *Ornithodoros,* two *Argas,* one *Dermacentor,* and one *Amblyomma* species have been investigated. Most experiments were designed to separate mechanical from chemical effects of the mating process. In general, mechanical stimuli were less effective in initiating egg development than were chemical stimuli, although mechanical stimuli frequently did influence the early stages of egg maturation in some species.

In metastriate ixodids egg development is dependent upon engorgement and the latter upon mating. *Amblyomma americanum* females were tested to determine the effects of various potential stimuli on the completion of feeding, oocyte development, and fat body growth (Oliver *et al.* 1975). The experimental groups include females receiving various mechanical stimuli to their vaginas, i.e. insertion of sand, agar, bits of rubber, and probe, as well as male accessory gland homogenate. All treatments including a 'trauma procedure' provided some stimulation (maximum egg development to stage 3 of Balashov) compared to untreated control females, but none of the treatments induced ovulation or oviposition. Interestingly, acidophilic fat body cell diameters increased in size and ranged from 48 to 70% of the fully differentiated size seen in ovipositing control females. This increase in the size of fat body cells probably represents synthesis of vitellogenins (Vg), some of which may have already been secreted into the hemolymph for final conversion to vitellin (Vn) in the oocyte. Scant yolk was seen in the oocytes, and the fat body cells were not fully developed; so, clearly, Vg synthesis was in the early stage.

Various parts of the mating act of *Dermacentor variabilis* were isolated and examined to determine which part promoted rapid engorgement and thereby oogenesis (Pappas & Oliver 1972). Females attached to a host were allowed to undergo the preliminary stages of mating (mechanical insertion of male mouthparts into female genital apertures and salivary secretions of male) with males which had had their genital apertures sealed. These females were slightly stimulated (4.5 times that of controls), but clearly needed to receive the spermatophore to fully feed and develop their eggs. In another experiment half of the females that mated with gamma irradiated sterile males (no spermatids) entered the rapid engorgement stage and oviposited normally. Thus it appears that female *D. variabilis* require a spermatophore and/or its contents, but not necessarily spermatids, for them to oviposit.

It is instructive to compare results obtained from the two ixodid species with those derived from the following argasid species. Fed female *Argas persicus* which received tactile and stretch stimuli from the placement of beads in their uteri produced some mature oocytes. Increased numbers of beads increased the number of mature oocytes. Moreover, spermless spermatophores enhanced fecundity, but no oviposition occurred. Male genital accessory gland complex injected into virgins caused more oogenesis, but did not result in oviposition.

In contrast to results obtained in experiments with *Argas persicus*, insertion of silica gel beads into the uteri of *Argas arboreus*, and the mechanical plus chemical stimulation of copulation with *A. persicus* males, did not stimulate ova development (Khalil & Shanbaky 1975). Injection of conspecific male accessory gland and/or testes extract into the hemocoel of fed virgin *A. arboreus*, however, induced ova development and ovulation similar to normally mated females. Nevertheless, the ovulated eggs degenerated in the ovary lumen, oviduct, and uterus, and no oviposition occurred. Testes and accessory gland extract both contain a common factor that induces development, yet the effect of this factor(s) is not enhanced when extracts of both tissues are injected into fed virgins.

The function of each stage in the mating process was also investigated in another argasid species, *Ornithodoros tholozani* (Galun & Warburg 1967). Fed virgin females were allowed to complete the mating act with males except they were denied spermatophore implantation (males' genital apertures were blocked). These females, as well as females which received plastic beads in their uteri, and females whose uteri were filled with NaCl solution, did not show any egg development. Females which received contents (but not capsules) of spermatophores from nonconspecific males (*O. moubata*) showed egg development at almost the same rate as normally mated females, but no oviposition occurred. Thus egg development must have been caused by factors present in sperm and/or seminal fluid. To differentiate between these factors of sperm and seminal fluid, females were mated with gamma irradiated sterile males. These males were spermless, but had functional accessory glands which produced spermatophores containing seminal fluid. Females which mated with these sterile males developed eggs at a normal rate, but only oviposited a few eggs. Thus, seminal fluid appears adequate to initiate oogenesis, but not quite enough to induce full oviposition. This interpretation is based on the assumption that X-ray treatment does not alter the chemistry of accessory glands and does not thereby destroy factors that might be present which would induce oviposition. The alternative interpretation does not seem to have been considered. If the X-ray treatment did not destroy any oviposition factors, however, then sperm are necessary for oviposition to occur and they must be from *O. tholozani* males because females mated by *O. moubata* males develop eggs, but do not oviposit. Interestingly, accessory gland homogenate of the roach *Periplaneta americana* can even cause some growth of oocytes when introduced into the uterus of *O. tholozani* (Galun & Warburg 1967).

Contrary to findings in *Ornithodoros tholozani,* mechanical stimuli had a great effect on oogenesis in *O. moubata* (see Connat *et al.,* Chapter 8 of this book, for a review of gonotropic cycles in *O. moubata*). Introduction of small metal beads into the vagina and uterus of fed virgin *O. moubata* not only provoked vitellogenesis, but caused oviposition. There was no significant difference in the proportion of 'bead' treated females which laid eggs and the mated control females, although there were fewer eggs oviposited per female weight of the treated females.

Chemical stimuli of egg development in *O. moubata* were derived from only mature sperm according to Sahli *et al.* (1985), but earlier results by tick workers in Neuchâtel, Switzerland, indicated that injection of seminal vesicle homogenates distended with immature spermatids was also effective (Connat *et al.,* Chapter 8 of this book). Nevertheless, Sahli *et al.* (1985) could not stimulate oogenesis or oviposition with an incubation medium of immature, nondevaginated spermatids. The medium became active only after the spermatids devaginated (capacitated). Evidence presented by the Neuchâtel group indicates that there is an active compound that causes egg maturation and oviposition in *O. moubata,* and that it is a non-specific heat labile protein of approximately 150,000 daltons (Germond & Aeschlimann 1977).

A third *Ornithodoros* species, *O. parkeri,* has been investigated regarding stimuli of egg maturation and oviposition (Oliver *et al.* 1984). As there were differences in the results of experiments on oogonial maturational and ovipositional stimuli between *O. tholozani* and *O. moubata,* so are there differences between *O. parkeri* and the other two species. Hemocoelic injection of homogenates of entire male reproductive system, testis, or genital accessory gland resulted in all *O. parkeri* female recipients ovipositing. Homogenates of salivary glands prompted 40% of the recipients to oviposit, but synganglial homogenates from unfed males showed no stimulation. When these same homogenates were introduced via the vagina instead of injected, the entire reproductive system and the accessory gland stimulated oviposition in only 20 and 40%, respectively, of the recipients. The testicular, salivary glandular, and synganglial homogenates did not elicit oviposition.

To unravel the complexity surrounding induction of oogenesis and oviposition, it helps to recall the need and interdependence of feeding and mating. With this in mind recall two interesting findings. DOPA initiated early stages of yolk deposition in 100% of the treated *O. tholozani* females as compared to 50% of the untreated controls (Galun & Warburg 1967); and DOPA, dopamine, epinephrine, and norepinephrine were detected in genital accessory glands of this species. Also recall that injection of homogenates of salivary glands caused 40% of recipient *O. parkeri* to oviposit (Oliver *et al.* 1984), and that saliva is abundant immediately prior to and during mating and transfer of spermatophores to females (Feldman-Muhsam *et al.* 1970). Mating involves repeated insertion of the male's mouthparts into the female's genital aperture, and saliva serves to lubricate the mouthparts as well as to augment spermatophore transfer. It is logical that some component(s) of saliva may also function in a physiological sequence of events in egg development. This seems even more likely when one recalls that prostaglandins, which have been reported to stimulate oviposition in two species of crickets (Destephano & Brady 1977, Loher *et al.* 1981), are present in the salivary glands of *Boophilus microplus* (Dickinson *et al.* 1976, Higgs *et al.* 1976) and in the salivary glands, reproductive organs, and egg-batches of *Hyalomma anatolicum excavatum* (Shemesh *et al.* 1979).

Given the information above regarding the catecholamines and prostaglandins in ticks, plus the general pharmacological action of these compounds, it seems likely that one or several of them might play a role in stimulating or mediating egg development and/or oviposition. Thus experiments were conducted involving these catecholamines and prostaglandins to determine their effectiveness in promoting egg development in *O. parkeri*. Tables 10.1 and 10.2 show the extent of oocyte maturation (scored

Table 10.1 — Effects of injecting 0.5 μl/♀ (see below for molar concentrations) L-DOPA on oocyte development in fed virgin female *O. parkeri*

		Controls			DOPA		
Days after injection	*n*	untreated	Fed/mated	Saline inject	10^{-2}M	10^{-3}M	10^{-4}M
	1	1–2	3	2	3	3	3
5	1	1–2	3	1–2	3	4	3
	1	1–2	4	1–2	3	3	3
	1	2	4	1–2	3	4	3
10	1	1–2	3	2–3	3	3	3
	1	1–2	4	1–2	3	3	3
	1	1–2	4	2	4	3	3
15	1	1–2	4	1–2	4	3	3
	1	1–2	4	1–2	4	4	3

Table 10.2 — Effects of injecting 0.5 μl/♀ (see below for molar concentrations) L-DOPA on oocyte development in fed virgin female *O. parkeri*

Days after injection	*n*	Saline inject control	DOPA 10^{-2}M	DOPA 10^{-3}M
27	1	2–3	4	3–4
	1	1–2	5	5
31	1	3	3	3
38	1	3	4	4

as stages given by Balashov 1972) caused by DOPA under the particular experimental conditions listed. Note that DOPA does indeed stimulate early oocyte development and that during the 15 days following treatment oocytes of the treated females are similar to those from fed mated controls, and more developed than those from the untreated and saline injected control females. The treated females did not oviposit however, and upon examination of their ovaries at days 27, 31 and 38 after injection, no further oocyte development occurred, even though fed mated controls had oviposited (Table 10.2). Broadly similar results were obtained after injection of dopamine (Table 10.3), epinephrine (Table 10.4) and norepinephrine

Table 10.3 — Effects of injecting 0.5 μl/♀ (see below for molar concentrations) dopamine on oocyte development in fed virgin female *O. parkeri*

Days after injection	*n*	Controls: Saline inject	Controls: Fed/mated	Dopamine 10^{-5}M	Dopamine 10^{-6}M	Dopamine 10^{-7}M
	1	1–2	2–3	2–3	2–3	3
6	1	1–2	3	3	1–2	3
	1	1–2	2–3	3	2–3	2–3
	1	1–2	5	2–3	3	3
13	1	1–2	2–3	3	3	3
	1	2–3	4	2–3	3	3
	1	1–2	5	3	3	3
20	1	1–2	5	3	4	3
	1	1–2	ovip	5	3	3
28	1	1–2	—	3	3	3–4
35	1	2–3	—	3	3–4	4

(Table 10.5). Although these results are preliminary and based on small samples, they do suggest that these catecholamines might play a stimulatory role in early oocyte development provided that they are transferred to the female as part of the seminal fluid or part of the spermatophore.

Two approaches were used to test the effects of prostaglandin E_2 (PGE_2) on oocyte development (Table 10.6). The direct method involved injection

Table 10.4 — Effects of injecting 0.5 μl/♀ (see below for molar concentrations) epinephrine on oocyte development in fed virgin female *O. parkeri*

Days after injection	*n*	Fed/mated control	Epinephrine 10^{-4}M	Epinephrine 10^{-5}M	Epinephrine 10^{-6}M
	1	3	3	3	3
6	1	3	3	3	3
	1	3	3	3	3
	1	3	3	3	3
13	1	3	3	4	4
	1	4	3	5	4
	1	5	4	4	4
20	1	5	4	4	4
	1	ovip	4	5	5

Table 10.5 — Effects of injecting 0.5 μl/♀ (see below for molar concentrations) norepinephrine on oocyte development in fed virgin female *O. parkeri*

Days after injection	*n*	Fed/mated control	Norepinephrine 10^{-4}M	Norepinephrine 10^{-5}M	Norepinephrine 10^{-6}M
	1	3	3	3	3
6	1	3	3	3	3
	1	3	3	3	3
	1	3	2–3	4	2–3
13	1	3	3	3–4	3
	1	4	3	4	2–3
	1	5	3–4	4	4
20	1	5	3–4	4	5
	1	ovip	4	5	5

of several concentrations of PGE_2 into the hemocoel of fed virgins as was done for the several catecholamines. The second indirect approach involved two protocols. In the first, DOPA and the prostaglandin synthetase inhibitor, acetaminophen (N-acetyl-p-aminophenol), were injected into virgin fed females (DOPA was injected approximately 1.5 hours after acetaminophen). The second protocol involved injection of acetaminophen followed two hours later by mating. The rationale for the two indirect protocols was that in the former the DOPA should stimulate the initial stages of oocyte maturation, and that if prostaglandin was necessary for subsequent oocyte development then oocyte growth would be blocked because the acetamino-

Table 10.6 — Effects of injecting prostaglandin E_2 (PGE_2), acetaminophen (N-acetyl-p-aminophenol) and DOPA on oocyte development in fed female *O. parkeri*

Days ater injection	*n*	Fed/ mated control	DOPA (μl 10^{-2}M)	Acetaminophen (1μl 10^{-2}M) +DOPA* (1μl 10^{-2}M)	Acetaminophen (1μl 10^{-2}M) +mated⁺	PGE_2 (1μl) 10^{-2}M	PGE_2 (1μl) 10^{-3}	PGE_2 (1μl) 10^{-4}M
6	1	5	3	2–3	3	3	3	3
	1	3	3	dead	3	3	3	3
13	1	5	dead	3	5	3	3	3
	1	4	3	dead	4	3	3	3
20	1	5	2–3	3	5	3	3	3
	1	3–4	3	3	5	3	3	4

*DOPA injected *ca.* 1.5 h after acetaminophen
⁺Mated 2 h post-inject.

phen would prevent prostaglandin formation. Prevention of prostaglandin formation through inhibition of prostaglandin cyclo-oxygenase is the most widely accepted mode of action of acetaminophen (Nickander *et al.* 1979). The rationale behind the second protocol was simliar to the first except that instead of using DOPA as an initial stimulus of oocyte development, mating was used to provide the stimulation.

Results of the three experiments involving PGE_2 (the one direct and two indirect) were disappointing (Table 10.6). The direct injection of PGE_2 resulted only in mild initial oocyte development similar to that observed in the DOPA and other catecholamine experiments. In the indirect experiment in which females received both DOPA and acetaminophen, results were inconclusive because it is unclear if PGE_2 plays a role in oocyte developmment past stage 3 of oocyte development. If PGE_2 is normally produced by females that have had an initial ovum stimulation via DOPA received in the seminal fluid during mating, and if PGE_2 stimulates later ovum development, then absence of ovum development past stage 3 suggests that acetaminophen blocked formation of PGE_2. Unfortunately, the direct PGE_2 injection experiment suggests that it does not stimulate ovum development past stage 3.

Results of the experiment in which females received acetaminophen followed by mating add additional evidence that prostaglandin is not necessary for later stages of oocyte growth, at least under those particular experimental conditions. If protaglandin were necessary for later stages of oocyte development then acetaminophen should have prevented prostalgandin formation and thereby blocked oocyte maturation in the mated females, but this did not happen. Ovum development in these females was similar to that in fed/mated control females (Table 10.6). Nevertheless, it is uncertain if the failure to block development was due to the lack of a proper dose of acetaminophen, breakdown of acetaminophen during the two-hour interval between injection and mating, or, in fact, that it was an indication that prostaglandin plays no role in oocyte maturation beyond perhaps very early stages. Given the unstable nature of prostaglandins (Brady 1983) and the unknown stability of injected acetaminophen in ticks, it is likely that the two-hour interval was too long and that the injected acetaminophen had chemically changed prior to the mating stimulus. Additional experiments are called for involving various experimental designs using different doses of exogenous PGE_2 and acetaminophen.

Identification of prostaglandins in the salivary glands of the two tick species cited above and the presence also in testes, ovaries, and eggs of *Hyalomma anatolicum excavatum* (Shemesh *et al.* 1979), plus the involvement of male saliva early in the mating process, provide the basis for the hypothesis of prostaglandin involvement in oocyte development. Reports of reproductive involvement of prostaglandins in insects indicate that the action of prostaglandins involves oviposition stimulation. No mention is made of a possible earlier role in oocyte development, but this may not have been examined. It is also possible that prostaglandins might play a regulatory role in the egg laying of ticks as it does in several insect species (Brady 1983). Prostaglandin synthetase is transferred in the spermatophore of the crickets *Acheta domesticus* (Destephano & Brady 1977, Destephano *et al.* 1976, 1982) and *Teleogryllus commodus* (Loher 1979, 1981, Loher & Edson 1973, Loher *et al.* 1981) to the female and is considered to be the 'mating factor' in *T. commodus*. Prostaglandin synthetase acts on arachidonic acid to produce PGE_2 in these crickets. If prostaglandins act as ovipositional stimuli in ticks, in which mating and egg laying are separated by several days, then there may be present some temporary endogenous inhibitor of prostaglandin synthesis. Endogenous inhibitors are known from certain tissues in crickets (Brady 1983) and bovine seminal vesicles (Takeguchi *et al.* 1971). Alternatively, delay between the hypothesized prostaglandin synthetase transfer during mating and oviposition in ticks could be the result of a temporal presence of the prostaglandin precursor, arachidonic acid.

10.5 CONCLUSIONS

In attempting to partition and define the roles of various aspects of feeding and mating in stimulating oocyte development and oviposition in ticks, attention should be concentrated on argasid rather than ixodid species. It is more difficult to separate parts of feeding and mating in ixodids which generally do not complete feeding until after mating, whereas feeding is separate from mating among argasids. In reconsidering experiments involving the latter, we note that experiments on induction of oogenesis and oviposition have involved two *Argas* (*A. persicus* and *A. arboreus*) and three *Ornithodoros* species (O. tholozani, O. moubata and *O. parkeri*).

Since these five species have similar ecological adaptations which allow them to be nest inhabitors and to undergo several gonotrophic cycles, we might suppose that they would share essentially the same stimuli for reproduction. Results of studies noted earlier in this chapter, however, make it clear that this supposition is erroneous. Mechanical stimulation of beads inserted into the vagina/uterus of *A. arboreus* or mating with *A. persicus* males did not stimulate ova development, whereas the number of beads received by female *A. persicus* was directly related to the number of mature oocytes. Spermless spermatophores also enhanced fecundity as did the introduction of male genital glands into *A. persicus* (Leahy & Galun 1972). However, none of these stimuli caused oviposition. Testicular and accessory glandular extracts injected into *A. arboreus* caused ova development as in normally mated females, but no oviposition occurred (Khalil & Shanbaky 1975). Moreover, egg retention in *A. arboreus* did not inhibit oocyte development or ovulation and thus did not serve as a 'feedback' signal to control oocyte development. In the two argasid species it appears that there are different stimuli needed for oocyte development and ovulation in each species, but neither species oviposits unless it receives sperm from conspecific males.

Among the *Ornithodoros* species, *O. tholozani* females do not show any egg development after receiving mechanical stimuli, but heterospecific spermatophores, conspecific mating with sterile males, and injections of the catecholamine DOPA provoke oogenesis. Thus seminal fluid is necessary for egg development, but sperm are required for oviposition.

Female *O. moubata* respond to the introduction of beads into the vagina/uterus by not only undergoing oogenesis, but also by oviposition, which is strikingly different from any other species tested. The percentage of 'bead' treated females undergoing oviposition was not significantly different from mated females (see Connat *et al.*, Chapter 8 in this book). These results along with those indicating that some virgin fed females oviposit a normal number of eggs without any mating or artificially introduced stimuli, suggest that *O. moubata* might be a species 'experimenting' with parthenogenesis, and that practially any stimulus might provoke oogenesis and oviposition. In opposition to this assumption, however, are results indicating no vitellogenesis or oviposition in females upon receiving injections of either conspecific accessory gland homogenates or testicular homogenates. Only spermatids and mature spermatozoa provoked oogenesis. Thus the strange situation exists whereby either mechanical stimulus of metal beads in the uterus or the presence of spermatids/spermatozoa are the only stimuli that provoke oogenesis, and that these same stimuli also elicit oviposition. In *O. parkeri* homogenates of entire male reproductive system and accessory glands cause 20% and 40%, respectively, of recipient females to oviposit if homogenates are inserted vaginally. If homogenates are injected into the hemocoel, then the entire reproductive system, accessory glands and testes provoke 100% oviposition, and even the salivary glands stimulate 40% of the females to oviposit.

In summary, it is uncertain if differences recorded concerning induction of oogenesis and oviposition among the seven species of investigated ticks represent true differences in the reproductive biology of these species, intraspecific genetic variation, or a failure to properly control all of the experimental variables. In nature and in laboratory colonies of argasids the innate variation in ovipositional response to 'normal' conditions expressed by most species is difficult to analyze and nearly impossible to control. Perhaps future laboratory experiments should be conducted on genetically more homogenous populations. Intraspecific variation of tendencies toward parthenogenesis, which clearly relate to 'mating' stimulus, are known, but not understood (Oliver 1971, 1983). Nevertheless, from existing data it seems likely that several summary conclusions are justified. First, mating and feeding are usually necessary for oogenesis and oviposition in ticks (excluding parthenogentic and autogenous spp.). Second, reproductive strategies differ between the Argasidae and Ixodidae, and a complex interplay among various stimuli involved in feeding, mating, and oviposition occurs in different species. There are probably valid differences, even among the argasid species, regarding necessary stimuli for the initiation of oogenesis and oviposition. Third, induction of oogenesis and oviposition are two distinct phenomena and are usually controlled by different stimuli. Fourth, probably catecholamines, perhaps prostaglandins and almost certainly a juvenile hormone-like gonadotrophic hormone, regulate egg maturation and oviposition. Vitellogenesis and its control may be very similar to that found in insects, whereby the fat body produces most of the vitellogenins under control of JH or a JH-like compound. Precisely where this JH-like compound is produced, how it is regulated, and its interrelationships with neurosecretory materials and ecdysones remain to be discovered. Fifth, there is a complicated interplay among nervous, endocrine, and reproductive systems prior to and during egg maturation and oviposition. And last, scarcity of data and the reported differences among species preclude any general reliable model at present regarding the induction and regulation of egg maturation and oviposition in ticks.

ACKNOWLEDGEMENTS

I thank Richard L. Osburn for critically reading this manuscript, and Cheryl M. Claiborne for clerical and editorial assistance. This work was supported in part by Public Health Service research grant AI-09556 from the National Institute of Allergy and Infectious Diseases and contract N-00014-80-C-0546 from the Office of Naval Research.

REFERENCES

Balashov, Yu. S. (1972) A translation of bloodsucking ticks (Ixodoidea) — vectors of diseases of man and animals, *Misc. Publ. Entomol. Soc. Am.* **8** 161–376.

Brady, U. E. (1983) Prostaglandins in insects, *Insect Biochem.* **13** 443–451.

Chinzei, Y. & Yano, I. (1985) Fat body is the site of vitellogenin synthesis in the soft tick, *Ornithodoros moubata, J. Comp. Physiol.* **155** 671–678.

Chinzei, Y., Chino, H., & Takahashi, K. (1983) Purification and properties of vitellogenin and vitellin from a tick, *Ornithodoros moubata, J. Comp. Physiol.* **152** 13–21.

Connat, J. L., Ducommun, J., & Diehl, P. A. (1983) Juvenile hormone-like substances can induce vitellogenesis in the tick *Ornithodoros moubata* (Acarina: Argasidae), *Int. J. Invert. Reprod.* **6** 285–294.

Coons, L. B., Tarnowski, B., & Ourth, D. D. (1982) *Rhipicephalus sanguineus:* Localization of vitellogenin synthesis by immunological methods and electron microscopy, *Exp. Parasit.* **54** 331–339.

Deeley, R. G. & Goldberger, R. F. (1979) Regulation of expression of the vitellogenin gene in avian liver, In: Hamilton, T. H., Clark, J. H., & Sadler, W. H. (eds) *Ontogeny of receptors and reproductive hormone action,* Raven Press, New York, p. 291–307.

Destephano, D. B. & Brady, U. E. (1977) Prostaglandin and prostaglandin synthetase in the cricket, *Acheta domesticus, J. Insect Physiol.* **23** 905–911.

Destephano, D. B., Brady, U. E., & Farr, C. A. (1982) Factors influencing oviposition behavior in the cricket, *Acheta domesticus, Ann. Entomol. Soc. Am.* **75** 111–114.

Destephano, D. B., Brady, U. E., & Woodall, L. B. (1976) Partial characterization of prostaglandin synthetase in the reproductive tract of the male house cricket, *Acheta domesticus, Prostaglandins* **11** 261–273.

Dickinson, R. G., O'Hagan, J. E., Schotz, M., Binnington, K. C., & Hegarty, M. P. (1976) Prostaglandin in the saliva of the cattle tick, *Boophilus microplus, Aust. J. Exp. Biol. Med. Sci.* **54** 475–486.

Diehl, P. A., Aeschlimann, A., & Obenchain, F. D. (1982) Tick reproduction: oogenesis and oviposition, In: Obenchain, F. D., & Galun, R. (eds) *Physiology of ticks,* Pergamon Press, New York, p. 277–350.

Feldman-Muhsam, B., Borut, S., & Saliternik-Givant, S. (1970) Salivary secretion of the male tick during copulation, *J. Insect Physiol.* **16** 1945–1949.

Galun, R. & Warburg, M. (1967) Studies on the reproductive physiology of the tick *Ornithodoros tholozani:* the effect of mating on oogenesis, *Vestnick Cs. spol. Zool. (Acta Soc. Zool. Bohemoslov.)* **31** 329–334.

Germond, J. E. & Aeschlimann, A. (1977) Influence of copulation on vitellogenesis and egg-laying in *Ornithodoros moubata* (Ixodoidea: Argasidae), In: Adiyodi, K. G., & Adiyodi, R. G. (eds) *Advances in invertebrate reproduction,* Peralam-Kenoth, Kerivellur, Kerala, India, p. 308–318.

Hagedorn, H. H. & Kunkel, J. G. (1979) Vitellogenin and vitellin in insects, *Ann. Rev. Entomol.* **24** 475–505.

Hänel, H. (1983) Effect of JH III on the reproduction of *Varroa jacobsoni, Apidologie* **14** 137–142.

Higgs, G. A., Vane, J. R., Hart, R. J., Potter, C., & Wilson, R. G. (1976)

Prostaglandins in the saliva of the cattle tick, *Booplilus microplus* (Canestrini) (Acarina, Ixodidae), *Bull. Entomol. Res.* **66** 665–670.

Himmelman, J. H. (1981) Environmental synchronization of reproduction, In: Clark, W. H., & Adams, T. S. (eds) *Advances in invertebrate reproduction,* Elsevier/North-Holland, New York, p. 3–19.

Khalil, G. M. & Shanbaky, N. M. (1975) The subgenus *Persicargas* (Ixodoidea: Argasidae: *Argas*) 21. The effect of some factors in the process of mating on egg development and oviposition in *A. (P.) arboreus* Kaiser, Hoogstraal and Kohls, *J. Med. Entomol.* **12** 47–51.

Leahy, M. G. & Galun, R. (1972) Effect of mating on oogenesis and oviposition in the tick *Argas persicus* (Oken), *Parasitology* **65** 167–178.

Loher, W. (1979) The influence of prostaglandin E_2 on oviposition in *Teleogryllus commodus, Entomol. Exp. Appl.* **25** 107–109.

Loher, W. (1981) The effect of mating on female sexual behavior of *Teleogryllus commodus* Walker, *Behav. Ecol. Sociobiol.* **9** 219–225.

Loher, W. & Edson, K. (1973) The effect of mating on egg production and release in the cricket *Teleogryllus commodus, Entomol. Exp. Appl.* **16** 483–490.

Loher, W., Ganjian, I., Kubo, I., Stanley-Samuelson, D., & Tobe, S. S. (1981) Prostaglandins: their role in egg-laying of the cricket, *Telegryllus commodus, Proc. Natl. Acad. Sci. U.S.A.* **78** 7835–7838.

Neumann, D. (1981) Synchronization of reproduction in marine insects by tides, In: Clark, W. H., & Adams, T. S. (eds) *Advances in invertebrate reproduction,* Elsevier/North-Holland, New York, p. 21–35.

Nickander, R., McMahon, F. G., & Ridolfo, A. S. (1979) Nonsteroidal anti-inflammatory agents, *Ann. Rev. Pharmac. Toxicol.* **19** 469–490.

Oliver, J. H., Jr. (1971) Parthenogenesis in mites and ticks (Arachnida: Acari), *Am. Zool.* **11** 283–299.

Oliver, J. H., Jr. (1974) Symposium on reproduction of arthropods of medical and veterinary importance. IV. Reproduction in ticks (Ixodoidea), *J. Med. Entomol.* **11** 26–34.

Oliver, J. H., Jr. (1983) Chromosomes, genetic variance and reproductive strategies among mites and ticks, *Bull. Entomol. Soc. Am.* **29** 8–17.

Oliver, J. H., Jr., Murphy, R. W., & Obenchain, F. D. (1975) Reproduction in ticks (Acari: Ixodoidea). 4. Effects of mechanical and chemical stimulation on oocyte development in *Amblyomma americanum, J. Parasit.* **61** 782–784.

Oliver, J. H., Jr., Pound, J. M., & Andrews, R. H. (1984) Induction of egg maturation and oviposition in the tick *Ornithodoros parkeri* (Acari: Argasidae), *J. Parasit.* **70** 337–342.

Oliver, J. H., Jr., Pound, J. M., & Severino, G. (1985) Evidence of a juvenile hormone-like compound in the reproduction of *Dermanyssus gallinae* (Acari: Dermanyssidae), *J. Med. Entomol.* **22** 281–286.

Pappas, P. J. & Oliver, J. H., Jr. (1972) Reproduction in ticks (Acari: Ixodoidea) 2. Analysis of the stimulus for rapid and complete feeding of female *Dermacentor variabilis* (Say), *J. Med. Entomol.* **9** 47–50.

Pound, J. M. & Oliver, J. H., Jr. (1979) Juvenile hormone: Evidence of its role in the reproduction of ticks, *Science* **206** 355–357.

Pound, J. M., Oliver, J. H., Jr., & Andrews, R. H. (1984) Effects of temperature and tick weight on expression of autogeny in the argasid tick *Ornithodoros parkeri* Cooley (Acari: Argasidae) *J. Parasit.* **70** 279–284.

Sahli, R., Germond, J. E., & Diehl, P. A. (1985) *Ornithodoros moubata:* Spermateleosis and secretory activity of the sperm, *Exp. Parasit.* **60** 383–395.

Shanbaky, N. M. & Khalil, G. M. (1975) The subgenus *Persicargas* (Ixodoidea: Argasidae: *Argas*). 22. The effect of feeding on hormonal control of egg development in *Argas (Persicargas) arboreus, Exp. Parasit.* **37** 361–366.

Shemesh, M., Hadani, A., Shklar, A., Shore, L. S., & Meleguir, F. (1979) Prostaglandins in the salivary glands and reproductive organs of *Hyalomma anatolicum excavatum* Koch (Acari: Ixodidae), *Bull. Entomol. Res.* **69** 381–385.

Takeguchi, C., Kohno, E., & Sih, J. (1971) Mechanism of prostaglandin biosynthesis. I-characterization and assay of bovine prostaglandin synthetase, *Biochemistry* **10** 2372.

Tata, J. R. (1978) Induction and regulation of vitellogenin synthesis by estrogen, In: Litwack, G. (ed) *Biochemical actions of hormones,* Vol. 5, Academic Press, New York, p. 397–431.

11

Bloodmeal digestion in ticks

Lewis B. Coons, Rosemarie Rosell-Davis, and **Betty I. Tarnowski**

11.1 INTRODUCTION

This chapter is divided into four sections: digestive tract anatomy, midgut histology and bloodmeal digestion, vitellogenin production, and physiology of bloodmeal digestion and excretion. The first section describes the gross anatomy and histology of the entire digestive tract. The sections on midgut histology and bloodmeal digestion review the changes that occur in the midgut epithelium during bloodmeal digestion, especially at the ultrastructural level. We use the adult ixodid *Dermacentor variabilis* (Say) and the adult argasid *Ornithodorus moubata* Murray as models for midgut histology and bloodmeal digestion. There are not enough data as yet to describe a model for immature ticks. The section on vitellogenin production describes two different methods that have evolved in the ticks and relates these to the different life histories of the two families of ticks. The final section covers the physiology of bloodmeal digestion and excretion.

11.2 DIGESTIVE TRACT ANATOMY

The digestive system in both families of ticks can be divided into three different regions: foregut, midgut, and hindgut. Each region differs in its histology and function, and each can be subdivided anatomically. The foregut is divided into a sucking pharynx and an esophagus. Descriptions of the feeding apparatus of ticks and the foregut are given by Kemp *et al.* (1982), Sonenshine & Gregson (1970), Gregson (1960), and Douglas (1943). The pharynx is lined with chitin and has a muscle arrangement enabling it to expand, contract, and imbibe the bloodmeal. The pharynx has a set of valves which probably function to prevent regurgitation of the bloodmeal while it is being propelled from the pharynx into the ventriculus. The esophageal lumen is lined by cuticle secreted by underlying epithelial cells. A well developed circular muscle band overlies these cells. Since there is no structural or experimental evidence to suggest that any water reabsorption occurs here the esophagus probably functions solely as a food transfer tube.

The midgut consists of a ventriculus, a variable number of blind diverticula or caeca, and the rectal tube. A valve is found at the junction of the esophagus and the ventriculus (Sonenshine & Gregson 1970, Till 1961, Gregson 1960). The ultrastructure of the valve has yet to be described. No valves between the ventriculus and the caeca, or between the ventriculus and the rectal tube have been described. Bloodmeal digestion takes place in the midgut. Epithelial cells of the midgut lie on a basal lamina which is covered externally by circular and longitudinal muscle layers that move the bloodmeal through the midgut by peristalsis. The muscle has an ultrastructure simliar to the muscles of the digestive system of the mite *Macrocheles muscaedomesticae* (Coons 1978). The histology of the ventriculus and the caeca appears to be identical in ticks. A peritrophic membrane is found in the midgut of *Ixodes dammini* (Rudzinska *et al.* 1982). Another study suggests the existence of a peritrophic membrane in the midgut of the soft tick *Ornithodorus moubata* (Grandjean 1984). It is not yet clear how common peritrophic membranes are in ticks, nor under what conditions they form.

The hindgut is divided into the rectal bulb and the rectum. The rectal bulb is identified by a variety of names (Grandjean & Aeschlimann 1973). Published electron micrographs show that in ixodids, the rectal bulb is an extension of the Malpighian tubes with respect to its ultrastructure (Raikhel 1983b). This investigation concluded that the rectal bulb is involved in water reabsorption. However, the ultrastructure of the rectal bulb does not resemble organs known to function in the movement of water such as the hindgut of insects (Wall & Oschman 1975), nor is this conclusion supported by any experimental evidence. Most likely both the Malpighian tubes and the rectal bulb have the same function. The rectal bulb also receives and temporarily stores guanine produced by the Malpighian tubes. During feeding and after the ticks drop from the host, both the Malpighian tubes and the rectal bulb enlarge in *Dermacentor variabilis* (Coons, unpublished observation) and other ixodid species (Enigk & Grittner 1952). In argasids, the Malpighian tubes alternately fill and empty with guanine during feeding cycles of adults (Efremova 1967, Hamdy 1973a). The large and well developed musculature of the rectal bulb aids in moving the fecal material during defecation. The rectum or anal canal is short and lined by chitin (Efremova 1967). The rectum's musculature, which originates on the margins of the anal plates and inserts into the wall of the rectum, opens and closes the anal valves during defecation (Douglas 1943, Till 1961, Sonenshine & Gregson 1970). The ultrastructure of the rectum is unknown. In *D. variabilis,* the rectum is not involved in water vapor uptake as in some insects (McEnroe 1972).

The chitinous lining of the foregut and hindgut of ticks is continuous with the integumental cuticle, suggesting that the foregut and the rectum are ectodermal in origin (Sonenshine & Gregson 1970). The ventriculus, the caeca, and the intestine lack a cuticular lining, and are probably endodermal in origin as are the rectal bulb and the Malpighian tubes (Aeschlimann 1958). Based on the presence or absence of a cuticular lining, on similarities and differences in cellular ultrastructure, and on gross anatomy, we conclude that the digestive system of ticks consists of the following: a foregut including pharynx and an esophagus; a midgut consisting of a ventriculus, caeca, and a rectal tube; and a hindgut consisting of a rectum. The rectal bulb should be considered as a part of the Malpighian tubes that has been incorporated into the digestive tract proper.

11.3 MIDGUT HISTOLOGY AND BLOODMEAL DIGESTION

In describing midgut histology and bloodmeal digestion, we will first describe the structure of the midgut cells found in unfed ticks and in the various stages of feeding ticks; second, correlate bloodmeal digestion with the different feeding periods of ticks; and third, apply what is known about the structure and function of eukaryotic cells as a means of differentiating cell types in the tick midgut.

In the Ixodidae all digestive processes occur in the ventriculus and caeca. In unfed male and female *D. variabilis,* the epithelial lining consists of one simple undifferentiated cell type that lies on a thick, infolded basal lamina (Fig. 11.1). In unfed adult *D. variabilis,* the apical plasma membrane which borders the small lumen has few microvilli and no glycocalyx. The midgut lumen is most commonly small and inconspicuous. There is no evidence of endocytosis occurring in the unfed adult midgut. Few cytoplasmic organelles are present in these cells. The most common cytoplasmic constituents are lipid inclusions, glycogen granules, and residual bodies remaining after the digestion of the nymphal bloodmeal.

Several authors have suggested that the midgut epithelium acts as a food reserve in unfed ixodid ticks (Raikhel 1983a, Tarnowski 1983). Recently it has been shown that endosomes decrease in size in the midgut epithelial cells of unfed laboratory colonies of *Amblyomma americanum* as the tick ages (Jaworski *et al.* 1983). This suggests that endosomes from the nymphal bloodmeal persist through to the adult form, where they are utilized as a food reserve. A further study on field reared ticks of the same species showed an initial increase in lipid inclusions and a decrease in endosomes as the tick ages, followed by a decrease in the lipid inclusions and an increase in lysosomal activity in 28-month old ticks (Williams *et al.* 1985). This suggests that some of the bloodmeal is converted to lipid in field reared ticks for later utilization as energy. This conversion to lipid was not found in the previous study on laboratory ticks and suggests an important difference in energy utilization between the two populations. The function of the midgut as a food reserve is supported by the absence of a well developed fat body in unfed male and female *D. variabilis* (Spires 1982). Nuclei in the digestive cells of unfed ticks are small and have no prominent heterochromatin or nucleolus. The basal plasma membrane is sometimes infolded (Tarnowski

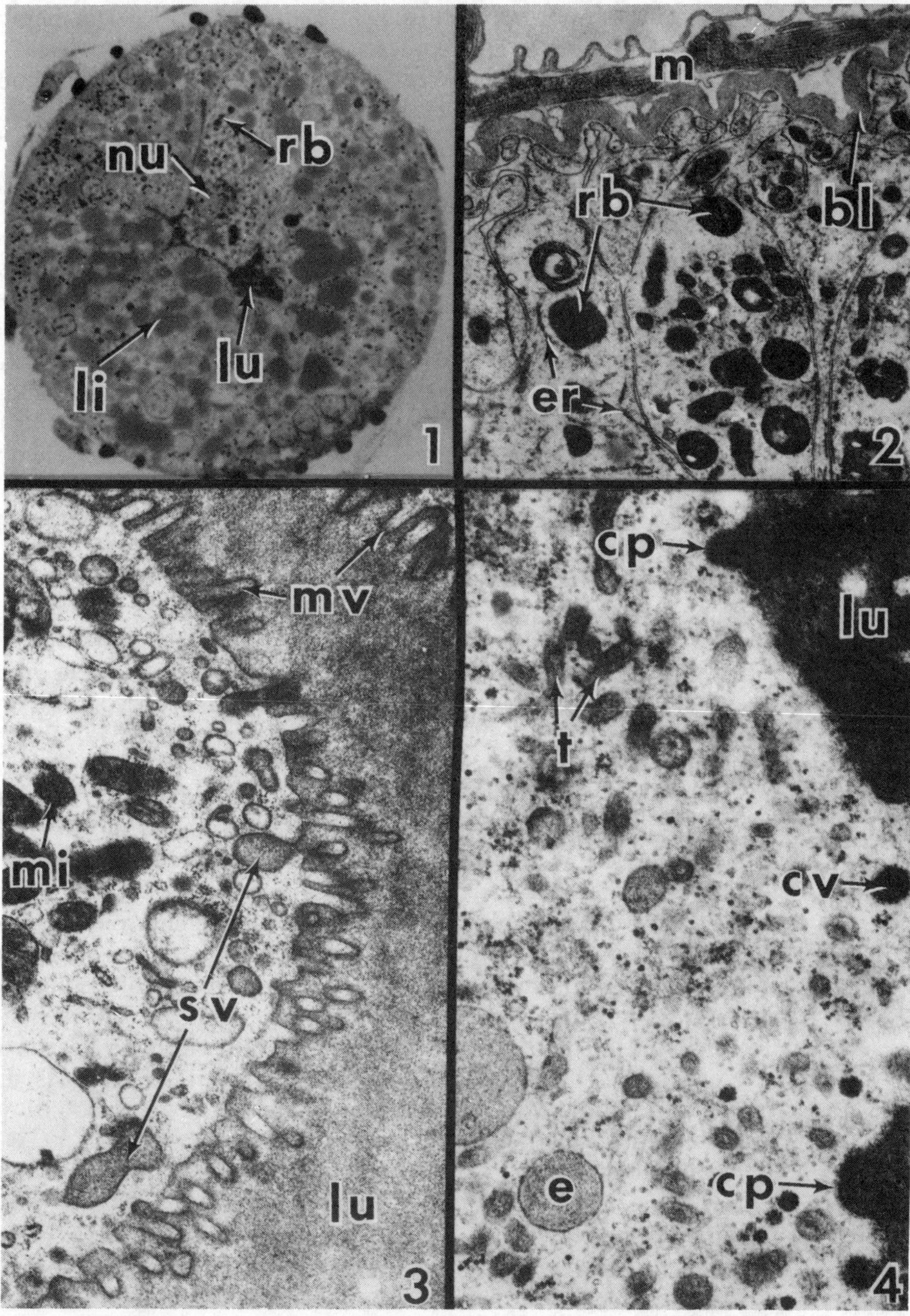

Fig. 11.1 — Light micrograph (LM) of a midgut of an unfed female *D. variabilis*. lu, lumen; rb, residual body; li, lipid inclusion; nu, nucleus. 453 ×. Fig. 11.2 — Transmission electron micrograph (TEM) of the basal areas of several cells from a midgut of an unfed female *D. variabilis*. m, visceral muscle; bl, basal lamina; rb, residual body; er, rough endoplasmic reticulum. 9865 ×. Fig. 11.3 — TEM of the apical area of a digestive cell of a feeding male *D. variabilis*. lu, lumen; mv, microvilli; sv, smooth flask-shaped vesicles; mi, mitochondria. 17 648 ×. Fig. 11.4 — TEM of the apical area of a digestive cell of a feeding female *D. variabilis*. lu, lumen; cv, clathrin coated vesicles; t, tubular elements; cp, clathrin coated pits; e, endosome. 35 748 ×.

1983), but often is not infolded (Fig. 11.2). The lateral plasma membrane does not enclose extracellular spaces.

The following description of cell junctions in the tick midgut epithelium is from Tarnowski (1983). Zonula adherens and septate junctions are found at the apical borders between digestive cells. During feeding, the electron-dense bloodmeal appears not to penetrate through the septate junctions, suggesting that this junction may act as a seal to separate the midgut lumenal space from the lateral spaces of the midgut epithelium and the hemocoel. However, several recent studies suggest that large molecules such as antibodies cross the gut epithelium of several species of ticks during feeding (Ackerman *et al.* 1981, Brossard & Rais 1984, Fujisaki *et al.* 1984, also see Ch. 14 of the present book). Whether these molecules move through or between cells or whether they are partly digested but still retain immunological properties after entry into the hemocoel, is unknown. In feeding ticks, gap junctions are commonly found along the lateral plasma membrane posterior to the septate junctions. Sclariform-like junctions, often associated with the gap junctions, are found along the lateral plasma membrane and the infoldings of the basal membrane. Sclariform-like junctions are also found in the fat body of *D. variabilis* (Spires 1982). The role of this junction is unknown, but it is believed to assist in keeping the extracellular spaces around it open, which would aid in the secretion of large proteins.

In ixodids, the lysed bloodmeal is taken up by digestive cells via fluid-phase endocytosis (Fig. 11.3) and via clathrin coated pits in a process termed 'receptor mediated endocytosis' (Fig. 11.4) (Raikhel 1983a, Tarnowski 1983). Once inside the cell the coated vesicles lose their clathrin and fuse to form endosomes (Figs. 11.4, 11.6). Organelles termed tubular elements are found associated with coated vesicles and endosomes (Fig. 11.4). They probably function to recycle the receptor and cell membrane involved in bloodmeal uptake back to the apical plasma membrane and are not involved in uptake as proposed by Raikhel (1983a). Intracellular digestion of the bloodmeal begins with the fusion of primary lysosomes and endosomes to form secondary lysosomes, and ends with the production of residual bodies (Figs. 11.5, 11.6) (Raikhel 1983a, Tarnowski 1983). As shown by X-ray microanalysis, the amount of iron increases in order from host erythrocyte, to endosomes and then to residual bodies (Tarnowski 1983) , suggesting that iron is concentrated into undigestible residues in residual bodies during intracellular digestion.

Female ixodids progress through two different feeding and two different post-feeding periods depending on whether they have mated. Slow feeding and an increase in weight occur first. During this time the weight of *Dermacentor andersoni* females increases from ten to about two hundred and fifty mg (Kaufman 1976). Slow feeding lasts for six to nine days, depending on the species and feeding conditions. A rapid engorgement period follows the slow feeding. *Dermacentor andersoni* females weigh about seven hundred and fifty mg following rapid engorgement (Kaufman 1976) which occurs twelve to twenty-four hours prior to detachment from the host (Balashov 1972). Mating is the stimulus for rapid engorgement in *D.*

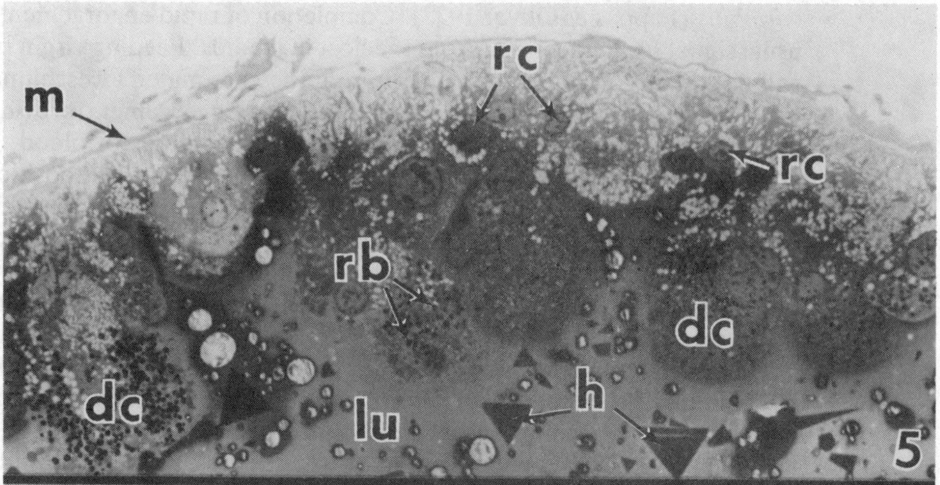

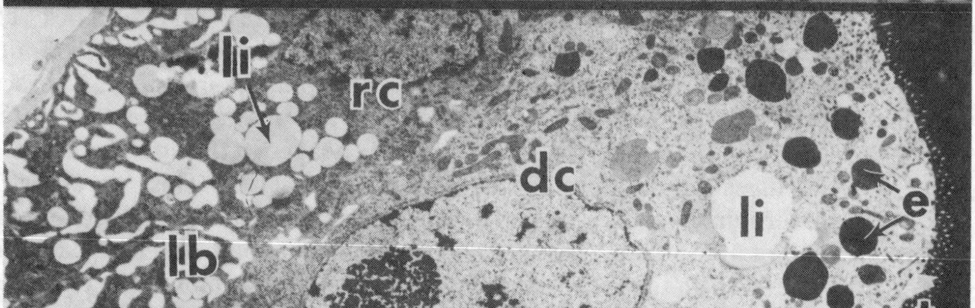

Fig. 11.5 — LM of a female *D. variabilis* midgut showing the first continuous digestive phase. rc, replacement cells; dc, digestive cells; h, hemoglobin crystals; lu, lumen; m, visceral muscle; rb, residual bodies. 445 ×. Fig. 11.6 — TEM of a female *D. variabilis* midgut showing the first continuous digestive phase. bl, basal lamina; rb, residual bodies; dc, digestive cell; rc, replacement cells; lb, basal labyrinth; es, extracellular spaces; li, lipid inclusion; lu, lumen, t, tubular element; e, endosomes. 3738 ×.

variabilis (Pappas & Oliver 1972). Completion of rapid engorgement is the usual signal for detachment from the host in ixodids. Feeding virgin ixodids will not undergo rapid engorgement until they are mated (Aeschlimann & Grandjean 1973, McSwain *et al.* 1982). Rapid engorgement provides the means for the mated female to take up large quantities of blood. Rapid engorgement is followed by two post-feeding periods: preoviposition and oviposition. The preoviposition period is the time between the detachment of the tick from the host and the appearance of the first egg; the oviposition period is the time during which eggs are laid (Nagar 1968). After oviposition, the female dies (Smith *et al.* 1946). Male ixodids do not have different feeding or post-feeding periods.

In mated female *Dermacentor variabilis,* the digestion of the bloodmeal is divided into three phases: continuous digestion, delayed or reduced digestion, and a second phase continuous digestion. Each phase coincides with a feeding or post-feeding period. Each digestive phase is dependent on the occurrence of a specific event in the life cycle of an ixodid tick. The first continuous phase is initiated by feeding and lasts from about 24 hours after attachment to just before the onset of rapid engorgement. Characteristically, two different populations of cells are observed during slow feeding. One group, termed digestive cells, are engaged in the uptake and intracellular digestion of the bloodmeal (Figs. 11.5, 11.6). This process is termed heterophagy and results in the production of usable products of digestion and of unusable by-products stored in residual bodies (Holtzman & Novikoff 1984). In these cells, residual bodies are not removed via exocytosis but accumulate in the cytoplasm. When large numbers are present, the cells detach from the basal lamina, are released into the lumen, and are then replaced by inactive digestive cells (replacement cells) found beneath the digestive cells (Figs. 11.5, 11.6). Although not actively involved in the uptake and digestion of the bloodmeal, replacement cells have a full component of cellular organelles, including lysosomes, rough endoplasmic reticulum, a well developed basal labyrinth, and lateral extracellular spaces. During this phase of digestion, two cyclic events occur: first, uptake of the bloodmeal, intracellular digestion and sloughing of digestive cells filled with residual bodies; and second, the sloughed cells are replaced by new cells. Replacement cells become digestive cells and begin a new cycle of uptake and intracellular digestion. Digestive cells accumulate residual bodies at different rates, so that all cells are not in the same stage of intracellular digestion. These two cyclic events and their characteristic ultrastructure continue throughout the first continuous digestion phase which lasts six or seven days after the onset of feeding in mated females. The first continuous digestive phase correlates with the two biological and biochemical feeding phases of 'preparatory' and 'growth' reported by Raikhel (1983a), Araman (1979), and Balashov (1972). In the former phase, the midgut epithelium develops, while in the latter phase the mobilized nutrients are utilized for cuticle development and organ growth. In *D. variabilis,* no ultrastructural evidence could be found for the preparatory and growth phases.

Mating is required to initiate the delayed or reduced phase of digestion.

The reduced phase of digestion is correlated with rapid engorgement, although cytological changes in the midgut epithelium occur slightly before the onset of rapid engorgement. This phase of digestion enables the mated female tick to acquire the largest possible bloodmeal by allowing the digestive cells of the midgut epithelium as well as the lumen of the midgut to function as storage sites for the bloodmeal. The reduced digestive phase is characterized by four events. First, intracellular digestion is greatly reduced and large endosomes rather than residual bodies accumulate in the cytoplasm of digestive cells (Figs. 11.7, 11.8). Few residual bodies are observed in digestive cells at this time. Second, digestive cells cease to be sloughed into, the lumen. Third, replacement cells are not observed. Fourth, a new type of cell, the vitellogenic cell appears in the midgut epithelium. This phase is correlated with the expansion phase of feeding described by Araman (1979). We suggest that the change in the active digestive cells prior to the onset of reduced digestion is the result of a great reduction or a cessation of lysosome production which would allow an accumulation of large endosomes.

The second continuous phase of digestion is initiated by detachment from the host (Fig. 11.9). This phase extends from the preoviposition through the oviposition periods, after which the female dies. During this period the products of digestion are used for the production of vitellogenin and egg maturation. Uptake of the bloodmeal slows during this phase, as evidenced by the reduction in the number of microvilli, vesicles, and tubular elements present in digestive cells. At the beginning of this phase, the lysosomal system once again becomes active. Fusion of primary lysosomes with the endosomes produces secondary lysosomes which are the sites of intracellular digestion, and residual bodies which contain non-digestible materials. As oviposition proceeds, the digestive cells accumulate large numbers of residual bodies in their cytoplasm. No cells detach into the lumen during the second continuous digestion phase. The basal labyrinth and extracellular spaces progressively shrink until little remains (Fig. 11.10). During this period there is a progressive growth in the vitellogenic cells and a concomitant shrinkage of the digestive cells (Figs. 11.9, 11.16). At the end of oviposition, the digestive cells and the vitellogenic cells have an organization suggestive of cell death (Fig. 11.10).

In unmated or virgin females that are allowed to continue feeding, the first continuous digestion phase persists, but bloodmeal uptake slows and the lumen becomes occluded with distended digestive cells (Figs. 11.11, 11.12). We found that cells were sloughed into the lumen during early feeding in virgins but not in the thirteen-day fed virgins. The reduced phase of digestion is not found in feeding virgin females, nor in feeding males. It is found in virgin females detached by hand that subsequently oviposit after first passing through a preoviposition period (Fig. 11.13).

Male *D. variabilis* do not produce vitellogenin and therefore do not require large quantities of blood. Consequently, they have a different feeding cycle than mated females. Males feed for only two to three days when females are present. Males detach and mate in response to a sex

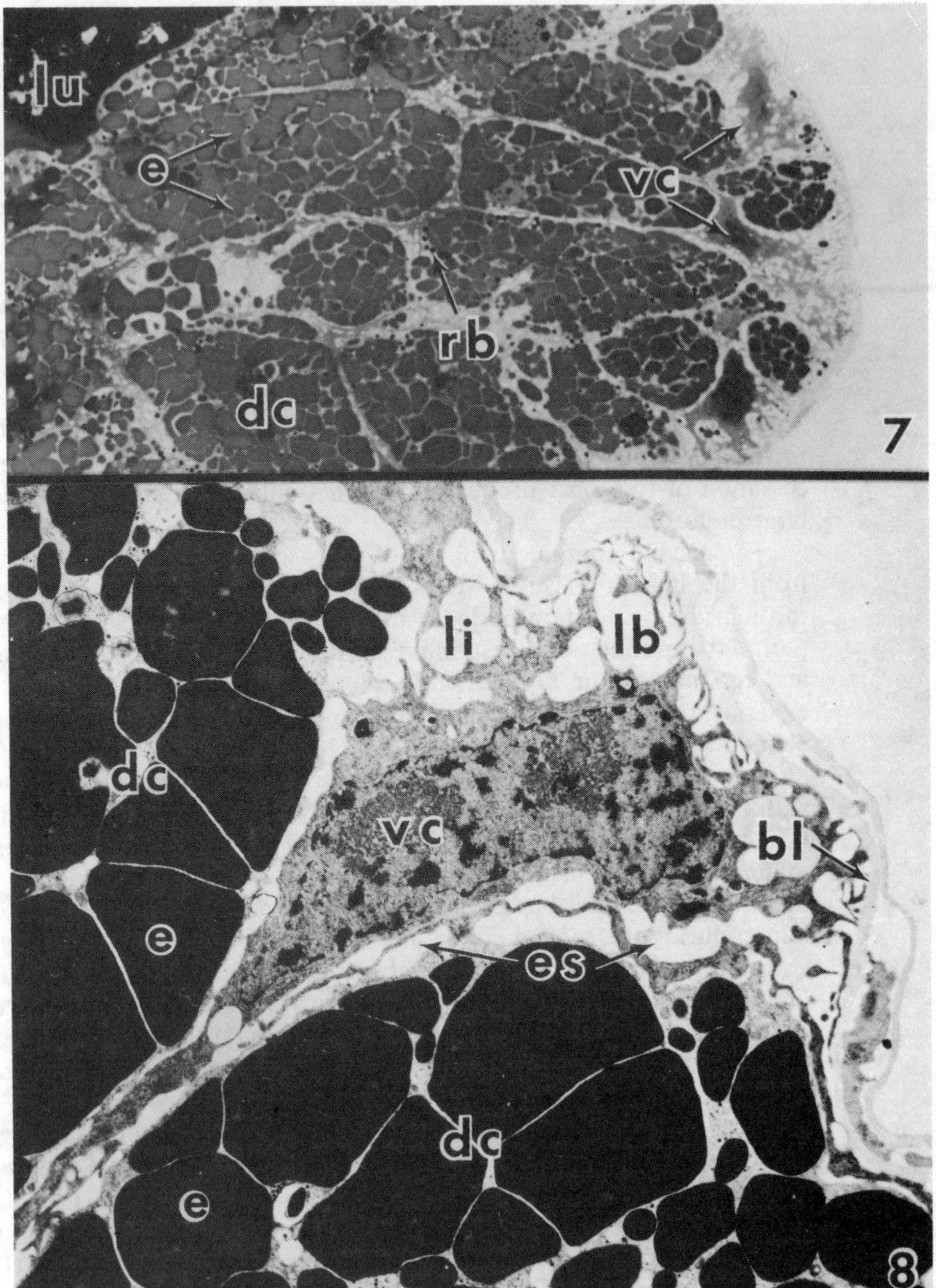

Fig. 11.7 — LM of a female *D. variabilis* midgut showing the reduced digestive phase. e, endosomes in digestive cells, dc; lu, lumen; vc, vitellogenic cells; rb, residual bodies. 445 ×. Fig. 11.8 — TEM of the basal region of a female *D. variabilis* midgut showing the reduced digestive phase. vc, vitellogenic cell; dc, digestive cells; e, endosomes in digestive cells; bl, basal lamina; es, extracellular spaces; lb, basal labyrinth; li, lipid inclusion. 5785 ×.

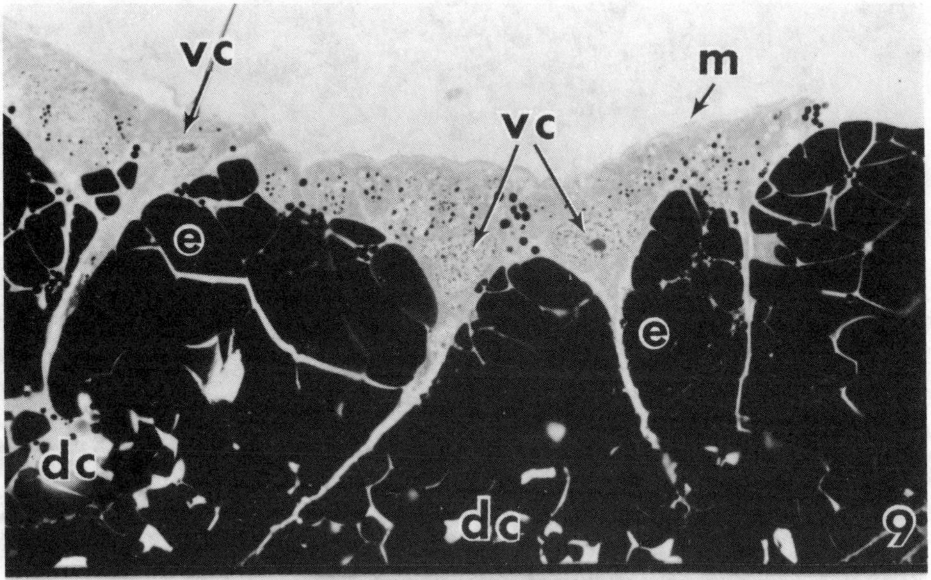

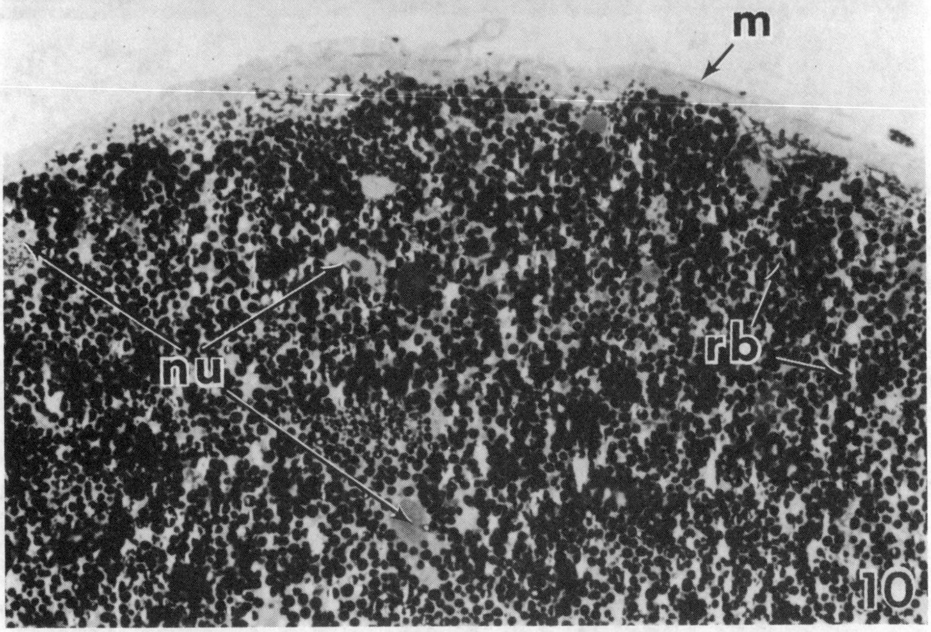

Fig. 11.9 — LM of an ovipositing *D. variabilis* midgut showing the second continuous digestive phase. e, endosomes in digestive cells; vc, vitellogenic cells; m, visceral muscle; dc, digestive cells. 438 ×. Fig. 11.10 — LM of a female *D. variabilis* midgut at the end of oviposition just before death. m, visceral muscle; rb, residual bodies; nu, nuclei. 438 ×.

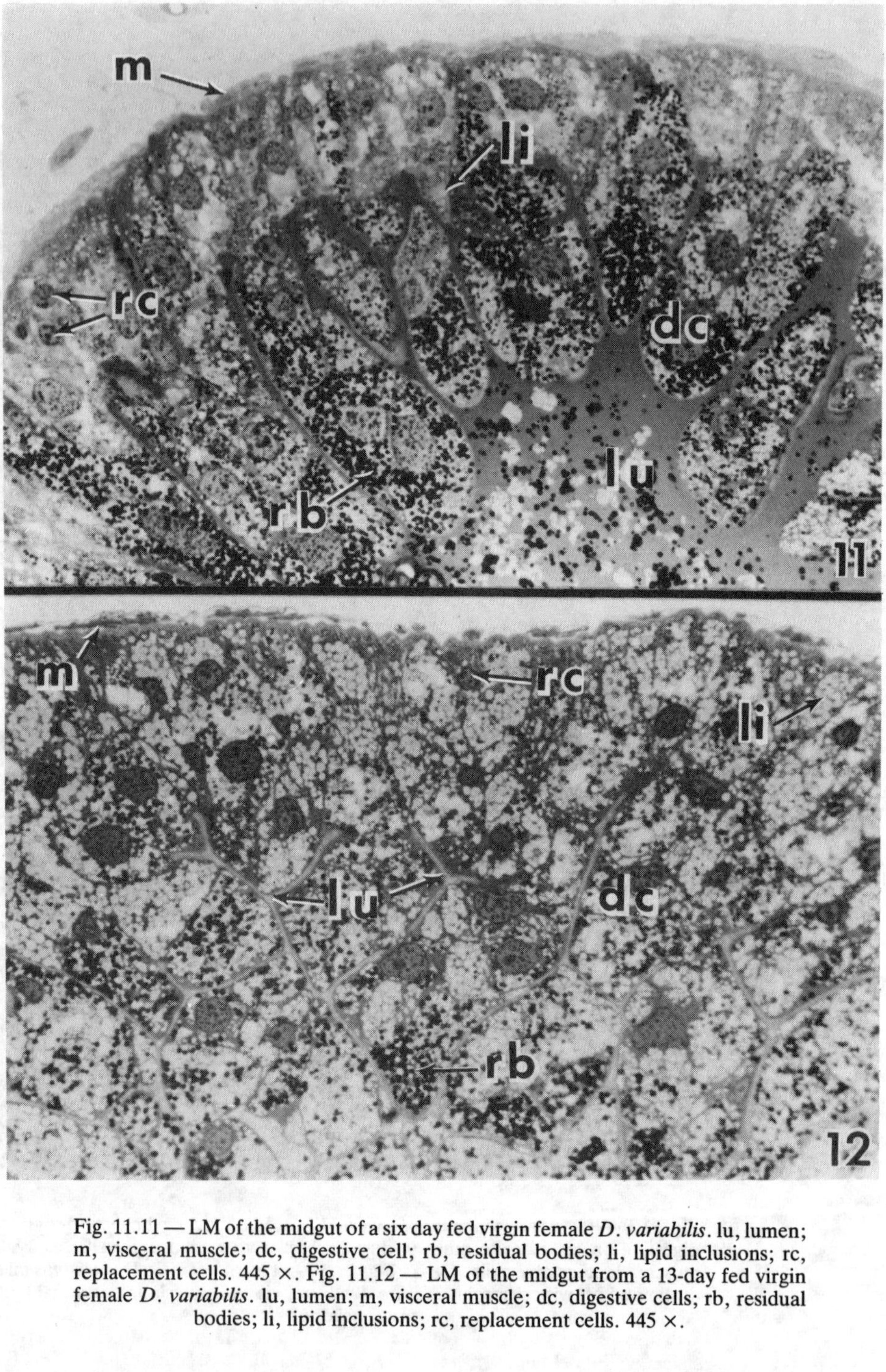

Fig. 11.11 — LM of the midgut of a six day fed virgin female *D. variabilis*. lu, lumen; m, visceral muscle; dc, digestive cell; rb, residual bodies; li, lipid inclusions; rc, replacement cells. 445 ×. Fig. 11.12 — LM of the midgut from a 13-day fed virgin female *D. variabilis*. lu, lumen; m, visceral muscle; dc, digestive cells; rb, residual bodies; li, lipid inclusions; rc, replacement cells. 445 ×.

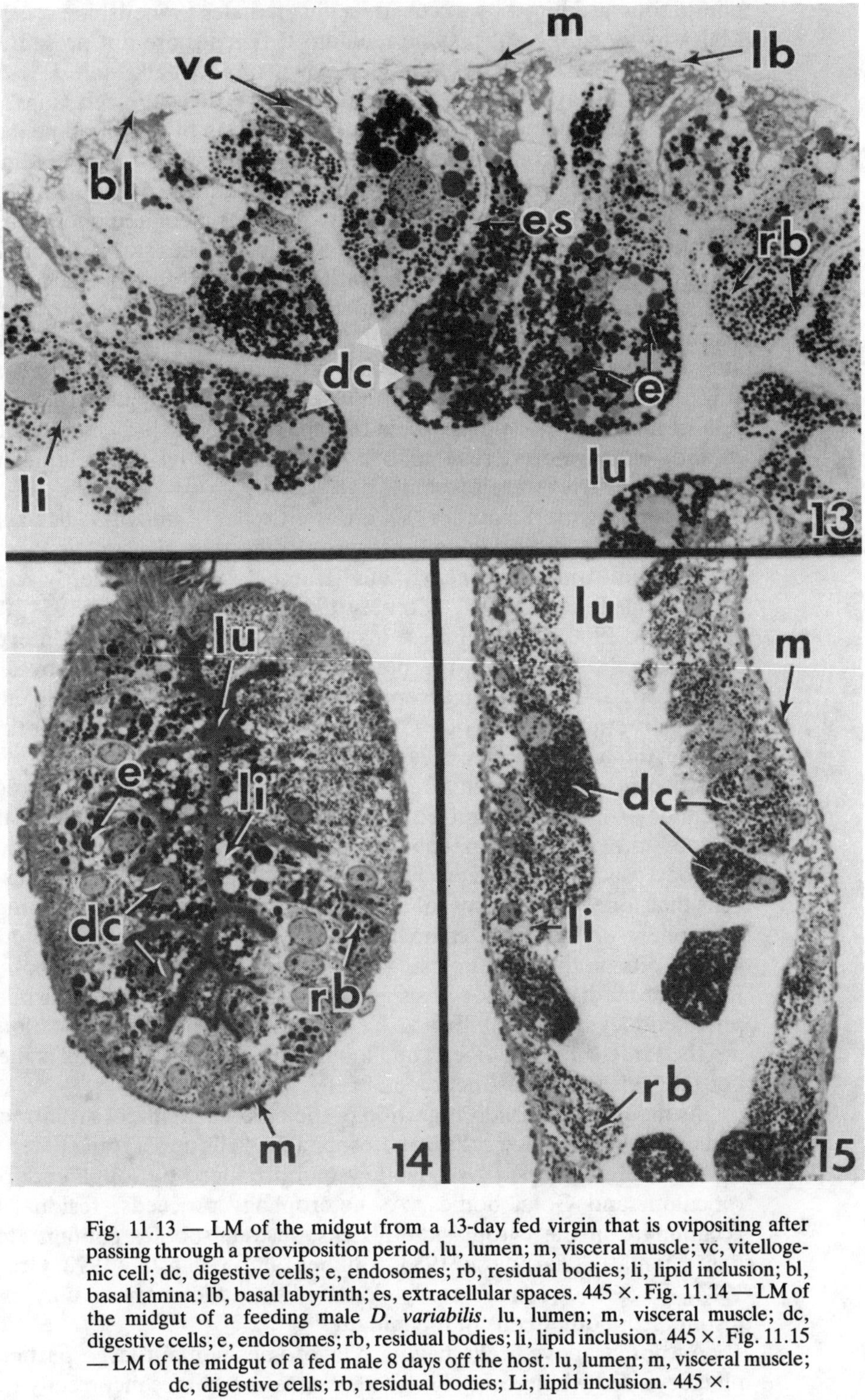

Fig. 11.13 — LM of the midgut from a 13-day fed virgin that is ovipositing after passing through a preoviposition period. lu, lumen; m, visceral muscle; vc, vitellogenic cell; dc, digestive cells; e, endosomes; rb, residual bodies; li, lipid inclusion; bl, basal lamina; lb, basal labyrinth; es, extracellular spaces. 445 ×. Fig. 11.14 — LM of the midgut of a feeding male *D. variabilis*. lu, lumen; m, visceral muscle; dc, digestive cells; e, endosomes; rb, residual bodies; li, lipid inclusion. 445 ×. Fig. 11.15 — LM of the midgut of a fed male 8 days off the host. lu, lumen; m, visceral muscle; dc, digestive cells; rb, residual bodies; Li, lipid inclusion. 445 ×.

pheromone produced by attached feeding females (Sonenshine *et al.* 1982). Males may or may not resume feeding. If females are not present, males remain attached to the host and feed slowly or not at all. Male *D. variabilis* have only one type of cell in the midgut, the digestive cell (Figs. 11.14, 11.15).No replacement cells are found. Digestion of the bloodmeal corresponds to the first continuous digestion phase in females. We could not find any digestive cells that were sloughed into the lumen of feeding males. The bloodmeal in males is also taken up by receptor mediated endocytosis and digested intracellularly by lysosomes. Residual bodies slowly accumulate in digestive cells (Fig. 11.14). Some profiles of rough endoplasmic reticulum and Golgi bodies are present which are involved in the production of lysosomes. In males which had fed for three days and then were sampled after eight days off the host, digestive but no vitellogenic cells were present (Fig. 11.15). At the ultrastructural level these cells resemble the digestive cells of three-day fed males except that the digestive cells in males eight days off the host have more residual bodies. No lipid inclusions or large areas of glycogen particles were present.

Depending on the author, six different types of midgut cells have been described in ixodid midguts: absorptive, digestive, phagocytic, secretory, replacement, and vitellogenic cells (Raikhel 1983a, Belozerov & Tymopheev 1971, Balashov 1972, Chinery 1964, Till 1961, Hughes 1954, Roesler 1934, Saito 1960). Most of these investigators relied heavily on interpreting light micrographs. Using the much better resolution offered by electron microscopy and relying on accepted principles of interpretation in eukaryotic cell structure, we rule out below the presence of three types: absorptive, phagocytic, and secretory cells.

Absorptive cells are involved in some form of endocytosis. Three types of endocytosis are recognized: receptor mediated endocytosis, characterized by the presence of coated pits on the apical plasma membrane; phagocytosis, characterized by the uptake of large particles such as whole cells that, once phagocytosed, are found inside the cell in phagosomes; and fluid-phase endocytosis, characterized by the presence of uncoated flask-shaped pits on the apical plasma membrane (Holtzman & Novikoff 1984). Receptor mediated endocytosis and phagocytosis are selective, while fluid-phase endocytosis is usually not. Cells that absorb or take up the bloodmeal are the same cells that digest the bloodmeal. An absorptive cell is therefore not recognized as a distinct type of cell.

As mentioned earlier, digestion of the bloodmeal in ticks is intracellular and occurs in digestive cells by heterophagy. Cells undergoing heterophagy have a well developed lysosomal system produced by rough endoplasmic reticulum and Golgi bodies. As heterophagy proceeds, residual bodies accumulate in the cytoplasm. In ticks, residual bodies contain electron-dense hematin (Tarnowski 1983, Raikhel 1983a, Balashov 1972, Grandjean 1984). Digestive cells are easily identified since they have an ultrastructure distinct from other cells in the midgut.

As noted above, cells having phagosomes contain large particles and phagocytosed whole cells. Phagosomes are sometimes incorrectly referred to as vacuoles. Phagosomes undergo heterophagy by fusion with a primary lysosome to produce a digestive vacuole or phagolysosome which then undergoes heterophagy to produce residual bodies. We know of no experimental evidence to support the conclusion of Raikhel (1983a) that phagocytosis is a property of specialized cells in the midgut. Until such evidence is demonstrated, we conclude that phagocytosis is a generalized property of digestive cells, and not a function of specific cells designated phagocytic cells.

Secretory cells have large amounts of rough endoplasmic reticulum, Golgi bodies, and membrane bound secretory vesicles for the storage of material to be secreted. The secretory process involves the fusion of the membrane of the secretory vesicles with the apical cell membrane and the subsequent release of its contents to the exterior (Palade 1975). Using these criteria, we have found no cell in the midgut of adult *D. variabilis* that could be described secretory, nor have we seen any evidence in the literature supporting the existence of such a cell. The relatively undifferentiated replacement cells have been found in the midguts of many different hard ticks and are probably a common cell type in the Ixodidae.

The cell types we recognize in adult *D. variabilis* midguts, digestive, replacement, and vitellogenic cells, all differentiate from one unspecialized type of cell found in the midguts of unfed ticks (Fig. 11.1). These cells differentiate at different times during feeding. Only the digestive cells and replacement cells are found during the first continuous phase of digestion. During the remaining two phases, reduced digestion and the second continuous digestion phase, only digestive and vitellogenic cells are present. In virgin feeding females, digestive and replacement cells are found during feeding, but several days after tick detachment only digestive and vitellogenic cells are present in the midgut. No replacement cells have been observed. In feeding males, only digestive cells are present.

It is not clear from the literature if all of the digestive processes are carried out in the ventriculus and caeca in the Argasidae. Other workers conclude that the posterior caeca are storage organs which do not participate in the three phases of digestion (Balashov 1972, Guirgis 1971). However, others do not agree that the posterior caeca are just storage organs. Grandjean (1983) found that all caeca participate in all digestive processes, but agreed with Khalil (1971) that the second or rapid digestion phase occurs more commonly in anterior caeca, while the third or slow digestion phase occurs more commonly in the posterior caeca. Soft ticks do store part of the concentrated bloodmeal as a food reserve (Tatchell 1964).

Most authors agree on the histology and ultrastructure of the argasid midgut. In *O. moubata*, the basal lamina is thicker than the basal lamina of other organ systems and is highly folded (Grandjean 1984). In addition, the relatively undifferentiated midgut cell of the unfed tick has an infolded basal plasma membrane (Grandjean & Aeschlimann 1973, Grandjean 1983). The thickened basal lamina and the infolded basal plasma membrane seem to be adaptations to accommodate the sudden stretching of the epithelium during the first digestive phase. Cell–cell junctional specializations in the midgut epithelium of *O. moubata* are the same as in *D. variabilis* and are ordered in similar fashion along the lateral plasma membranes. The junctional specializations of *O. moubata* midgut cells were described by Grandjean (1984). This author described a special cell membrane that resembles the sclariform-like junction found in *D. variabilis* midguts. In addition to having a similar structure, the two junctions are found in the same area of the cells. Digestive cells in the midgut of *O. moubata* take up the lysed bloodmeal via receptor mediated endocytosis, and take up cellular components via phagocytosis (Grandjean 1984). Inside the digestive cell, the endocytosed bloodmeal appears in coated vesicles that lose their clathrin coat and fuse to form endosomes. Most likely, the tubules thought to be involved in bloodmeal uptake (Grandjean 1984) are, in fact, involved in the recycling of the receptor and/or plasma membrane as discussed earlier. Phagocytosed materials, such as erythrocytes or leukocytes, are found in phagosomes inside the digestive cell. Intracellular digestion in the Argasidae, like that in the Ixodidae, is by heterophagy. Digestion begins with the fusion of a primary lysosome and an endosome or a phagosome. Digestion ends with the production of electron-dense residual bodies containing hematin, which is the waste product of digestion (Grandjean 1984).

Utilization of the bloodmeal in soft ticks can be divided into three phases: no digestion, rapid digestion, and slow digestion (Balashov 1972, Tatchell 1964, Guirgis 1971, Grandjean 1983, Grandjean & Aeschlimann 1973). The following description of three digestive phases is taken from the latter two studies on *O. moubata.* These are the only ultrastructural studies on an argasid midgut in the literature. It is not known how typical digestion is in *O. moubata* compared to other argasids, since the midgut is not continuous with the rectal sac and therefore the tick does not defecate during the adult stage (Enigk & Grittner 1952). The first digestive phase involves a stretching of the undifferentiated midgut epithelium by the bloodmeal. There is no evidence of digestion occurring during this phase. The first phase continues for about two days after uptake of the bloodmeal. During this period, the bloodmeal is concentrated by the removal of excess water and ions which are moved across the midgut epithelium and out of the tick through the coxal glands. The next phase, rapid digestion, continues from the end of the first phase until oviposition. The rate of digestion in virgin female *O. moubata* proceeds until about six days after imbibing the bloodmeal, when the rate either decreases or stops for several days (Tatchell 1964). More active digestive cells appear in a shorter time in mated female *O. moubata* compared to virgin female *O. moubata* (Galun & Warburg 1967). After introduction of recently fed males into the colony, females enter a second rapid digestion phase that is associated with egg production.

During rapid digestion many active digestive cells appear. These take up the bloodmeal and digest it, as previously described. Red blood cells are lysed during the first phase and during the early part of the second phase. The rate of lysis is not different when virgin *O. savignyi* females are compared to mated females (Osterhoff & Gothe 1966). During rapid digestion, crystallization of hemoglobin occurs in the midgut lumen. The final digestive phase is characterized by the presence of relatively few active digestive cells in the midgut epithelium and by a slower digestion of the bloodmeal. This phase allows soft ticks to survive long periods without feeding. Digestion in male *A. (P.) arboreus* is similar to digestion in females but proceeds more slowly (Guirgis 1971).

No ultrastructural study of the digestive system of the immature stages of argasids have been reported, but at the light microscopic level, the digestive epithelium in immature *A. (P.) arboreus* is simlar to that of adults (Guirgis 1971). The same study showed that in larval *A. (P.) arboreus* digestion of yolk is completed within five days after hatching. Only part of the bloodmeal is digested during the premolting period in larvae and nymphs. Digestion is then completed slowly after molting to the next stage. Unlike ixodids, the digestive cells in argasids are not synchronized — an adaptation to promote slow digestion and a long life span.

The reported number of cell types in the midgut of the Argasidae varies with the investigator. Balashov (1972) discussed three distinct cell types in *O. papillipes,* reserve or replacement, digesting, and secreting. However, in *O. moubata* (Grandjean 1983), *Argas arboreus* (Guirgis 1971), and *Argas persicus* (Tatchell 1964), a single digestive cell was interpreted as having many functions. We have not seen any convincing ultrastructural evidence in the literature of the existence of a secretory cell in the midgut of an argasid. We conclude, as previously discussed, that phagocytosis is a general property of cells and not a property of a specialized cell. For these reasons, we support the conclusion of Grandjean & Aeschlimann (1973) that the midgut epithelium in the Argasidae consists of only digestive and reserve or replacement cells.

Cell detachment from the midgut epithelium is common in both families of ticks (Balashov 1972, Guirgis 1971, Grandjean & Aeschlimann 1973, Roesler 1934, Hughes 1954, Chinery 1964, Raikhel 1983a, Tarnowski 1983). Tatchell (1964) concluded that cell detachment is an artifact. Some investigators supporting cell detachment in the midgut epithelium describe a partial detachment or rupture of the apical area of digestive cells as a natural means of cell detachment. This type of detachment is not supported by any known process at the cellular level. It is possible that partial detachment or rupture of the apical area of digestive cells is an artifact caused by specimen preparation. This is supported by Tatchell (1964) who found that mechanical disruption of the midgut in *Argas persicus* caused cells to be ruptured. Even when cells were rapidly fixed with cold formol–saline, attached cell stalks with disruptured apical borders were observed.

We suggest that detachment of whole cells from the midgut epithelium into the lumen is a mechanism designed to remove spent digestive cells that are filled with residual bodies so that replacement cells can begin bloodmeal digestion. In those midgut cells undergoing heterophagy during the first continuous phase to digestion in *D. variabilis,* the apical membrane is heavily involved in endocytosis of the bloodmeal. Exocytosis of the residual bodies into the lumen or into the hemocoel has not been reported. The hematin found in the midgut lumen comes from the residual bodies released from ruptured cells after they are sloughed into the lumen. In all ticks except adult *O. moubata,* the hematin is removed by defecation. Hematin in *O. moubata* is retained in the midgut lumen since there is no continuity between the midgut and hindgut. There seems to be no mechanism for the active digestive cell to rid itself of residual bodies which accumulate during digestion. The ability of discard cells that fill with residual bodies allows replacement cells to begin taking up and digesting the bloodmeal.

In ticks, an important event in digestion is the elimination of excess water and ions from the bloodmeal. In both families of ticks, the first step in the elimination process is to move the water and ions from the bloodmeal through the midgut into the hemolymph. Elimination of the water and ions from the hemolymph differs in the two tick families. Argasids eliminate the water and ions through coxal glands (Kaufman & Sauer 1982, Hecker *et al.* 1969), while ixodids eliminate these substances through the salivary glands (Lees 1946a,b, Tatchell 1967, 1969, Kaufman & Phillips 1973, Meredith & Kaufman 1973, Kaufman & Sauer 1982). The amount of water eliminated depends on the amount of bloodmeal taken up, the life stage, the species, and the sex of the individual. About two-thirds of the water from the bloodmeal in *Ixodes ricinus* was eliminated (Lee 1946a). While replete larvae, nymphs, and adult female *Amblyomma americanum* concentrated the bloodmeal, water elimination was most evident in nymphs and adult females (Sauer & Hair 1972). *Dermacentor andersoni* males take up only small amounts of blood and therefore do not have the large amounts of excess water that females have (Kaufman & Philips 1973).

We propose the following structural basis for the movement of water and ions in adult female *D. variabilis* midgut epithelium. A basal labyrinth and lateral extracellular spaces develop early during the first continuous phase of digestion (Figs. 11.5, 11.6). These reach maximum size during rapid engorgement (Figs. 11.7, 11.8). This parallels the maximum secretory activity of the salivary gland. After detachment from the host, the basal labyrinth and the lateral extracellular channels become reduced and finally absent (Fig. 11.9, 11.10). In ovipositing females, the vitellogenic cells have numerous basal foot-like processes which produce a well developed series of extracellular spaces in this region (Fig. 11.16). Sclariform-like junctions are common in the basal region during oviposition. In feeding males, the labyrinth and channels are poorly developed (Figs. 11.14, 11.15). Likewise, in virgin feeding females they are poorly developed (Figs. 11.11, 11.12). Neither feeding males nor virgin feeding females transport the large amount of water and ions moved by mated feeding females. If unmated fed females are removed by hand and allowed to proceed through preoviposition, some will begin to oviposit. In these ticks, reduced digestion occurs along with the development of a basal labyrinth and lateral extracellular channels (Fig. 11.13). Extracellular channels are closely associated with septate junctions which are believed to function as tight junctions in arthropods. This structural arrangement suggests involvement of paracellular routes in fluid removal (Ellis *et al.* 1977).

11.4 VITELLOGENIN PRODUCTION

The midgut is one of two sites of vitellogenin production in two species of ixodids (Coons *et al.* 1982, Tarnowski 1983). In *Rhipicephalus sanguineus,* immunodiffusion using Ouchterlony's double-diffusion technique and immunoelectrophoresis demonstrated that vitellogenin is produced by the fat body and the midgut (Fig. 11.17). Vitellogenin from either source is immunologically identical. It is transported to the ovary via the hemolymph and taken up unchanged to become vitellin. Likewise, vitellogenin from the same organs in *Dermacentor variabilis* is immunologically identical and is taken up from the hemolymph unaltered by the ovary (Tarnowski 1983). Vitellogenin from either tick had no immunological identity with host hemoglobin. This is also true in *Dermacentor andersoni* (Boctor & Kamel 1976). In *R. sanguineus,* immunoelectrophoresis shows that vitellin from the eggs and vitellogenin from the midgut and hemolymph have a similar electrophoretic mobility (Figs. 11.18, 11.19). In *Dermacentor andersoni,* the ovary is thought to contribute a component to the yolk in addition to extraovarial sources (Brinton & Oliver 1971). Vitellogenin from *D. variabilis, R. sanguineus,* and *O. moubata* is a glycolipohemeprotein (Tarnowski 1983, Araman 1979, Boctor & Kamel 1976, Diehl 1969, 1970, Chinzei *et al.* 1983). All ticks probably have this type of vitellogenin. Using immunodiffusion techniques, mated and unmated adult females and various stages of feeding adult *D. variabilis* were tested for the presence of vitellogenin in the fat body and midgut, and for vitellin in the ovary (Tarnowski 1983). Production of vitellogenin by the fat body and the midgut begins before detachment from the host (unpublished observation).

Virgins fed for up to nine days show no precipitation lines when extracts of midgut and fat body are tested against polyclonal antivitellogenin antibody (Tarnowski 1983). Lines appear for midgut, fat body, and the ovary only if the virgins are removed by hand from the host and they begin to lay eggs after continuing through a preoviposition period. This supports histological data discussed earlier that virgins lack the vitellogenic cell until after they are removed by hand from the host. Parthenogenesis occurs in some races of *D. variabilis* (Oliver 1971). Oliver found that an average of 1.5% of eggs from parthenogenic *D. variabilis* hatched. Only two out of 200 parthenogenically derived larvae attached and engorged on a host. One of the two fed larvae attained adulthood and was female. Parthenogenesis is rare and biologically unimportant in ticks (Hoogstraal 1978).

Vitellogenic cells are found in the midgut of mated *D. variabilis* females just before the onset of rapid engorgement to the completion of oviposition (Figs. 11.7, 11.8, 11.9). They are located between each digestive cell and extend to the lumen of the midgut. The basal area of the vitellogenic cell has extracellular spaces resulting from foot-like processes. Sclariform-like junc-

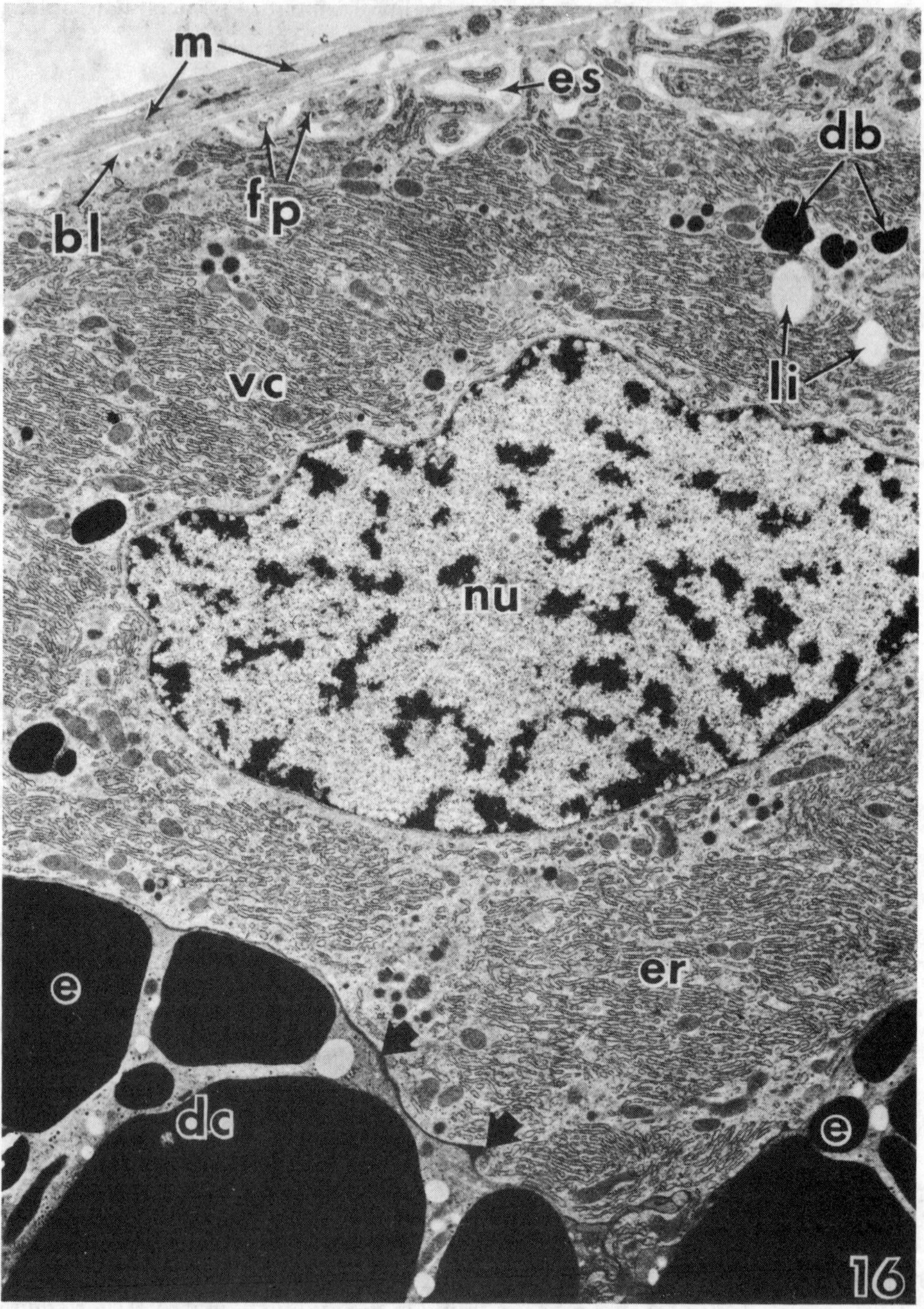

Fig. 11.16 — TEM of the basal area of the midgut of an ovipositing *D. variabilis*. dc, digestive cell; e, endosomes in digestive cells; m, visceral muscle; vc, vitellogenic cell; db, dense bodies; arrows, electron-dense material in intercellular space; fp, foot-like processes; li, lipid inclusions; nu, nucleus; er, rough endoplasmic reticulum; bl, basal lamina; es, extracellular spaces. 5775 ×.

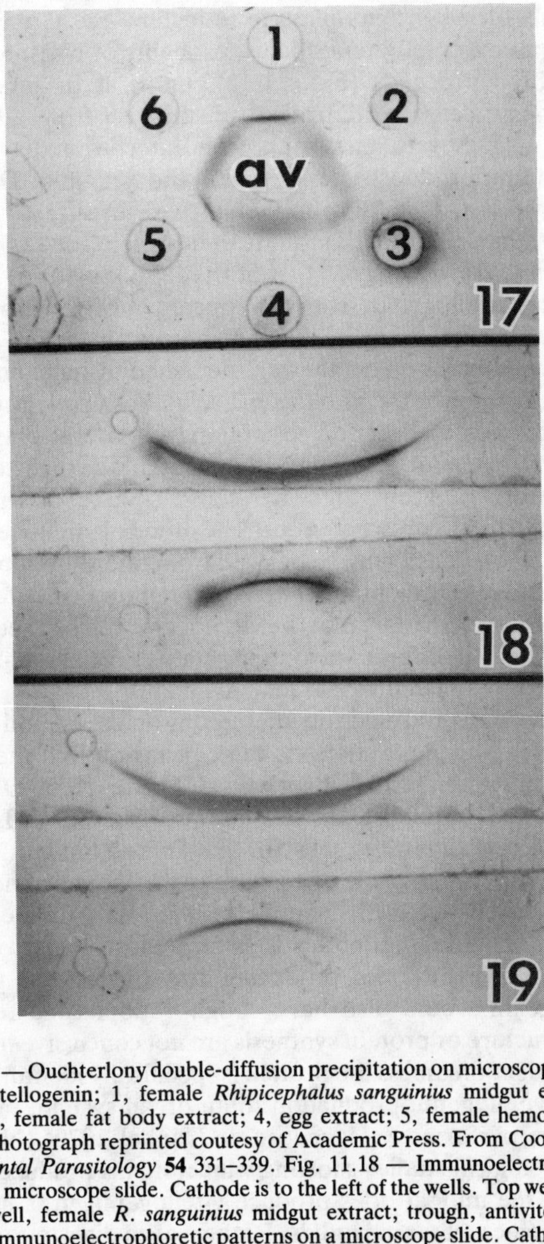

Fig. 11.17 — Ouchterlony double-diffusion precipitation on microscope slide. Wells: av, antivitellogenin; 1, female *Rhipicephalus sanguinius* midgut extract; 2, egg extract; 3, female fat body extract; 4, egg extract; 5, female hemolymph; 6, egg extract. Photograph reprinted coutesy of Academic Press. From Coons *et al.* (1982) *Experimental Parasitology* **54** 331–339. Fig. 11.18 — Immunoelectrophoretic patterns on a microscope slide. Cathode is to the left of the wells. Top well, egg extract; bottom well, female *R. sanguinius* midgut extract; trough, antivitellogenin. Fig. 11.19 — Immunoelectrophoretic patterns on a microscope slide. Cathode is to left of the wells. Top well, egg extract; bottom well, female *R. sanguinius* hemolymph; trough, antivitellogenin.

tions are found here, and on the lateral border near the extracellular spaces. We suggest that the sclariform-like junctions function to maintain constant extracellular spaces. The cytoplasm of the vitellogenic cell contains large amounts of rough endoplasmic reticulum and Golgi bodies (Fig. 11.16). Dense membrane bound bodies, probably of lysosomal origin, increase as oviposition proceeds (Fig. 11.16), until, at the end of oviposition, the vitellogenic cells are difficult to distinguish from adjacent digestive cells (Fig. 11.10). Several facts support the conclusion that the vitellogenic cell is the cell that produces vitellogenin in the midgut of *D. variabilis*. First, it is the only cell in the midgut epithelium with an ultrastructure characteristic of cells synthesizing large amounts of a protein for export. Second, it has an identical ultrastructure to that of the fat body during the time the fat body produces vitellogenin. Third, it appears only when the midgut is synthesizing vitellogenin. Fourth, it is not present in males and is not present in unmated females unless they are detached by hand and subsequently begin to lay eggs. Fifth, the vitellogenic cell takes up electron-dense material via receptor mediated endocytosis along its lateral plasma membrane (Fig. 11.16). This material is believed to contain the digested bloodmeal components from adjacent digestive cells. Uptake of the bloodmeal by the vitellogenic cells does not occur. The origin of vitellogenic cells in the midgut is not known. They may be derived from undifferentiated replacement cells in the midgut epithelium which differentiate just before the onset of rapid engorgement. The fact that the vitellogenic cell has been found in two ixodid ticks of widely different taxonomic status suggests that the cell is a common feature of the mated ixodid female midgut.

Use of light and electron microscopy on the fat body of adult *O. moubata* during vitellogenesis supports the conclusion that the fat body is active in vitellogenin synthesis (Obenchain & Oliver 1973, Obenchain, Dhadialla, and Ahmed, unpublished observations, reviewed by Diehl *et al.* 1983). The existence of digestive cells in the argasid midgut which also produce vitellogenin is suggested by the presence of rough endoplasmic reticulum in these cells (Diehl 1970, Grandjean 1983), and by the uptake and incorporation of tritiated amino acids as an indicator of protein synthesis (Jenni 1971). However, cells producing lysosomes or a protein precursor of vitellogenin would also have rough endoplasmic reticulum. Therefore, ultrastructure or protein synthesis are not conclusive indicators of vitellogenin production unless the protein is specifically identified as vitellogenin. To assign vitellogenin production to an organ system, immunological studies must be performed (Hagedorn & Kunkel 1979). No one has published immunological studies showing which organs produce vitellogenin in argasids. If the midgut does produce vitellogenin, it is not within specialized vitellogenic cells as in ixodids. Rather, each digestive cell at some time in its cycle seems capable of producing vitellogenin (Grandjean 1983, 1984).

The available evidence points to the development of two distinctive patterns of vitellogenin production in ticks: the argasid pattern which lacks a discrete vitellogenic cell in the midgut and lacks the ability to produce large amounts of vitellogenin, and the ixodid pattern which has a discrete vitellogenic cell in the midgut and produces large amount of vitellogenin. Ixodids which feed on multiple hosts have little contact with the nest, burrow, or home of the host, necessitating the production of thousands of eggs to ensure successful parasitism. Most argasids live in close proximity to the burrow or nest of their host and require only small batches of eggs per feeding to ensure successful parasitism. The ixodid type of life history is therefore dependent on an ability to ingest large quantities of blood and convert this to vitellogenin. The development of the ability to produce large amounts of vitellogenin within a relatively short time seems to be a major evolutionary strategy for reproductive success in the Ixodidae.

11.5 PHYSIOLOGY OF BLOODMEAL DIGESTION AND EXCRETION

Bloodmeal digestion in ticks occurs intracellularly at an acid pH and involves lysosomes in a process termed heterophagy. In most hematophagous insects, bloodmeal digestion occurs extracellularly at neutral or slightly alkaline pH, and occurs faster than in ticks (Gooding 1972). The bug *Rhodnius prolixus* has three digestive enzymes that are maximally active at an acidic pH and one that is maximally active at an alkaline pH (Houseman & Downe 1983). These three enzymes, cathepsin B, cathepsin D, and lysosomal carboxypeptidase B, are associated with lysosomes (Colowick & Kaplan 1981), and are found in different locations in the bug's gut (Houseman 1978, Houseman & Downe 1981, 1982). Only a few studies have described the properties of digestive enzymes in ticks, and these have dealt with the *in vitro* activity of proteases (Akov *et al.* 1976, Bogin & Hadani 1973, Reich & Zorzopulos 1978, Tatchell 1964, Tatchell *et al.* 1972). Akov (1982) concludes that the properties of the proteolytic enzyme suggest that it is cathepsin D, an enzyme characteristic of lysosomes.

The bloodmeal consists of different mixtures of tissues and cells at different life stages of ixodids. Because of the different bloodmeal constituents, it is difficult to estimate its size and concentration. The size of the bloodmeal also varies considerably between species (Kitaoka 1961b). In *Haemophysalis bispinosa,* Kitaoka found that the bloodmeal was three to five times the final weight of the engorged ticks, but in *Boophilus microplus* it was only three times the final weight of the engorged tick. Bloodmeal size also varies within a given species with the duration of starvation. The significance of this variation is unclear except that a larger bloodmeal in *D. variabilis* results in a larger egg mass (Amin 1969). In *Amblyomma americanum,* the size of the bloodmeal can be generalized with respect to sex and life stage (Sauer & Hair 1972). Larvae take in mostly lysed tissue and the lowest volume of erythrocytes of any life stage. Adult females take in the most erythrocytes; adult males take up a smaller meal than adult females. Concentration of the bloodmeal caused by the removal of ions and water varies with the life stage in four species of ixodids (Koch *et al.* 1974). The bloodmeal is least concentrated in larvae and most concentrated in adult females, with nymphs intermediate. Male *Argas arboreus* ingest about one-third to two-thirds as much blood as females (Hafez *et al.* 1972).

The physiology of bloodmeal digestion involves the degradation of proteins, most notably hemoglobin. In engorged female *Boophilus micro-*

plus, the digestion of hemoglobin proceeds at a steady rate for thirteen days, during which time all the hemoglobin was metabolized to hematin (O'Hagan 1974). In the same tick, Bremner (1959) found that only 5.1% of the iron ingested by the adult female is transferred to the eggs, with most remaining in the female tick, presumably in the lumen or the cells of the midgut. Hemoglobin crystals were observed in the lumens of male and female *O. moubata* (Smit *et al.* 1977). Hemoglobin molecules crystallized as oxyhemoglobin which later metabolized to methemoglobin and hemochrome. Phagocytosis of hemoglobin crystals by digestive cells occurs in the same tick (Grandjean 1984). As discussed by Akov (1982), the significance of these crystals is unknown since ticks fed on hosts whose blood does not crystallize have the same rate of survival as ticks fed on hosts whose blood does crystallize.

In ixodids, the hemoglobin within the gut contents in fed *Ixodes ricinus* females was measured using spectrophotometric methods. Ticks of equal weight ingest different amounts of hemoglobin (Sutton & Arthur 1962). Five to ten percent of ingested iron is transferred to eggs in *Haemaphysalis bispinosa* females (Kitaoka 1961a). Only a few other components of the blood were followed through digestion. Quantitative measurements in feeding and preovipositing *Haemaphysalis bispinosa* females showed that nitrogen in the form of guanine increased and was not excreted (Kitaoka 1961a). Also the amount of sterols in the tick was dependent on the concentration of sterols in host blood. Only indirect evidence exists for an essential need of sterols in ticks. Sterols are excreted in small amounts, which suggests that they are probably essential for tick development (Kitaoka 1961a). In comparison, dietary sterols play an essential role as factors for the normal growth, development, and survival of insects (Trager 1953, Lipke & Fraenkel 1956, House 1958).

Hemoglobin digestion seems to be correlated with the varying length of the preoviposition period. When temperature and relative humidity are held constant, the duration of the preoviposition period is a fixed time in *D. variabilis* and *R. sanguineus,* while the oviposition period varies between individuals (Nagar 1968). Two studies, one using *D. variabilis* (Tarnowski 1983) and another using *Rhipicephalus sanguineus* (Araman 1979) found that both ticks digest about one half of their hemoglobin by the end of the preoviposition period even though the time of the period varied from twelve days in *D. variabilis* to five days in *R. sanguineus*. The digestion of half of the hemoglobin in the bloodmeal may be a method of marking the onset of oviposition in different species of ixodid ticks. This also helps explain the observation that the preoviposition period of *Boophilus decoloratus* is temperature-, not humidity-dependent (Loudt 1974), since temperature is the environmental variable that would most affect the digestion of the bloodmeal. In *D. variabilis*, the concentration of hematin, which is the final metabolite of hemoglobin digestion in the midgut, increased as hemoglobin concentrations decreased (Tarnowski 1983).

The Ixodidae have a distinct cycle of excretion (Enigk & Grittner 1952, Till 1961). Ixodids excrete guanine from their Malpighian tubes and, at certain times, hematin from the digested bloodmeal. Ixodids excrete up to 15% of their excreta as a second unidentified purine compound (Hamdy 1972). Some ixodid species may excrete undigested blood from the bloodmeal. In addition, ixodids excrete up to 68% protein by weight (Hamdy 1973b, 1977). Ten different amino acids have been identified in the excreta of ixodids (Hamdy 1977). Non-feeding ixodids do not excrete amino acids (Hamdy 1973b). For this reason, it is suggested that the amino acids are products of digestion rather than products of intermediary metabolism (Hamdy 1977). Female *Hyalomma dromedarii* excrete 13% of their dry body weight in guanine during oogenesis and oviposition (Bassal 1977). The death of the female after oviposition may be due in part to an accumulation of toxic purines and hematin (Hamdy 1973b).

Studies on excretion in *Argas (P.) arboreus* (Hamdy 1973a) and *Alveonasus lahorensis* (Efremova 1967) show that a definite cycle of excretion also exists in the Argasidae. Hamdy found that the maximum amount of excretion was produced three to four days after feeding, and that the production of excreta is terminated in nymphs five to six days before molting. No proteins or amino acids are found in fecal material. Argasids excrete only guanine as a nitrogenous waste product.

11.6 SUMMARY AND CONCLUSIONS

The digestive system of ticks consists of a chitinous lined foregut and hindgut, and a non-chitinous lined midgut. The foregut consists of a sucking pharynx and an esophagus. The midgut consists of a ventriculus and a variable number of caeca and the rectal tube. The rectal tube leads to the rectal sac which is an extension of the paired Malpighian tubes. The hindgut is lined with chitin and consists of the rectum which leads to the anus.

Digestion of the bloodmeal takes place in the midgut and is taken up by the digestive cells via endocytosis. Digestion is intracellular and is termed heterophagy. The end product that remains in the digestive cell is a residual body. Digestive cells fill with residual bodies and detach into the midgut lumen. Other cells take their place and begin to take up and digest the bloodmeal.

While the mechanism of bloodmeal uptake and digestion is similar if not identical, in both families of ticks, the time of digestion and the cells involved are different and reflect the different strategies for reproductive success found in soft and hard ticks. As a model for describing bloodmeal digestion in adult ixodids, we use *D. variabilis,* and for adult argasids, we use *O. moubata.* There are not enough data to describe a model of bloodmeal digestion in immature ticks. Adult female *D. variabilis* have three distinct phases of bloodmeal digestion. Each phase occurs during a specific feeding or post-feeding period. The first continuous digestion phase begins after the first 24 hours of attachment and lasts until just prior to the onset of the rapid engorgement. The midgut epithelium changes from a single undiffentiated cell in the unfed tick to two different cell types in a feeding tick: digestive cells and replacement cells. The latter cells become digestive cells as the digestive cells fill with residual bodies and are sloughed into the lumen. During the first continuous digestive phase, the products of digestion are used mostly for non-reproductive organ and tissue growth.

The second digestive phase, delayed or reduced digestion, is initiated by mating and occurs during the rapid engorgement period of feeding. During this feeding period large amounts of blood are taken up by the tick. Some of the blood is stored in the midgut lumen and some is stored in the digestive cells which have greatly reduced their digestive activities and are filled with large endosomes. A new cell type, the vitellogenic cell, is found between each digestive cell. The final digestive phase, the second continuous phase, is initiated by detachment from the host and occurs during the preoviposition and oviposition periods. During this phase, digestion is accelerated and no cells are detached into the lumen of the midgut. As this phase of digestion proceeds, vitellogenic cells grow in size, and digestive cells decrease in size. At this time, the products of digestion are used mostly for the production of vitellogenin and egg production. Males and virgin females have only one continuous digestion phase.

Three phases occur during feeding in the argasid *Ornithodoros moubata:* no digestion, rapid digestion, and slow digestion. During the first phase, the undifferentiated epithelium of the midgut is stretched by the bloodmeal. Excess water and ions are removed at this time which lasts about two days. The second rapid digestion phase lasts until oviposition. During this period, many active digestive cells appear. During the last slow digestion phase only a few active digestive cells are present. This phase allows soft ticks to survive long periods witout a bloodmeal. In soft ticks, as opposed to hard ticks, the digestive cells are not synchronized, which allows the bloodmeal to be digested slowly. The midguts of other species of ticks should be studied at the ultrastructural level to ascertain whether or not these descriptions of bloodmeal digestion apply throughout the Ixodidae and Argasidae.

Vitellogenin in ticks is a glycolipohemeprotein and shows no immunological identity with host hemoglobin. In *D. variabilis,* vitellogenin production is triggered by mating and occurs in the fat body and midgut. In the latter organ, it is produced by the vitellogenic cells. Immunologically identical vitellogenin from both sources is released into the hemolymph, and taken up unaltered by the ovary to become vitellin. No definitive studies have been published which localize vitellogenin production in soft ticks, but there is good evidence to suggest the existence of two distinctive patterns of vitellogenin production in the two families of ticks. The simplest pattern is found in the Argasidae which lack a specific vitellogenic cell in the midgut and produce only a small number of eggs per feeding. Ixodids have a specific cell in the midgut that produces vitellogenin and such ticks produce thousands of eggs per feeding. This in turn is related to bloodmeal size, frequency of feeding, and digestion. Argasids take in a relatively small amount of blood in a short time but feed many times as adults. Ixodids take in a large amount of blood over a relatively long feeding period but do so only once as adults.

The bloodmeal of hard ticks consists of a variety of constituents. In general, larvae take in the most lysed tissues and the least number of erythrocytes, while adult females take in the most erythrocytes of any life stage. Nymphs are intermediate in this repsect. Adult males take in a smaller bloodmeal than adult females. The bloodmeal of soft ticks consists mostly of erythrocytes, which are lysed in the midgut. Very little is known about the physiology of bloodmeal digestion in ticks other than the breakdown of hemoglobin into hematin. Hematin is found in residual bodies and in the lumen of the midgut as a result of the lysis of detached cells. The only digestive enzyme in tick midguts studied in some detail is a partially characterized enzyme similar to cathepsin D. Much work needs to be done on the biochemistry of bloodmeal digestion in ticks.

ACKNOWLEDGEMENTS

The authors wish to acknowledge William J. Lamoreaux and Maureen Robinson-Prather for their assistance in the preparation of the manuscript.

REFERENCES

Ackerman, S., Brian Clare, F. McGill, T. W., & Sonenshine, D. E. (1981) Passage of host serum components, including antibodies, across the digestive tract of *Dermacentor variabilis* (Say). *J. Parasitol* **67** 373–380.

Aeschlimann, A. (1958) Development embryonnaire d'*Ornithodorus moubata* (Murray) et transmission transovarienne de *Borrelia duttoni*. *Acta Trop*. **15** 15–64.

Aeschlimann, A. & Grandjean, O. (1973) Influence of natural and artificial mating on feeding, digestion, vitellogenesis and oviposition in ticks (Ixodoidea). *Folia Parasitologica (Praha)*. **20** 67–74.

Akov, S. (1982) Blood digestion in ticks. In: *Current themes in tropical sciences*. (gen. ed. T. Odhiambo) I. Physiology of Ticks, (eds. F. Obenchain & R. Galun), pp. 197–211. Pergamon Press, Oxford.

Akov, S., Samish, M., & Galun, R. (1976). Protease activity in female *Ornithodoros tholozani* ticks. *Acta Trop*. **33** 37–52.

Amin, O. (1969) Growth of the dog tick *Dermacentor variabilis* (Say) (Acarina: Ixodidae) II. The effect of starvation and host species on its growth and fecundity. *J. Med. Entomol.* **6** 305–316.

Araman, S. (1979) Protein digestion and synthesis in ixodid females. In: *Recent Advances in Acarology*. (ed. J. Rodriguez), pp. 385–395. Academic Press, New York.

Balashov, Y. (1972) Bloodsucking ticks (Ixodoidea), vectors of disease to man and animals (in Russian). *Akad. Nauk. SSSR, Zool. Inst. Leningrad.* 1967. (in English): *Misc. Pub. Entomol. Soc. Amer.* **8** 161–376.

Bassal, T. T. M. (1977) Demonstration of guanine biosynthesis in *Hyalomma dromedarii* ticks using ^{14}C-labeled glycine and glyoxylate. *J. Parasitol.* **63** 758–759.

Belozerov, V. & Tymopheev, V. (1971) An electron-microscope investigation of the midgut epithelium of the adult female of *Dermacentor marginatus* Sulz. *Proc. 3rd International Congress of Acarology*. pp. 707–714. Prague.

Boctor, F. & Kamel, M. (1976) Purification and characterization of two lipovitellins from eggs of the tick *Dermacentor andersoni. Insect Biochem.* **6** 233–240.

Bogin, E. & Hadani, A. (1973) Digestive enzymes in "hard ticks" (Ixodoidea, Ixodidae). I. Proteolytic enzyme activity in the gut of *Hyalomma excavatum* female ticks. *Z. Parasitenk* **41** 139–146.

Bremner, K. (1959) Studies on haemixodovin, the pigment in the eggs of the cattle tick *Boophilus microplus* (Acarina: Ixodidae). *Australian Journal of Biological Sciences* **12** 263–273.

Brinton, L. P. & Oliver, J. H. (1971) Fine structure of oogonial and oocyte development in *Dermacentor andersoni* Stiles (Acari: Ixodidae). *J. Parasitol.* **57** 720–747.

Brossard, M. & Rais, O. (1984) Passage of hemolysins through the midgut epithelium of female *Ixodes ricinus* L. fed on rabbits infested or reinfested with ticks. *Experientia* **40** 561–563.

Chinery, W. (1964) The midgut epithelium of the tick *Haemaphysalis spinigera* Neumann. *J. Med. Entomol.* **1** 206–212.

Chinzei, Y., Chino, H., & Takahashi, K. (1983) Purification and properties of vitellogenin and vitellin from a tick, *Ornithodoros moubata. J. Comp. Physiol.* **152** 13–21.

Colowick, S. & Kaplan, N. (1981) Proteases of diverse origin and function. In: *Methods in enzymology* (ed. L. Lorand), Vol. 80, pp. 535–736. Academic Press, New York.

Coons, L., (1978) Fine structure of the digestive system of *Macrocheles muscaedomesiticae* (Scopoli) (Acarina: Mesostigmata). *Int. J. Insect Morphol. Embryol.* **7** 137–153.

Coons, L., Tarnowski, B., & Ourth, D. (1982) *Rhipicephalus sanguineus:* localization of vitellogenin synthesis by immunological methods and electron microscopy. *Exp. Parasitol.* **54** 331–339.

Diehl, P. (1969) Hämolymphenproteine und vitellogenese bei *Ornithodorus moubata* Murray (Ixodoidea Argasidae). Mitteilungen der Schweizerischen Entomologischen Gesellschaft. *Bull. Soc. Entomol. Suisse.* **XLII,** 117–125.

Diehl, P. (1970) Zur Oogenese bei *Ornithodorus moubata* Murray (Ixodoidea: Argasidae) unter besonderer Berucksichtigung der Vitellogenese. *Acta Trop.* **27** 302–355.

Diehl, P., Aeschlimann, A., & Obenchain, F. (1983) Tick reproduction: oogenesis and oviposition. In: *Current themes in tropical sciences.* (gen. ed. T. Odhiambo) I. Physiology of Ticks (eds. F. Obenchain & R. Galun), pp. 277–350. Pergamon Press, Oxford.

Douglas, J. R. (1943) The internal anatomy of *Dermacentor andersoni* Stiles. *Univ. Calif. Publ. Entomol.* **7** 207–72.

Efremova, L. K. (1967) Functioning of Malpighian tubules in the tick

Alveonasus lahorensis (Ixodidae, Argasidae). *Zool. Zh.* **46** 46–54. (In Russian) (In English: NAMRU3–T464).

Ellis, R. A., Goertemiller, C. G., & Stetson, D. L. (1977) Significance of extensive 'leaky' cell junctions in the avian salt gland. *Nature* **268** 555–56.

Enigk, K. & Grittner I. (1952) Die Exkretion der Zecken. *Z. Tropenmed. Parasitol.* **4** 77–94.

Fujisaki, K., Kamio, T., & Kitaoka, S. (1984) Passage of host serum components, including antibodies specific for *Theileria sergenti*, across the digestive tract of argasid and ixodid ticks. *Ann. Trop. Med. Parasitol.* **78** 449–450.

Galun, R. & Warburg, M. (1967) Studies in the reproductive physiology of the tick *Ornithororus tholozani:* the effect of mating on oogenesis. *Vestn. Cs. Spol. Zool.* **31** 329–334.

Gooding, R. (1972) Digestive processes of haematophagous insects. I. A literature review. *Quaest. Entomol.* **8** 5–60.

Grandjean, O. (1983) Blood digestion in *Ornithodorus moubata* Murray *sensu stricto* Walton females (Ixodoidea: Argasidae). II. Modifications of midgut cells related to the digestive cycle and to the triggering action of mating. *Ann. Parasitol. Hum. Comp.* **58** 493–514.

Grandjean, O. (1984) Blood digestion in *Ornithodorus moubata* Murray *sensu stricto* Walton (Ixodoidea: Argasidae) females. I. Biochemical changes in the midgut lumen and ultrastructure of the midgut cell, related to intracellular digestion. *Acarologia* **25** 147–165.

Grandjean, O. & Aeschlimann, A. (1973) Contribution to the study of digestion in ticks: histology and fine structure of the midgut epithelium of *Ornithodoros moubata,* Murray (Ixodoidea, Argasidae). *Acta Trop.* **30** 193–212.

Gregson, J. D. (1960) Morphology and functioning of the mouthparts of *Dermacentor andersoni* Stiles. Part I. The feeding mechanism in relation to the tick. *Acta Trop.* **17** 48–72.

Guirgis, S. (1971) The subgenus *Persicargas* (Ixodoidea, Argasidae, *Argas*). 13. Histological studies on *A.* (P.) *arboreus* Kaiser, Hoogstraal & Kohls. *J. Med. Entomol.* **8** 648–667.

Hafez, M., Abdel-Malek, A., & Guirgis, S. (1972) The subgenus *Persicargas* (Ixodoidea, Argasidae, *Argas*). 14. Biological studies on the adult stage of *A.* (P.) *arboreus* Kaiser, Hoogstraal & Kohls in Egypt. J. Med. Entomol. **9** 19–29.

Hagedorn, H. & Kunkel, J. (1979) Vitellogenin and vitellin in insects. *Annual review of entomology* **24** 475–505.

Hamdy, B. H. (1972) Biochemical and physiological studies of certain ticks (Ixodidae). Nitrogenous excretory products of *Argas (Persicargas) arboreus* Kaiser, Hoogstraal, and Kohls, and of other argasid and ixodid species. *J. Med. Entomol.* **9** 346–350.

Hamdy, B. (1973a) Biochemical and physiological studies of certain ticks (Ixodoidea). Cycle of nitrogenous excretion in *Argas (Persicargas) arboreus* Kaiser, Hoogstrtaal, and Kohls (Argasidae). *J. Med. Entomol.* **10** 53–57.

Hamdy, B. H. (1973b) Biochemical and physiological studies of certain ticks (Ixodoidea). Cycle of nitrogenous excretion in *Hyalomma dromedarii* Koch (Ixodidae). *J. Med. Entomol.* **10** 345–348.

Hamdy, B. H. (1977) Biochemical and physiological studies of certain ticks (Ixodoidea). *J. Med. Entomol.* **14** 15–18.

Hecker, H., Diehl, P., & Aeschlimann, A. (1969) Recherches sur l'ultrastructure et l'histochimie de l'organe coxal d'*Ornithodoros moubata* (Murray) (Ixodoidea; Argasidae). *Acta Trop.* **26** 346–360.

Holtzman, E. & Novikoff, A. B. (1984) *Cells and organelles.* Saunders College Pub. Co. New York. 660 pp.

Hoogstraal, H. (1978) Biology of ticks. In: *Tickborne diseases and their vectors.* (gen. ed. J. K. H. Wilde). pp. 3–14. *Proc. Internat. Conf., Edinburgh, Sep.–Oct. 1976.* University of Edinburgh, Centre for Tropical Veterinary Medicine; Edinburgh.

House, H. (1958) Nutritional requirements of insects associated with animal parasitism. *Exp. Parasitol.* **7** 555–609.

Houseman, J. (1978) A thiol-activated digestive proteinase from adults of *Rhodnius prolixus. Can. J. Zool.* **56** 1140–1143.

Houseman, J. & Downe, A. (1981) Exoproteinase activity in the posterior midgut of *Rhodnius prolixus* Stal (Hemiptera: Reduviidae). *Insect Biochem.* **11** 579–582.

Houseman, J. & Downe, A. (1982) Characterization of an acidic proteinase from the posterior midgut of *Rhodinius prolixus* Stal (Hemiptera: Reduviidae). *Insect Biochem.* **12** 651–655.

Houseman, J. & Downe, A. (1983) Activity cycles and the control of four digestive proteinases in the posterior midgut of *Rhodnius prolixus* Stal (Hemiptera: Reduviidae). *J. Insect Physiol.* **29** 141–148.

Huges, T. (1954) Some histological changes which occur in the gut epithelium of *Ixodes ricinus* females during engorging and up to oviposition. *Ann. Trop. Med. Parasitol.* **48** 397–404.

Jaworski, D. C., Barker, D. M., Williams, J. P., Sauer, J. R., Ownby, C. L., & Hair, J. A. (1983). Age-related changes in midgut ultrastructure and surface tegument of unfed adult lone star ticks. *J. Parasitol.* **69** 701–708.

Jenni, L. (1971) Synthese und Aufnahme von Proteinen wahrend der Vitellogenese in Ovocyten von *Ornithodorus moubata* Murray (Ixodoidea: Argasidae). *Acta Trop.* **28** 153–163.

Kaufman, W. (1976) The influence of various factors on fluid secretion by *in vitro* salivary glands of ixodid ticks. *J. Exp. Biol.* **64** 727–742.

Kaufman, W. & Phillips, J. (1973) Ion and water balance in the ixodid tick *Dermacentor andersoni.* I. Routes of ion and water excretion. *J. Exp. Biol.* **58** 523–536.

Kaufman, W. & Sauer, J. (1982) Ion and water balance in feeding ticks: mechanisms of tick excretion. In: *Current themes in tropical sciences.* (gen. ed. T. Odhiambo) I. Physiology of ticks (eds. F. Obenchain & R. Galun), pp. 197–211. Pergamon Press, Oxford.

Kemp, D. H., Stone, B. F., & Binnington, K. C. (1982) Tick attachment and feeding: Role of the mouthparts, feeding apparatus, salivary gland secretions and the host response. In: *Current themes in tropical sciences.* (Gen. ed. T. Odhiambo). I. Physiology of ticks (eds. F. Obenchain & R. Galun), pp. 119–168. Pergamon Press, Oxford.

Khalil, G. W. (1971) Biochemical and physiological studies of certain ticks (Ixodoidea). Incorporation of tritiated tyrosine in the digestive system of nymphal *Argas (Periscargas) arboreus* (Argasidae). *Ann. Entomol. Soc. Amer.* **64** 1149–1154.

Kitaoka, S. (1961a) Physiological and ecological studies on some ticks. IV. Physiological stage and lipid deposit during the blood-sucking process in the tick. *Natn. Inst. Anim. Health. Q.* **1** 85–95.

Kitaoka, S. (1961b) Physiological and ecological studies on some ticks. V. Nitrogen and iron excretion and amount of bloodmeal ingested during the blood sucking process in the tick. *Natn. Inst. Anim. Health Q.* **1** 96–104.

Koch, H. G., Sauer, J. R., & Hair, J. A. (1974) Concentration of the ingested meal in four species of hard ticks. *Ann. Entomol. Soc. Amer.* **67** 861–866.

Lees, A. (1946a) Chloride regulation and the function of the coxal gland in ticks. *Parasitology* **37** 172–184.

Lees, A. (1946b) The water balance in *Ixodes ricinus* L. and certain other species of ticks. *Parasitology* **37** 1–20.

Lipke, H. & Fraenkel, G. (1956) Insect nutrition. *Ann. Rev. Entomol.* **1** 17–44.

Loudt, H. G. H. (1974) The preoviposition period of *Boophilus decoloratus* (Koch 1844) (Acarina: Ixodidae). *J. Entomol. Soc. South Afr.* **37** 405–412.

McEnroe, W. D. (1972) The rectum is not the site of water vapor uptake by *Dermacentor variabilis* Say (Acarina: Ixodidae). *Acarologia* **14** 542–543.

McSwain, J. L., Essenberg, R. C., & Sauer, J. R. (1982) Protein changes in the salivary glands of the female lone star tick, *Amblyomma americanum* during feeding. *J. Parasitol.* **68,** 100–106.

Meredith, J. & Kaufman, W. (1973) A proposed site of fluid secretion in the salivary gland of the ixodid tick *Dermacentor andersoni. Parasitology* **67** 205–217.

Nager, S. (1968) On the significance of the duration of preoviposition and oviposition periods in ixodid ticks. *Acarologia* **10** 621–29.

Obenchain, F. & Oliver, J. (1973) A qualitative analysis of the form, function and interrelationships of fat body and associated tissues in adult ticks (Acari–Ixodoidea). *J. Exp. Zool.* **186** 217–236.

O'Hagan, J. (1974). *Boophilus microplus:* digestion of hemoglobins by the engorged female tick. *Exp. Parasitol.* **35** 110–118.

Oliver, J. (1971) Parthenogenesis in mites and ticks (Arachnida: Acari). *Am Zool.* **11** 283–299.

Osterhoff, D. R., & Gothe R. (1966) The use of red cell antigens for timing break-down of erythrocytes during digestion in *O. savignyi. Parisitology* **56** 613–618.

Palade, G. (1975) Intracellular aspects of the process of protein synthesis. *Science* **189** 347–358.

Pappas, P. & Oliver, J. (1972) Reproduction in ticks (Acari: Ixodoidea) 2. Analysis of the stimulus for rapid and complete feeding of female *Dermacentor variabilis* (Say). *J. Med. Entomol.* **9** 47–50.

Raikhel, A. (1983a) The intestine. In: *An atlas of ixodid tick ultrastructure.* pp. 59–97. (ed. Y. Balashov). In English: Special Publication, Entomological Society of America.

Raikhel, A. (1983b) The excretory system. In: *An atlas of ixodid tick ultrastructure.* pp. 129–146. (ed. Y. Balashov) In English: Special Publication, Entomological Society of America.

Reich, C. R., & Zorzopulos, J. (1978) *Boophilus microplus:* Characterization of larval proteases. *Exp. Parasitol.* **44** 1–6.

Roesler, R. (1934) Histologische, physiologische und serologishe Untersuchungen uber die Verdauung bie der Zeckengattung *Ixodes* Latr. *Z. Morph u. Ökol. Tiere* **28** 297–317.

Rudzinska, M. A., Spielman, A., Lewengrub, S., Presman, J., & Karakashian, S. (1982) Penetration of the peritrophic membrane of the tick by *Babesia micoti. Cell Tissue Res.* **221** 471–481.

Saito, Y. (1960) Studies on ixodid ticks. IV. The internal anatomy in each stage of *Haemaphysalis flava* Neumann 1897. *Acta Medica et Biologica* **8** 189–239.

Sauer, J. & Hair, J. (1972) The quantity of blood ingested by the lone star tick (Acarina: Ixodidae). *Ann. Entomol. Soc. Amer.* **65** 1065–1068.

Smit, J., Grandjean, O., Guggenheim, R., & Winterhalter, K. (1977) Haemoglobin crystals in the midgut of the tick *Ornithodorus moubata* Murray. *Nature* **255** 536–538.

Smith, C., Cole, M., & Gouck, H. (1946) Biology and control of the American dog tick. *Tech. Bull.* 905. United States Dept. of Agriculture.

Sonenshine, D. E. & Gregson, J. D. (1970) A contribution to the internal anatomy and histology of the bat tick *Ornithodoros kelleyi* Colley & Kohls, 1941. I. The alimentary system, with notes on the food channel in *Ornithodoros denmarki* Kohls, Sonenshine and Clifford, 1965. *J. Med. Entomol.* **7** 46–64.

Sonenshine, D., Silverstein, R., & Rechav, Y. (1982) Tick pheromone mechanisms. In: *Current themes in tropical science.* (gen. ed. T. Odhiambo) Vol. I. Physiology of ticks. (eds. F. Obenchain & R. Galun) pp. 439–468. Pergamon Press, Oxford.

Spires, L. (1982) Fine structure of the fat body and nephrocytes of the American dog tick *Dermacentor variabilis* (Say). Unpublished MS thesis. Memphis State University.

Sutton, E. & Arthur, D. (1962) Aspects of disease transmission by ticks. In *Symposia of the Zoological Society of London* **6** 223–252.

Tarnowski, B. (1983) Bloodmeal digestion and vitellogenesis in the tick

Dermacentor variabilis Say. Unpublished PhD. Dissertation. Memphis State University.

Tatchell, R. (1964) Digestion in the tick *Argas persicus* Oken. *Parasitology* **54** 423–40.

Tatchell, R. J. (1967) Salivary secretion in the cattle tick as a means of water elimination. *Nature* **213** 940–941.

Tatchell, R. J. (1969) The ionic regulatory role of the salivary secretion of the cattle tick *Boophilus microplus*. *J. Insect Physiol.* **15** 1421–1430.

Tatchell, R. J., Araman, S. F., & Boctor, F. N. (1972) Biochemical and physiological studies of certain ticks (Ixodoidea). Protease activity cycles in *Argas (Periscargas) persicus* (Oken) and *A. (P.) arboreus* Kaiser, Hoogstraal, and Kohls (Argasidae). *Z. Parasitked.* **39** 345–50.

Till, W. (1961) A contribution to the anatomy and histology of the brown ear tick *Rhipicephalus appendiculatus* Neumann. *Mem. Entomol. Soc. S. Afr.* **6** 1–124.

Trager, W. (1953) Nutrition. In: *Insect physiology* (ed. K. Roeder), pp. 350–380. John Wiley & Sons, New York.

Wall, B. J. & Oschman, J. L. (1975) Structure and function of the rectum of insects. *Fortschrifte der Zoologie* **23** 193–222.

Williams, J. P., Barker, D. M., Sauer, J. R., Hair, J. A., Ownby, C., & Koch, H. (1985) Ultrastructural changes in the midgut epithelium of unfed lone star ticks with increasing age. *Ann. Entomol. Soc. Am.* **78** 62–69.

NOTE ADDED IN PROOF

Additional references in vitellogenin and host immunoglobulin G are:

Chinzei, Y. (1983) Quantitative changes of vitellogenin and vitellin adult emale ticks *Ornithodoros moubata* during vitellogensis. *Mie Med. J.* **32** 117–127.

Chinzei, Y. and Yano, I. (1985) Fat body is the site of vitellogenin synthesis in the soft thick, *Ornithodoros moubata*. *J. Comp. Physiol. B.* **155** 671–678.

Chinzei, Y. and Yano, I. (1985) Vitellin is the nutrient reserve during starvation in the nymphal stage of a tick. *Experientia* **41** 948–950.

12

Some aspects of carbohydrate biochemistry in ticks

Taha T. M. Bassal

12.1 INTRODUCTION

Several reviews of carbohydrate biochemistry in insects are available (Gilmour 1961, 1965, Chefurka 1965, Wyatt 1967, Sacktor 1974, 1975, Chippendale 1978). In contrast, much less is known about tick biochemistry despite an upsurge of work in the past decade.

Carbohydrates as well as lipids and proteins are the principal classes of organic substances in the diet of animals, and they are well known contributors to the structure and function of animal tissues. The general principles of carbohydrate metabolism, known from most living groups, are applicable — with some specific exceptions — to ticks. In this chapter the limited amount of work concerning the carbohydrate biochemistry of ticks is considered and discussed, with hypotheses of the author.

12.2 OCCURRENCE OF SUGARS

12.2.1 Hemolymph

In ticks, as in other animals, glucose has a central place in carbohydrate metabolism. Glucose is the major neutral free sugar of tick hemolymph, e.g. in the argasids: *Argas* (*persicargas*) *arboreus* and *A.* (*P.*) *persicus* (Bassal 1970, Aboul-Nasr & Bassal 1971), *A.* (*A.*) *lagenoplastis* (Rehacek & Brzostowski 1969), *Ornithodoros moubata* and *O. savignyi* (Bassal 1973b); in the ixodids: *Boophilus microplus* (Tatchell 1969), *Hyalomma dromedarii* and *H. excavatum* (Bassal 1970, Aboul-Nasr & Bassal 1971), and *Dermacentor andersoni* (Parker *et al.* 1972, Bassal 1973a, Levenbook *et al.* 1980). Some polyols such as *m*-inositol are found in the hemolymph of *A.* (*A.*) *legenoplastis* (Rehacek & Brzostowski 1969) and *B. microplus* (Tatchell 1969), or *m*-inositol as well as glycerol in the hemolymph of *D. andersoni* (Levenbook *et al.* 1980).

Mannose was detected, using GLC, in hemolymph samples collected from fed *D. andersoni* females, but only at 5% that of the glucose content (Levenbrook *et al.* 1980). This non-physiological amount may be a digestion product absorbed into the hemolymph from the glycoprotein of lysed host tissues and from the bloodmeal. Alternatively, it could be a result of hexose interconversion, with glucose, particularly in the developing ovary (see section 12.4). Artifacts caused by the oozing of ovarian contents into the hemolymph during sample collection should be considered as a possible source as well.

The disaccharide, trehalose, has been detected in the bodies of non-tick arachnids (Fairbairn, 1958). Insects, with a few exceptions, have relatively low levels of glucose along with a higher level of trehalose (Wyatt 1961, 1967, Florkin & Jeuniaux 1974). Trehalose could not be detected in the hemolymph of larval *Phormia regina* (Evans & Dethier 1957) and *Celerio euphorbiae* during feeding (Mochnaka & Petryszen 1959). In other cases, the sugar content of the insect diet may affect the types of sugars in its hemolymph. Thus, fructose is present in the hemolymph of horse bot fly larvae *Gastrophilus intestinalis* (Levenbook 1950) and locusts (Hansen 1964). Fructose is probably derived from the high levan content of the horse's food in *Gastrophilus* and from a fructose-rich diet in locusts. Both fructose and sucrose are present in small amounts in the hemolymph of *Bombyx mori* (Wyatt *et al.* 1956) and *Megoura viciae* (Ehrhardt 1962) which feed on plant sap.

Sugar levels of tick hemolymph are generally lower than those of insects, and those of argasid species lower than ixodids. The hemolymph sugar levels in ticks range between 100 and 200 mg/100 ml in nymph and/or adult *A.* (*A.*) *lagenoplastis* (Rehacek & Brzostowski 1969), *A.* (*P.*) *arboreus* and *A.* (*P.*) *persicus* to 200 and 250 mg/100 ml in *H. excavatum* (Bassal 1970, Aboul-Nasr and Bassal 1971) and *B. microplus* (Tatchell 1969). It is worth noting that the sugar level in the plasma of the rabbit-host of *Hyalomma* ticks is about 110 mg/100 ml and 300 mg/100 ml in the plasma of the pigeon-host of *Argas* ticks (Bassal 1970). In contrast, the sugar level in both the hemolymph and gut-fluid is lower in argasids and higher in ixodids, probably because of a difference in the mechanisms of water elimination and concentration of the bloodmeal, i.e. coxal fluid filtrate versus salivary secretion in the argasids and ixodids respectively. Thus, the sugar level in the argasid hemolymph and gut-fluid is 0.16–0.66 while in ixodids it is 2.5–2.8 times that of host blood (Bassal 1970). Ixodid ticks, in general, concentrate the blood meal 2–3 times (Kitaoka & Fujisaki 1976). In insects (Wyatt 1967), the total sugar concentration in hemolymph is usually about 500 mg/100 ml, and in some cases it may exceed 5000 mg/100 ml.

12.2.2 Eggs

The carbohydrate biochemistry in the eggs of ticks has been studied (Roshdy 1967, Levenbook *et al.* 1980, Kamel *et al.* 1983, Ahmed & Bassal 1984). Glucose and mannose were identified as the free neutral sugars of *D. andersoni* eggs (Levenbook *et al.* 1980). Besides these two sugars, fructose, ribose, and deoxyribose were identified in the eggs of *H. dromedarii* (Kamel *et al.* 1983). In addition, *m*-inositol and glycerol were detected in both *D. andersoni* and *H. dromedarii* eggs. Glycogen was demonstrated histochemically in *A. persicus* eggs as finely dispersed particles in the cytoplasm between the yolk bodies (Roshdy 1967). Glycogen in the eggs of *H. dromedarii* is 1.5 times that of the neutral sugar content (Kamel *et al.* 1983). When *H. dromedarii* eggs were examined for amino- and non-amino sugar content, the amino-sugars of newly laid eggs (50.8 mg/gm eggs) were about 2.7 times that of the non-amino sugars (19 mg/gm eggs) (Ahmed & Bassal 1984). Prior to hatching, the amino- and non-amino sugar contents were 40.3 and 11.9 mg/gm eggs respectively.

Glucose was found in the eggs of ticks, but other sugars were detected as well (Levenbook *et al.* 1980, Kamel *et al.* 1983) probably because homogenates of whole eggs were analyzed. This included yolk and its digestion products as well as the intermediate and end products of the metabolic pathways occurring within the egg and the developing embryo. Thus, the nucleic acid sugar-moieties, ribose and deoxyribose, were observed possibly because they are intermediates of the pentose phosphate cycle (PPC), a predominant metabolic pathway in ticks (Bassal & Ahmed 1983). Nucleic acids were found in growing eggs in the ovary of *A. persicus* (Roshdy 1967) and embryonated eggs of *H. dromedarii* (Kamel *et al.* 1983). Fructose, mannose, *m*-insoitol, and glycerol could have been formed through metabolic interconversions, as in insects (Chefurka 1965). Mannose and *m*-inositol in particular may be liberated from conjugated glycoproteins. The considerable amount of amino-sugars (50.8 mg/gm eggs) in *H. dromedarii* eggs (Ahmed & Bassal 1984) is thought to be in the form of glycoproteins. In *A. persicus*, the yolk bodies contain large amounts of glycoproteins as estimated histochemically (Roshdy 1967). It is worth noting that the egg glycoproteins in insects have considerable amounts of mannose and glucosamine (Chippendale 1978, Kunkel, *et al.* 1984). For example, 77% of the dry weight of *Locusta migratoria* eggs are glycoproteins, 14% of its carbohydrate is in the form of mannose and glucosamine in a molar ratio of 1:7 (Yamasaki 1974).

12.3 METABOLISM

Sugars are absorbed from the tick gut into the hemolymph, transported to various tissues and metabolized. These sugars are converted to other molecules, and/or degraded, and the derived energy is used by tissues in various metabolic reactions. Glucose has a central role in carbohydrate metabolism in living organisms, and intermediate steps seem to be remarkably similar. Although glycolysis provides the main pathway for the anaerobic conversion of glucose to pyruvate in insects, the PPC cycle is also important (Chefurka 1965, Chefurka *et al.* 1970, Ela *et al.* 1970. Friedman 1970).

The order of appearance of carbon atoms in CO_2 when glucose is oxidized indicates whether the PPC or glycolytic-citric acid cycle (GCAC) is operating. When glucose is metabolized by the GCAC, carbons 3 and 4 are the first to appear in CO_2 as they appear in the carboxyl group of two equivalent pyruvate molecules. After two complete turns of the citric acid cycle (CAC), carbons 2, 5, 1, and 6 of glucose — i.e. carbons 2 and 3 of the two pyruvate molecules — appear as CO_2. If, on the other hand, glucose is metabolixed via PPC, carbon 1 appears first in CO_2 — resulting from decarboxylation of the formed gluconate-6-phosphate. Carbon 6 of glucose metabolized via the latter cycle may contribute to CO_2 if the resulting triose phosphate is converted to pyruvate and then decarboxylated and completely oxidized through the CAC. The CO_2 formed from carbon 6 will appear after that resulting from carbon 1.

12.3.1 Contribution of GCAC and PPC in glucose metabolism

In insects, extensive enzymatic studies have implicated the GCAC and PPC as major catabolic routes for the metabolism of carbohydrates (Chefurka 1965, Bailey, 1975, Crabtree & Newsholme 1975, Sacktor 1975, Chippendale 1978). However, the presence of the enzymes of any metabolic pathway merely suggests the possibility of any pathway being operative (Chefurka 1965).

Knowledge of the contribution of both GCAC and PPC to the metabolism of glucose by ticks is limited. The relative contribution of the two pathways was estimated, using the ratio of $^{14}CO_2$ output during *in vivo* metabolism of glucose-6-^{14}C (G_6) and glucose-1-^{14}C (G_1) by *A.* (*P.*) *arboreus* and *D. andersoni* (Bassal 1973a). Application of the radiorespirometric technique after injection of G_1, glucose-3,4-^{14}C ($G_{3,4}$) or G_6 into the tick hemocoel enabled determination of the relative contributions of the GCAC and PPC in *B. microplus* and *D. andersoni* (Tatchell & Schuntner 1972; Bassal & Ahmed 1983). Results showed that the PPC contributes about 25, 50 and 85% and the GCAC about 75, 50, and 15% of the glucose metabolized in unfed, partly fed, and engorged *D. andersoni* females respectively (Bassal & Ahmed 1983). In *B. microplus* females, the PPC contribution was about 50% on the day of repletion and 20% seven days after dropping from the host (Tatchell & Schuntner 1972). Thus, there is a general trend of high PPC activity during physiological states of ticks when metabolic rates are high, such as oogenesis.

The PPC includes reaction sequences which are of major importance to all living organisms. It is a source of NADPH, i.e. the metabolic hydrogen required for reductive biosynthetic processes, and the pentose phosphate required for nucleic acid and nucleotide biosynthesis. Other intermediates i.e. two-carbon units resulting from rearrangements catalyzed by transketolases and transaldolases are considered to be a source of glyoxylate (Bassal 1973a, Bassal & Ahmed 1983, 1984).

12.3.2 End products of glycolysis

End products of glycolysis in insects include lactate and α-glycerophosphate (Chefurka 1965, Sacktor 1975). The oxidation of reduced NAD produced during conversion of glyceraldehyde-3-phosphate to 3-phosphoglycerate in the glycolytic sequence is accomplished by two well-known reductions, i.e. pyruvate to lactate catalyzed by lactate dehydrogenase and dihydroxyacetone phosphate to α-glyerophosphate catalyzed by α-glycerophosphate dehydrogenase.

Insect muscle with low lactate dehydrogenase activity usually has high α-glycerophosphate dehydrogenase activity; however, some insect tissues contain a lactate dehydrogenase that is equally active (Zebe & McShan 1957). The relative activities of these two dehydrogenases may be related to the function and energy-yielding demands of the tissues (Chefurka 1965). Thus, flight muscles of flying insects and leg muscles which have a high, rapid demand for energy are rich in α-glycerophosphate dehydrogenase activity (Chefurka 1958, Zebe & McShan 1957, Sacktor 1970, 1975, Crabtree & Newsholme 1972, 1975). In ticks, isozymes of α-glycerophosphate dehydrogenase were demonstrated in *Ixodes ricinus* which were related to the zygocity, sex, and activity of individuals within the population (Healy 1979). Based on these results, it is likely that ticks have the enzymes necessary to reoxidize the reduced NAD of the glycolytic pathway.

12.3.3 Citric acid cycle (CAC)

The operation of the CAC is preceded by a series of reactions which form either its metabolites or acetyl groups. In carbohydrate metabolism, pyruvate is the last member of the glycolytic sequence preceeding the CAC. Pyruvate can also be derived from the triose phosphate end product of the PPC. Prior to pyruvate entry into the CAC, it must be oxidatively decarboxylated to acetyl-CoA ('active acetate') by pyruvate dehydrogenase. The derived acetyl-CoA combines with oxaloacetate to form citrate in a condensing reaction catalyzed by citrate synthase. Citrate is subsequently oxidized by a series of oxidations to oxaloacetate liberating CO_2 and reducing equivalents.

12.3.3.1 *Formation of acetyl-CoA*

The oxidative decarboxylation of pyruvate to acetyl-CoA is catalyzed by several different enzymes that work sequentially and are collectively designated the pyruvate dehydrogenase complex. The reaction is summarized as follows:

$$\text{Pyruvate} + \text{CoA} + \text{NAD}^+ \rightarrow \text{acetyl-CoA} + \text{NADH} + \text{H}^+ + \text{CO}_2$$

This reaction has been demonstrated in insects (Watanabe & Williams 1951, Rees 1954, Hoskins *et al.* 1956). Acetyl-CoA itself can be formed in insects by the direct activation of acetate through the acetate-activating enzyme (Rees 1954, Hoskins *et al.* 1956, Levenbook 1961).

Indirect evidence for the operation of pyruvate dehydrogenase in ticks was demonstrated *in vivo* (Bassal 1973a). The fate of radioactive pyruvate-1-^{14}C (PY) was followed after its injection into the hemocoel of *A. arboreus* and *D. andersoni* females. One-two times more radioactivity was released into CO_2 than that incorporated into carbohydrates, amino acids, and lipids.

The operation of the acetate-activating enzyme was indicated by incorporation of acetate-1-^{14}C (AC) into CAC intermediates of *A. arboreus* and *D. andersoni* females (Bassal 1973a, Bassal & Ahmed 1984) and *H. dromedarii* nymphs (Dajani *et al.* 1969). Acetate incorporations are known to be preceeded by acetate 'activation' to acetyl-CoA by acetate-activating enzyme:

$$\text{Acetate} + \text{CoA} + \text{ATP} \xleftrightarrow{Mg^{2+}} \text{acetyl-CoA} + \text{AMP} + \text{PP} \qquad (2)$$

12.3.3.2 *Formation of tricarboxylic acids*

Acetyl-CoA condenses with the last intermediate member of the CAC to form citrate through a reaction catalyzed by the condensing enzyme citrate synthase:

$$\text{Acetyl-CoA} + \text{oxaloacetate} + H_2O \longleftrightarrow \text{citrate} + \text{CoA} \qquad (3)$$

At equilibrium the reaction favors citrate formation (Kosicki & Srere 1961).

The formed citrate is dehydrated to *cis*-aconitate and then rehydrated to isocitrate. Aconitase catalyzes the reversible interconversion of the three tricarboxylic acids: citrate, *cis*-aconitate, and isocitrate. Citrate predominates at equilibrium (Krebs 1953). An unknown intermediate (UI) was supposed:

$$\begin{array}{ccccc} \text{citrate} & \longleftrightarrow & \text{UI} & \longleftrightarrow & \textit{cis}\text{-aconitate} \\ & & \updownarrow & & \\ & & \text{isocitrate} & & \end{array} \qquad (4)$$

Therefore, it is possible that *cis*-aconitate may not be an obligatory intermediate between citrate and isocitrate but may in fact be a side reaction from the main pathway (Speyer & Dickman 1956).

In ticks, citrate and isocitrate have been demonstrated in females of *A.* (*P.*) *arboreus* and *D. andersoni* (Bassal 1973a). Citrate, isocitrate, and *cis*-aconitate have been demonstrated in nymphs of *H. dromedarii* (Danjani *et al.* 1969), and citrate in eggs of *H. dromedarii* (Fahmy 1983). These investigations revealed that the concentrations of the tricarboxylic acids are less than those of dicarboxylic acid members of the CAC (Dajani *et al.* 1969, Fahmy 1983). Incorporations from AC into tricarboxylic acids via the CAC operation are also less in comparison to those into dicarboxylic acids (Dajani *et al.* 1969, Bassal 1973a). The enzymes concerned with the formation of the three tricarboxylic acids of the primal part of the CAC, i.e. citrate synthase and aconitase, have been demonstrated in embryonated eggs of *H. dromedarii* (Fahmy 1983). In addition, fluoroacetate (an inhibitor of aconitase) did not cause the accumulation of radioactive citrate or minimization of radioactive organic acid members of the CAC in nymphal homogenates of *H.*

dromedarii (Dajani *et al.* 1969). These data indicate that the CAC operates in the forward direction at low activity, and raises the possibility of additional auxiliary metabolic reactions and routes (see eq. (14)).

Citrate synthase and aconitase have been demonstrated in insects (Rees 1954, Hoskins *et al.* 1956, McGinnis *et al.* 1956, Hearfield & Kilby 1958, Levenbook 1961). The citrate concentration in insects is considerably higher than those of other organisms (Levenbook & Hollis 1961). In contrast, the concentration in ticks is less than that of other organisms. Therefore, the operation of a citrate cleavage enzyme and/or reversible reaction of citrate synthase in ticks can be assumed. Finally, it is worth noting that citrate synthase is controlled by the availability of oxaloacetate (Lowenstein 1967), an intermediate in reactions of an assumed pyruvate-dicarboxylic acid cycle in ticks.

12.3.3.3 *Formation of 5-carbon dicarboxylic acids*

Through the usual pathway of the CAC, isocitrate is converted to oxalosuccinate in a reversible reaction catalyzed by isocitrate dehydrogenase. In most organisms, an NADP-requiring enzyme is located in the cytoplasm and mitochondria (Lowenstein & Smith 1962), and an NAD-requiring enzyme is located in the mitochondria (Murthy & Rappoport 1963). Reversibility of the reaction involving the NAD-specific type is difficult to demonstrate because it is strongly inhibited by NADH (Lowenstein 1967). Isocitrate dehydrogenase is involved in a two step reaction: dehydrogenation which requires nicotinamide adenine (NA^+) that yields oxalosuccinate:

$$\text{isocitrate} + NA^+ \longleftrightarrow [\text{oxalosuccinate}] + NAH + H^+ \qquad (5)$$

and decarboxylation which requires a divalent metal ion (M^{2+}):

$$[\text{oxalosuccinate}] + H^+ \xleftrightarrow{M^{2+}} \alpha\text{-ketoglutarate} + CO_2 \qquad (6)$$

It appears that oxalosuccinate remains bound to the enzyme as an intermediate in the overall reaction.

Tick isocitrate dehydrogenases in embryonated eggs of *H. dromedarii* require both NADP and NAD (Fahmy 1983). Maximum activities of these two enzyme forms are 8.4 and 4.4 units/mg protein respectively. Both the concentration of and incorporation of AC into α-ketoglutarate in *H. dromedarii* nymphs (Dajani *et al.* 1969), *A.* (*P.*) *arboreus*, and *D. andersoni* females (Bassal 1973a) lie within the same range of those of dicarboxylic acids of the CAC. Also, the concentration of α-ketoglutarate is 2 times that of fumarate, 3 times that of succinate, and 8 times that of malate in embryonated eggs of *H. dromedarii* (Fahmy 1983). Therefore, α-ketoglutarate is probably the point at which a rise in the CAC-organic acid intermediates occurs in ticks i.e. the change from tri-to dicarboxylic acids. Hence, future studies of cytoplasmic NADP- isocitrate dehydrogenase, may support or contradict the hypothesis of an isocitrate hydrogenase operating in a reverse direction.

Insect isocitrate dehydrogenase seems to be mostly of the NADP-specific type (Rees 1954, Hoskins, *et al.* 1956, Levenbook 1961); however, a NAD-specific type has also been reported (McGinnis *et al.* 1956). Both types of isocitrate dehydrogenase have been found in insect muscle (Goebell & Klingenberg 1964, Ku and Cochran 1971, Hansford 1972, Zahavi & Tahori 1972).

12.3.3.4 *Formation of 4-carbon dicarboxylic acids*
α-Ketoglutarate results in the formation of succinyl-CoA in an oxidative decarboxylation reaction catalyzed by the α-ketoglutarate dehydrogenase complex (Sanadi *et al.* 1958). The equilibrium of the reaction strongly favors formation of succinyl-CoA, and hence the reaction should be considered as physiologically unidirectional. The reaction is catalyzed by four enzymes, each requiring specific cofactors, i.e. α-ketoglutarate decarboxylase, lipoyl reductase, transuccinylase, and dihydrolipoic dehydrogenase:

$$\alpha\text{-ketoglutarate} + \text{CoA} + \text{NAD}^+ \rightarrow \text{succinyl-CoA} + \text{NADH} + \text{H}^+ + \text{CO}_2 \qquad (7)$$

Succinyl-CoA is next converted to succinate by succinate thiokinase in the presence of GDP or IDP (i.e. 'X'DP) and P_i:

$$\text{Succinyl-CoA} + \text{XDP} + \text{P}_i \rightarrow \text{succinate} + \text{XTP} + \text{CoA} \qquad (8)$$

An alternative pathway converts succinyl-CoA to succinate coupled with conversion of acetoacetate to acetoacetyl-CoA, catalyzed by succinyl-CoA-acetoacetate-CoA transferase:

$$\text{Succinyl-CoA} + \text{acetoacetate} \rightarrow \text{succinate} + \text{acetoacetyl-CoA} \qquad (9)$$

α-Ketoglutarate dehydrogenase has been demonstrated in mitochondrial preparations of insects (Rees 1954, Hoskins *et al.* 1956, Levenbook 1961, Sacktor 1975).

The concentration and the incorporation of AC into succinate is higher than other members of the CAC in *H. dromedarii* nymphs (Dajani *et al.* 1969) and females of *A.* (*P.*) *arboreus* and *D. andersoni* (Bassal 1973a). However, in eggs of *H. dromedarii*, the maximum succinate concentration was only 39% of that of α-ketoglutarate (Fahmy 1983).

Further metabolism of succinate to fumarate is accomplished through a reversible dehydrogenation reaction catalyzed by succinate dehydrogenase which contains FAD and non-heme iron:

$$\text{Succinate} + \text{FAD} \longleftrightarrow \text{fumarate} + \text{FADH}_2 \qquad (10)$$

This enzyme is widespread in insects (Gilmour 1961, Sacktor 1974, 1975).

The concentration of fumarate in *H. dromedarii* eggs is usually about that of succinate (Fahmy 1983) or in *H. dromedarii* nymphs within the same range as that of CAC dicarboxylic acids (Dajani *et al.* 1969). The incorporation of AC into fumarate in *H. dromedarii* nymphs (Dajani *et al.* 1969) and females of *A.* (*P.*) *arboreus* and *D. andersoni* (Bassal 1973a) is about the same as that of the other dicarboxylic acids of the CAC. Succinate dehydogenase was demonstrated, in the forward direction, in embryonated eggs of *H. dromedarii* at a low maximal activity of 0.34 unit/mg protein (Fahmy 1983). Inhibition of succinate dehydrogenase by malonate in *H. dromedarii* nymphs (Dajani *et al.* 1969) leads to the accumulation of succinate but does not reduce organic acids following the succinate step. Therefore, the suggested auxiliary routes of the CAC probably operate, at least partly, to introduce acetate into the CAC after the succinate step.

Under the influence of fumarase which catalyzes the reversible interconversion of fumarate and malate, water is added to fumarate to form malate:

$$\text{fumarate} + H_2O \longleftrightarrow \text{L-malate} \tag{11}$$

This enzyme has been identified in insects (Gonda *et al.* 1957, Hearfield & Kilby 1958, Ito *et al.* 1958, Levenbook 1961, Sacktor 1974) and in ticks (Fahmy 1983). In *H. dromedarii* embryonated eggs, it was measured in the reverse direction and found to be a highly active enzyme (54 units/mg protein at maximum). However, there are no data available concerning its activity in the forward direction.

Malate is converted to oxaloacetate in a freely reversible reaction catalyzed by malate dehydrogenase. However, the reaction equilibrium favors the reduction of oxaloacetate:

$$\text{L-malate} + NAD^+ \longleftrightarrow \text{oxaloacetate} + NADH + H^+ \tag{12}$$

The malate dehydrogenase is both intra- and extra-mitochondrial and different at the two sites (Grimm & Doherty 1961). The two enzyme forms were demonstrated in insects (Delbruck *et al.* 1959). In embryonated eggs of *H. dromedarii* (Fahmy 1983), the activities of mitochondrial and cytoplasmic malate dehydrogenase were studied in the reverse direction, i.e. catalyzing the reduction of oxaloacetate to malate. Comparison of activities of these two forms showed that the cytoplasmic form is always higher than the mitochondrial enzyme (70 and 50 units/mg protein maximum activities respectively).

12.3.4 Specific aspects of organic acid metabolism

12.3.4.1 *The fate of organic acids*

The fate of PY, AC, citrate-1,5-^{14}C (CT), and glyoxylate-u-^{14}C (GX) in *D. andersoni* females was estimated by Bassal & Ahmed (1984). The estimated individual incorporations of radioactivity from one precursor into possible metabolites is referred to as the summed total incorporations (TI). Their specific characteristics have been clarified. Thus, 65% of the TI from PY was incorporated into CO_2, indicating that decarboxylation of PY is a principal route for its metabolism.

Production of $^{14}CO_2$ from AC in *A.* (*P.*) *arboreus* and *D. andersoni* females was elevated 0.3–0.5 times when this precursor was injected into the ticks together with glyoxylate (Bassal 1973a). This observation and the high incorporation of AC into dicarboxylic rather than tricarboxylic acids of the CAC (Bassal 1973a) can be explained by the operation of a malate synthase reaction as an auxiliary route for AC introduction into the CAC. This enzyme has been demonstrated in embryonated eggs of *H. dromedarii* (Fahmy 1983). However, its activity is low (0.34–0.50 unit/mg protein at maximum).

The first observation of the ticks' ability to metabolize extrinsic glyoxylate was in females of *A.* (*P.*) *arboreus*, *D. andersoni* (Bassal 1973a), and *H. dromedarii* (Bassal 1977). Participation of GX, through a malate synthesis catalyzed reaction, is a means for its contribution to the general metabolic pool of ticks. Other alternative routes such as amino transfer and participation in the succinate-glycine cycle (SGC) were considered before as a source of precursors for guanine and general purine biosynthesis (Bassal & Ahmed 1984).

About 25% of the TI from each of PY, AC, CT, and GX was traced to neutral sugars. This indicates that these acids participate, at least through the CAC, in metabolic reactions leading to phosphoenolpyruvate (PEP), and thereafter in gluconeogenesis reactions.

Formation of radioactive alanine, glycine, aspartate, glutamate, and serine from PY, AC, CT, and GX indicates the presence of either transaminases or enzymes that indirectly form ketoacid-metabolites which are then transaminated. Transaminases were demonstrated in *H. dromedarii* nymphs (Frayha *et al.* 1972, Dajani *et al.* 1973). Glutamate-oxaloacetate and glutamate-pyruvate transaminases were found in embryonated eggs of *H. dromedarii* (Fahmy 1983) and *D. andersoni* (Boctor & Kamel 1977). With respect to GX, about half was metabolized to amino acids of which one third was glycine. This finding could indicate direct transamination of glyoxylate to glycine (Bassal & Ahmed 1984).

12.3.4.2 *Metabolic reactions and routes of some organic acids*

Isocitratase was demonstrated in embryonated eggs of *H. dromedarii* (Kamel & Fahmy 1982) with a maximum activity of 7 units/mg protein. This isocitratase catalyzes the aldol cleavage of isocitrate to succinate and glyoxylate:

$$\text{Isocitrate} \longleftrightarrow \text{succinate} + \text{glyoxylate}\,. \qquad (13)$$

In insects, the presence of weak isocitratase activity was demonstrated in *Prodenia eridania* (Carpenter & Jaworski 1962) and *Tribolium confusum* (Beauddoin & Demonde 1972). Isocitratase of ticks (Kamel & Fahmy 1982) requires Mg^{2+} (K_m 2.1 mM) and sulfhydryl compounds for maximal activity with a pH optimum of 7.4. The K_m for isocitrate is 2.4 mM. At optimal pH, the enzyme is inhibited competitively by oxaloacetate (K_i 0.7 mM) and pyruvate (K_i 0.63 mM); and noncompetitively by acetyl-CoA (K_i 1.6 mM) and succinate (K_i 1.35 mM). It is also inhibited competitively and noncompetively by PEP, depending upon the pH.

Malate synthesis was demonstrated in the embryonated eggs of *H. dromedarii*, with a maximum activity of 0.34–0.50 unit/mg protein (Fahmy 1983). This enzyme catalyzes the condensation of acetyl CoA with glyoxylate to form malate:

$$\text{Acetyl-CoA} + \text{glyoxylate} + H_2O \rightarrow \text{L-malate} + \text{CoA} . \qquad (14)$$

Previous indirect evidence suggested the presence of this enzyme in *A.* (*P.*) *arboreus* and *D. andersoni* females. This evidence included production of more dicarboxylic than tricarboxylic acid intermediates of the CAC and more CO_2 when AC and GX were injected into the ticks (Bassal & Ahmed 1984).

Both isocitratase and malate synthesis are key enzymes of the 'glyoxylate cycle' in nematodes (Barrett *et al.* 1970) and microorganisms (Kornberg & Elsden 1961). This cycle operates in microorganisms capable of growing aerobically on acetate, ethanol, or fatty acids with an even number of carbon atoms. It is also important for converting fatty acids to carbohydrates. The cycle was apparent in ticks during embryogenesis of *H. dromedarii* when carbohydrates and proteins (Ahmed & Bassal 1984) and lipids (Kamel *et al.* 1983) were consumed for energy production. However, the presence of isocitratase and malate synthesis does not necessarily mean a complete glyoxylate cycle is operating, but rather a combined modified form of citric acid, glyoxylate, and dicarboxylate acid cycles. Thus, isocitratase can be considered as necessary for producing glyoxylate, supposedly an amino nitrogen carrier for glycine synthesis. Malate synthesis, on the other hand, is considered as an auxiliary route for biosynthesis of malate from glyoxylate and acetyl-CoA (eq. (14)).

As discussed earlier the highest percentage of the TI from GX enters amino acids, the most accessible being glycine. Glycine is a precursor for carbons 4 and 5 and nitrogen 7 and partly carbons 2 and 8 of the purine bases of nucleic acids. Guanine is the principal nitrogenous excretory material of ticks (Schultz 1955). It is synthesized from glyoxylate and glycine in fed *H. dromedarii* females and excreted in amounts equalling 13% of the female dry weight (Bassal 1977). Glycine provides fuel for generating the succinate-glycine cycle (SGC) (Nemeth *et al.* 1957). This cycle was demonstrated in ticks by Bassal (1973a, 1977) and Hamdy *et al.* (1974).

Some of the intermediates of the SGC are important metabolites for ticks. These include methylenetetrahydrofolate which supplies carbons 2 and 8 during the biosynthesis of guanine; δ-aminolevulinate for porphyrin synthesis; and α-ketoglutarate that is diverted into the CAC. The operation of the dicarboxylic acid part of the CAC in the reverse direction favors fumarate and later succinate formation which can then enter the SGC and thus provide an additional link between the two cycles. The reverse direction of comparatively active malate dehydrogenase and fumarase (72 and 54 units/mg protein at maximum respectively) was demonstrated in embryonated eggs of *H. dromedarii* (Fahmy 1983). The formation of succinate from fumarate could be predicted by the low activity of succinate dehydrogenase in the forward direction (i.e. 0.3 unit/mg protein at maximum). In other animals, succinate dehydrogenase, in the opposite direction, catalyzes the reduction of fumarate to succinate. However, another enzyme, fumarate reductase, found in microorganisms (Hirsch *et al.* 1963), is probably working in ticks to complement the highly active reversed malate dehydrogenase and fumarase, and demand for succinate in the SGC.

The concentration of α-ketoglutarate is high in comparison to tricarboxylic acids in ticks. This may show that α-ketoglutarate is produced not only from isocitrate but also from other precursors such as succinate via the SGC that leads to the formation of γ-ketoglutaraldehyde. It is worth noting that the high concentration of glutamate in the haemolymph and eggs of ticks (Boctor 1972, Boctor & Kamel 1977) could serve as a reserve molecule for the synthesis of α-ketoglutarate.

The pyruvate dehydrogenase reaction which produces acetyl-CoA is a principal route of pyruvate metabolism in ticks (eq. (1)). Pyruvate and PEP were shown to be suitable substrates for CO_2 fixation to produce dicarboxylic acids (Bassal 1973a, Dajani *et al.* 1973, Bassal and Ahmed 1984). Pyruvate is known to produce malate and/or oxaloacetate through malic enzyme in ticks (Fahmy 1983):

$$\text{Pyruvate} + \text{NADPH} + \text{H}^+ + \text{CO}_2 \longleftrightarrow \text{L-malate} + \text{NADP} \quad (15)$$

or through pyruvate carboxylase:

$$\text{Pyruvate} + \text{ATP} + \text{CO}_2 \rightarrow \text{oxaloacetate} + \text{ADP} + \text{P}_i \ . \quad (16)$$

The fixation of CO_2 to form oxaloacetate from PEP, on the other hand, is accomplished through other enzymes e.g. PEP carboxykinase:

$$\text{PEP} + \text{IDP} + \text{CO}_2 \longleftrightarrow \text{oxaloacetate} + \text{ITP} \ . \quad (17)$$

Dajani *et al.* (1973) concluded that PEP is preferred over pyruvate for CO_2 fixation in *H. dromedarii* nymphs.

Therefore, both CO_2 fixation and reversed malate dehydrogenase reactions include PEP, oxaloacetate, and malate as substrates. It is likely that a cycle including PEP and/or pyruvate, oxaloacetate, and malate is operating between the cytoplasm and mitochondria of ticks. This cycle starts in the cytoplasm with the PEP and/or pyruvate up to malate formation, then enters the mitochondria and is converted to oxaloacetate and later pyruvate. This proposed cycle may be useful for the transport of reducing equivalents across the mitochondrial membrane. It may also lead to the conversion of CAC intermediates from one metabolic site to pyruvate and acetyl-CoA to fuel the CAC at another site. An analogous shuttle including malate, oxaloacetate-aspartate, and glutamate-α-ketoglutarate was reported in insects (Crabtree & Newssholme 1975).

12.4 SUMMARY AND CONCLUSIONS

Glucose is a physiologically-circulating neutral free sugar of tick hemolymph. Polyols such as *m*-inositol and glycerol are also present in tick hemolymph. The concentration of glucose in tick hemolymph is less than that of the host blood in argasids but more than twice that of the host blood in ixodids. Glucose, mannose, fructose, ribose, deoxyribose, *m*-inositol, glycerol, and glycogen have been estimated in tick eggs. Glucose and glycogen represent the free forms of carbohydrates, while the others are considered metabolites. Amino-sugars are 2–3 times that of the non-amino sugars in tick eggs. The amino-sugars are probably in the form of glycoproteins and are utilizable reserve carbohydrates during embryogenesis. Ticks metabolize glucose via the glycolytic-citric acid and pentose phosphate cycles. The contribution of the pentose phosphate cycle increases during physiologically active states of the tick's metabolism, utilizing 85% of the glucose during maximum oogenesis of ixodids.

Gluconeogenesis proceeds with the following precursors: pyruvate, acetate, citrate, and glyoxylate. Enzymes and intermediates of the citric acid cycle were demonstrated in ticks. Malate dehydrogenase and fumarase (in the reverse direction) and *cis*-aconitase are the most active enzymes of the citric acid cycle (up to 72-140 units/mg protein). Citrate synthase and isocitrate dehydrogenase are moderately active enzymes of this cycle (up to 6–12 units/mg protein). The concentrations of dicarboxylic acids of the citric acid cycle are always higher than those of the tricarboxylic acids. An auxiliary route for the introduction of acetyl-CoA into the citric acid cycle occurs through a malate synthase reaction.

Pyruvate decarboxylation is a functional metabolic route. Also, fixation of CO_2 to form dicarboxylic acids from pyruvate and/or phosphoenolpyruvate was demonstrated. In this context, malic enzyme operation was reported. The data on carbon dioxide fixation and reversed active malate dehydrogenase and fumarase have lead us to propose a limited cycle including pyruvate and/or phosphoenolpyruvate, oxaloacetate, and malate operating between the cytoplasm and mitochondria. The cycle may transport reducing equivalents between these two cellular compartments, and convert the 4-carbon intermediates of the citric acid cycle to metabolizable pyruvate molecules.

Ticks are able to metabolize extrinsic glyoxylate, and produce intrinsic glyoxylate as a result of the activity of isocitratase, and 2-carbon units accompanying rearrangement reactions of the pentose phosphate cycle. Glycine, glutamate, aspartate, alanine, and serine were found to be metabolites of glyoxylate, acetate, pruvate, and citrate. Glyoxylate is a principal source of glycine. Glycine is an indispensable precursor of carbons 4 and 5 and nitrogen 7 and also to some extent carbons 2 and 8 of purines such as guanine.

A succinate-glycine cycle is operating in ticks which utilizes succinate and glycine for the production of (a) δ-aminolevulinate for porphyrin biosynthesis, (b) γ-ketoglutaraldehyde, a precursor of α-ketoglutarate, and

(c) methylenetetrahydrofolate (from the α-carbon of glycine), a source of carbons 2 and 8 for biosynthesis of purines, and regeneration of succinate. The citric acid cycle and succinate-glycine cycle are linked through succinate and α-ketoglutarate. Functions of isocitratase and malate synthase, the pyruvate and/or phosphoenolpyruvate-dicarboxylic acid cycle, the production of glycine from glyoxylate from different sources, and linkage of the succinate-glycine cycle with the citric acid cycle are considered to be specific aspects of organic acid metabolism in ticks.

ACKNOWLEDGEMENTS

This manuscript is dedicated to the world's leader of tick science — Harry Hoogstraal. I am grateful to Dr Afaf S. Fahmy, Faculty of Science, Al-Azhar University, and to members of the Biochemistry Department, National Research Centre, for providing me with the available thesis, and cooperation.

REFERENCES

Aboul-Nasr, A. E., & Bassal, T. T. M. (1971) Biochemical and physiological studies of certain ticks (Ixodoidea). The sugar content and concentration in *Argas* and *Hyalomma* biological fluids, *J. med. Ent.* **8** 521–524.

Ahmed, S. H. & Bassal, T. T. M. (1984) The general energy metabolism and economy, and morphogenetic observations on the embryo of *Hyalomma dromedarii*, *J. Egypt Soc. Parasit.* **14** 455–461.

Bailey, E. (1975) Biochemistry of insect flight. 2. Fuel supply. In: *Insect biochemistry and function* (Candy, D. J. & Kilby, A. A., Eds.), Chapman and Hall, London. p. 89–176.

Barrett, J., Ward, C. W., & Fairbairn, D. (1970) Glyoxylate cycle and the conversion of triglycerides to carbohydrates in developing eggs of *Ascaris lumbricoides*, *Comp. Biochem. Physiol.* **35** 577–586.

Bassal, T. T. M. (1970) Physiological and biochemical studies on carbohydrates in the biological fluids of certain Egyptian ticks, M.Sc. Thesis, Cairo University.

Bassal, T. T. M. (1973a) Carbohydrate metabolism in *Argas* (*Persicargas*) *arboreus* (*Argasidae*) and *Dermacentor andersoni* (*Ixodidae*), *in vivo* studies using ^{14}C-substrates, Ph.D. Thesis, Cairo University.

Bassal, T. T. M. (1973b) Detection on silicic acid chromatograms of trehalose and other sugars in arthropod fluids, *J. med. Ent.* **10** 228.

Bassal, T. T. M. (1977) Demonstration of guanine biosynthesis in *Hyalomma dromedarii* ticks using ^{14}C-labelled glycine and glyoxylate, *J. Parasit.* **65** 758–759.

Bassal, T. T. M. & Ahmed, S. H. (1983) Radiorespirometric determination of contribution of pentose phosphate cycle to metabolism of glucose in *Dermacentor andersoni*, *J. Egypt. Soc. Parasit.* **13** 101–107.

Bassal, T. T. M. & Ahmed, S. H. (1984) Some aspects of organic acid metabolism in *Dermacentor andersoni* female, *J. Egypt. Soc. Parasit.* **14** 411–420.

Beaudoin, A. R. & Lemonde, A. (1972) Evolution de acides organiques et cycle du glyoxylate chez *Tribolium confusum*, *J. Insect Physiol.* **18** 1427–1434.

Boctor, F. N. (1972) Biochemical and physiological studies of certain ticks (Ixodoidea). Free amino acids in female *Argas* (*Persicargas*) *arboreus* (*Argasidae*) analyzed by gas-liquid chromatography, *J. med. Ent.* **9** 201–204.

Boctor, F. N. & Kamel, M. Y. (1977) Biochemical studies of tick embryogenesis. Free amino acid pools during embryogenesis of *Dermacentor andersoni*, *Comp. Biochem. Physiol.* **56B** 169–173.

Carpenter, W. D. & Jaworski, E. G. (1962) The presence of isocitrate lyase activity in the southern army-worm, *Prodenia eridania* (Cramer), *Biochim. Biophys. Acta* **58** 369–371.

Chefurka, W. (1958) On the importance of α-glycerophosphate dehydrogenase in glycolysing insect muscle, *Biochim. Biophys. Acta* **28** 660–661.

Chefurka, W. (1965) Intermediary metabolism of carbohydrates in insects. In: *The physiology of insecta* (Rockstein, M., Ed.), Vol. 2, Academic Press, New York. p. 581–667.

Chefurka, W., Horie, Y., & Robinson, J. R. (1970) Contribution of the pentose cycle to glucose metabolism by insects, *Comp. Biochem. Physiol.* **37** 143–165.

Chippendale, G. M. (1978) The functions of carbohydrates in insect life processes. In: *The biochemistry of insects* (Rockstein, M., Ed.), Academic Press, New York, p. 1–55.

Crabtree, B. & Newsholme, E. A. (1972) The activities of phosphorylase, hexokinase, phosphofructokinase, lactate dehydrogenase and glycerol-3-phosphate dehydrogenase in muscles from vertebrates and invertebrates, *Biochem. J.* **126** 49–58.

Crabtree, B. & Newsholme, E. A. (1975) Comparative aspects of fuel utilization and metabolism by muscle. In: *Insect muscle* (Usherwood, P. N. R., Ed.), Academic Press, London. p. 405–500.

Dajani, R. M., Frayha, G. J., & Sweatman, G. K. (1969) A study of the citric acid cycle in the tick *Hyalomma dromedarii*, *Comp. Biochem. Physiol.* **28** 347–357.

Dajani, R. M., Frayha, G. J., Samia, L. H., & Sweatman, G. K. (1973) Carbon dioxide fixation by the nymphs of the tick, *Hyalomma dromedarii*, *J. Parasit* **59** 897–899.

Delbruck, A., Zebe, E., & Bucher, T. (1959) Uber Verteilungsmuster von Enzymen des Engergic lieferden Stoffwechsels in Flugmuskel. Sprungmuskel und Fettkorper von *Locusta migratoria* und ihre cytologische Zuordnung, *Biochem. Z.* **331** 273–296.

Ehrhardt, P. (1962) Untersuchungen zur Stoffwechselphysiologie von *Megoura viciae* Burkt. einer phloemsaugenden Aphide, *Z. vergl. Physiol.* **46** 169–211.

Ela, R., Chefurka, W., & Robinson, J. R. (1970) *In vivo* glucose metabolism in the normal and poisoned cockroach, *Periplaneta americana*, *J. Insect Physiol.* **16** 2137–2156.

Evans, D. R. & Dethier, V. G. (1957) The regulation of taste threshold for sugars in the blowfly, *J. Insect Physiol.* **1** 3–17.

Fahmy, A. S. (1983) Biochemical studies on organic acids metabolism in ticks, Ph.D. Thesis, Al-Azhar University.

Fairbairn, D. (1958) Trehalose and glucose in helminths and other invertebrate, *Can. J. Zool.* **36** 787–795.

Florkin, M. & Jeuniaux, Ch. (1974) Hemolymph: Composition. In: *The physiology of insecta* (Rockstein, M., Ed.), Vol. 5, Academic Press, New York. p. 255–307.

Frayha, G. J., Dajani, R. M., Samia, L. H., & Sweatman, G. K. (1972) Dicarboxylic and amino acid formation by nymphs of the tick *Hyalomma dromedarii*, *Insect biochem.* **2** 78–86.

Friedman, S. (1970) Metabolism of carbohydrates in insects. In: *Chemical zoology* (Florkin, M., and Scheer, B. T., Eds.), Vol. 5A, Academic Press, New York. p. 167–197.

Gilmour, D. (1961) *The biochemistry of insects*, Academic Press, New York.

Gilmour, D. (1965) *The metabolism of insects*, Oliver and Boyd, Edinburgh.

Goebell, H. & Klingenberg, M. (1964) DPN-spezifische isocitrate-dehydrogenase dermitochondria. 1. Kinetische Eigenschaften, Vorkommen und Funktion der DPN-spezifichen isocitrate dehydrogenase, *Biochem. Z.* **340** 441–464.

Gonda, O., Traub, A., & Avi-Dor, Y. (1957) The oxidative activity of particulate fractions from mosquitoes, *Biochem. J.* **67** 487–493.

Grimm, F. C. & Doherty, D. G. (1961) Properties of the two forms of malic dehydrogenase from beef heart, *J. biol. Chem.* **236** 1980–1985.

Hamdy, B. H., Taha, A. A., & Sidrak, W. (1974) Haemolymph and egg pigment of *Dermacentor andersoni* (Ixodidae) with reference to δ-aminolevulinic acid catabolism, *Insect Biochem.* **4** 205–213.

Hansen, O. (1964) Effect of diet on the amount and composition of locust blood carbohydrates, *Biochem. J.* **92** 333–337.

Hansford, R. G. (1972) Some properties of pyruvate and 2-oxoglutarate oxidation by blowfly flight-muscle mitochondria, *Biochem. J.* **127** 271–283.

Healy, J. A. (1979) Analysis of α-glycerophosphate dehydrogenase variability in the tick *Ixodes ricinus* (Acari: Ixodidae), *Genetica* **50** 19–30.

Hearfield, D. A. H. & Kilby, B. A. (1958) Enzymes of the tricarboxylic acid cycle and cytochrome oxidase in the fat body of the desert locust, *Nature* **181** 546–547.

Hirsch, C. A., Raminsky, M., Davis, B. D., & Lin, E. C. C. (1963) A fumarate reductase in *Escherichia coli* distinct from succinate dehydrogenase, *J. biol. Chem.* **238** 3770–3774.

Hoskins, D. D., Cheldelin, V. H., & Newburgh, R. W. (1956) Oxidative enzyme systems of the honey bee *Apis mellifera* L., *J. gen. Physiol.* **39** 705–713.

Ito, T., Horie, Y., & Ishikawa, S. (1958) Oxidative enzymes of the midgut of the silkworm *Bombyx mori*, *J. Insect Physiol.* **2** 313–323.

Kamel, M. Y. & Fahmy, A. S. (1982) Biochemical studies of tick embryogenesis. V. Purification and partial characterization of isocitrate lyase from eggs of the tick *Hyalomma dromedarii*, *Comp. Biochem. Physiol.* **72B** 107–115.

Kamel, M. Y., Shalaby, F. Y., Ghazy, A. M., & Farid, N. M. (1983) Variations in the amount of glycogen, free sugars, lipids and phospholipids during embryonic development of *Hyalomma dromedarii* (Acarina: Ixodidae), *Insect Biochem.* **13** 115–121.

Kitaoka, S. & Fujisaki, K. (1976) Accumulating process and concentration ratios of ingested blood meals in larvae and nymphs of ten species of ticks. *Natn. Inst. Anim. Hlth. Q.* **16** 114–121.

Kornberg, H. L. & Elsden, S. R. (1961) The metabolism of 2-carbon compounds by microorganisms, *Adv. Enzymol.* **23** 401–470.

Kosicki, G. W. & Srere, P. A. (1961) Kinetic studies on the citrate-condensing enzyme, *J. biol. Chem.* **236** 2560–2565.

Krebs, H. A. (1953) The equilibrium constants of the fumarase and aconitase systems, *Biochem. J.* **54** 78–82.

Ku, T. Y. & Cochran, D. G. (1971) Isocitrate dehydrogenase from fly and cockroach thoracic muscle mitochondria, *Insect Biochem.* **1** 81–96.

Kunkel, J. G., Estes, P. A., Gochoco, C. A., Purcell, J., Wojchowski, D. M., & Nordin, J. H. (1984) Vitellogenin of the german cockroach, *Blatella germanica*: structure and function of high mannose oligosaccharide in secretion, uptake and storage, *Proc. XVII International Congress of Entomology*, 275, August, 20–26, Hamburg.

Levenbook, L. (1950) The composition of horse bot fly (*Gastrophilus intestinalis*) larva blood, *Biochem. J.* **47** 336–346.

Levenbook, L. (1961) Organic acids in insects. II. Tricarboxylic acid cycle and related enzymes in the larvae of the southern armyworm, *Prodenia eridania*, *Arch. Biochem. Biophys.* **92** 114–121.

Levenbook, L. & Hollis, V. W. (1961) Organic acids in insects. I. Citric acid, *J. Insect Physiol.* **6** 52–61.

Levenbook, L., Boctor, F. N., & Fales, H. M. (1980) Biochemical studies of tick embryogenesis. Free sugars in adult hemolymph and during embryogenesis of *Dermacentor andersoni*, *J. Insect Physiol.* **26** 381–383.

Lowenstein, J. M. (1967) The tricarboxylic acid cycle. In: *Metabolic pathways* (Greenberg, D. M., Ed.) **1**, Academic Press, New York. p. 146–270.

Lowenstein, J. M. & Smith, S. R. (1962) Intra- and extra-mitochondrial isocitrate dehydrogenase, *Biochim. Biophys. Acta* **56** 385–387.

McGinnis, A. J., Cheldelin, V. H., & Newburgh, R. W. (1956) Enzyme studies of various stages of the blowfly *Phormia regina* (Meig.), *Arch. Biochem. Biophys.* **63** 427–436.

Mochnacka, I. & Petryszyn, C. (1959) Trehalose in *Celerio euphorbiae*, *Acta biochim. Pol.* **6** 307–311.

Murthy, M. R. V. & Rappoport, D. A. (1963) Biochemistry of the developing rat brain. II. Neonatal mitochondrial oxidations, *Biochim. Biophys. Acta* **74** 51–59.

Nemeth, A. M., Russel, C. S., & Shemin, D. (1957) The succuinate-glycine cycle. II. Metabolism of δ-aminolevulinic acid, *J. biol. Chem.* **229** 415–422.

Parker, R., Ribi, E., & Yunker, C. E. (1972) Application of centrifugal chromatography to the separation of sugars in arthropod hemolymph and tissue culture. In: *Insect and mite nutrition* (Rodriguez, J. G., Ed.), North-Holland Publishing Company, Amsterdam, p. 397–406.

Rees, K. R. (1954) Aerobic metabolism of the muscle of *Locusta migratoria*, *Biochem. J.* **58** 196–202.

Rehacek, J. & Brzostowski, H. W. (1969) A tick tissue culture medium based on analyses of tick hemolymph, *J. Insect Physiol.* **15** 1431–1436.

Roshdy, M. A. (1967) Histochemical studies in the developing oocytes of the tick, *Argas persicus* Oken, *Proc. zool. Soc. Egypt* **2** 11–26.

Sacktor, B. (1970) Regulation of intermediary metabolism, with special reference to the control mechanisms in insect flight muscle, *Adv. Insect Physiol.* **7** 267–347.

Sacktor, B. (1974) Biological oxidations and energetics in insect mitochondria. In: *The physiology of insecta* (Rockstein, M., Ed.), Vol. 4. Academic Press, New York. p. 271–353.

Sacktor, B. (1975) Biochemistry of insect flight. 1. Utilization of fuels by muscle. In: *Insect biochemistry and function* (Candy, D. J., & Kilby, B. A., Eds.), Chapman and Hall, London. p. 1–88.

Sanadi, D. R., Langley, M., & White, F. (1958) On the mechanisms of oxidative decarboxylation of α-keto acids, *Biochim. Biophys. Acta* **29** 218–219.

Schulze, P. (1955) Uber Ausscheidungsvorgange und Ablagerung von Stoffwechselprodukten bei den Zecken, *Z. Parasitenk.* **17** 217–236.

Speyer, J. F. & Dickman, S. R. (1956) On the mechanism of action of aconitase, *J. biol. Chem.* **220** 193–208.

Tatchell, R. (1969) The ionic regulatory role of the salivary secretion of the cattle tick *Boophilus microplus*, *J. Insect Physiol.* **15** 1421–1430.

Tatchell, R. & Schuntner, C. A. (1972) Glucose metabolism in the cattle tick *Boophilus microplus*, *J. Insect Physiol.* **18** 283–288.

Watanabe, M. E. & Williams, C. M. (1951) Mitochondria in flight muscles of insects. I. Chemical composition and enzymatic content, *J. gen. Physiol.* **34** 675–689.

Wyatt, G. R. (1961) The biochemistry of insect hemolymph, *Ann. Rev. Ent.* **6** 75–102.

Wyatt, G. R. (1967) The biochemistry of sugars and polysaccharides in insects, *Adv. Insect Physiol.* **4** 287–360.

Wyatt, G. R., Loughheed, T. C., & Wyatt, S. S. (1956) The chemistry of insect hemolymph. I. Organic components of the hemolymph of the silkworm, *Bombyx mori*, and two other species, *J. gen. Physiol.* **39** 853–868.

Yamasaki, K. (1974) Yolk protein from eggs of *Locusta migratoria* and glycopeptides from the proteins, *Insect Biochem.* **4** 411–422.

Zahavi, M. & Tahori, A. S. (1972) Activity of mitochondrial NAD-linked isocitric dehydrogenase in alatiform and apteriform larvae of *Myzus persicae*, *J. Insect Physiol.* **18** 609–614.

Zebe, E. C. & McShan, W. H. (1957) Lactic and α-glycerophosphate dehydrogenases in insects, *J. gen. Physiol.* **40** 779–790.

ADDENDUM

After preparation of the manuscript of the present chapter, a work was undertaken in our laboratory to study the nature and metabolism of carbohydrates in the biological fluids of the argasid tick *O. moubata*. The GLC chromatograms show that the hemolymph contains both the free sugars glucose and mannose, and the polyols *m*-inositol and glycerol. The concentration of mannose is more than half that of glucose (Sallam, M. H. & Bassal, T. T. M., unpublished work).

Part 2

Ecology, behavior, and tick-host relationship

13

Characterization of *Amblyomma americanum* — derived salivary gland proteins responsible for the elicitation of host immunity.

Stephen J. Brown and Philip W. Askenase

13.1 INTRODUCTION

Several laboratories concerned with the problem of developing an efficient and safe tick control program have turned to examining the feasibility of artificially immunizing cattle against ticks using various types of crude tick extracts (Riek 1958, Willadsen *et al.* 1978, McGowan *et al.* 1981). Similar studies using guinea pigs as model hosts for bovines have been conducted in the laboratory (Trager 1939, Garin & Grabarev 1972, Allen & Humphreys 1979, McGowan *et al.* 1980, Wikel 1981, Brown *et al.* 1984b, Brown & Askenase 1986). However, few reports have been published describing the isolation and purification of tick products capable of eliciting skin responses in animals previously exposed to ticks, or, more importantly, capable of inducing host immunity to ticks.

Strong immediate skin responses were evoked in the dermis of tick-sensitized cattle by the injection of a protein isolated from the saliva of *Boophilus microplus* (Geczy *et al.* 1971). This skin reactive factor was noted to have esterase activity and estimated to have a molecular weight of 30 000.

Similarly, a skin reactive substance of approximately 60 000 mw, also described as an esterase, was isolated from crude *B. microplus* extracts (Willadsen & Williams 1976, Willadsen & Riding 1979). In a more recent study, antibodies against *B. microplus*-derived phosphomonoesterases were found in tick-infested cattle (Reich & Zorzopulos 1980). However, none of the described studies have demonstrated that the isolated material is the immunogen responsible for the induction of host immunity.

Our studies on the mechanism of immunity to *Amblyomma americanum* ticks in guinea pig hosts have concentrated on defining the effector cells and the mechanism of their recruitment, as reviewed recently (Brown & Askenase 1983). This chapter will review our studies aimed at the identification and isolation of tick antigen(s) responsible for the induction of host immune resistance. These studies addressed the following questions. First, can guinea pigs be protected against *A. americanum* feeding by artificial immunization using tick-derived substances, and conventional adjuvants? Second, can the immunogen(s) in the tick extract be identified, as to the site of production, and characterized as to their molecular weight, by employing anti-tick antibodies that transfer immune resistance to ticks? Third, what are the kinetics of immunogen production? Fourth, what are some of the biochemical properties of the immunogen(s)?

13.2 MATERIALS AND METHODS

Adult *A. americanum* ticks served as salivary gland donors for the production of antigen, and both larval and adult ticks were used in feeding assays. Larval ticks were 4–6 weeks of age at the time of infestation, and adults were 1–3 months of age. Both stages were maintained at 85 ± 5% relative humidity, 25°C, and 12 h photophase. Outbred female albino Hartley guinea pigs (CrL:HA:BR) ranging in weight from 250 to 450 g were used.

13.2.1 Tick infestation procedure

Larval ticks were enumerated into groups of 100 in the bottom half of Petri culture dishes (35 × 10 mm, Costar, Cambridge, MA., USA) on ice (Brown & Askenase 1981). The dishes were inverted and fastened to the shaved flank of guinea pigs with adhesive tape. Adult ticks (10 females and 10 males/host) were confined to hosts within the screw-top portion (3.7 cm in diam.) of polyethylene reagent bottles fastened with adhesive tape to the shaved flank of gunea pigs (Brown 1982). Primary infestations were on the right flank and challenge infestations were on the opposite flank 26 days later. At the end of the tick feeding period, the percentage tick yield (number fed/number applied × 100) and mean tick weight were determined. In addition, the level of tick rejection and the percentage weight reduction of fed ticks from challenged hosts was calculated (1 — tick yield or weight from challenged hosts/tick yield or weight from controls × 100). All data are presented as the mean ± standard error and differences were considered significant if $p < 0.05$ (*Student's t* test).

13.2.2 Antigen preparation

Since the salivary gland represents the site of production of enzymes and other substances required for feeding, extracts of this gland from partially fed adult female ticks were used as antigen. The salivary gland of female ticks increases in size and total protein content as feeding progresses (McSwain *et al.* 1982); the feeding period lasting 10–12 days. To optimize our ability to artificially immunize guinea pigs, we selected females that fed for 6 days, since salivary glands begin to degenerate as the feeding period terminates.

Salivary gland harvest and antigen preparation was according to Brown *et al.* (1984b). Partially fed ticks were removed from guinea pig hosts and the ventral surface placed on a small area of wax, in a wax-filled Petri dish, melted by the heated tip of a spatula. Then, viewing through a dissecting microscope, the ticks dorsal surface was separated and removed from the body by cutting the lateral edge of the cuticle with a #11 scalpel blade. The exposed organs were immersed in 10 mM phosphate buffered saline (0.7 mM NH_2PO_4, 4 mM K_2HPO_4, 0.15 M NaCl) (PBS) pH 7.4 to facilitate observation and prevent dessication of the paired salivary glands. Glands were removed intact with fine tipped forceps and placed in 200 μl PBS pH 7.4, triturated and centrifuged at 15 000 g for 15 min. The supernatant was decanted, stored at −70°C in the absence of protease inhibitors, and thereafter referred to as salivary gland antigen (SGA).

Cement material is derived from the salivary glands and, unlike other salivary secretions that remain soluble, it solidifies around the mouthparts to aid in anchoring the tick to the host. Cement accumulates in large quantities around the mouthparts of *A. americanum* female ticks allowed to feed for more than 6 days, so that it can be collected. Thus, the immunogenicity of this material was also examined. Collected cement was removed from contaminating host tissue by physical separation and rinsing in PBS. This material was triturated in 10 mM PBS pH 7.4, centrifuged at 15 000 g for 5 min, the supernatant decanted, stored at −70°C in the absence of protease inhibitors, and referred to as cement antigen (CA).

13.2.3 Immunization with crude extracts

Naive guinea pigs were immunized with SGA in saline or emulsified with Freund's complete (CFA) or incomplete (IFA) adjuvant (Difco Laboratories, Detroit, MI., USA) and injected into each footpad at 0.1 ml/footpad (antigen:adjuvant = 1:1 v/v). The immunizing dose of protein per guinea pig was arbitrarily selected at 280 μg based upon the report that guinea pigs parasitized by one adult *Ixodes holocyclus* tick were protected from a challenge infestation with larvae (Askenase *et al.* 1982). Since the total salivary gland protein in a single engorged adult female *A. americanm* tick is approximately 150 μg (McSwain *et al.* 1982), a dose of 280 μg represented approximately a 2-fold increase in total salivary gland protein, and therefore should be sufficiently immunogenic. Also included in this experiment were animals given IFA or CFA alone followed by tick infestation to control for the effect of these adjuvants on tick feeding.

Animals were immunized with a total of 280 μg of SGA in 0.4 ml IFA or CFA emulsion (0.1 ml/footpad). Additional salivary gland pairs were triturated in PBS alone. Animals immunized with SGA in CFA were skin tested with purified protein derivative (PPD) (Connaugat Laboratories, Ontario, Canada) from *Mycobacterium tuberculosum* at the time of tick challenge 17 days later, to verify the efficacy of the CFA preparation. Negative controls consisted of naive guinea pigs, and positive controls consisted of animals actively immunized by tick feeding and were the same age, sex, and weight as animals immunized with tick extracts. Cement material was administered at a single dose of 50 μg/0.4 ml in IFA. This dose was selected because after 6 days of tick feeding, the ratio of total cement to salivary gland protein is approximately 11 μg:53 μg respectively, or 1:5.

In a subsequent experiment, the minimum effective dose of SGA required to protect guinea pigs from larval *A. americanumn* infestation was examined. SGA extract was administered in 5-fold dilutions of a 500 μg stock in IFA only, based upon the results of the first experiment. Guinea pigs were challenged with larval ticks 17 days postimmunization.

13.2.4 Antigen identification

13.2.4.1 Preparation of radiolabeled tick salivary gland extract

Radiolabeling methods were according to Brown *et al.* (1984b). SGA from partially fed female ticks were radiolabeled by two different techniques. In the first procedure, 6-day fed female ticks were removed from their hosts, immobilized in solidified paraffin on their ventral surface, and inoculated through the cuticle using a 20 μl Hamilton syringe (model 702-N, Reno, NV., USA) with 0.3 mCi each of ^{35}S methionine and cysteine (New England Nuclear, Boston, MA., USA). The ticks sat at room temperature for 6 h, and then their salivary glands were removed and subjected to a rigorous extraction procedure involving several quick freeze–thaws in the presence of TEN buffer (20 mM Tris-HCl pH 7.4, 1 mM EDTA, 100 mM NaCl), 1% Triton X-100 and PMSF at 10^{-3} M followed by centrifugation at 15 000 g for 15 min. The pellet was subjected to the same procedure and the two supernatants were pooled, dialyzed against TEN buffer pH 7.4 for 48 h, centrifuged at 100 000 g for 1 h and stored at −30°C until used in immunoprecipitation with anti-tick serum for the detection of antigen.

In the second radiolabeling technique, salivary gland extracts and cement material from female ticks were iodinated *in vitro* with ^{125}I using two procedures. In the first procedure, 10 to 80 μg of SGA in 1 ml of PBS was added to tubes (12 × 75 mm) containing 100 to 500 μCi of Na ^{125}I. The iodination reaction was initiated by the addition of 3 to 6 Iodo-beads (N-Chlorobenzenesulfonamide) (Pierce Chemical Co., Rockford, IL., USA) and allowed to proceed in capped tubes for 15 min at room temperature (Markwell 1982). In the second procedure, 20 to 60 μg of SGA was added to Iodo-gen (1,3,4,6-tetrachloro-3-α-, 6-α-diphenylglycouril) coated test tubes

(12 × 75 mm) (Fraker & Speck 1978). The reaction was started by adding 125 μCi of Na ^{125}I and was allowed to proceed for 15 min at room temperature.

In both iodination procedures, ^{125}I labeled SGA proteins were separated from free iodide by passage through 5 ml G-25 coarse columns (Pharmacia, Piscataway, N.J., USA) made in 10 mM NH_4HCO_3 pH 8.0. Samples comprising the protein-containing peak were pooled, centrifuged at 100 000 g for 1 h at 50°C, lyophilized, and reconstituted in TEN buffer (20 mM Tris-HCl pH 7.4, mM EDTA, 100 mM NaCl) pH 7.4 in preparation for immunoprecipitation.

13.2.4.2 Immunoprecipitation of salivary gland proteins by guinea pig anti-tick serum

Immune serum that transferred resistance against *A. americanum* was obtained from guinea pigs exposed to larval infestations (Brown & Askenase 1981). Immunoprecipitation procedures were according to Shapiro & August (1976). In 12 × 75 mm polypropylene test tubes, a 1 ml aliquot of each salivary gland preparation (10–80 μg protein) was mixed with either 10 μl of antiserum or 10 μl of control serum from uninfested guinea pigs, and then incubated on ice for 30 min. Immune complexes were isolated by the addition of 50 μl of a 10% (v/v) suspension of fixed *Staphylococcus aureus* (Cowan strain, Calbiochem–Behrig, San Diego, CA.) and 50 μl of a 10% (v/v) fixed *S. aureus* bound with rabbit anti-guinea pig IgG. After 30 min on ice, the suspension was centrifuged at 15 000 g for 2 min at 22°C. The pellet was resuspended in TEN buffer (20 mM Tris-HCl pH 8.5, 1 mM EDTA, 100 mM NaCl) and centrifuged as before. This wash procedure was then repeated with TEN buffer containing 0.1% Triton X-100 and 2 M NaCl, and finally with TEN buffer containing 2 M NaCl.

13.2.4.3 Identification of tick antigens using sodium dodecyl sulfate polyacrylamide gel electrophoresis (SDS/PAGE)

To the final pellet from immunoprecipitation was added sample buffer (62 mM Tris-HCL pH 6.8, 2% sodium dodecylsulfate, 7.5% glycerol, 5% 2-mercaptoehanol) containing 0.02% bromophenol blue. The preparation was heated at 100°C for 6 min and centrifuged at 15 000 g for 20 min at 22°C. Immunoprecipitated salivary gland proteins were resolved by SDS/PAGE using a 1.5 mm thick discontinuous gel system (Laemmli 1970). The stacking gel was 3% acrylamide in 0.5 M Tris pH 6.8 containing 0.25% SDS, and the resolving gel was either 13% acrylamide or a 7.5% to 20% acrylamide gradient in 1.5 M Tris pH 8.8 containing 0.25% SDS. Electrophoresis was conducted for 1 h at 70 V, at which time the voltage was increased to 120 constant voltage for the next 8 hours. Gels were stained with Coomassie Brilliant blue R-250 or silver nitrate (Merril *et al.* 1981). Marker proteins were a standard mixture ranging from 94 000 to 14 000 mw (Pharmacia,

Piscataway, N.J.). After destaining, gels were fixed (5% glycerol-10% acetic acid), dried, and autoradiograms obtained after 8 h at 70°C, or after 72 h at 4°C using cassettes with two Cronex Quanta III intensifying screens (Dupont Co., Willington, DE.) and X-OMAT AR film (Eastman Kodak, Rochester, NY).

13.2.5 Protein extraction from SDS polyacrylamide gels and skin testing for specificity in tick-sensitized guinea pigs

The SDS/PAGE of SGA proteins immunoprecipitated by anti-tick antibody demonstrated a single unique band at 20 000 molecular weight. This finding suggested that a single protein of 20 000 mw was the crucial antigen. The initial test for antigenicity of this protein involved an Active Cutaneous Anaphylaxis (ACA) assay using protein eluted from SDS gels and injected into the dermis of tick exposed guinea pigs versus naive uninfested guinea pigs.

The gel area occupied by the 20 000 mw band, as indicated by autoradiography and the position of the molecular weight standard, was excised, cut into fine pieces, put into PBS PH 7.4 containing TEN buffer pH 7.4 and 0.1% Triton X-100, and incubated at 37°C for 24 h. The solution was decanted and replaced by buffer, followed by incubation at 56°C for 24 h. The two solutions were then pooled, concentrated to 1 ml by vacuum dialysis, and dialyzed against PBS pH 7.4.

In the ACA assay, guinea pigs exposed to larval ticks 26 days earlier were used as sensitized controls, and uninfested normal guinea pigs were used as normal controls. Respective animals were injected intravenously with 0.1% Evans Blue dye followed immediately by an intradermal injection of SGA from unfed (1 μg/.1 ml) and 6-day fed (0.1 μg/.1 ml) female ticks, and the band extraction preparation (0.1 μg/.1 ml). After 30 min, animals were anesthesized with ether and the diameter of blueing was measured on the inside of the excised flank tissue. The skin test doses of unfed and 6-day SGA were determined, based upon a) the quantity of protein extracted from the gels and available for skin testing; 0.1 μg/0.1 ml at each site, and b) the relative abundance of the 20 000 mw protein to total salivary gland protein in unfed 6-day preparations (Table 13.1). According to the estimated data, unfed SGA contained approximately 12% of the concentration of antigen that 6 day SGA contained. Therefore, the concentration of protein from unfed SGA used was approximately 8 times greater than 6 day SGA and equivalent to the antigen concentration eluted from the gels.

13.2.6 Antigen purification

13.2.6.1 Gel filtration chromatography

A 1 mg sample of SGA in 1 ml of PBS pH 7.4 was applied to a 2 × 38 cm column of Sephadex G-75 (Pharmacia Chemical, Piscataway, NJ) at room temperature and eluted with 9 ml/h of PBS containing 0.02% sodium azide. The absorbance at 280 nm was determined for each 9 ml fraction collected. Sample composed of fractions from each of the five peaks obtained were

Table 13.1 — Summary of protein kinetics and determination of relative abundance of the 20 000 mw salivary gland antigen (SGA) to total protein salivary gland protein in female *Amblyomma americanum* ticks.

Days of tick feeding at time of SGA harvest (No. glands)[a]	Mean protein per salivary gland (μg ± S.E.)	Amount of the 20 000 mw protein (ng)	Estimated antigen/total protein (%)
Unfed (60)	4 ± 1	1.72[b]	0.043[c]
1 day (45)	3 ± 2	0.6	0.02
2 day (45)	6 ± 2	0.12	0.002
3 day (80)	19 ± 4	1.9	0.01
6 day (88)	55 ± 9	203.5	0.37
9 day (116)	66 ± 13	792.0	1.2

a) Each gland is from a single tick. Females fed in the absence of males.
b) 1000 × (% antigen/total protein × mean salivary gland protein)
c) (cpm of SGA proteins precipitated by immune anti-tick serum — cpm of SGA proteins precipitated by normal guinea pig serum)/cpm of SGA proteins prior to immunoprecipitation

pooled separately and concentrated to 1 ml, corresponding to the volume applied to the column. Fractions were then dialyzed against PBS pH 7.4 that did not contain azide, and stored at −20°C. The column was cleaned with 3 bed volumes of PBS and 1 mg of dextran blue 2000 was applied to determine the exclusion volume.

13.2.6.2 Ion-exchange chromatography

A 2 mg sample of SGA in 3 ml of PBS pH 7.4 was dialyzed extensively against 20 mM Tris-HCl pH 7.4 and applied to a 1.6 × 70 cm column of diethylaminoethyl (DEAE) cellulose (Whatman DE-52) that had been equilibrated with 20 mM Tris-HCl pH 7.4 at room temperature. Samples were eluted with a continuous gradient of 0.05 to 0.5 M KCl and fractions collected every 10 min at a flow rate of 60 ml/h. The absorbance at 280 nm was determined for each 6 ml fraction collected. Samples composed of fractions from each of the peaks were pooled separately and concentrated to 1 ml, dialyzed against PBS pH 7.4 that did not contain azide, and stored at −20°C.

13.2.6.3 Immunoaffinity chromatography

Heavy chain specific guinea pig anti-*Amblyomma americanum* IgG_1 was prepared by applying whole immune anti-tick guinea pig serum to a 3.5 × 15 cm column of Sepharose rabbit anti-guinea pig IgG_1 as already reported (Brown *et al.* 1982b). Anti-tick IgG_1 antibody was eluted with 10 ml of 0.2 M sodium carbonate (Na_2CO_3) pH 11.3, dialyzed against PBS, concentrated, and bound to Sepharose by cyanogen bromide (99% coupling) according to Graziano *et al.* (1983).

Immunoaffinity purified IgG_1 from whole anti-tick immune serum has already been shown to transfer resistance to ticks, whereas immune serum depleted of IgG_1 does not (Brown *et al.* 1982b).

Initial applications of tick SGA extracts to the guinea pig anti-tick IgG_1 column resulted in the binding of 30% of the total protein. Estimations of antigen concentration to total protein in the salivary gland indicated a maximum concentration of 1.2% (see Table 13.1). Therefore, guinea pig immunoglobulin was nonspecifically binding SGA protein causing a high level of background in the eluate. To alleviate this, SGA extract was applied first to a Sepharose anti-ovalbumin IgG_1 column. In addition, both the anti-ovalbumin and anti-tick columns were pretreated with a 10% fetal calf serum preparation to bind 'sticky' sites on the guinea pig IgG_1 molecule, and then equilibrated with PBS pH 7.4 at room temperature.

SGA (0.5 mg in 1 ml of PBS) was applied to the anti-ovalbumin column followed by PBS, and fractions were collected every hour at a flow rate of 12 ml/h. The absorbance at 280 nm was determined for each 12 ml fraction, and only one peak resulted. This preparation (anti-OVA IgG_1 pass) was concentrated to 1 ml and applied to the guinea pig anti-tick IgG_1 column. Once the SGA had entered the column matrix, flow was stopped for 30 min to facilitate antigen–antibody interaction. SGA was followed by 90 ml of PBS, and fractions were collected every hour at a flow rate of 12 ml/h. The absorbance at 280 nm was determined for each 12 ml fraction collected, and only one peak resulted (anti-tick IgG_1 pass). The column was then eluted with 10 ml 0.2 M Na_2CO_3 pH 11.3 followed by 80 ml of PBS pH 7.4. The eluate was monitored by absorbance at 280 nm and with pH indicator paper for, first, the elution of high base (pH 11.3) indicating potential elution of 20,000 mw antigen, and, second, for the elution of a neutral (pH 7.4) solution indicating complete rinse of the column. The column was then treated with 0.1 M Glycine-HCl pH 2.3 as a final elution of bound proteins. The high base and low acid fractions were pooled separately, dialyzed against PBS, concentrated to 1 ml, and stored at −20°C.

13.2.7 Skin testing of tick-sensitized guinea pigs with column eluted salivary gland proteins

Fractions derived from gel filtration, ion-exchange, and immunoaffinity chromatography were used in Active Cutaneous Anaphylaxis skin test assays to determine what fraction from each procedure yielded skin reactivity in guinea pigs sensitized by tick feeding 26 days earlier; naive uninfested guinea pigs served as controls. This assay has been described in a previous section.

13.2.8 Immunization with tick proteins eluted from the immunoaffinity column

Each fraction obtained from the passage of SGA down the guinea pig anti-tick IgG_1 column was assayed for its ability to immunize naive guinea pigs against feeding by *A. americanum*. Each of the following preparations were prepared in Incomplete Freund's adjuvant (IFA) and administered in the footpads of guinea pigs at a volume of 0.1 ml/footpad: a) pass from the guinea pig anti-tick column (9 μg/animal), b) Na_2CO_3 eluate of the anti-tick column (200 ng/animal), and c) glycine-HCl eluate of the anti-tick column (1 μg/animal). The dose used was a function of the availability of material. IFA only was used because our initial studies demonstrated Complete Freund's Adjuvant (CFA) was not a permissive adjuvant (see Table 13.2). Controls consisted of naive unimmunized guinea pigs and animals immunized with crude SGA (380 μg/animal). The minimum effective dose of crude SGA was shown to be 280 μg. Animals were challenged with larval ticks 17 days later, and the level of resistance or tick rejection and the mean weight of the engorged ticks were determined.

Table 13.2 — Feeding success of *Amblyomma americanum* larval ticks on guinea pigs 17 days after immunization with 280 μg of salivary gland antigen (SGA).

Immunization regimen (animals/group)	Tick yield (%)	Tick rejection[a] (%)	Tick weight (mg)	Weight decrease[a] (%)
Naive (6)	92±4[b]	—	0.95±02	—
Active immunization (3)	47±[c]	48.9	0.65±04[c]	31.5
SGA + IFA (6)	57±5[c,d]	38.4	0.76±.02[c,e]	20.0
SGA + CFA (6)	86±3[g]	6.5	0.84±.02[h]	11.6
SGA + SALINE (6)	92±3	0.0	0.89±04[g]	6.3
IFA (6)	89±3	3.3	0.99±.01	0.0
CFA (6)	88±6	4.4	0.95±.02	0.0

a) $100\left(1 - \frac{\text{Tick yield or weight from experimentals}}{\text{Tick yield or weight from controls}}\right)$.
b) Mean ± S.E.
c) $p < 0.0001$ (versus naive hosts).
d) $p > 0.4$ (versus active sensitization).
e) $p > 0.06$ (versus active sensitization).
f) $p < 0.05$ (versus active sensitization).
g) $p > 0.2$ (versus naive hosts).
h) $p < 0.005$ (versus naive).

13.2.9 Protein determination

Protein concentrations were determined by the method of Bradford (1976).

13.2.10 Determination of the relative abundance of antigen to total salivary gland protein

In an effort to estimate the amount of the 20 000 mw protein to total salivary gland protein, the following assumptions were made. First, that all the various salivary gland proteins contained tyrosines in relatively equal proportion, and, therefore, all proteins were iodinated to comparable specific activities; an analagous assumption is made in the Lowry *et al.* (1951) protein assay. Second, that immune serum precipitated 100% of the detectable amount of antigen since no remaining antigen could be detected on SDS/PAGE of supernatants from immunoprecipitation reactions. The following calculations were made. For each SGA preparation, the total counts per minute (cpm) after column separation from free iodide yielded the post-iodination cpm. After immunoprecipitation, the cpm from the normal serum preparation was subtracted from the cpm from the immune serum preparation. A positive value reflected the presence of antigen, since immune serum recognized the same background proteins as normal serum, but in addition recognized antigen. A zero value indicated an absence of antigen. This value was divided by the post-iodination cpm to yield the percentage antigen in the initial preparation. This value was multiplied by the mean salivary gland protein in female and male ticks at each feeding time examined, to yield the estimated level of antigen at that time. Any error in the absolute amount of the antigen is a systematic error since all samples were treated and analyzed in the same way.

13.2.11 Feeding success of *Amblyomma americanum* females on guinea pigs sensitized by the partial feeding of female ticks

In vivo examination of antigen secretion by female *Amblyomma* ticks was conducted in feeding assays to determine the accuracy of the extrapolated data presented in Table 13.1. Sensitization periods of 3 and 9 days by female ticks were selected to see if partial feeding was adequate to appropriately sensitize guinea pigs against a challenge by female ticks allowed to feed to term (10–14 days) in the presence of males. Such a design would indicate if and when the antigen is secreted.

This experiment consisted of:

Group A — guinea pigs infested with 6 female and 6 male ticks for the first and only time (naive controls) at the time of challenge of the other groups;
Group B — guinea pigs sensitized with 6 female and 6 male ticks followed 26 days later by a challenge on the opposite flank with the same number of ticks (sensitized controls);
Group C — guinea pigs sensitized with 6 female ticks, allowed to feed for 3 days and challenged 26 days later with 6 female and male ticks;
Group E — guinea pigs sensitized with 6 female ticks allowed to feed for 9 days and challenged 26 days later with 6 female and male ticks.

Ticks were confined to their hosts, using the capsule technique as described in detail in a previous section. Female ticks were collected and weighed as they became replete with blood and dropped from the host. Male ticks feed sporadically and increase in weight by only 1.5 to 2 times the unfed weight and are not usually assayed for changes in weight, but are used in female feeding trials because females must be mated before normal engorgement occurs.

The number of females that fed to completion and dropped from the host, compared to the number added, constituted the yield expressed as yield ± standard error. Successful sensitization by tick feeding resulted in the expression of resistance when challenged. This was exhibited as a decrease in yield expressed as rejection, and a decrease in tick engorgement weight expressed as percentage weight reduction. Means between groups were compared using the Student's *t* test and differences were considered significant if $p < 0.05$.

13.3 RESULTS

13.3.1 Induction of resistance by immunization with crude tick extracts

Guinea pigs immunized with 280 μg SGA+IFA manifested significant tick rejection and reduction in the mean weight of fed ticks (Table 13.2). Induction of resistance by artificial immunization with SGA+IFA was nearly equivalent to active sensitization by tick feeding. Artificial immunization with SGA emulsified with CFA did not protect against tick feeding, but did result in a significant decrease in the mean weight of engorged individuals. At the time of tick challenge, hosts immunized with SGA+CFA responded to PPD skin tests with strong erythematous and indurated delayed reactions, demonstrating the efficacy of the CFA preparation. In a subsequent immunization experiment, the limiting dose of SGA was shown to be between 100 and 280 μg. Interestingly, cement material (CA) collected from the mouthparts of partially fed ticks and given at a dose of only 50 μg in IFA resulted in a level of protection (tick rejection) similar to active immunization or immunization with 280 μg SGA. Since the cement material originates from the salivary gland, this result suggests that cement contains increased amounts of salivary gland antigens responsible for induction of immune resistance, or that these antigens are present in a particularly immunogenic form in the cement.

13.3.2 Analysis of antigens in SGA and CA preparation by immunoprecipitation and SDS/PAGE employing anti-tick sera.

In vivo (^{35}S) and *in vitro* (^{125}I) radiolabeled salivary gland proteins were immunoprecipitated, solubilized, and separated by SDS/PAGE on discontinuous 13% gels, and then subjected to autoradiography. A single unique protein of 20 000 mw was identified with polyclonal anti-larval *A. americanum* immune guinea pig serum employing the two different isotopes and labeling techniques (Figs 13.1, 13.2). Fig. 13.1 also illustrates the efficiency of guinea pig anti-tick serum to eliminate irrelevant protein leading to the identification of a single unique band at 20 000 mw. This band is not seen after Coomassie blue or silver staining of the gel suggesting that this protein is at a concentration less than 20 ng in the gel.

Serum components from all guinea pigs bound a similar protein of 14 000 mw; this is considered to be background. However, sera obtained from actively sensitized guinea pigs that transferred anti-tick immune resistance

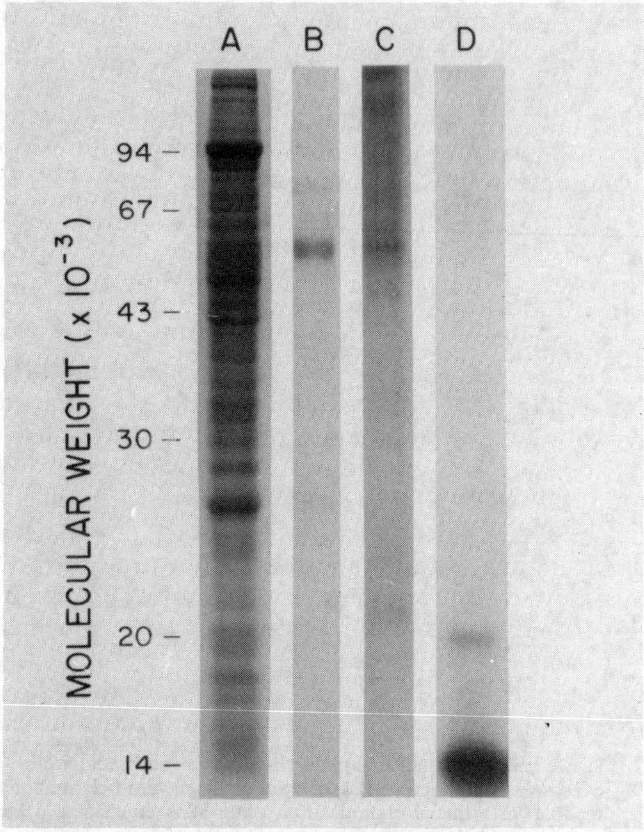

Fig. 13.1 — Analysis of tick salivary gland protein antigens (SGA) by SDS/PAGE (13%, reducing conditions) and visualized by staining with Coomassie brilliant blue after electrophoresis of the crude preparation (Lane A) or after immunoprecipitation with guinea pig anti-tick antibody from guinea pigs expressing resistance to ticks (Lane B). Lane C is a silver stain of SGA immunoprecipitated by immune serum. Lane D is autoradiograph of Lanes B and C.

responsiveness, and sera from guinea pigs immunized with SGA+IFA and that expressed resistance, both recognized an additional salivary gland protein of 20 000 mw. Furthermore, anti-larval *A. americanum* serum from actively sensitized hosts recognized 20 000 mw proteins from ^{125}I labeled salivary gland and cement extracts (Fig. 13.3) that was similar to those recognized in the gel prepared with internally ^{35}S-labeled SGA. These results suggest that a 20 000 mw protein from the salivary gland of adult female *A. americanum* is an important antigen responsible for the induction of anti-tick immune resistance responses in guinea pigs. The observation that this protein is in the cement material demonstrates that it is a component of the tick's salivary gland secretions that are required for normal feeding.

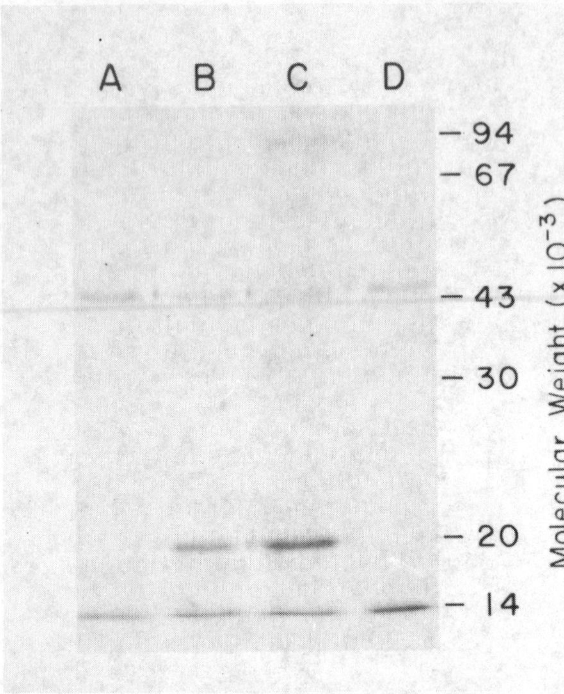

Fig. 13.2 — Analysis of tick salivary gland protein antigens (SGA) by SDS/PAGE (13%, reducing conditions) and visualized by autoradiography. ^{35}S *in vivo* labeled SGA was immunoprecipitated with guinea pig anti-tick antibody induced by active feeding, or artificial immunization using SGA emulsified in Incomplete (IFA) or Complete (CFA) Freund's adjuvant. Samples in Lane A: SGA proteins recognized by serum from normal uninfested guinea pigs; Lane B: SGA recognized by serum from actively sensitized guinea pigs expressing resistance to ticks, this serum transferred immune resistance responsiveness to naive guinea pigs; Lane C: SGA proteins recognized by serum from guinea pigs immunized with SGA emulsified with IFA and expressing resistance to ticks; Lane D: SGA proteins recognized by serum from guinea pigs immunized with SGA emulsified with CFA and not expressing resistance. (From Brown *et al.* 1984b).

It should be mentioned that SGA in these SDS/PAGE studies was prepared without protease inhibitors. Therefore, it is possible that the 20 000 mw protein is a cleavage product of a larger molecular weight species.

13.3.3 Kinetics of total protein production by the salivary gland

Total protein production by female *A. americanum* salivary glands increased as the 12-day attachment phase progressed, even in the absence of being mated (see Table 13.1). Mean protein concentration ranged from 4 μg per unfed tick to 116 μg per 12-day fed tick. During the attachment period, the rate of mean female weight gain increased markedly through day 7, and

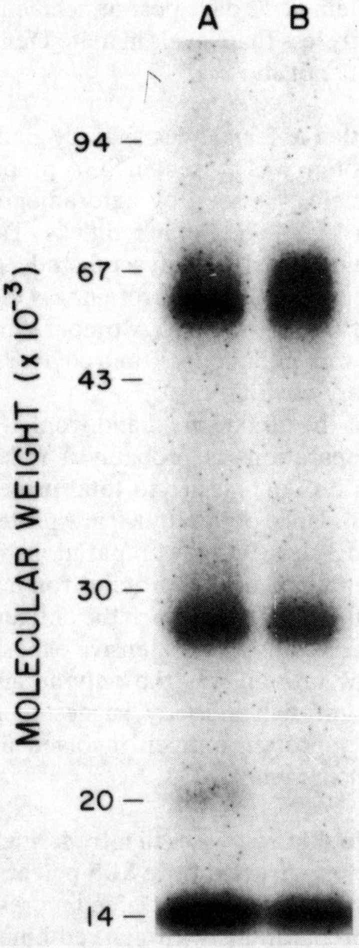

Fig. 13.3 — Analysis of salivary gland derived cement antigen (CA) by SDS/PAGE (13%, reducing conditions) and visualized by autoradiography after extensive (2 weeks) storage in a cassette with 2 Quanta III intensifying screens. ^{125}I *in vitro* labeled CA was immunoprecipitated with guinea pig anti-tick antibody induced by active feeding. Samples in Lane A: CA proteins recognized by serum from guinea pigs expressing resistance to ticks and that transferred immune resistance responsiveness; Lane B: CA proteins recognized by serum from normal uninfested guinea pigs. (From Brown *et al.* 1984b).

then increased minimally through day 12, resulting in a mean weight 10 times greater than unfed ticks. *A. americanum* female ticks do not reach normal engorgement weight (500 to 700 mg) until mated (Gladney & Dawkins 1971). If unmated, females remain attached to the host and apparently feed at a low rate, as evidenced by their small size and minimal tissue destruction at the attachment site compared to mated females (Brown & Knapp 1980b). Even in the unmated state, female salivary glands increase in total protein (see Table 13.1) to a level similar to mated females since salivary gland protein at 12 days post-attachment in unmated females (116 μg) was only slightly less than levels in mated females (127 μg) harvested at the same time (data not shown).

13.3.4 Identification of female tick salivary gland derived antigens

The 20 000 mw protein was present in SGA preparations from unfed and 6- and 9-day fed females, as seen by autoradiography, but absent in SGA preparations from 1-, 2-, and 3-day fed ticks (Table 13.1). The 20 000 mw protein was not visualized in polyacrylamide gels of 3 day SGA preparations, even though substantial protein was iodinated (60 μg). This is in contrast to gels of 6 and 9 day SGA preparations where similar concentrations of protein were iodinated (30 and 60 μg, respectively) and the 20 000 mw protein was very obvious.

The absence of the 20 000 mw band from 1-, 2-, and 3-day fed female salivary gland preparations is probably a reflection of the low relative abundance of this protein species to total protein (Table 13.1). The estimated relative abundance of the 20 000 mw protein in the 1-day (0.02%), 2-day (0.002%), and 3-day (0.01%) preparations were at least 50% less than the estimated abundance of this protein from salivary gland preparations from unfed ticks (0.043%) from which the antigen was faint after autoradiography and visualized only after extensive exposure (7 days) at −70°C.

The 20 000 mw protein was the only antigen identified from female salivary glands by anti-tick antibody, suggesting further that the 20 000 mw species is the only protein antigen involved in host immune resistance responses to *A. americanum*.

13.3.5 Immediate skin responses to intradermal injection of crude SGA and proteins extracted from SDS polyacrylamide gels

Guinea pigs sensitized by larval tick feeding responded to both the unfed and 6-day SGA preparations with marked bluing responses. In addition, cutaneous responses to protein extracted from SDS gels were also of comparable size. This extract was prepared from gels of immunoprecipitates from 6- and 9-day fed female SGA where the 20 000 mw protein was the most abundant (Table 13.1) and easily seen by autoradiography (Figs 13.1, 13.2). These results demonstrate that very little antigen was present in the gels. Only 10 μg total was extracted from the 3 gels, but there was a closely migrating protein as seen by silver staining. Attempts to immunize guinea pigs with gel extracted protein (2 μg/host) emulsified with Incomplete Freund's adjuvant were unsuccessful. However, the presence of polyacrylamide and SDS in the gel extract may have acted as complete adjuvant (Stumph *et al.* 1974) which is not acceptable for SGA immunization (see Table 13.2).

13.3.6 Gel filtration chromatography

SGA proteins eluted from a G-75 Sephadex column yielded 5 peaks as shown in Fig. 13.4, with 86% recovery of applied protein. Intradermal skin testing of each peak (0.19 to 0.22 μg protein/0.1 ml) into naive and tick-

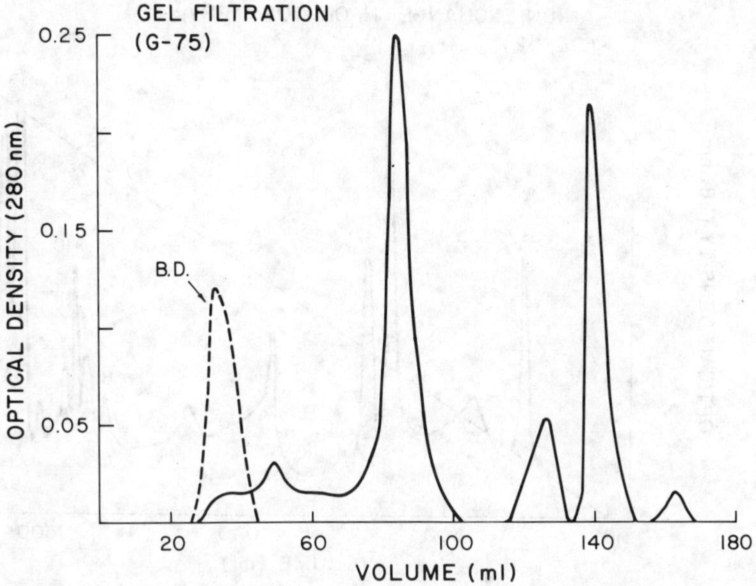

Fig. 13.4 — Elution profile of salivary gland proteins from Sephadex G-75 gel filtration (0.01 M PBS pH 7.4). The void volume is represented by Blue Dextran elution (B.D.) (From Brown & Askenase 1986).

sensitized hosts demonstrated the presence of skin reactive protein (antigen) in fraction 1 as evidenced by strong (4+) cutaneous responses. However, the presence of antigen in the first fraction suggests it was in excess of 70 000 mw since the inclusion range of the column was 3000 to 70 000 mw. This is contrary to that reported previously where guinea pig anti-tick IgG_1 recognized a 20 000 mw protein only. Immunoprecipitation of fraction 1 with anti-tick antibody demonstrated recognition of several proteins, the band of interest at 20 000 with an additional band at 67 000 mw. Since the proteins other than the 20 000 mw species have not previously been recognized, it appears that the two proteins aggregated on the column causing them to elute with the large molecular weight proteins in the exclusion volume.

13.3.7 Ion-exchange chromatography

SGA eluted from DEAE by a 0.05 to 0.5 M KCl gradient yielded 72% recovery divided among 16 fractions as shown in Fig. 13.5. The relative concentration of protein in each fraction ranged from 0.1 to 16%. Active cutaneous anaphylaxis using 0.11 to 0.17 μg protein/0.1 ml demonstrated that only fraction 13 induced significant skin reactions (7+), suggesting the presence of antigen. Fraction 13 eluted at approximately 0.45 M KCl and contained 1.4% of the total protein. We were unable to obtain sufficient quantities of protein to visualize immunoprecipitated proteins by SDS/PAGE because of the high level of dilution resulting from this procedure.

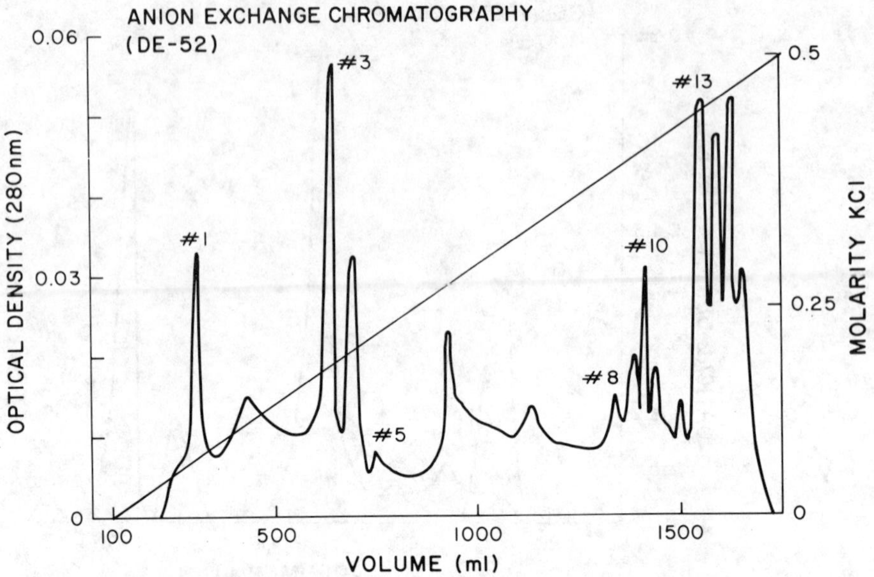

Fig. 13.5 — Elution profile of salivary gland proteins from DEAE ion-exchange column (0.02 M Tris-HCl pH 7.5) after treatment with 0.05 to 0.5 M KCl pH 7.5 gradient. (From Brown & Askenase 1986).

13.3.8 Immunoaffinity chromatography

Whole immune anti-tick serum, immune serum depleted of anti-tick IgG_1, and anti-tick IgG_1 recognized the 20 000 mw protein (Fig. 13.6). These results demonstrate the specificity of guinea pig anti-tick IgG_1 as a solid phase ligand for isolating antigen from SGA. Furthermore, the results also suggest that not all of the IgG_1 is removed by passage through the anti-IgG_1 column, or that IgG_2 also recognizes antigen but is unable to transfer resistance (Brown *et al.* 1982b).

Fifty-two percent of the SGA protein applied to the anti-ovalbumin IgG_1 control column passed, whereas 33% was retained and subsequently eluted with Na_2CO_3. The SGA protein in the pass was applied to the anti-tick IgG_1 column where 100% of the 260 μg of protein applied was recovered. 14% (36 μg) passed in the void column, whereas 24% and 14% were eluted by Na_2CO_3 pH 11.3 and Glycine-HCl pH 2.3, respectively. The largest protein-containing eluate was obtained with the PBS wash of the column after Glycine-HCl treatment. The fact that this material did not give positive skin reactions (Table 13.3) suggested it contained irrelevant proteins, such as antibody and fetal calf serum leached from the column. Interestingly, both Na_2CO_3 and Glycine-HCl eluates yielded comparable skin reactions in tick-sensitized guinea pigs (Table 13.3), suggesting the presence of two antigens, or two species of the same protein.

Analysis of Na_2CO_3 and Glycine-HCl eluates by SDS/PAGE resulted in

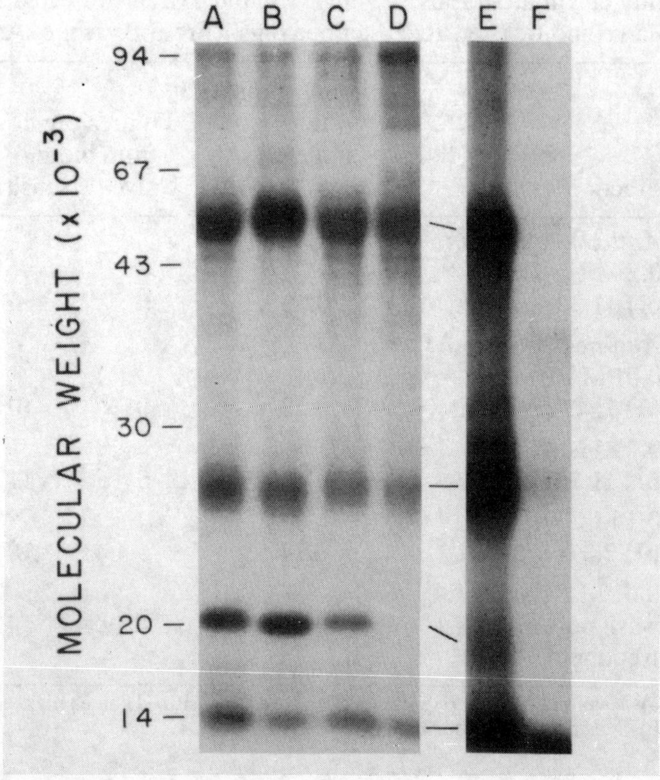

Fig. 13.6 — Analysis of tick salivary gland protein antigens by SDS/PAGE (%13, reducing conditions), and visualized by autoradiography. ^{125}I *in vitro* labeled SGA was immunoprecipitated with: Samples in Lane A: serum from guinea pigs expressing resistance to ticks that transferred resistance; Lane B: serum depleted of IgG_1 by passage through rabbit anti-guinea pig IgG_1 column; Lane C: anti-tick immunoaffinity purified IgG_1; Lane D: serum from normal uninfested guinea pigs. Samples in Lane E: SGA proteins eluted from the anti-tick column in Na_2CO_3; Lane F: SGA proteins eluted from the anti-tick column in Glycine-HCl.

the detection of only one protein in the Na_2CO_3 eluate (20 000 mw) with only several faint bands in the Glycine-HCl eluate (Fig. 13.6) none of which were at 20 000 mw; comparable concentrations of protein were iodinated (40 to 60 μg). The prominent bands at 67 000, 30 000, and 14 000 were also recognized by normal guinea pig serum. These results demonstrate that the 20 000 mw protein shown previously to be the only antigen recognized by guinea pig anti-tick antibodies was bound by IgG_1 in the fluid phase (immunoprecipitation) and solid phase (immunoaffinity column) where it was eluted by Na_2CO_3 only. Interestingly, a small amount of highly skin reactive protein was eluted by Glycine-HCl treatment as evidenced by the comparable skin reactions to Na_2CO_3; however, the 20 000 mw protein could not be detected in SDS/PAGE after immunoprecipitation. The low

Table 13.3 — Cutaneous responses to fractions obtained from immunoaffinity chromatography of salivary gland extracts injected intradermally into naive and tick sensitized guinea pigs. (From Brown & Askenase 1986).

Fraction	Amount of protein injected (μg/0.1 ml)	Guinea pig host (mm bluing)[a] Naive	Sensitized	Adjusted reaction[b]
Anti-OVA column:				
0.2 M Na_2CO_3 pH 11.3 treatment	0.13	0	0	0
Anti-tick column:				
0.01 M PBS pH 7.4 Treatment	0.13	10	10	0
0.2 M Na_2CO_3 pH 11.3 treatment	0.15	5	11	6
0.1 M Gly-HCl pH 2.3 treatment	0.14	5	10	5
0.01 M PBS pH 7.4 wash post Gly-HCl treatment	0.14	4	4	0

a) Mean reactions at one site per 2 naive and sensitized guinea pigs.
b) Sensitized minus naive.

pH of this eluate may have denatured the protein to such an extent that only determinants were present, as opposed to the polypetide from the base elution which was not denaturing. A small concentration of determinants could yield skin reactions, but because of their small molecular weight it could not be detected by SDS/PAGE after immunoprecipitation which had a lower limit of 14 000 molecular weight.

13.3.9 Immunization with immunoaffinity purified antigen

Each fraction obtained from immunoaffinity purification of antigen was prepared in IFA and administered to naive guinea pigs challenged 17 days later by tick feeding. However, only the Na_2CO_3 pH 11.3 eluate, at 200 ng/animal, conferred resistance (Table 13.4). The level of tick rejection achieved with the Na_2CO_3 eluate (23.4%) was significant ($p<0.001$), but less ($p>0.01$) than that achieved with crude SGA (38.3% rejection). However, both crude SGA and the Na_2CO_3 eluate reduced tick engorgement weight to similar levels (21.4%), a significant drop compared to ticks from controls ($p<0.001$). Interestingly, the pass and glycine-HCl pH 2.3 fractions had no ability to inhibit tick attachment, as tick yield was normal, but did have a slight inhibitory effect on the ticks' ability to acquire a normal meal. The observed reductions in tick weight were significant compared to ticks from naive controls ($p<0.001$), but statistically similar ($p>0.05$)

Table 13.4 — Feeding success of larval *Amblyomma americanum* ticks on naive guinea pigs and guinea pigs immunized with salivary gland extract (SGA) or fractions of SGA after passage through a guinea pig anti-tick IgG_1 sepharose column. (From Brown & Askenase 1986).

Treatment (No. animals)	Dose (μg)	% Tick yield	% Tick rejection	Tick weight (mg)	% Tick weight reduction
Naive (4)	—	94 ± 2	—	0.98 ± .01	—
Salivary gland antigen (3)	380	58 ± 2[a]	38.3	0.77 ± .05[a]	21.4
Pass from the guinea pig anti-tick IgG_1 column (3)	9	90 ± 1	4.3	0.87 ± .01[ac]	11.2
Na_2CO_3 eluate from guinea pig anti-tick IgG_1 column (3)	0.2	72 ± 3[ab]	23.4	0.77 ± .03[a]	21.4
Glycine-HCl eluate from guinea pig anti-tick IgG_1 column (3)	1	90 ± 1	4.3	0.88 ± 03[d]	10.2

a) $p < 0.001$ (versus naive)
b) $p < 0.01$ (versus SGA immunized)
c) $p > 0.05$ (versus SGA immunized)
d) $p < 0.01$ (versus naive)

compared to ticks from successfully immunized animals. These results suggest that tick weight may be a more sensitive assay for tick resistance, further suggesting that salivary gland-derived materials other than 20 000 mw protein may be involved in interfering with the acquisition of a normal meal. However, it is obvious that the 20 000 mw protein is the primary antigen because it not only reduces tick weight, but has a significant effect on tick yield. This conclusion seems appropriate since the glycine-HCl and Na_2CO_3 eluates yielded comparable skin test reactions, whereas the pass did not. It may have been that a small amount of antigen was in the pass (this preparation was not analyzed by immunoprecipitation and SDS/PAGE), but a large proportion of inflammatory proteins were also present that masked any antigenic skin reactive factor. The skin test data appear to substantiate this conclusion since this preparation caused large areas (10 mm) of bluing in naive animals, demonstrating the presence of highly inflammatory proteins.

13.3.10 Feeding success of female Amblyomma americanum on guinea pigs with different sensitization histories

Female *A. americanum* ticks allowed to feed on guinea pigs sensitized by a full-term feeding of females and males together, 26 days earlier, expressed a 60% level of rejection and an 89% decrease in engorgement weight (Table

13.5). Similarly, female infestation of guinea pigs sensitized by female feeding for only 3 or 9 days resulted in comparable reductions in yield and weight. The findings of this experiment corroborate the results of the gels and the extrapolated data in Table 13.1. That is, unfed female ticks contain antigen and sensitize the host during the first 3 days of feeding.

13.4 DISCUSSION

The ability to induce immune resistance to ticks by immunization with salivary gland extracts has been described in guinea pigs and cattle using *Dermacentor andersoni* (Allen & Humphreys 1979, Wikel 1981). In the present study, we report the successful immunization of guinea pigs against larval *Amblyomma americanum* feeding using salivary gland extracts (SGA) and solubilized cement substance (CA) obtained from adult female ticks. Furthermore, we have identified a unique protein of 20 000 mw from both crude salivary gland extracts and secreted salivary substances (cement) using guinea pig anti-tick serum. This protein is believed to be responsible for the induction of host resistance.

Tick-derived antigens induced resistance when emulsified with incomplete Freund's adjuvant (IFA), but had no such ability when emulsified with complete Freund's adjuvant (CFA), or administered in saline. Furthermore, serum from resistant animals immunized with SGA+IFA recognized a 20 000 mw protein from salivary gland extracts, as shown by SDS/PAGE, similar to serum that transferred resistance. Serum from normal uninfested guinea pigs and non-resistant animals immunized with SGA-CFA did not recognize this protein. These results suggest further that the substance responsible for the induction and/or elicitation of host immune resistance responses is a protein of 20 000 mw.

The inability of soluble SGA preparations to be significantly immunogenic in saline may be due to the rapid removal from the site of inoculation, compared to the slow release effect provided by IFA. The success of IFA compared to CFA may be due to the type of immune-inflammatory response induced by these respective adjuvants. We have shown previously that the immune resistance response to ticks is dependent upon an accompanying cutaneous basophil response (Brown *et al.* 1982b). Immunization of guinea pigs employing CFA results in preferential IgG_2 antibody production (Beneacerraf *et al.* 1963) and suppression of cutaneous basophil responses (Mitchell & Askenase 1982). In contrast, immunization with antigens emulsified with IFA results in preferential IgG_1 antibody production (Benacerraf *et al.* 1963) and promotes cutaneous basophil responses (Dvorak *et al.* 1971). Therefore, successful anti-tick immunization with SGA not only requires antigen recognition and production of antibodies and T cells of appropriate specificity, but also depends on the use of the proper adjuvant. The adjuvant of choice must allow for the induction of the correct antibody isotype and/or T cell-dependent response, that allows expression of a cutaneous cellular response that calls in particular effector cells, such as

Table 13.5 — Female *Amblyomma americanum* feeding success on guinea pigs sensitized by female infestation of varying durations.

Infestation[a] (2 guinea pigs/group)		Female tick feeding parameters			
Sensitization (day 0)	Challenge (day 26)	Tick yield (%)	Tick rejection (%)	Tick weight (mg)	Weight decrease (%)
Female + male (12–14 days)	Nil	84 ± 17[b]	—	327 ± 31[b]	—
Female + male (12–14 days)	Female + male (12–14 days)	34 + 16[c]	59.5	36 ± 6	88.9[d]
Female (3 days)	Female + male (12–14 days)	42 ± 8[c]	50.0	50 ± 9[e]	84.7
Female (9 days)	Female + male (12–14 days)	33 ± 17[c,f]	60.7	68 ± 7[e,f]	79.2

a) Six female ticks per animal; 2 animals/group.
b) Mean ± S.E.
c) $p < 0.001$ (versus naive controls)
d) $1 - \dfrac{\text{Tick yield or weight from challenged animals}}{\text{Tick yield or weight from naive controls}} \times 100$
e) $p > 0.05$ (versus sensitized controls)
f) $p < 0.05$ (versus female 3 day)

basophils and eosinophils required for the rejection of ticks (Brown *et al.* 1982a).

Biochemical analyses of tick salivary secretions have been few and have depended on skin test assays in tick-sensitized animals. However, all studies to date agree that the important tick-derived component is less than 60 000 molecular weight. The allergen from the oral secretion of *Boophilus microplus* was found to be 30 000 mw after extensive purification on G-100, DEAE, and isoelectric focusing (Geczy *et al.* 1971). Gel filtration (G-25) of saliva obtained from adult *B. microplus* ticks yielded a protein estimated to be 60 000 mw and reported to have carbohydrase activity (Willadsen & Williams 1976). However, subsequent chromatography yielded three distinct subunits, two of which had skin reactivity in tick-sensitized cattle. Similarly, skin-reactive proteins with an estimated molecular weight of 60 000 were obtained from extracts of larval *B. microplus* by gel filtration on Sephadex G-200 (Reich & Zorzopulos 1980). Subsequent separation on DEAE-cellulose yielded two protein species, one that co-chromatographed with esterase activity suggesting it was an esterase. These three studies represent the most detailed biochemical analyses of tick-derived substances believed to be antigens. However, none of the studies demonstrated that anti-tick antibody recognized these proteins, or more importantly that these proteins could protect animals from subsequent tick feeding. In the present study, we have demonstrated that a 20 000 mw protein is the only species recognized by anti-tick antibody, it induces skin reactions in tick-sensitized animals, and protects naive animals from tick infestation.

Salivary gland proteins from adult *A. amercanum* ticks were separated by gel filtration and ion-exchange chromatography and yielded 5 and 16 fractions, respectively. However, only one fraction from each purification procedure elicited significant skin reactions in tick-sensitized guinea pigs. Furthermore, the positive fraction from G-75 chomatography contained the 20 000 mw protein, as well as several other proteins. However, since these other proteins were not demonstrated in antigen kinetics studies in female ticks, it is believed that they aggregated with the 20 000 mw protein and were removed from suspension when this protein was precipitated by anti-tick antibody. This would explain the apparent recognition of these extraneous proteins by antibody, and the observation of skin reactivity and the presence of the 20 000 mw protein in the first fraction from G-75 chromatography that contained substances in excess of 70 000 mw.

Elution of SGA proteins from the anion exchange column with a KCl gradient yielded a skin reactive fraction in peak 13 at approximately 0.45 M KCl. This result suggests that the protein has a high proportion of negative charged amino acids. Insufficient quantities of protein from this peak prevented visualization of any banding pattern after SDS/PAGE.

The final attempt to purify the 20 000 mw protein from crude salivary gland extract was with the use of an immunoaffinity column containing guinea pig *anti-A. americanum* IgG_1. This antibody effectively recognized the 20 000 mw protein from the SGA extract as shown by SDS/PAGE. However, passage of crude SGA through this column yielded essentially the same background proteins being more diffuse and relatively faint compared to the G-75 gel. These results suggest that the column was probably not very specific, containing low titer anti-tick IgG_1 in a heterogeneous population of IgG_1 molecules yielding more background protein by virtue of the 'stickiness' of this molecule, with less antigen. Also, guinea pig IgG_1 in general is a low affinity immunoglobulin that presents further complications in antigen purification. Therefore, it appears that for the immunoaffinity approach to be successful, a monoclonal antibody may be required, or at least a polyclonal serum with a high anti-tick titer and high avidity.

This study demonstrates that in the female salivary gland, the 20 000 mw protein is the only occurring species recognized by anti-tick antibodies. This protein was observed from unfed female ticks, and from females that fed for 6 and 9 days, but could not be seen by autoradiography in glands from ticks that fed for 1, 2, and 3 days. This protein appears to be both an initial and latent secretion of the tick, since it is found in the unfed gland and after the first 3 days of feeding. In the female ixodid tick, which feeds only once, this protein may be required for the initial insertion of the mouthparts into the host's tissue, and then again several days after attachment since feeding *per se* does not increase in magnitude until 4–5 days after the female is mated (Gladney & Drummond 1971). Female *A. americanum* are receptive to mating 3 to 4 days after attachment (Brown & Stenner 1982), therefore, active feeding is initiated at about day 7 to 8 post-attachment.

Males were not kept with the females, therefore mating, the primary stimulus for induction of the rapid engorgement phase, did not occur. However, salivary glands in unmated females increased in total protein content (116 μg/tick) to a level similar to mated females (127 μg/tick) harvested at the same time, and as reported elsewhere (McSwain *et al.* 1982). In addition, most females increased only slightly in weight (50 mg) whereas others were markedly engorged (200 to 300 mg), suggesting unmated females entered the fast feeding phase in the absence of mating. Maximum engorgement (500–700 mg) occurs generaly only after mating. Since the ticks used were of the same age and were equal in size, it appears that salivary gland development may function independently of the mating process when ticks are attached, whereas body expansion to the normal engorgement state is generally strictly regulated by the act of mating.

In addition to enzymes that remain soluble after injection into the host, the salivary gland of most ixodid ticks also secretes a cement substance during the attachment phase (Chinery 1973), and the feeding phases as well (Brown & Knapp 1980a,b), that solidifies around the mouthparts. This cement substance lies between the tick's mouthparts and the host's tissue and is apparently used as an important mechanism to secure and maintain the attachment to the host. The primary holdfast structure is the hypostome that bears numerous recurved teeth. Histopathological analysis of tick attachment sites in guinea pigs experiencing a primary infestation indicate a strong neutrophil response characteristic of nonspecific inflammation (Brown & Knapp 1980a,b). However, previously tick-exposed hosts that are challenged, express resistance manifested by a significant decrease in tick feeding success (Brown & Askenase 1983). Histological analysis of tick feeding sites in resistant animals reveals the presence of epidermal vesicles, packed with basophils, in intimate contact with the cement around the tick's mouthparts (Allen 1973, Brown *et al.* 1983, 1984a).

Cement from *B. microplus* was found to be a glycoprotein–lipoprotein complex, and believed to be immunologically inert (Moorhouse & Tatchell 1966). This conclusion was based upon the weak cellular infiltrate adjacent to the cement material compared to the intense cellular infiltrate at the tip of the mouthparts where soluble secretions were released. However, human peripheral blood lymphocytes from individuals sensitized to *Dermacentor andersoni* expressed greater *in vitro* mitogenesis to cement than to salivary gland extracts (Gregson 1970). In the current study, cement from *A. americanum* was found to be a potent immunogen, and contained a 20 000 mw protein similar to that identified from SGA. This observation suggests further that antigens in the cement are important not only in the induction of the immune resistance response, but in elicitation of this response.

Basophil-filled epidermal vesicles adjacent to attached ticks in resistant guinea pigs are reminiscent of those occurring in contact hypersensitivity responses (Askenase 1977). In tick reactions, basophil-rich epidermal vesicles may develop in response to cement substances or salivary gland-derived enzyme secretions or both, after processing and/or presentation by local antigen presenting cells such as the Langerhans cells. Langerhans cells are Ia^+ cells of the epidermis and upper dermis. Langerhans cells are thought to act as crucial antigen presenting cells in the induction of contact hypersensitivity reaction (Ptak *et al.* 1980), and have been shown by indirect immunofluorescence to bind SGA in the skin of guinea pigs infested with ticks (Allen *et al.* 1979). There is evidence to suggest that depletion of Langerhans cells from the epidermis of *D. andersoni* sensitized guinea pigs inhibits their ability to express resistance to tick challenge (Allen, personal communication); this is analogous to findings indicating that Langerhans cells are required for induction of contact hypersensitivity responses (Ptak *et al.* 1980, Streilein & Bergstresser 1981, Elments *et al.* 1983). Basophils are also required for the resistance response to ticks (Brown *et al.* 1982), but the exact relationship between Langerhans cells and basophils is not known. If Langerhans cells are required for expression of these responses, then it is likely that Langerhans cells take up and present SGA to specifically sensitized T cells. The latter are then activated to produce chemotactic factors that recruit circulating basophils to infiltrate the local site and effect the immune resistance response.

In summary, the results of this study demonstrate four important points. First, the potency of crude salivary gland and cement extracts as immunogens that, when administered with IFA, can induce resistance to subsequent tick feeding as manifested by a significant tick rejection and reduced tick weight. Second, that polyclonal IgG_1 antibodies from serum of guinea pigs resistant to larval *A. americanum* ticks, and which transfer immune resistance to ticks, are apparently monospecific recognizing a single 20 000 mw protein in salivary gland and cement extracts. Third, polyclonal anti-tick IgG_1 is a highly specific probe for isolating this crucial tick antigen that induces host immune resistance. Fourth, protein production by female salivary glands increased as feeding progressed, and that a single protein of 20 000 mw was consistently recognized by anti-tick antibodies.

Subsequent studies are directed towards isolating this protein in pure form and characterizing its biochemical properties that may elucidate its function. Knowledge of the biochemical properties of this protein are also important to enhance isolation procedures involving the production of high affinity anti-tick antibody for m RNA isolation leading to recombinant DNA technology and *in vitro* expression of this protein.

ACKNOWLEDGEMENTS

We would like to thank the following individuals for their contribution to various immunological and biochemical phases of our research on 'Host immunity to tick feeding' that has resulted in the present review; Dr Stephen Galli (Harvard University, Boston, MA, USA), Dr Gerald Gleich (Mayo Clinic, Rochester, MN, USA), Dr Frank Graziano (University of Wisconsin, Madison, WI, USA), Dr Stuart Shapiro (International Laboratory for Research on Animal Diseases, Nairobi, Kenya), and Dr Grant Shimamoto

(Yale University, New Haven, CT, USA). In addition, we acknowledge the excellent technical assistance of Donna Cipriano, Robin Meade, and Lisa Langlois.

This work was supported in part by grants from the United States Public Health Service AI-19657, AI-17555, and from the Agricultural and Biomedical Sciences Divisions of The Rockefeller Foundation.

REFERENCES

Allen, J. R. (1973) Tick resistance: basophils in skin reactions of resistant guinea pigs. *Intl. J. Parasitol.* **3**:195-200.

Allen, J. R. & Humphreys S. J. (1979) Immunization of guinea pigs and cattle against ticks. *Nature* **280**:491–495.

Allen, J. R., Khalil, H. M. & Wikel, S. K. (1979). Langerhans cells trap tick salivary antigens in tick resistant guinea pigs. *J. Immunol.* **122**:563–565.

Askenase, P. W. (1977) Role of basophils, mast cells and vasoamines in hypersensitivity reactions with a delayed time course. *Prog. Allergy.* **23**:199–320.

Askenase, P. W., Bagnall, B. G., & Worms, M. J. (1982) Cutaneous basophil associated resistance to ectoparasites (ticks). I. Transfer with immune serum or immune cells. *Immunology* **45**:501–511.

Benacerraf, B., Ovary, Z., Bloch, K. J., & Franklin, E. C. (1963) Properties of guinea pig 7S antibodies: I. electrophoretic separation of two types of guinea pig 7S antibodies. *J. Exp. Med.* **117**:937–950.

Bradford, M. (1976) A rapid and sensitive method for the quantitation of microgram quantities of protein utilizing the principle of protein-dye binding. *Anal. Biochem.* **72**:248.

Brown, S. J. (1982) Antibody and cell mediated immune resistance by guinea pigs to adult *Amblyomma americanum* ticks. *Amer. J. Trop. Med. Hyg.* **32**:1285–1290.

Brown, S. J. & Askenase, P. W. (1981) Cutaneous basophil responses and immune resistance of guinea pigs to ticks: passive transfer with peritoneal exudate cells or serum. *J. Immunol.* **127**:2163–2167.

Brown, S. J. & Askenase, P. W. (1983) Immune rejection of ectoparasites (ticks) by T-cell and IgG_1 antibody recruitment of basophils and eosinophils. *Fed. Proc.* **42**:1744–1747.

Brown, S. J. & Askenase, P. W. (1986) *Amblyomma americanum:* physiochemical isolation of a salivary gland-derived protein capable of inducing immune resistance in guinea pigs. *Exp. Parasitol.* (In press).

Brown, S. J. & Knapp, F. W. (1980a) *Amblyomma americanum:* sequential histological analysis of larval and nymphal feeding sites on guinea pigs. *Exp. Parasitol.* **49**:188–205.

Brown, S. J. & Knapp, F. W. (1980b) *Amblyomma americanum:* sequential histological analysis of adult feeding sites on guinea pigs. *Exp. Parasitol.* **49**:303–318.

Brown, S. J. & Stenner, T. (1982) Regulation of feeding and ovipositional success of *Amblyomma americanum* ticks. *Experientia* **138**:1060–1061.

Brown, S. J., Galli, S. J. Gleich, G. J. & Askenase, P. W. (1982a) Ablation of immunity to *Amblyomma americanum* by anti-basophil serum: cooperation between basophils and eosinophils in expression of immunity to ectoparasites (ticks) in guinea pigs. *J. Immunol.* **129**:790–795.

Brown, S. J., Graziano, F. M., & Askenase, P. W. (1982b) Immune serum transfer of cutaneous basophil associated resistance to ticks: mediation by 7S IgG_1 antibodies. *J. Immunol.* **129**:2407–2412.

Brown, S. J., Worms, M. J. & Askenase, P. W. (1983) *Rhipicephalus appendiculatus:* larval feeding sites in guinea pigs actively sensitized and receiving immune serum. *Exp. Parasitol.* **54**:111–120.

Brown, S. J., Bagnall, B. G. & Askenase, P. W. (1984a) *Ixodes holocyclus:* Cutaneous basophil responses in naive and sensitized guinea pigs and recipients of immune serum and cells. *Exp. Parasitol.* **57**:40–47

Brown, S. J., Shapiro, S. Z. & Askenase, P. W. (1984b) Characterization of tick antigens inducing host immune resistance. I. Immunization of guinea pigs with *Amblyomma americanum* derived salivary gland extracts and identification of an important salivary gland antigen with guinea pig anti-tick antibodies. *J. Immunol.* **132**:3319–3325.

Chinery, W. A. (1973) The nature and origin of the "cement" substance at the site of attachment and feeding of adult *Haemaphysalis spinigera* (Ixodidae). *J. Med. Entomol.* **10**:355–362.

Dvorak, H. F., Simpson, B. A., Bast, R. C. Jr., & Leskowitz, S. (1971). Cutaneous basophil hypersensitivity. III. Participation of the basophil in hypersensitivity to antigen antibody complexes, delayed hypersensitivity and contact allergy. Passive Transfer. *J. Immunol.* **107**:138–144.

Elments, C. A., Bergstresser, P. R., Tigelaar, R. E., Wood, P. J., & Streilein, J. W. (1983) Analysis of the mechanism of unresponsiveness produced by haptens painted on skin exposed to low dose ultraviolet radiation. *J. Exp. Med.* **158**:781–785.

Fraker, P. J. & Speck, J. C. Jr. (1978) Protein and cell membrane iodinations with a sparingly soluble chloramide, 1,3,4,6-tetrachloro 3-alpha, 6-alpha-diphenylglcoluril. *Biochem. Biophys. Res. Commun.* **80**:849–857.

Garin, N. S. & Grabarev, P. A. (1972) Protective reactions in rabbits and guinea pigs upon repeated feeding by ixodid ticks, *Rhipicephalus sanguineus* (Latr., 1806). *Med. Parazitol. Muskra.* **41**:274–279 (In Russian).

Geczy, A. F., Naughton, M. A., Clegg, J. B., & Hewetson, R. W. (1971) Esterases and a carbohydrate splitting enzyme in the saliva of the cattle tick, *Boophilus microplus*. *J. Parasitol.* **57**:437–438.

Gladney, W. J. & Dawkins, C. (1971). Parthenogenetic reproduction by *Dermacentor variabilis* Say. *Ann. Entomol. Soc. Amer.* **64**:1286–1289.

Gladney, W. J. & Drummond, R. O. (1971) Spermatophore transfer and fertilization of Lone Star ticks off the host. *Ann. Entomol. Soc. Amer.* **64**:378–381.

Graziano, F. M., Gunderson, L., Larson, L., & Askenase, P. W. (1983) IgE antibody mediated cutaneous basophil hypersensitivity reactions in guinea pigs. *J. Immunol.* **131**:2675–2678.

Gregson, J. D. (1970) Antigenic properties of tick secretions. *J. Parasitol.* **56**:1038–1039.

Laemmli, U. K. (1970) Cleavage of structural proteins during the assembly of the head of Bacteriophage T4. *Nature* **227**:680.

Lowry, O. H., Rosebrough, N. J., Farr, A. L., & Randall, R. J. (1951) Protein measurement with the Folin Phenol reagent. *J. Biol. Chem.* **193**:265–275.

McGowan, M. J., Homer, J. T., O'Dell, G. V., McNew, R. W. & Barker, R. W. (1980) Performance of ticks fed on rabbits inoculated with extracts derived from homgenized ticks, *Amblyomma maculatum* Koch (Acarina: Ixodidae). J. Parasitol. **66**:42–48.

McGowan, M. J., Barker, R. W., Homer, J. T. & Holscher, K. H. (1981) Success of tick feeding on calves immunized with *Amblyomma americanum* extract (Acari:Ixodidae). *J. Med. Entomol.* **18**:328–332.

McSwain, J. L., Essenberg, R. C., & Sauer, J. R. (1982) Protein changes in the salivary gland of the femle Lone Star tick, *Amblyomma americanum,* during feeding. *J. Parasitol.* **68**:100–106.

Markwell, M. A. K. (1982) A new solid-state reagent to iodinate proteins. I. Conditions for the efficient labeling of antiserum. *Anal. Biochem.* **125**:427–432.

Merril, C. R., Goldman, D., Sedman, S. A., & Ebert, M. H. (1981). Ultrasensitive stain for proteins in polyacrylamide gels shows regional variation in cerebrospinal fluid proteins. *Science.* **211**:1437–1438.

Mitchell, E. B. & Askenase, P. W. (1982) Suppression of T cell mediated cutaneous basophil hypersensitivity by serum from guinea pigs immunized with mycobacterial adjuvant. *J. Exp. Med.* **156**:159–172.

Moorhouse, D. E. & Tatchell, R. J. (1966) The feeding process of the cattle tick *Boophilus microplus* (Canestrini): a study in host/parasite reactions. I. Attachment to the host. *Parasitology.* **56**:623–632.

Ptak, W., Rozycka, D., Askenase, P. W., & Gershon, R. K. (1980) Role of antigen presenting cells in the development and persistence of contact hypersensitivity. *J. Exp. Med.* **151**:362–375.

Riek, R. F. (1958) Studies on the reactions of animals to infestation with ticks. III. The reactions of laboratory animals to repeated sublethal doses of egg extracts of *Heamaphysalis bispinosa* Neumann. *Aust. J. Agr. Res.* **9**:830–841.

Reich, C. I. & Zorzopulos, J. (1980) *Boophilus microplus*: cattle antienzymes to larval phosphomonoesterases. *Exp. Parasitol.* **50**:272–277.

Shapiro, S. Z. & August, J. T. (1976) The use of immune-precipitation to study the synthesis and cleavage processing of viral proteins. *J. Immunol. Meth.* **13**:153–158.

Streilein, J. W. & Bergstresser, P. R. (1981) Langerhans cell function dictates induction of contact hypersensitivity or unresponsiveness to

DNFB in Syrian hamsters. *J. Invest. Dermatol.* **77**:272–276.

Stumph, W. E., Elgin, S. C. R., & Hood, L. (1974) Antibodies to proteins dissolved in sodium dodecyl sulfate. *J. Immunol.* **113**:1752–1755.

Trager, W. (1939) Further observation on acquired immunity to the tick *Dermacentor variabilis* (Say). *J. Parasitiol.* **25**:137–139.

Wikel, S. K. (1981) The induction of host resistance to tick infestation with a salivary gland antigen. *Am. J. Trop. Med. Hyg.* **30**:284–288.

Willadsen, P. & Riding, G. A. (1979) characterization of a proteolytic enzyme inhibitor with allergenic activity. *Biochem. J.* **177**:41–47.

Willadsen, P. & Williams, P. G. (1976) Isolation and partial characterization of an antigen from the cattle tick, *Boophilus microplus*. *Immunochem.* **13**:591–597.

Willadsen, P., Williams, P. G., Roberts, J. A., & Kerr, J. D. (1978) Responses of cattle to allergens from *Boophilus microplus*. *Int. J. Parasitol.* **8**:89–95.

14

Quantitative studies of host immunoglobulin G passage into the hemocoel of the ticks *Amblyomma americanum* and *Dermacentor variabilis*

David Ben-Yakir, J. Carl Fox, John T. Homer, and R. W. Barker

14.1 INTRODUCTION

The main method for tick control has been the treatment of hosts and their environment with chemical acaricides. Recent emphasis on environmental protection and pure food laws, as well as the emergence of tick strains resistant to acaricides, indicate a need to develop alternative methods for tick control. Vaccination of hosts with antigenic tissues and secretions of ticks may provide a new method to control ticks. Extracts of whole tick body homogenates and of various organs have been tested as to their ability to induce tick resistance in the host (McGowan *et al.* 1980, and Brown *et al.* 1984, and reviews by Willadsen 1980 and Wikel 1982). Antigens from tick salivary glands have been studied the most extensively. For some tick species the antigenic component responsible for the induction of resistance has been identified (Brown *et al.* 1984). Many interactions between the salivary antigens and the host immune system have been described (Wikel & Allen 1982). In contrast, little is known about the composition of immunogens from other internal tissues of ticks. It is not known how these antigens induce host resistance to ticks. Some investigators suggested that specific antibodies may be able to inhibit an enzyme or neutralize a hormone in the tick hemocoel (Roberts 1968, Galun 1978, Allen & Humphreys 1979, Ackerman *et al.* 1980, McGowan *et al.* 1980), but no evidence for such reactions has been found.

Recently, several investigators reported that immunoglobulin G (IgG) of mammalian hosts crossed the gut epithelium of ticks without losing its ability to bind the sensitizing antigen. When *Dermacentor variabilis* was fed on rabbits immunized with tick salivary glands and ovaries the specific IgG crossed into the hemocoel and attached specifically to these organs (Ackerman *et al.* 1981). Intact rabbit hemolysin antibodies were found in the hemolymph of *Ixodes ricinus* fed on rabbits immunized with sheep red blood cells (Brossard & Rais 1984). Bovine IgG against *Theileria sergenti* was found in the hemolymph of *Ornithodores moubata* and *Haemaphysalis longicornis* fed on a calf infected with the parasite (Fujisaki *et al.* 1984). The above reports did not establish that the passage of antibodies occurred in all individuals of the experimental population. Hemolysin antibodies were found only in a small fraction of the ticks tested, and other investigators used hemolymph pooled from several ticks. No information about the actual concentration of IgG in the hemolymph was given.

Knowledge of the concentration of specific IgG in hemolymph could be used to evaluate the capacity of antibodies to deactivate hemolymph components. Both the rate and extent of antigen reaction with antibodies are greatly dependent on the concentrations of these molecules in the reaction medium.

The study described in this chapter was designed to determine the concentrations of specific and total host IgG in the hemolymph of female *Amblyomma americanum* and *D. variabilis*. Specific antibodies were produced in hosts against ovalbumin, a simple immunogenic protein (mw *c.* 40 000), which will simulate a hemolymph component to be deactivated by antibodies. The variations in the concentration of host IgG in the hemolymph of the two tick species and between individuals of the same species were investigated. The changes in the concentrations of host IgG in the hemolymph during the bloodmeal were studied. Because the serum antibody concentrations of the hosts are often used to evaluate their immunological status, the IgG in the tick hemolymph was also expressed as a fraction of their host serum IgG concentration. The variations in these fractions in ticks fed on rabbits, sheep, and calves were noted.

14.2 MATERIALS AND METHODS

The *A. americanum* used in this study originated from Eastern Oklahoma and were maintained in the Medical Entomology Laboratory at Oklahoma State University for the last seven years. *Dermacentor variabilis* ticks were collected in Central Oklahoma and maintained in the laboratory for the last five years. Ticks were reared on rabbits and sheep as described by Patrick & Hair (1975). Adult ticks 3–5 months post-molt were used in this study. Offspring of one or two females were used in each experiment.

New Zealand White rabbits were the principal hosts in this study. None of the rabbits had been infested with ticks or mites before they were used. The rabbits were 4–8 months old and weighed 2–3 kg. Three sibling rabbits of both sexes formed an experimental group. Each tick species was fed on two groups of rabbits. Two rabbits in each group were hyperimmunized with ovalbumin, and the third rabbit served as the nonimmunized control. Three mg of ovalbumin (grade VIII, Sigma Chemical Co., St. Louis, MO) were used per injection. Ovalbumin was dissolved in 1 ml saline and mixed with 1 ml of adjuvant. Freund's complete adjuvant (DIFCO Laboratories, Detroit, MI.) was used for the first injection given intramuscularly in both legs. Freund's incomplete adjuvant was used for the following two injections given subcutaneously in the neck area, at 3-week intervals. Immunized rabbits were infested with ticks 2 weeks after the last injection. Special care was taken to prevent mechanical damage to the ticks. Ticks were fed in six cells made of polystyrene sample vials affixed with adhesive to the shaved backs of the rabbits (Ben-Yakir 1985). Plastic collars prevented the rabbits from disturbing the cells (Watts *et al.* 1972). Two cells were infested on day 1 and one cell was infested on days 3, 5, 7, and 9. Seven pairs of ticks (males and females) were placed in each cell. When 50% of the adult females had engorged and detached in the cells infested on day 1, the rabbit was bled by heart puncture and then sacrificed and skinned. The skins with attached ticks were left for 24 h at 23°C and 90% RH to allow the ticks to freely detach themselves and thus prevent potential damage from forced detachment of the partially engorged ticks. About 90% of the ticks detached within 24 h. Those ticks that did not detach were removed by excising a small piece of skin around their mouthparts.

Two 6-month-old Hereford calves weighing about 100 kg and two 8-month-old sibling Suffolk sheep weighing about 40 kg were used as hosts. None of these animals had been used to rear ticks prior to this study. In each pair one animal was immunized with ovalbumin (15 mg per injection) as described above for the rabbits. Each animal was fitted with eight cells made from tubular orthopedic stockinette affixed with adhesive to the shorn upper side of the animal (Ben-Yakir 1985). The animals were held in headgate stanchions when ticks fed on them. Both species of ticks were fed on each animal. Thirty pairs of ticks were placed in each cell on days 1, 3, 5, and 7. When c. 50% of the females had engorged and detached in the cells infested on day 1, all ticks were carefully removed with fine-tipped forceps. Animals were bled by venapuncture at that time.

All detached ticks were weighed and assigned to a weight-group as described in Table 14.1. Ticks were rinsed for 1 min with cold water to wash away serum exudate, blood or tick feces from their exterior. Ticks were cooled for 30–60 min at 4°C. Hemolymph was collected into precalibrated capillary tubes after amputating the distal portion of one or more legs with fine-tipped forceps while ticks were held on a cold table (5–10°C). Hemolymph with reddish tint, indicating it may have been contaminated with gut content, was not used. The proteins in the hemolymph and the gut content were compared, using polyacrylamide gel electrophoresis (Ben-Yakir 1985) to confirm the purity of the hemolymph samples. Table 14.1 shows the mean quantities of hemolymph collected from each weight-group. Because it was difficult to measure quantities less than 1 μl, samples from weight-groups III and IV were often pooled from several ticks. Hemolymph samples were

Table 14.1 — Weight-groups of feeding ticks and quantities of hemolymph collected*

Group number	Days post-infestation	Weight range (mg) *A. americanum*	Weight range (mg) *D. variabilis*	Hemolymph collected/tick (μl) Mean	Hemolymph collected/tick (μl) Range
I	8–12	400–1200	300–800	5.0	0.5–60
II	6	60–400	80–300	3.0	0.5–20
III	4	10–60	15–80	1.0	0.2–2.0
IV	2	4–10	5–15	0.3	0.1–0.5

*Ticks were fed on rabbits, sheep, and calves. There were no significant weight difference between ticks fed on the different host species.

immediately diluted 1:50 in cold 0.01 M phosphate buffered saline at pH 7.4 with 0.5% Tween 20 (PBS-Tween 20). The diluted hemolymph was kept frozen in 250 μl polyethylene microcentrifuge tubes until assayed for host IgG content.

An enzyme-linked immunosorbent assay (ELISA) was used to measure the amount of host IgG in the hemolymph. A double antibody sandwich method as outlined by Voller *et al.* (1979) was found to be the most sensitive and consistent. NUNC 96F Immunoplate I plates (Irvine Scientific, Santa Anna, CA) were used to bind the IgG fraction of antihost IgG serum at 500 ng/ml (Cooper Biomedical, Malvern, PA) or Ovalbumin at 50 μg/ml. Alkaline phosphatase was conjugated to the antihost IgG antibodies. All assayed materials were diluted with PBS-Tween 20 and were run in duplicate wells. Working volume for all reagents was 100 μl per well except for the NaOH solution (50 μl/well). Absorbance was measured with a Bio-Tek EIA Reader (Model EL 307 Bio-Tek Instruments, Burlington, VT). The absorbance readings for host IgG at concentrations of 0.5–5.0 ng/ml were clearly positive compared with the negative controls. The detection range for rabbit antiovalbumin IgG was 1–10 ng/ml. These ranges were determined by using reference solutions of host IgG and of rabbit antiovalbumin IgG, respectively. A linear correlation ($r > 0.96$) existed between the absorbance values and the IgG concentration. During preliminary studies (Ben-Yakir 1985) interference by nontarget hemolymph components reduced the sensitivity of the assay. Centrifugation of the diluted hemolymph at 12 000 g for 3 min did not eliminate the interference. Attempts to 'block' the interfering components with various concentrations of bovine serum albumin were unsuccessful. Therefore, reference solutions of rabbit IgG were assayed with various concentrations of hemolymph from sheep-fed ticks, which was negative for rabbit IgG. Ovine and bovine IgG reference solutions were assayed with hemolymph of rabbit-fed ticks for the same reason. Ten separate pools of hemolymph were used for these assays. The mean absorbance values of the above assays were compared to the absorbance values of the reference solutions without hemolymph to determine the percent reduction in absorbance due to interference (Table 14.2). Reference IgG solutions and negative control hemolymph were assayed with every plate. Hemolymph from sheep-fed ticks was used as a negative

Table 14.2 — The percent reduction (mean±SD) in absorbance values of reference IgG due to interference by nontarget hemolymph proteins.

Hemolymph dilution	Percentage*	
	A. americanum	*D. variabilis*
1:250	50±13	80±15
1:500	40±10	65±12
1:1000	30±7	50±10
1:2000	20±4	40±8
1:4000	15±2	25±4
1:8000	10±1	15±2

*Based on assays with 10 different pools of hemolymph.

control in the rabbit IgG studies, and hemolymph from rabbit-fed ticks served as the negative control for the bovine and ovine IgG studies. For the antiovalbumin IgG assay, hemolymph from ticks fed on nonimmunized hosts of the same species was used as a negative control. For consistency of results the enzyme reaction was stopped when the reference solution of 2.0 ng IgG/ml reached the absorbance value of 1.0 (on a scale of 0.0–2.0). This took about 30 min at room temperature. Hemolymph samples were diluted to give absorbance values of 0.4–1.8 and these values were corrected for the effect of interference by dividing them by (100-% reduction)$\times 10^{-2}$. The concentrations of host IgG in the samples were determined by comparing the corrected absorbance values to a linear regression curve (the least squares method) calculated from the absorbance values for the reference IgG solutions tested on the same plate. The serum IgG concentrations were determined in a similar manner, but no interference was observed at the high dilutions at which they were assayed (*c.* $1{:}10^6$). The fraction of host IgG that was found in the hemolymph was determined by dividing the mean concentration of host IgG in the hemolymph by the serum concentration of IgG. The fraction of antiovalbumin IgG found in the hemolymph of sheep- or calf-fed ticks was estimated by comparing the corrected mean absorbance values of the hemolymph samples with the absorbance values of the serum.

14.3 RESULTS

The sites of tick feeding on most hosts in this study became intensely irritated about one week after the initial infestation. The special feeding cells were very effective in protecting the ticks from host grooming.

Most hemolymph collected had a brown-green tint, that became lighter as feeding progressed. In weight-groups I and II about 10–20% of the ticks had colorless hemolymph that coagulated very slowly and allowed collection of large volumes. The protein patterns of the hemolymph samples and the gut content were clearly different (Ben-Yakir 1985). When hemolymph was diluted with PBS-Tween 20 a brown precipitate formed near the tip of the capillary tube and it disappeared upon mixing. This precipitate also formed in 0.05 M Tris buffer solution (pH 7.2) and distilled water, and it was not affected by the addition of phenyl-thiourea (prevents melanization of insect hemolymph) to the diluent solutions.

The absorbance values for the control samples were all below 0.2. Hemolymph was diluted 1:250–1:1000 to measure antiovalbumin IgG and 1:1000–1:8000 for measuring the total host IgG. The lowest detectable concentrations of IgG in the hemolymph were 0.5 μg/ml for antiovalbumin IgG and 0.3 μg/ml for total IgG.

Table 14.3 — The concentration (mean±SD) of total rabbit IgG in the hemolymph of *A. americanum* females fed on rabbits.

Tick weight-group*	Hemolymph IgG in μg/ml (number of samples)					
	Rabbit-1	Rabbit-2	Rabbit-3	Rabbit-4	Rabbit-5	Rabbit-6
I	7.8±4.2 (11)	3.4±1.5 (10)	12.1±2.7 (10)	3.9±1.6 (10)	9.0±3.0 (11)	5.6±1.2 10)
II	1.3±1.0 (9)	0.7±0.3 (9)	0.8±0.3 (10)	1.0±0.4 (7)	2.6±1.6 (8)	1.3±0.5 (10)
III	0.6±0.2 (5)	0.4±0.1 (6)	0.4±0.1 (6)	0.5±0.3 (6)	0.8±0.2 (5)	0.6±0.3 (6)
IV	0.7±0.4 (5)	0.5±0.1 (6)	0.4±0.1 (4)	0.4±0.2 (4)	1.3±0.5 (5)	0.5±0.3 (4)
Serum IGg† (μg/ml)	15.0×10^3	9.0×10^3	17.0×10^3	8.0×10^3	18.0×10^3	16.0×10^3

*See Table 14.1.
†Concentration in host.

Table 14.4 — The concentration (mean±SD) of total rabbit IgG in the hemolymph of *D. variabilis* females fed on rabbits.

Tick weight-group*	Hemolymph IgG in μg/ml (number of samples)					
	Rabbit-7	Rabbit-8	Rabbit-9	Rabbit-10	Rabbit-11	Rabbit-12
I	13.7±7.3 (11)	16.4±6.7 (10)	8.2±4.3 (11)	10.4±4.2 (11)	6.9±2.9 (11)	9.7±3.7 (11)
II	7.1±3.2 (9)	13.4±8.2 (9)	6.2±4.2 (8)	7.2±3.6 (10)	5.0±2.8 (9)	4.4±2.3 (9)
III	1.2±0.5 (7)	2.5±1.7 (8)	1.2±0.5 (7)	1.0±0.4 (8)	1.6±0.7 (7)	1.8±0.8 (6)
IV	0.9±0.7 (6)	1.7±1.0 (4)	1.7±0.9 (4)	0.8±0.4 (5)	0.8±0.5 (4)	1.0±0.3 (3)
Serum IgG† (μg/ml)	12.0×10^3	20.0×10^3	8.0×10^3	10.0×10^3	12.0×10^3	9.0×10^3

*See Table 14.1.
†Concentration in host.

Tables 14.3 to 14.6 summarize the concentrations of total host IgG in the hemolymph. Tables 14.7 and 14.8 summarize the concentrations of antioval

Table 14.5 — The concentration (mean±SD) of total ovine and bovine IgG in the hemolymph of *A. americanum* females fed on sheep and calves.

Tick weight-group*	Hemolymph IgG in μg/ml (number of samples)			
	Sheep-1	Sheep-2	Calf-1	Calf-2
I	10.0±4.3 (17)	9.8±4.1 (16)	5.8±4.4 (16)	4.7±2.6 (13)
II	3.6±2.5 (9)	4.2±1.4 (10)	4.0±3.7 (12)	2.6±1.8 (11)
III	1.2±0.9 (6)	1.5±1.1 (5)	1.5±1.4 (7)	0.7±0.6 (12)
IV	2.8±1.2 (4)	1.4±0.6 (4)	3.5±1.1 (4)	2.6±0.8 (4)
Serum IgG† (μg/ml)	36.0×10^3	34.0×10^3	18.0×10^3	12.0×10^3

*See Table 14.1.
†Concentration in host.

Table 14.6 — The concentration (mean±SD) of total ovine and bovine IgG in the hemolymph of *D. variabilis* females fed on sheep and calves.

Tick weight-group*	Hemolymph IgG in μg/ml (number of samples)			
	Sheep-1	Sheep-2	Calf-1	Calf-2
I	9.8±3.4 (15)	11.3±3.5 (15)	8.4±5.1 (20)	7.6±3.8 (15)
II	1.0±0.5 (9)	4.8±3.1 (10)	1.3±1.0 (9)	4.5±3.6 (9)
III	0.8±0.6 (7)	1.8±1.3 (7)	1.3±0.5 (6)	2.2±1.4 (8)
IV	2.1±1.6 (4)	2.5±1.5 (4)	3.7±2.7 (5)	2.7±1.1 (4)
Serum IgG† (μg/ml)	23.0×10^3	24.0×10^3	18.0×10^3	12.0×10^3

*See Table 14.1.
†Concentration in host.

bumin rabbit IgG in the hemolymph. Several samples in group III and all samples in group IV were pooled hemolymph from 2–5 ticks, therefore, a smaller number of samples is shown for these groups. Within each group there was no apparent relation between the tick body weights and the IgG concentration in the hemolymph. Ticks with colorless hemolymph usually had the lowest IgG concentrations. These ticks greatly increased the variation in the results in weight-groups I and II.

The serum IgG concentrations, as established by the ELISA, were slightly higher than the reported values for rabbits (Kozma *et al.* 1967) and sheep, and slightly lower than the reported values for calves (Tizard 1977). In the hyperimmunized rabbits the specific antiovalbumin IgG was about

Table 14.7 — The concentration (mean±SD) of antiovalbumin rabbit IgG in the hemolymph of *A. americanum* females fed on rabbits hyperimmunized with ovalbumin.

Tick weight-group*	Hemolymph IgG in μg/ml (number of samples)			
	Rabbit-1	Rabbit-2†	Rabbit-4	Rabbit-5†
I	2.8±1.7 (9)	0.8±0.6 (10)	0.9±0.5 (8)	2.6±1.3 (9)
II–IV	<0.5	<0.5	<0.5	<0.5
Serum IgG‡ (μg/ml)	5.0×10^3	4.0×10^3	3.0×10^3	4.5×10^3

*See Table 14.1.
†Serum and hemolymph of control rabbits #3 and #6 were negative.
‡Concentration in host.

Table 14.8 — The concentration (mean±SD) of antiovalbumin rabbit IgG in the hemolymph of *D. variabilis* females fed on rabbits hyperimmunized with ovalbumin.

Tick weight-group*	Hemolymph IgG in μg/ml (number of samples)			
	Rabbit-7	Rabbit-8†	Rabbit-10	Rabbit-11†
I	3.1±0.5 (10)	4.6±0.9 (8)	3.4±1.2 (9)	4.3±0.9 (9)
II	1.9±1.5 (8)	3.5±0.7 (7)	2.5±1.5 (8)	3.0±0.6 (8)
III	0.9±0.2 (5)	1.6±0.6 (6)	1.3±0.5 (6)	1.4±0.5 (5)
IV	1.3±1.0 (3)	0.8±0.2 (3)	0.8±0.3 (3)	1.7±1.1 (3)
Serum IgG‡ (μg/ml)	4.0×10^3	6.0×10^3	4.5×10^3	5.0×10^3

*See Table 14.1.
†Serum and hemolymph of control rabbits #9 and #12 were negative.
‡Concentration in host.

20–40% of the total serum IgG. A large increase (about 50%) in serum IgG occurred in both sheep from the time of *D. variabilis* removal to the time of *A. americanum* removal, 4 days later. Tables 14.9 and 14.10 summarize the fractions of total and antiovalbumin host IgG that were found in the hemolymph.

14.4 DISCUSSION

This is the first report of host IgG in the hemolymph of female *A. americanum*. The lack of similarity between the protein patterns of the hemolymph samples and the tick gut content, and the consistency of the

Table 14.9 — The fraction* of total host serum IgG that was found in the hemolymph of female ticks of weight-group I.†

Host††	No. of samples	*A. americanum*	No. of samples	*D. variabilis*
Rabbit-1;7	11	1:1887	11	1:909
Rabbit-2;8	10	1:2632	10	1:1250
Rabbit-3;9	10	1:1408	11	1:1000
Rabbit-4;10	10	1:2083	11	1:1000
Rabbit-5;11	11	1:2000	11	1:1724
Rabbit-6;12	10	1:2341	12	1:909
Sheep-1	17	1:3571	15	1:2326
Sheep-2	16	1:3448	15	1:2128
Calf-1	16	1:3125	20	1:2128
Calf-2	13	1:2564	15	1:1587

$$* = \frac{\text{Mean concentration of host IgG in the hemolymph}}{\text{concentration of host IgG in the serum}}.$$

†See Table 14.1.

††See Tables 14.3–14.6.

Table 14.10 — The fraction* of *antiovalbumin,* host serum IgG that was found in the hemolymph of female ticks of weight-group I.†

Host††	No. of samples	*A. americanum*	No. of samples	*D. variabilis*
Rabbit-1;7	9	1:1786	10	1:1282
Rabbit-2;8	10	1:5000	8	1:1299
Rabbit-4;10	8	1:3333	9	1:1316
Rabbit-5;11	9	1:1695	9	1:1163
Sheep-1	12	1:3030	11	1:1493
Calf-2	16	1:4000	20	1:2128

*For rabbits:

$$= \frac{\text{Mean concentration of } \textit{antiovalbumin} \text{ IgG in the hemolymph}}{\text{concentration of } \textit{antiovalbumin} \text{ IgG in the serum}}$$

For sheep and calf:
Estimated by comparing the corrected mean absorbance values of the hemolymph to the absorbance values of the sera.

†See Table 14.1.

††See Tables 14.7–14.8.

results in this study, both indicate that the IgG detected in the hemolymph had passed naturally into the hemocoel of these ticks. Based on this study and previous reports (Ackerman *et al.* 1981, Brossard & Rais 1984, Fujisaki *et al.* 1984) it appears that the crossing of serum proteins, including antibodies, into the hemocoel is a widespread phenomenon among ticks. The host IgG in the hemocoel maintains its ability to bind specifically to the sensitizing antigen. It was not determined in this study whether intact IgG molecules or only the antigen binding fragments (Fab') were present in the hemolymph. This information has little bearing on our goal to inhibit or neutralize an antigen in the hemolymph. The hemolysin activity in the hemolymph of *I. ricinus* (Brossard & Rais 1984), which requires an intact antibody molecule, indicates that at least some intact antibodies are able to pass through the gut wall of ticks. It is not known how these large molecules (mw 180 000 daltons) are able to avoid digestion and pass through an intact layer of gut epithelium. According to Akov (1982) the digestive enzymes of ticks act rapidly on hemoglobin but slowly upon other blood proteins. There is a slight possibility that some of the undigested blood proteins are transported into the hemocoel to aid in osmoregulation, as was reported for host albumin in the hemolymph of tsetse flies (Nogge & Giannetti 1980).

A gradual increase in host IgG concentration was generally observed as feeding progressed. The concentration of host IgG in the hemolymph was the highest during the rapid feeding phase and around detachment time (weight-group I). This may be due to the accumulation of IgG in the hemocoel or to an increase in the gut permeability as it stretches to accommodate the bloodmeal. All ticks in weight-group I had a detectable concentration of IgG in their hemolymph. The concentration of IgG in the host serum had a direct relationship to the IgG concentration in the hemolymph. The variation among ticks that fed on the same host may be due in part to differences in the fluid content of the hemolymph at the time of collection. For example, the ticks with colorless hemolymph and relatively low IgG concentrations probably failed to excrete the excess fluid via their salivary glands. The findings of greater fractions of host IgG in the hemolymph of *D. variabilis* may be due to their ability to concentrate the blood meal about 3-fold compared with only 2-fold concentration by *A. americanum* (Koch & Sauer 1984). Slightly larger fractions of rabbit IgG were found in the tick hemolymph compared with the fractions of serum IgG of sheep and calves. This may be due to an actual difference between these hosts or to an experimental artifact, since each host was tested with a different set of reagents.

From these results it might be expected that 0.5–3.0 μg/ml of specific IgG will be available for reaction with a hemolymph antigen in ticks feeding on a vaccinated host. This reaction may lead to inhibition of an enyzme, neutralization of a hormone, or agglutination and precipitation of hemolymph components which would interfere with their normal function. Only hemolymph components at low concentration are expected to be affected by specific antibodies to a significant extent. Although the total concentration of soluble proteins in the hemolymph of replete female ticks was 85 mg/ml for *A. americanum* and 100 mg/ml for *D. variabilis* (Ben-Yakir 1985), enzymes and hormones in the hemolymph are likely to be present at low concentrations. The results of this study also provide a possible explanation of why ticks fed on rabbits hyperimmunized with cell-free hemolymph minor deleterious effects (Ben-Yakir 1985). Rather than immunizing a host with an entire tissue, like hemolymph, there is a need to identify a single antigenic molecule or a particular antigenic determinant in the hemolymph whose deactivation with specific antibodies will produce deleterious changes in ticks.

Most of the research toward developing an anti-tick vaccine has focused on the use of salivary gland antigen (SGA). This approach is based on the detailed study of the immunological mechanisms responsible for the natural development of tick resistance in the host. These studies have led to the relatively successful induction of resistance in naive animals by immunization with SGA (Wikel & Allen 1982, Brown *et al.* 1984). When extracts of whole tick homogenates or specific internal organs were used as immunogens (McGowan *et al.* 1980, 1981 and review by Wikel 1982) the resistance that was induced was probably due to salivary antigens or their homologues in other tick tissues (Wikel, personal communications). Resistant hosts usually exhibit intense cutaneous basophilic hypersensitivity reaction to tick saliva which is the apparent mechanism that prevents ticks from feeding successfully on them. This form of resistance is most effective against the larvae because of their short mouthparts. Usually, only the nymphs and adults of the two tick species used in this study feed on domesticated animals. Several problems may be caused by hyperimmunization with SGA. Exaggerated hypersensitivity often results in harmful effects in the host (Riek 1958). Intense cutaneous inflammation may produce skin lesions that affect the hide quality of cattle, and in some instances may be exploited by secondary pathogens. On the other hand, hosts may become immunologically toleratnt to SGA following repeated exposure to tick saliva (Berdyev & Khudainazarova 1976).

Immunization with a hemolymph component that the host never encounters naturally eliminates the risk of hypersensitivity or immunotolerance. This type of immunization would use the tick hemocoel as the site where the host immune system harms the tick. In addition, natural resistance will continue to play its role without being affected by artificial immunization. It was established in this study that female ticks might be particularly susceptible to this type of immunization, provided that an appropriate antigen can be found in the hemolymph.

A study of the proteins in the hemolymph of *A. americanum* and *D. variabilis* has been initiated with the aim of identifying proteins that are present at low concentrations in the hemolymph of female ticks during the rapid-feeding phase. When identified, the effects of deactivating these proteins with specific IgG will be investigated with the hope that these studies will provide material for an anti-tick vaccine.

14.5 SUMMARY

The concentration of specific (antiovalbumin) and total host immunoglobulin G (IgG) were measured in the hemolymph of female *Amblyomma americanum* and *Dermacentor variabilis* fed on rabbits, sheep, and calves.

An enyme-linked immunosorbent assay (ELISA) was used to determine the concentration of IgG in the hemolymph of individual females every 2 days during their bloodmeal. Results were compared with the IgG concentrations in the host serum to determine the fraction of serum IgG that was found into the hemolymph. In most ticks the IgG concentration in the hemolymph increased as feeding progressed, and reached a maximum level around the time of repletion. The fraction of host serum IgG that was found in the hemolymph at repletion time was 1:1000–1:2000 for *D. variabilis* and 1:2000–1:4000 for *A. americanum.* The concentration of antiovalbumin specific IgG in the hemolymph was about 2 μg/ml at repletion. Results were similar for ticks fed on all host species used in this study. The information from this study is being used for selecting an appropriate hemolymph component for an anti-tick vaccine.

ACKNOWLEDGEMENT

We thank Ms Thea Yellin and the rest of the staff of the Medical Entomology Laboratory. We are also thankful to Ms Dawn Neely for her patience and care in typing this chapter.

REFERENCES

Ackerman, S., Floyd, M., & Soneshine, D. E. (1980). Artificial immunity to *Dermacentor variabilis* (Acari: Ixodidae): Vaccination using tick antigens. *J. Med. Entomol.* **17:** 391–97.

Ackerman, S., Brian Clare, F., McGill, T. W., & Soneshine, D. E. (1981). Passage of host serum components, including antibodies, across the digestive tract of *Dermacentor variabilis* (Say). *J. Parasitol.* **67:** 737–40.

Akov, S. (1982). Blood digestion in ticks *in Physiology of ticks* (Obenchain, F. D., & Galun, R., eds.) Pergamon Press. Elmsford, N. Y. p. 197–211.

Allen, J. R., & Humphreys, S. J. (1979). Immunization of guinea pigs and cattle against ticks. *Nature* **280:** 491–93.

Ben-Yakir, D. (1985). Evaluation of tick hemolymph as a potential material for anti-tick vaccine. PhD. thesis, Oklahoma State University, Stillwater, Ok.

Berdyev, A., & Khudainazarova, S. N. (1976). A study of acquired resistance to adults *Hyalomma asiaticum asiaticum* in experiments on lambs. *Parasitologiya* **10:** 619–25. (English transl.: NAMRU3-T1340).

Brossard, M., & Rais, O. (1984). Passage of hemolysins through the midgut epithelium of female *Ixodes ricinus* L. fed on rabbits infested or reinfested with ticks. *Experientia* **40:** 561–63.

Brown, S. J., Shapiro, S. Z., & Askenase, P. W. (1984). Characterization of tick antigens inducing host immune resistance. I. Immunization of guinea pigs with *Amblyomma americanum* derived salivary gland extract and identification of an important salivary gland antigen with guinea pig anti-tick antibodies. *J. Immunol.* **133:** 3319–25.

Fujisaki, K., Kamio, T., & Kitaoka, S. (1984). Passage of host serum components, including antibodies specific for *Theileria sergenti,* across the digestive tract of argasid and ixodid ticks. *Ann. Trop. Med. Parasitol.* **78:** 449–50.

Galun, R. (1978). *Control of livestock pest by interference in their development.* Bogata, Colombia, UC/AID Pest Management and Relative Environmental Protection Progress Report.

Koch, H. G., & Sauer, J. R. (1984). Quantity of blood ingested by four species of hard ticks (Acari: Ixodidae) fed on domestic dogs. *Ann. Entomol. Soc. Am.* **77:** 142–6.

Kozma, C. K., Pelas Ann, & Salvador, R. A. (1967) Electrophoretic determination of serum proteins of laboratory animals. *J. Am. Vet. Med. Assoc.* **151:** 865–9.

McGowan, M. J., Homer, J. T., O'Dell, G. V., McNew, R. W., & Barker, R. W. (1980). Performance of tick fed on rabbit inoculated with extracts derived from homogenized ticks *Ambylomma maculatum* Koch (Acarina: Ixodidae) *J. Parasitol.* **66:** 42–48.

McGowan, M. J., Barker, R. W., Homer, J. T., McNew, R. W., & Holscher, K. H. (1981). Success of tick feeding on calves immunized with *Ambylomma americanum* (Acari: Ixodidae) extract. *J. Med. Entomol.* **18:** 328–32.

Nogge, G., & Giannetti, M. (1980). Specific antibodies: A potential insecticide. *Science* **209:** 1028–29.

Patrick, C. D., & Hair, J. A. (1975). Laboratory rearing procedures and equipment for multi-host ticks (Acarina: Ixodidae) *J. Med. Entomol.* **12:** 389–90.

Riek, R. F. (1958). Studies on the reaction of animals to infestation with ticks. III. The reaction of laboratory animals to repeated sublethal doses of egg extract of *Haemaphysalis bispinosa* (Newman). *Aust. J. Agric. Res.* **9:** 830–41.

Roberts, J. A. (1968). Acquisition by the host of resistance to the cattle tick *Boophilus microplus* (Canestrini) *J. Parasitol.* **54:** 657–62.

Tizard, I. R. (1977). *An introduction to veterinary immunology.* W. B. Saunders Co. Philadelphia. p. 43.

Voller, A., Bidwell, D. E., & Bartlett, A. (1979). *The enzyme linked immunosorbent assay.* Dynatech Laboratories, Alexandria, Virginia.

Watts, B. P., JR., Pound, J. M., & Oliver, J. H. Jr. (1972). An adjustable plastic collar for feeding ticks on ears of rabbits. *J. Parasitol.* **58:** 1105.

Wikel, S. K. (1982). Immune response to arthorpods and their products. *Ann. Rev. Entomol.* **27:** 21–8.

Wikel, S. K., & Allen, J. R. (1982) Immunological basis of host resistance to ticks In *Physiology of ticks* (Obenchain, F. O., & Galun, R. eds.) Pergamon Press. Elmsford, N. Y. pp. 169–96.

Willadsen, P. (1980). Immunity to ticks. *Adv. in parasitol.* **18:** 293–313.

15

Tick pheromones: an overview

Daniel E. Sonenshine

15.1 INTRODUCTION

Knowledge of tick phermones has grown rapidly since the pioneering discovery of 2,6-dichlorophenol, the first reported tick pheromone, by Berger (1972). Recent reviews (Sonenshine 1984a, 1985) have described the complexity of this subject and commented on all but the latest discoveries. Consequently, this overview of tick pheromones will highlight only the most recent research and correlate it with earlier findings in an effort to enhance our understanding of chemical communication in ticks. Similarities with other arthropod groups will be noted, and the evolution of pheromone systems will be explored.

15.2 TYPES OF TICK PHEROMONES

Behavior-modifying chemicals are termed semiochemicals. Pheromones, perhaps the most common and best known of the various types of semiochemicals, may be defined as compounds secreted by individuals of a species that alter the behaviour of other individuals of that same species in such a manner as to benefit the survival of the population. Pheromones may comprise a single compound or a mixture of different compounds. Moreover, phermonal compounds may be volatile, inducing an attractant effect, or non-volatile, eliciting a response only when contact is made. The same compound may be used by many species, or may be highly species specific.

The classification of these information-bearing substances is based on their biological effects, rather than their chemistry. Indeed, an incredible variety of compounds have been found serving as pheromones in different animals, e.g., phenols, sesqueterpenes, terpenoids, and hydrocarbons, to name but a few of the chemical classes. Great diversity exists within each of the classes among the various animals that use these chemical compounds.

However, the behavioral patterns, i.e., pheromone-mediated behavior, are limited to only a few types, e.g., sexual attraction, food-locating behavior, assembly, or alarm behavior.

Ticks utilize three general types of pheromones, namely, (1) assembly pheromones: non-specific compounds which induce clustering in the natural habitat, (2) aggregation–attachment pheromones: species-specific volatile compounds or a mixture of compounds (Schöni *et al.* 1984), which induce clustering of ticks of the same species at a particular feeding site on the host body, and, finally, (3) sex pheromones: including both species specific and non-specific pheromones, which attract males to sexually active females and excite copulation.

In addition to the pheromones, there is evidence of the use of natural repellants (allomones) in regulating tick behavior in certain mixed species populations (Khalil *et al.* 1983a).

We will examine some of the most recent contributions that have been made in the study of each of these different pheromone types.

15.3 ASSEMBLY PHEROMONES

Many species in both the Argasidae and Ixodidae deposit unknown compounds on the substratum that they contact as they move about their environment, compounds which serve to arrest the movement of other individuals that encounter them. This response to the presence of such compounds leads to cluster of quiescent individuals, i.e., assembly. The interspecific nature, slow speed of response, low volatility, and high degree of chemical stability are attributes of these pheromones that were summarized in recent reviews by Leahy (1979) and Sonenshine (1985). These pheromones may be related to excretory products deposited by ticks as they move. Gothe & Kraiss (1982) suggest that the perhomone emerges via the anus and, possibly, the coxal glands (in argasids). Workers at the International Center for Insect Physiology and Ecology, Nairobi, Kenya, have considered purines in the tick excreta as the pheromonal compounds, although formal reports of their findings have not been published to date (Obenchain, personal communication). The need for identifying the chemical composition of these important compounds is one of the more urgent priorities in the study of tick pheromones.

15.4 AGGREGATION–ATTACHMENT PHEROMONES

These pheromones are unique to the genus *Amblyomma,* and only in a relatively small number of species that parasitize ungulates and other large herbivores. The pheromones are produced only by males, but are attractive to both sexes and even nymphal ticks in one of the species studied (Rechav 1978). Although males will attach and feed readily when exposed to suitable hosts, females will attach only in the presence of this male pheromone. This pheromone-regulated feeding behaviour contributes to the formation of well defined conspecific clusters of feeding males and females.

Feeding males also clasp moving, unfed females that approach them, using their hind legs to grasp them (Obenchain 1984). Thus, the aggregation–attachment response is actually a hierarchy of behavioral responses, i.e., attraction to the feeding site, grasping by the attached males, and attachment. This behavior also greatly improves the mating success of this species, since conspecific clusters are formed at the commencement of feeding.

Extensive, detailed studies of these pheromones were carried out by Gladney, Rechav, and Norval (reviewed by Rechav & Whitehead 1979 and Sonenshine 1984a,b, 1985). The first description of the chemical composition of an aggregation–attachment phermone was reported by Schöni *et al.* (1984). These workers observed *o*-nitrophenol, methyl salicylate, and pelargonic acid in the pheromone of *Amblyomma variegatum*. *o*-nitrophenol was found to be responsible for searching and aggregating behavior, while the other compounds elicited clasping and attaching behavior. This discovery represents the first report of the use of phenolic compounds serving in a role other than the regulation of sex-attractant behavior. It is also the second example of a blend of compounds serving as a pheromone. The first is the mixture of phenol and *p*-cresol described by Wood *et al.* (1975) in African rhipicephalids. Of course, an alternative explanation is that these three compounds are actually regulating two or more distinct behaviors, namely (1) searching and aggregation (*o*-nitrophenol), and (2) clasping and attachment (Methyl salicylate and pelargonic acid). Thus, one may consider the aggregation–attachment phenomenon as several related behavioral events, each mediated by its own pheromone. A noteworthy finding by Schöni *et al.* (1984) is that *o*-nitrophenol by itself would not elicit clasping and mounting behavior in the *in vitro* bioassay (akin to clasping and attachment behavior on the host). Similarly, when methyl salicylate or pelargonic acid were administered alone, they were ineffective.

New research is needed to determine whether different chemicals serve as the aggregation–attachment pheromones of other *Amblyomma* species, whether they are single compunds or a blend of compounds. Another important need is to determine the source of aggregation –attachment pheromone. The foveal glands, which produce 2,6-dichlorophenol in many other species, would appear to be the most likely sites of secretion.

15.5 SEX PHEROMONES

This subject has been reviewed in several recent papers (Sonenshine 1984a,b, 1985). Consequently, this paper will highlight recent findings and interpretations that may contribute to an understanding of the role of sex pheromones in the survival strategies of different species.

15.5.1 Courtship procedures in ticks

Mating behavior is a complex process which ensures successful meeting of the sexes as efficiently as possible while avoiding wasteful interspecific exchanges. Diverse behaviors have evolved to regulate this process, often involving a hierarchy of specific steps in a complex ritual. In most cases, chemical signals serve to guide specific events leading to copulation. This process is termed courtship. It provides a filtering mechanism, based on the ability of conspecific mates to recognize and respond positively to appropriate chemical or physical signals from the other partners, while rejecting all others.

In ticks, courtship patterns are diverse, although several genera utilize a very similar model (see Chapter 9). Other reviews (Sonenshine 1984a,b, 1985) have summarized the earlier literature upon which much of our current understanding of these courtship patterns is based, and the reader is referred to these reviews for additional details. Several principles emerge from a study of tick courtship behavior. One is that the process consists of at least two or more discrete events guided by one or more chemical signals, i.e., sex pheromones. Perhaps the simplest courtship process is that of the soft ticks (Argasidae). In *Ornithodoros tholozani* and *O. erraticus,* males are attracted to moving, walking females which they mount. The males palpate the female body surface and detect unknown compounds believed to be secreted in the female coxal fluid. Males do not respond to inactive females, but will mount and attempt to copulate with females of other species or nymphs coated with female coxal fluid (Schlein & Gunders 1981). The identity of this interspecific contact sex pheromone is unknown. No evidence of species specific selectivity in the courtship behavior of these two species has emerged. Presumably, ecological isolating mechanisms are sufficient to minimize interspecific meetings in nature, and elaborate courtship rituals are not needed. However, little is known about mating behavior in these rather secretive ticks, and speculations regarding their sex pheromones or courtship are premature.

Another principle that may be deduced from a study of tick courtship behavior is the evolution of complex, ritualistic mate-selection procedures in species that have sympatric (or parapatric) distributions and share common hosts. These courtship patterns utilize the filtering principle noted above, encouraging and reinforcing the courtship efforts of conspecific partners but rejecting heterospecific mating attempts. These systems are widespread throughout the metastriate Ixodidae, but no single model occurs in all species. Perhaps the most thoroughly investigated model is the one I proposed for *Dermacentor variabilis* (Sonenshine 1985). In this example, courtship consists of 9 discrete stages, as shown in Fig. 15.1. The first 6 phases (a–e in the diagram) are believed to be controlled by the volatile sex attractant pheromone, 2,6-dichlorophenol. This pheromone serves to excite sexually active feeding males, inducing them to detach, orient to the pheromone-emitting females, and mount them. Subsequently the courting males turn to the female venter and probe for the gonopore. The final steps, i.e., gonopore location and recognition of species identity, are controlled by a second sex pheromone, the genital sex pheromone (Sonenshine *et al.* 1982a). The courting male must detect the genital pheromone in the vulva with chemosensory sensilla on its cheliceral digits, or it will abort the process and depart. Thus, filtering occurs at the very end of the courtship process in this species, and requires the ability of the male to recognize a specific

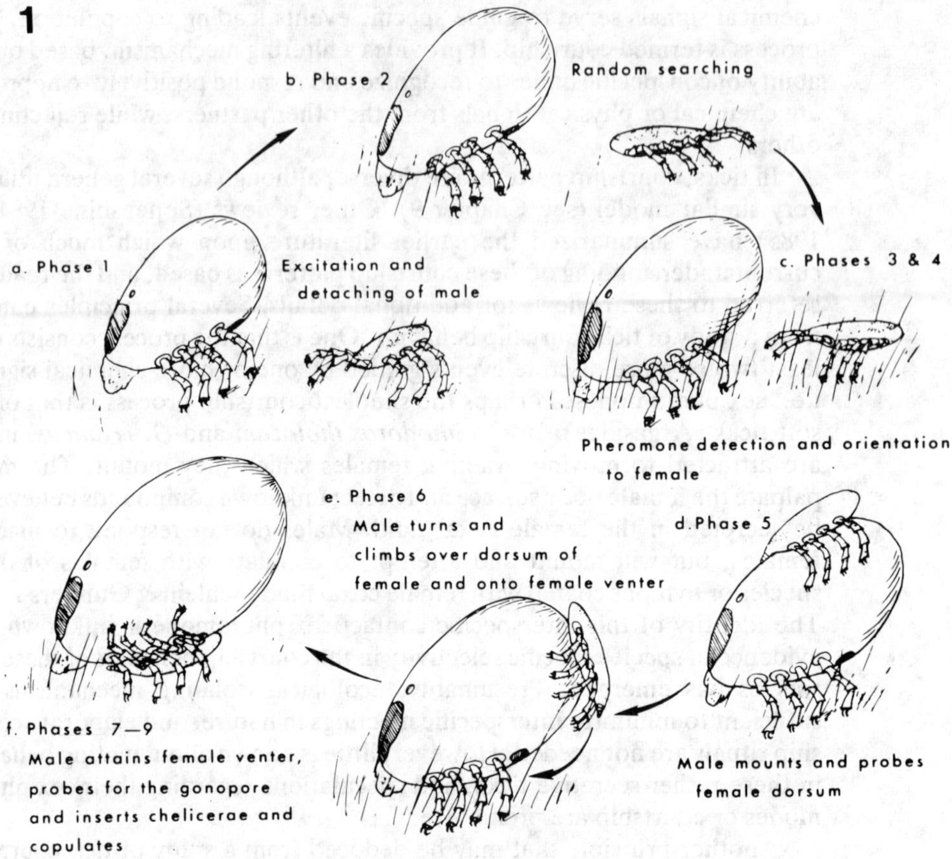

Fig. 15.1 — Hypothetical reconstruction of the behavioural stages that occur during courtship in *Dermacentor variabilis* and *Dermacentor andersoni* (a) Phase 1: feeding female secretes volatile sex-attractant pheromone, 2,6-dichlorophenol, exciting attached feeding male to detach. (b) Phase 2: detached male commences searching behavior. (c) Phases 3 and 4: sexually excited male detects the pheromone, orients to the emitting sources, and approaches the pheromone-secreting female. (d) Phase 5: male contacts the female, mounts, and probes the dorsal surface. The capitulus is flexed and the legs partially folded. (e) Phase 6: the male turns posteriorly, crawls over the female's opisthosoma, and proceeds to the ventral surface. The female lifts its body to facilitate these movements by the male. (f) Phases 7–9: the male positions itself under the female's podosoma, places its legs between those of the female, and flexes its capitulum to probe for the genital aperture. The chelicerae are inserted and copulation ensues. (From *Ann. Review Entomol*. 1985 **30:** 1–28, with permission of the editor.)

identifying code (i.e., pheromone). Males of other species, e.g., *Dermacentor andersoni,* appear to be incapable of distinguishing the *D. variabilis* females as heterospecific until they probe the gonopore with their chelicerae.

The courtship ritual in ticks such as *D. variabilis* is rigidly controlled. Males encountering females positioned with their ventral surface facing away from the host or substratum will still attempt to mount the female's dorsum, probe it, and then return to the ventral surface. Males placed with their mouthparts over the female gonopore will also attempt to repeat the earlier sequence, or they will depart. The female is relatively passive during courtship in these species, although some individuals will raise their bodies away from the host to expose their ventral surface when contacted by a courting male.

Many variations of this model occur in the Ixodidae, even in those species that share 2,6-dichlorophenol as a common, interspectific sex attractant. The decision point may occur at different periods in the courtship process. For example, in the camel ticks, *Hyalomma dromedarii* and *Hyalomma anatolicum excavatum,* differences in the amounts of 2,6-dichlorophenol emitted by the female enables mate-seeking males to determine the identity of their prospective partners. *H. dromedarii* males are insensitive to the small quantities secreted by *H. anatolicum excavatum* and frequently fail to orient to them. In contrast, *H. anatolicum excavatum* males are repelled by the quantities of this phermone secreted by *H. dromedarii* females (Khalil *et al.* 1983). Thus, in this example, filtering occurs at the very beginning of the courtship process, in contrast to the *D. variabilis* model, where it is delayed to the very end. Therefore, it is not surprising that *H. dromedarii* males copulate readily with *D. variabilis* females, once they have contacted and mounted them, whereas *D. variabilis* males rarely copulate with *H. dromedarii* females (Table 15.1). Other examples of different pheromone-guided courtship mechanisms occur in the Australian reptile ticks (Andrews & Bull 1982, Petney & Bull 1981) or in those species of the genus *Amblyomma* that have an aggregation–attachment pheromone (Rechav & Whitehead 1979).

Recent studies by Sonenshine, Silverstein, and their co-workers have been directed toward determining the chemical characteristics of the genital sex pheromone in *Dermacentor* species. Sonenshine *et al.* (1982) recognized the origin of this pheromone in the female reproductive tract, but did not attempt to identify it chemically. In recent studies (Sonenshine, Silverstein, and Brossut, unpublished), extracts were prepared from the anterior reproductive tract tissues, i.e., the vagina, seminal receptacle, connecting tube, uterus, and the various accessory glands, but excluding the ovary and oviducts. To test these extracts, females were 'neutered' by severing the vestibular vagina and washing the external genital surface as described in the previous report (Sonenshine *et al.* 1982). These neutered females were screened with sexually active males before extracts or authentic chemicals were administered to them for testing. Males mounted and probed these females, but most males withdrew without completing copulation, i.e., without depositing their spermatophores. Females that did elicit spermatophore formation were discarded (not neutered). Next, samples of the anterior reproductive tract extract were pipetted onto the genital surface or, in smaller quantities, directly into the remnant of the vestibular vagina. Controls consisted of neutered females without extract or chemical treatment. The results demonstrated that males would copulate readily with the treated females. Next, the extract was fractionated by gel permeation and the fractions tested in the same manner as with the crude extract. Prelimi-

Table 15.1 — Responses of mate-seeking males of *Hyalomma dromedarii* and *H. excavatum* to conspecific and transspecific CAV females, and comparisons with *D. variabilis* and *D. andersoni*.

Type of mating (♂+♀)	No. males	No. females	No.* trials	Percent of males responding by O	M	P	C	WD	A
H. dromedarii × *H. dromedarii*	14	12	14	100.0	100.0	85.7	85.7	0.0	14.3
H. dromedarii × *H. excavatum*	50	10	89	100.0	52.8	25.8	0.0	60.7	39.3
H. excavatum × *H. excavatum*	15	12	16	100.0	100.0	81.2	81.2	0.0	18.8
H. excavatum × *H. dromedarii*	50	10	87	95.4	77.0	12.6	0.0	48.3	47.1
H. dromedarii × *D. andersoni*	11	5	18	95.7	95.7	95.7	21.1	52.5	21.1
H. dromedarii × *D. variabilis*	15	8	18	94.7	84.2	84.2	40.0	21.1	23.1
D. andersoni × *H. dromedarii*	4	4	48	97.9	81.3	75.0	0.0	56.3	22.9
D. variabilis × *H. dromedarii*	34	7	78	100.0	92.3	92.3	0.0	57.7	34.0
D. variabilis × *D. variabilis*	10	10	10	100.0	100.0	100.0	100.0	0.0	0.0
D. variabilis × *D. andersoni*	38	8	103	97.1	96.1	96.1	0.0	86.0	10.7
D. andersoni × *D. andersoni*	13	13	13	100.0	100.0	100.0	84.6	0.0	15.4
D. andersoni × *D. variabilis*	54	14	131	93.9	83.2	83.2	2.3	64.9	26.7

*Each male was allowed 3 trials unless it copulated or attached.
1/From Sonenshine *et al.* (1982) and Khalil *et al.* (1983), with the permission of the editors.

nary results (unpublished) indicate the presence of at least two fractions which elicit copulation, i.e., spermatophore transfer, by the probing male. Electrophysiological studies (Davis & Sonenshine, unpublished) demonstrated highly significant increases in nerve spike activity when the inner cheliceral digits of *D. variabilis* and *D. andersoni* males were exposed to the crude extract, to washings of the tick vagina, and to 20-hydroxyecdysone (20-OH ecdysone). The cheliceral digits were also able to respond to phagostimulants commonly found in host blood. Morphologic studies demonstrated the existence of chemosensory and mechanosensory sensilla on the cheliceral digits of *D. variabilis* (Figs 15.2, 15.3) (Sonenshine *et al.* 1984a), similar to those described in *Boophilus microplus* (Waladde & Rice 1982). In addition to their role in detecting phagostimulants in host blood, experimental evidence demonstrated that these sensilla were required for copulation. Males deprived of their cheliceral digits mounted and probed sexually active females, but failed to copulate or even form their spermatophores (Table 15.2). The absence of a spermatophore suggests disruption of a required signal needed for its formation, e.g, the genital sex pheromone.

The assays described above were repeated using a cross-species test. Neutered *H. dromedarii* females were treated with *D. variabilis* female extract and then exposed to *D. variabilis* males. A highly significant response was obtained with the extract-treated females, but little response occurred with males exposed to the neutered controls; none copulated with normal females. This experiment demonstrates that the excitatory stimulis is chemical (i.e., a pheromone) since the compound or blend of compounds can be extracted, implanted into another, normally non-stimulatory environment, and still retain its activity. In effect, the *H. dromedarii* females were 'disguised' and a highly significant number of *D. variabilis* males responded as if these treated females were conspecific.

Studies on the identity of the genital sex pheromone suggest at least two active components (Sonenshine, Silverstein, Brossut, and Davis, unpublished). One of these components may be 20-hydroxyecdysone, or a closely related steroid. Significant responses were obtained when males were exposed to neutered females treated with 20-OH ecdysone, even in the cross-species test. However, the responses were invariably weaker than that obtained with the complete extract. Radioimmunoassay (RIA) and chemical assays suggest the occurrence of small but significant quantities of ecdysteroids in the anterior reproductive tract of *D. variabilis* females but not in *H. dromedarii* females. When analyzed by HPLC, 20-OH ecdysone was found in the anterior reproductive tract of *D. variabilis* females but none was found in extracts of these tissues from *H. dromedarri* females. Even after allowing for traces of ecdysteroid due to hemolymph contamination, this difference between the two species appears to be highly significant and supportive of the experimental findings.

The identity of the second component is unknown. Apparently, it is a somewhat larger molecule than 20-OH ecdysone, and does not absorb in the ultraviolet range (presumably excluding ecdysteroids). Nothing else is known about its chemistry. Treatment of ticks with this component of the

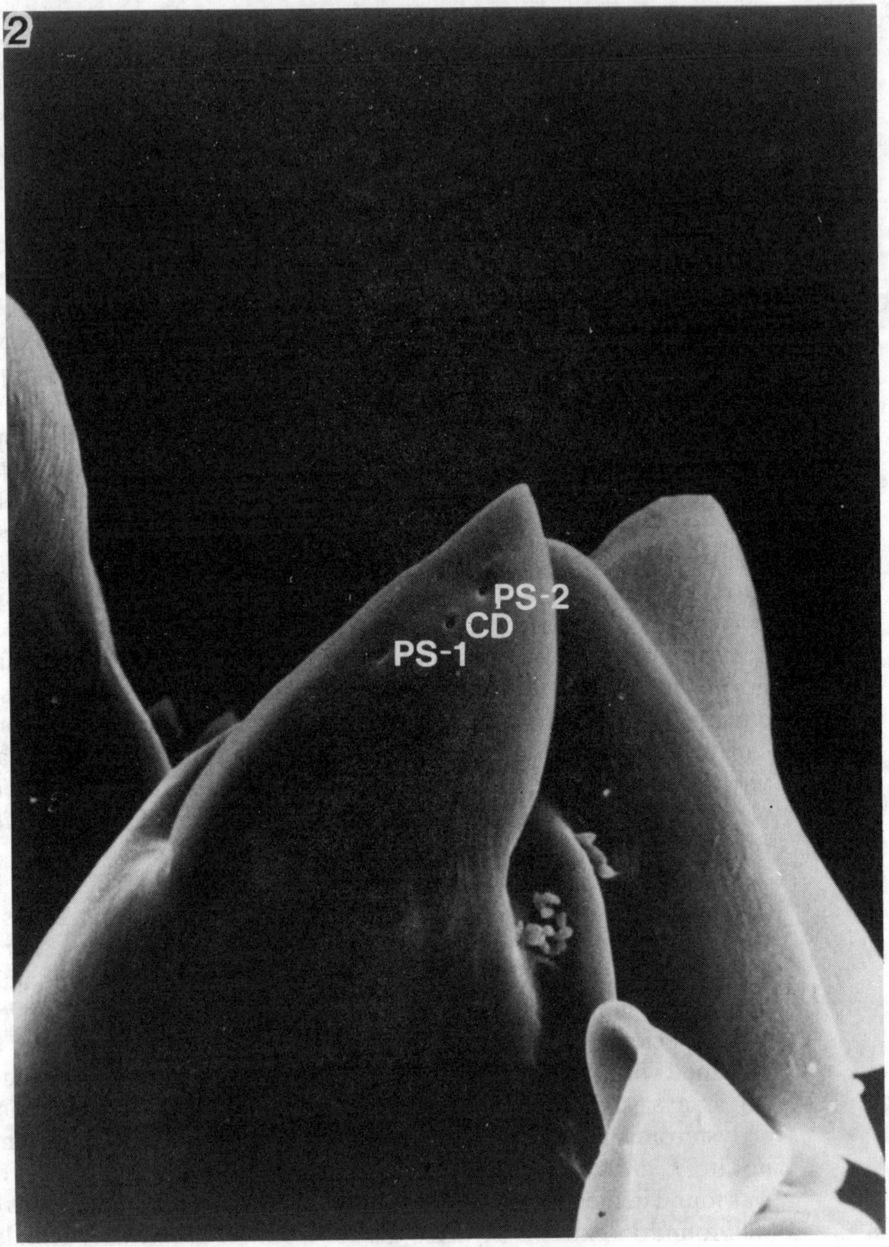

Fig. 15.2 — Scanning electron micrograph illustrating the sensilla on the distal denticle of the inner cheliceral digit of a male *D. variabilis*. Other portions of this digit and the tip of the outer cheliceral digit are visible in the background. The denticle bears the pit sensilla (PS-1, PS-2) and a cone like degression (CD). The cone-shaped depression (CD) is not continuous with the digit lumen. 6417 ×. (Reprinted, by permission of the editor, from the *Journal of Medical Entomology* **21** (3): 296–306.

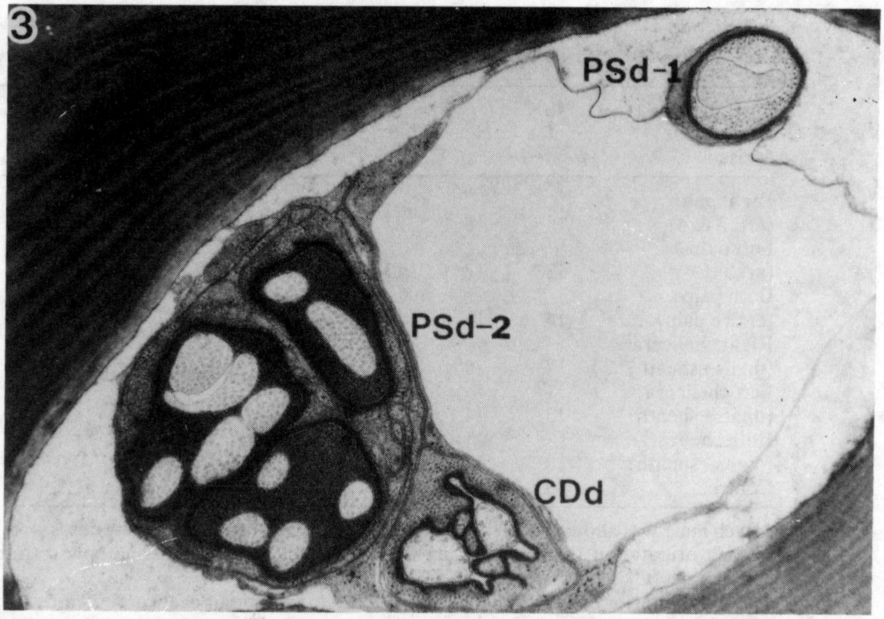

Fig. 15.3 — Transmission electron micrograph through the inner cheliceral digit near its tip, illustrating the central lumen and the three clusters of dendrites, PSd-1, PSd-2, and CDd. PSd-1, with only 1 dendrite, appears enclosed in a shaft of cuticle separate from the digit cuticle, while the other dendritic clusters are free in the main lumen. The PSd-2 dendrites are surrounded by dense granular material enveloped by a scolopale sheath which is surrounded by glial cell processes. The membranes of the latter are joined by other glial cell processes. The two CDd dendrites are surrounded by dense scolopale sheaths and are contained within glial cell processes. In addition to the dendrites, the lumen contains fine filamentous material. 23920 ×. Reprinted, by permission of the editor, from the *Journal of Medical Entomology*, **21** (3): 296–306.

original extract elicits very strong copulatory responses from the males, more or less indistinguishable from those elicted by the crude extract.

Studies to determine the specific site(s) of production of the genital sex pheromone are in progress. Preliminary tests implicate the lobular accessary glands, which expand greatly during feeding of the female and secrete an unknown substance into the glandular channels surrounding the vaginal lining cuticle.

These studies, it is hoped, will lead to an identification of the chemical strucure of the genital sex pheromone in the near future. Other studies are planned to determine the differences in the genital sex pheromone of other *Dermacentor* species. Knowledge of the identity of these pheromones will provide the necessary chemical basis for understanding the process of mating in these ticks in its entirety.

15.5.2 Hormonal regulation of sex pheromone activity

There can be little doubt that sex pheromone activity is highly regulated. Female metastriate ticks lack these pheromones until long after the nymphal–adult molt, and they secrete them only during feeding (Sonen-

Table 15.2 — Mating responses of sexually active *Dermacentor variabilis* males following excision of the cheliceral digits or palpal segments.

Parts of appendages excised	No. males treated	No.	No.* trials	% Response O	P	C	% copulating
Right palp (art. 3 & 4)	18	18	22	100	77.3	50.0	61.1
Left palp (art 3 & 4)	12	12	13	92.3	92.3	84.5	92.3
Both palps (entire palp)	14	19	30	100	96.7	26.7	57.1
Right chelicera (digits+sheath)**	17	20	35	88.6	40.0	2.9	5.9
Left chelicera (digits+sheath)**	13	14	49	100	83.7	1.7	7.7***
Both chelicerae (digits+sheath)**	10	10	10	100	100	0.0	0.0
Controls	9	9	13	100	100	61.5	88.9

*Each male was allowed three trials unless it copulated or attached; responses were characterized as orientation (O), positioning for copulation (P), and spermatophore transfer (i.e., copulation) (C).

**Digits, anterior end of cheliceral shaft, anterior portion of cheliceral sheaths removed.

***Examination of this male after copulation revealed excision of the right cheliceral sheath, but the cheliceral digits were intact.

Reprinted by permission of the editor, from *The Journal of Medical Entomology*, **21**(3): 296–306.

shine *et al.* 1984b). Biosynthesis and secretion of 2,6-dichlorophenol, the most common of the ixodid tick sex pheromones, appears to be stimulated by catecholamines, particularly dopamine. Moreover, the stimulatory activity of these neurosecretory substances appears to require the presence of the hormone 20-OH ecdysone to be effective. A brief review of the evidence in support of this hypothetical mechanism of sex pheromone regulation follows.

Electron micrographs of transections through the foveal nerve demonstrate abundant accumulations of neurosecretory substance, including material characteristic of catecholamine transport vesicles (Sonenshine *et al.* 1981, 1983). Histochemical tests also demonstrated the presence of monoamines, presumably catecholamines, in the foveal glands. When female ticks were exposed to abnormally high doses of dopamine, the demonstrable 2,6-DCP content (assayed by gas chromatography) increased in most cases, as compared to saline-treated controls. Treatment with the catecholamine-depleting drug reserpine or the synthesis-inhibiting drug α-methyl tyrosine ethyl ester led to significant declines in 2,6-DCP content in the treated ticks as compared to the controls. X-ray microanalysis also revealed increased chlorine content in the foveal glands of dopamine-treated females, and significantly lower chlorine content in the glands of reserpine-treated females. Female ticks treated with sublethal doses of reserpine were significantly less attractive to males than controls or dopamine treated individuals.

Hormonal regulation of mating behaviour has been reported in diverse insects (Barth & Bell 1969, Endo 1973), in crustaceans (Kittredge & Takahashi 1972), and in other arthoropods. However, the precise mechanism by which this is accomplished is unclear. In some insects, sex pheromone activity appears to be stimulated by gonadotropic hormone, i.e., coordinated with the stimulation of reproductive activity in the adults. In others, the ovaries appear to be the source of the stimulatory hormone (Endo 1973). The ovaries are the source of ecdysone in the adult females of several insect species (Riddiford & Truman 1978), and it is possible that this hormone is the responsible agent for exciting sex pheromone synthesis or the timing of its release. Studies by Blomquist *et al.* (1984) and Adams *et al.* (1984) demonstrated that the synthesis of the contact sex pheromone of the housefly, Z-(9) tricosene, a hydrocarbon secreted by the epidermis of these insects, was regulated by 20-OH ecdysone. Ovariectomized insects produced alkenes with chain lengths C_{27} and longer, but were unable to synthesize Z-(9) tricosene or other shorter chain hydrocarbons. When ovarian homogenates or 20-OH ecdysone are administered to ovariectomized flies, they respond by synthesizing Z-(9) tricosene. The authors speculate that 20-OH ecdysone causes a change in the chain length specificity of the decarboxylase enzyme which regulates the synthesis of these hydrocarbons.

Evidence of ecdysteroid stimulation of sex pheromone activity has also been found in ticks. Dees *et al.* (1984a, 1985) reported that unfed *H. dromedarii* females, treated with exogenous 20-OH ecdysone as nymphs, were attractive to male ticks. Moreover, when these treated ticks were examined by gas chromatography, significantly greater quantities of 2,6-DCP were found than in untreated or sham-treated controls. 2,6-DCP was even found in treated males. This last observation is an especially useful indicator of hormone excitation of sex pheromone synthesis, since unfed males do not normally produce this compound. X-ray microanalysis of the perhomone glands also revealed significantly greater quantities of chlorine in the treated ticks than in the controls. Autoradiography revealed unusually intense silver grains over the adult foveal glands, but not over other organs or tissues, following treatment of nymphs with ^{3}H ecdysone. In contrast with these findings with *H. dromedarii, D. variabilis* adults proved relatively unresponsive to exogenous hormone, although the pheromone glands of treated unfed females showed evidence of secretory activity. The natural occurrence of ecdysone and 20-OH ecdysone has been reported in several species of ticks (Delbecque *et al.* 1978, Bouvier *et al.* 1982, Diehl *et al.* 1982, Germond *et al.* 1982). More recently, Dees *et al.* (1984b, 1985) demonstrated 20-OH ecdysone in both *D. variabilis* and *H. dromedarii.* Howvever, comparisons between ecdysteroid titers, determined by radioimmunoassay (RIA) or 20-OH ecdysone, determined by HPLC, did not show a clear, concise correlation with changes in the amounts of 2,6-DCP in these ticks during maturation, feeding, and mating. Nevertheless, ecdysteroids were always present during adult development, and the measured concentration of 20-OH ecdysone never fell below 2.5 ng/tick.

Sonenshine (1985) proposed that the role of 20-OH ecdysone in regulating sex pheromone activity is indirect and tied to neurotransmitter function. This hormone is a known excitant of dopa decarboxylase activity (Schlaeger *et al.* 1974). This enzyme may also occur in ticks, since catecholamines have been demonstrated in the synganglion and salivary glands of *Boophilus microplus* (Binnington & Obenchain 10983, for a review). Histochemical evidence of monoamines, presumably dopamine or a related catecholamine, was reported in 'nerve endings associated with epidermal structures' (Binnington & Stone 1977) and, more recently, in the foveal nerve (Sonenshine *et al.* 1985a). None of these findings need preclude the role of neuropeptides as stimulatory and/or regulatory agents in sex pheromone activity. Neurosecretory vesicles with size and staining attributes characteristic of peptidergic neurosecretory molecules have been reported in the foveal nerves (Sonenshine *et al.* 1981, 1983).

An important limiting factor in the study of pheromone gland physiology is the lack of a reliable *in vitro* assay comparable to that described by Kaufman & Sauer (1983) for the salivary glands. The studies on neurosecretory and ecdysteroid regulation of sex pheromone activity reported above utilized *in vivo* experimental methods, e.g., flooding the tick's tissues prior to or even during feeding with above normal physiological doses of the compound to be tested. Clearly, an urgent priority is the development of methods for cultivation or at least maintenance of the pheromone glands *in vitro* in a phsyiologically active state and for sufficient time to permit experimentation.

The body organ or tissue that serves as the source of the tick ecdysteroid is unknown. Cox (1960) described evidence that the synganglion is the source of the molting hormone in the argasid tick *Ornithodoros turicata.* Transplanted synganglia from engorged nymphs induced supernumerary molting in adult recipients. In addition, Cox found that ligation of engorged nymps allowed molting only in the body region anterior to the ligature, i.e., in the region where the synganglion was located. However, the endocrine gland may not be in the synganglion *per se,* but in glandular structures located in adjacent nerve plexi. The ultrastructure of the lateral segmental organs, tiny glandular clusters located in the lateral nerve plexus, suggests an endocrine function, although the specific endocrine gland has not been clarified (Binnington 1981). Stuctures resembling neurohaemal organs of the insect brain have also been reported in tick synganglia (Obenchain & Oliver 1975, Pound 1981). Sonenshine *et al.* (1985b) noted unusually high concentrations of ecdysteroids in extracts of the synganglion/lateral nerve plexus of *D. variabilis* and *H. dromedarii* as compared to other tissues in these ticks (Table 15.3). Removal of the peripheral nerves and the lateral nerve plexi before extraction resulted in drastically reduced ecdysteroid content. Unusually high concentrations of ecdysteroids were also found in the foveal glands, especially when considered in terms of tissue mass (i.e., pg/mg). Samples of epidermis and fat body containing the foveal glands always had significantly more ecdysteroid than samples of these same tissues without foveal glands, suggesting that the ecdysteroid was actually in the

Table 15. 3 — Estimated ecdysteroid content of selected body organs and tissues of partially fed virgin *Hyalomma dromedarii* and *Dermacentor variabilis*

	H. dromedarii				*D. variabilis*			
Organ/ tissue	*N*	mg. tissue	pg/tick	pg/mg	*N*	mg. tissue	pg/tick	pg/mg
Synganglion	250	5.16 ±1.31	2479 ±1134	120,223 ±24,361	250	5.14 ±1.11	1901 ±1384	92,416 ±20406
Hemolymph	—	1.50	1934 ±85	85	—	1.50	1362** ±87	86
Salivary gland	25	26.93 ±8.55	312	290 ±92	25	22.45 ±2.88	957	1065 ±120
Ovary	50	12.82 ±2.60	115	448 ±74	50	21.45 ±2.22	57	133 ±13
Muscle	25	9.58 ±4.13	98	226 ±46	25	4.42 ±1.63	190	1074 ±289
Epidermis/ fat body/ foveal gland	10	12.59 ±0.91	500	397 ±27	10	4.87 ±0.24	1120	2300 ±122
Epidermis/ fat body	10	12.04 ±0.90	82	68 ±5	10	4.95 ±0.25	360	727 ±39
Foveal	49	0.67	158	11,731	50	0.39	133	17,183
Anterior reproductive tract	100	3.99 ±0.56	13	313	50	1.90 1.90	162	4263
Midgut***	14	100.00	947	1350				

* Average/tick=22.7±0.9 μL.
** Average/tick=15.9±6.9 μl.
*** Average weight of tick midgut=7.14 mg/tick.
Reprinted, by permission of the editor, from the *Journal of Medical Entomology*, **22** (3): 303–311.

glands and not merely in adjacent tissues. Substantial concentrations of ecdysteroids were also found in the hemolymph, as expected. Smallamounts of ecdysteroid were found associated with all of the other tissues organs examined, but this was regarded as reflecting unavoidable contamination with hemoloymph. High-pressure liquid chromatography confirmed the presence of 20-OH ecdysone in the extracts made from the synganglion/ lateral nerve plexus and hemolymph tissues, but not in any other body organs or tissues.

Other evidence of central nervous system involvement in ecdysteroid synthesis was obtained by inoculating ticks with ^{14}[C] cholestrerol. When the synganglion/lateral nerve plexus system was excised and examined by autoradiography, intense accumulations of silver grains were found over these organs. Although the highest accumulations occured over the cortical zone, significant accumulations also occurred over the lateral nerve plexus. Evidence of the use of ^{14}[C] cholesterol as a precursor for ecdysteroid synthesis was obtained by radioassay of 20-OH ecdysone from ^{14}[C] cholesterol inoculated ticks. Radioactivity was also found in several polar peaks, believed to be ecdysteroids as a result of their binding with anti-ecdysone antibody, their incorporation of ^{14}[C] radioactivity, and the similarity of their infra-red and NMR spectra with that of ecdysone.

Additional research is needed to determine conclusively where ecdysteroids are produced in ticks, as well as the steps involved in their biosynthesis and metabolism. At present, the available evidence implicates the central nervous system as the site for this endocrine activity, but the findings remain to be confirmed.

Much less is known about the role of juvenile hormones, if any, in regulating sex pheromone activity. Juvenile hormones have been reported to regulate sex pheromone activity in several species of insects (Barth & Bell 1969). That such hormones, acting as gonadotropic hormones, affected reproduction in ticks was demonstrated experimentally by Pound & Oliver (1979) in their studies with soft ticks. However, tests using direct application (topical treatment and inoculation) of authentic juvenile hormone or juvenile hormone analogues did not affect sex pheromone activity to any measurable degree when administered to *H. dromedarri* (Abdelmonem *et al.* in press). Similar results were noted in my laboratory in tests with *D. variabilis* (Sonenshine, unpublished). The hemolymph of these ticks contains esterases which readily degrade juvenoids, yielding inactive polar by-products which co-chormatograph with juvenile hormone acids and diols. Thus, it is not surprising that exogenous juvenile hormone, even when adminstered in microgram quantities, does not produce any demonstrable effects, since it is doubtful that it would remain active in the presence of these enzymes. In summary, the role of juvenoids in regulating sex pheromone activity in ticks remains uncertain.

15.6 SUMMARY

This brief overview of progress in tick pheromone research has attempted to highlight some of the new discoveries and to call attention to areas where further study is urgently needed. The mechanisms by which ticks use different chemical signals, often in a complex and precisely ordered sequence of stimulus and response events, are beginning to emerge. Within the Metastriata, a basic model appears to center around the use of a volatile sex-attractant pheromone to excite males and enable them to locate prospective mates. However, numerous variations exist, such as the formation of conspecific aggregations prior to mating or the addition of a second pheromone (= genital sex pheromone) for species specific recognition. These variations appear to have evolved in response to multiple species infestations of common hosts, and serve to reject hereospecific mating attempts. Although the mating strategies may appear imperfect, they are nevertheless important to the maintenance of the integrity of individual species in nature. Another important finding is the widespread use of phenolic compounds as pheromones, at least in the Metastriata. The prophetic view of Wood *et al.* (1975) regarding this class of compounds appears to have been confirmed. Other important discoveries relate to hormonal regulation of sex pheromone activity, especially the role of ecdysteroids.

ACKNOWLEDGEMENT

This work as supported by a contract, N-00014-80-C0546, from the Office of Naval Research, Microbiology Program, Naval Biology Project, Arlington, Virgina, and by a grant, AI 10986, from the National Institute of Allergy and Infectious Diseases, National Institutes of Health, U.S. Public Health Service, Department of Health and Human Services, Bethesda, Maryland. The opinions and assertions contained herein are the private ones of the author and are not to be construed as official or as reflecting the view of the Department of the Navy or of the naval service at large.

NOTE ADDED IN PROOF

Otieno, D. A., Hassanali, A., Obenchain, F. D., Sternberg, A., & Galun, R. (1985) *Insect Sci. Appl.,* **6:** 667–670 have identified guanine as the assembly pheromone of several argasid and ixodid species.

REFERENCES

Abdelmonem, A., Khalil, G. M., Sonenshine, D. E. & Sallam, O. A. (1986) Effects of two insect growth regulators on the camel ticks, *Hyalomma dromedarii* (Acari: Ixodoidea: Ixodidae). *J. Med. Entomol.* (In press).

Adams, T. S., Dillwith, J. W. & Blomquist, G. J. (1984) The role of 20-hydroxyecdysone in housefly sex pheromone biosynthesis. *J. Insect Physiol.* **30:** 287–294.

Andrews, R. H. & Bull, C. M. (1982) Mating behavior and reproductive isolation of three species of reptile ticks. *Anim. Behav.* **30:** 514–24.

Barth, R. H. & Bell, W. J. (1969) Effect of juvenile hormone and JH analogues on the reproductive biology of females of the cockroach, *Bryostria fumigata* Guerin. *Amer. Zool.* **9:** 1085.

Berger, R. S. (1972) 2,6-dichlorophenol, sex pheromone of the lone star tick. *Science* **177:** 704-705.

Binnington, K. C. (1981) Ultrastructural evidence for the endocrine nature of the lateral organs of the cattle tick *Boophilus microplus. Tissue & Cell* **13:** 475–490.

Binnington, K. C. & Stone, B. F. S. (1977) Distribution of catecholamine in the nervous system of the cattle tick *Boophilus microplus* Canestrini *Comp. Biochem. Physiol. C,* **58:** 21–28.

Binnington, K. C. & Obenchain, F. D. (1983) Structure and function of the circulatory, nervous, and neuroendocrine system of ticks. In: *Physiology of ticks.* Obenchain, F. D. & Galun R. (eds.) Pergamon Press, Ltd Oxford. p. 351–398.

Blomquist, G. J., Adams, T. S. & Dillwith, J. W. (1984) Induction of female sex pheromone production in male houseflies by ovary implants or 20-hydroxyecdysone. *J. Insect Physiol.* **30:** 295–302.

Bouvier, J., Diehl, P. A. & Morici, M. (1982) Ecdysone metabolism in the tick, *Ornithodoros moubata* (Argasidae, Ixodoidea). *Rev. Suisse Zool.* **89:** 967–976.

Cox, B. L. (1960) Hormonal involvement in the molting process in the soft tick *Ornithodoros turicata* Duges. Unpublished PhD dissertation, University of Oklamona.

Dees, W. H., Sonenshine, D. E. & Breidling, E. (1984a) Ecdysteroids in *Hyalomma dromedarii* and *Dermacentor variabilis* and their effect on sex pheromone activity. In: *Acarology VI,* Griffiths D. A. & Bowman, C. E. (eds.) Ellis Horwood, Ltd., Chichester, p. 405–413.

Dees, W. H., Sonenshine, D. E. & Breidling, E. (1984b) Ecdysteroids in American dog tick, *Demacentor variabilis* (Say) during different periods of tick development (Acari: Ixodidae). *J. Med. Entomol.* **21:** 514–523.

Dees, W. H., Sonenshine, D. E. & Breidling, E. (1985) Ecdysteroids in the camel tick, *Hyalomma dromnedarii* during different periods of adult tick development (Acari: Ixodidae). *J. Med. Entomol.* **22** 2–27.

Delbecque, P. A., Diehl, P. A. & O'Connor J. D. (1978) Presence of ecdysone and ecdysterone in the tick, *Amblyomma hebraeum* Koch (Acarina: Ixodidae). *Rev. Suisse Zool.* **89:** 859–68.

Diehl, P. A. & Aeschlimann, A. (1982) Tick reproduction: oogenesis and oviposition. In: *Physiology of ticks.* Obenchain, F. D. & Galun, R. (eds.). Pergamon Press, Ltd., Oxford, p. 277–350.

Diehl, P. A., Germond, J. E. & Morici, M. (1982) Correlations between ecdysteroid titers and integument structure in nymphs of the tick *Amblyomma hebraeum* Kock (Acarina: Ixodidae). *Rev. Suisse Zool.* **89:** 859–68.

Endo, K. (1973) Hormonal regulation of mating in the butterfly, *Polygonia C-aureum* L. *Dev. Growth Differ.* **15:** 1–18.

Germond, J. E., Diehl, P. A. & Morici, M. (1982) Correlations between integument structure and ecdysteroid titers in fifth-stage nymphs of the tick, *Ornithodoros moubata* (Murray 1877; sensu Walton 1962). *Gen. Comp. Endocrinol.* **46:** 255–266.

Gothe, R. & Kraiss, A. (1982) Zur Lokalisation der Pheromonemission und Persepstion bei *Argas (Persicargas) Walkerae* Kaiser und Hoogstraal 1969. *Zentralbl. Veterinäumlaut. Reihe B* **29:** 573–582.

Kaufmann, W. R. & Sauer, J. R. (1983) Ion and water balance in feeding ticks: Mechanisms of tick excretion. In: *Physiology of ticks.* Obenchain, F. D. & Galun, R. (eds.). Pergamon Press, Oxford, p. 213–244.

Khalil, G. M., Sonenshine, D. E., Sallam, O. & Homsher, P. J. (1983). Mating regulation and reproductive isolation in the camel ticks, *Hyalomma dromedarii* and *Hyalomma anatolicum excavatum. J. Med. Entomol.* **20:** 136–145.

Kittredge, J. S. & Takahashi, F. T. (1972) The evolution of sex pheromone communication in the Arthropoda. *J. Theor. Biol.* **35:** 467–71.

Leahy, M. G. (1979) Pheromones of argasid ticks. In: *Recent advances in acarology 2.* Rodriguez, J. G. (ed). Academic Press, New York, p. 297–309.

Obenchain, F. D. (1984) Behavioral alterations between the sexes and aspects of species specificity pheromone mediated aggregation and attachment in *Amblyomma.* In: *Acarology VI.* 1. Griffiths, D. A. & Bowman, C. E. (eds.). Ellis Horwood, Ltd. Chichester. 387–392.

Obenchain, F. D. & Oliver, J. H. Jr. (1975) Neurosecretory system of the American dog tick *Dermacentor variabilis* (Acari: Ixodidae) II. Distribution of secretory cell types, axonal pathways and putative neurohaemal-neuroendocrine associations; comparative histological and anatomical implications. *J. Morph.* **145:** 269–294.

Petney, T. N. & Bull, C. M. (1981) A non-specific aggregation pheromone in two Australian reptile ticks. *Anim. Behav.* **29:** 181–85.

Pound, M. J. (1981) Hormonal relationships in reproduction of female *Ornithodoros parkeri.* Unpublished PhD dissertation, University of Gerogia, Athen, Georgia, 103 pp.

Pound, M. J. & Oliver, J. H. Jr. (1979) Juvenile hormone: evidence of its role in the reproduction of ticks. *Science* **206:** 355–357.

Rechav, Y. (1978) Specificity in assembly pheromones of the tick *Amblyomma hebraeum* (Acarina: Ixodidae). *J. Med. Entomol.* **15:** 81–83.

Rechav, Y. & Whitehead, G. B. (1979) Male produced pheromones of Ixodidae. In: Rodriquez J. G. (ed.) *Advances in acarology,* **2:** Academic Press, New York, p. 291–96.

Rechav, Y., Norval, R. A. I. & Oliver, J. H. Jr. (1982) Inter-specific mating of *Amblyomma hebraeum and Amblyomma variegatum (Acari: Ixodidae) J. Med. Entomol.* **19:** 139–142.

Riddiford, L. M. & Truman, J. W. (1978) Biochemistry of insect hormones and insect growth regulators. In: *Biochemistry of insects* **1:** (ed.) Rockstein, M. Academic Press, N. Y. p. 307–357.

Schlaeger, D. A., Fuchs, M. S. & Kang, S. H. (1974) Ecdysone mediated stimulation of dopa-decarboxylase activity and its relationship to ovarian development in *Aedes aegypti. J. Cell. Biol.* **6:** 454:65.

Schlein, Y. & Gunders, A. E. (1981) Pheromone to *Ornithodoros* spp. (Argasidae) in the coxal fluid of female ticks. *Parasitol.* **83:** 467–71.

Schöni, R., Hess, E., Blum, W. & Ramstein, K. (1984) The aggregation–attachment pheromone of the topical bont tick, *Amblyomma variegatum* Fabricius (Acari: Ixodidae). Isolation, identification, and action of its active components. *J. Insect Physiol.* **30:** 613–618.

Sonenshine, D. E. (1984a) Tick pheromones In: *Pathogen-host-vector research.* Harris, K. F. (ed.) Praeger Scientific Press, Inc., New York, Vol. 2, 225–263 .

Sonenshine, D. E. (1984b) Pheromones of Acari and their potential use in control strategies. In: *Acarology VI 1,* Griffiths, D. A. & Bowman, C. E. (eds.) Ellis Horwood, Chichester, p. 100–108.

Sonenshine, D. E. (1985) Pheromones and other semiochemicals of the Acari. *Ann. Rev. Entomol.* **30:** 1–28.

Sonenshine, D. E., Homsher, P. J., VandeBerg, J. S. & Dawson, D. (1981) Fine structure of the foveal glands and foveae dorsales of the American dog tick, *Dermacentor variabilis* (Say) *J. Parasitol.* **67:** 627–46.

Sonenshine, D. E., Khalil, G. M., Homsher, P. J. & Mason, S. N. (1982) *Dermacentor variabilis* and *Dermacentor andersoni:* genital sex pheromones. *Exp. Parasitol.* **54:** 317–330.

Sonenshine, D. E., Khalil, G. M., Homsher, P. J., Dees, W. H., Carson, K. A. & Wang, V. (1983) Development, ultrastructure, and activity of thefoveal glands and foveae dorsales of the camel tick *Hyalomma dromedarii* (Acari: Ixodidae). 2. Maturation and pheromone activity. *J. Med. Entomol.* **20:** 424–439.

Sonenshine, D. E., Homsher, P. J., Carson, K. A. & Wang, V. B. (1984a) Evidence of the role of the cheliceral digits in the perception of genital sex pheromones during mating in the American dog tick, *Dermacentor variabilis* (Say) (Acari: Ixodidae). *J. Med. Entomol.* **21:** 296–306.

Sonenshine, D. E., Silverstein, R. M. & West, J. R. (1984b) Occurrence of the sex attractant pheromone, 2,6-dichlorophenol, in relation to age and feeding in the American dog tick, *Dermacentor variabilis* (Say) (Acari: Ixodidae). *J. Chem. Ecol.* **10:** 95–100.

Sonenshine, D. E., Silverstein, R. M., West, J. R., Carson, K. A. Homsher, P. J., Bennett, S. & Taylor, D. (1985a) Studies on the possible role of catecholamines in the regulation of sex pheromone activity in the American dog tick, *Dermacentor variabilis* (Say). *J. Chem. Ecol.* **11:** 363–382.

Sonenshine, D. E., Homsher, P. J., Beveridge, M. & Dees, W. H. (1985b) Occurrence of ecdysteroids in specific body organs of the camel tick, *Hyalomma dromedarii,* and the American dog tick, *Dermacentor variabilis,* with notes on their synthesis from cholesterol. *J. Med. Entomol.* **22:** 303–311.

Waladde, S. M. (1982) Tip-recording from ixodid tick olfactory sensilla: responses to tick related odours. *J. Comp. Physiol.* **148:** 411–418.

Waladde, S. M. & Rice, M. J. (1982) The sensory basis of tick feeding behavior. In: *Physiology of ticks,* Obenchain, F. D. & Galun, R. (eds.) Pergamon Press, Ltd., Oxford, p. 71–118.

Wood, W. F., Leahy, M. G., Galun, R., Prestwich, G. D., Meinwald, J. (1975) Phenols as pheromones of ixodid ticks: a general phenomenon? *J. Chem. Ecol.* **1:** 501–509.

16

Leg sense organs of ticks

E. Hess and M. Vlimant

16.1 INTRODUCTION

The forelegs of ticks are important sensory appendages bearing sensory hairs (SH), some of which are concentrated in Haller's organ. Numerous morphological and fine structural studies on tarsus 1 have revealed specific differences in tick chaetotaxy and morphology (Balashov & Leonovich 1976, 1978, Clifford & Anastos 1960, Homsher & Sonenshine 1975, 1977, Hoogstraal *et al.* 1983, Zumpt 1949). However, classical chaetotaxy does not seem to be a very powerful tool for discriminating tick species.

We have studied the foreleg tarsal sensory system from a functional point of view. By analyzing sensory hairs quantitatively with respect to the modality-specific structures, and the total input of the tarsal sensory system in terms of receptor neurons, we have assessed the relative importance of the different sensory modalities in ticks. This approach, called functional chaetotaxy, has shown that the contribution of the different sensory modalities to the total tarsal sensory input varies in the instars of *Amblyomma variegatum* (Hess & Vlimant 1983b). We have also compared the functional chaetotaxy of species differing in their systematic position, life cycle, and ecology. This approach, called comparative functional chaetotaxy, is useful in helping answer questions such as: Are the systematic position, life cycle, and ecology reflected in the morphology and fine structure of the tarsal sensory system of a tick species? Is the foreleg sensory system of a reptile-specific parasite such as the adult *Amblyomma nuttalli* structurally different from that of a closely related species, e.g. *A. variegatum* which feeds exclusively on warm-blooded vertebrates? Or, do adult one-host ticks such as *Boophilus microplus* have a reduced tarsal sensory system because they choose their hosts during the larval stage? Answers to questions such as these should help us solve problems associated with the physiology of host seeking and host specificity. The results of our comparative study on *Amblyomma variegatum*, *Amblyomma nuttalli*, *Dermacentor marginatus*, *Boophilus microplus*, *Ixodes ricinus*, and *Ornithodoros moubata*, are summarized and reviewed in this chapter.

Two other phenomena related to the sensory physiology of ticks are considered: i.e. the function of leg sense hairs located on other leg segments than the foreleg tarsi, and our discovery of intracuticular sensilla located ubiquitously on tick legs. It is unusual that researcher's have thus far limited their research to Haller's organ, tarsus 1, the palpal organ, and most recently the cheliceral sense organs, and have neglected the great number of sensory hairs on the legs and elsewhere. Even more astonishing is the absense of descriptions of the intracellular sensilla of tick legs prior to our study (Hess & Vlimant 1984) in spite of the known abundance of slit sense organs and lyriform organs in spiders, scorpions, and mites. It should be noted that authors erroneously interpreted gland openings on tick legs as campaniform sensilla or similar organs. Thus we have examined the structure of the sensory hairs on all segments of the four pairs of legs in *A. variegatum* and the intracuticular sensilla in the legs of *A. variegatum*, *I. ricinus*, and *O. moubata.*

16.2 STRUCTURE, FUNCTION, AND DISTRIBUTION OF SENSORY HAIRS OF TARSUS 1

Previously, the structure of the foreleg tarsal sensory hairs was described in detail, by means of transmission electron microscopy (TEM) or TEM and scanning electron microscopy (SEM) in the following species: *Alveonasus lahorensis*, *Amblyomma americanum*, *Amblyomma variegatum*, *Argas arboreus*, *Argas tridentatus*, *Boophilus microplus*, *Hyalomma asiaticum*, *Ixodes persulcatus*, *Ixodes ricinus*, and *Ornithodoros moubata* (Axtell 1979, Chu-Wang & Axtell 1973, Foelix 1972, Foelix & Axtell 1971, 1972, Hess & Vlimant 1980, 1982, 1983a,b, Jorgensen 1985, Leonovitch 1977, 1978, 1979, 1980, Roshdy *et al.* 1972, Thonney 1980, 1986, Waladde 1979). Two important facts emerged from these studies: a) There is a restricted number of structural types of sensory hairs present including 11 in argasids, 12 in the Metastriata, and 13 in the Prostriata. b) In the different species, the sensory hairs are distributed over the tarsus in a comparable manner.

16.2.1 Structure and function of sensory hairs

The structure and microanatomy of tick sensilla follows the general scheme as defined for arthropods (Altner 1977, Altner & Prillinger 1980, Hess & Vlimant 1982, 1983a,b). Thus four categories based on modality-specific structural features can be distinguished: a) wall pore single-walled sensilla (wp-sw). b) wall pore double-walled sensilla (wp-dw). c) terminal pore sensilla (tp). d) no pore sensilla (np).

These categories may contain sensory hairs with an innervated socket (sensilla chaetica) or with a noninnervated socket (sensilla basiconica or sensilla trichodea). In the former, the innervation of the base is always double, i.e. two tubular bodies are present). The lack of socket-innervation does not exclude the flexibility at the base of the sensillum.

wp-sw Sensilla (Fig. 16.1) are characterized by a shaft pierced by numerous pores and the presence of pore tubules or pore strands representing stimulus conducting structures. The dendrites of the sensory cells are branched or unbranched. These sensory hairs are mono- or multimodal olfactory sensilla; additional possible modalities include thermo- and hygrosensitivity.

wp-dw Sensilla (Fig. 16.2) have a double setal wall with the receptor lymph cavities separated in the shaft. Stimulus conduction occurs through 'spoke canals' opening into surface grooves. The dendrites of the shaft innervating units (SIU) are unbranched. A specialized innervated mechanosensory socket may be present or absent. These sensory hairs are mono- or multimodal olfactory sensilla; additional possible modalities are mechano-, hygro and thermosensitivity.

tp Sensilla (Fig. 16.3) have a terminal or subterminal pore or slit which connects the inner receptor lymph cavity to the outside. The inner and outer lymph cavities are separated by a sheath (scolopale) or a cuticular wall. Most of the unbranched dendrites reach to the tip of the hair. The base is generally innervated and forms a mechanosensitive socket. Tp sensilla are mechano-gustatory receptors (contact chemoreceptors), with other possible modalities being thermo- and hygrosensitivity.

np Sensilla (Fig. 16.4) are characterized by an absence of pores or slits. They possess either an innervated socket or an innervated shaft, or both. They are mono- or multimodal with possible modalities being mechano-, thermo, and hygrosensitivity.

All tick sensory hairs fit this classification scheme, and all four categories are present on tarsus 1. As in other arthropods, the four categories of sensory hairs include several different types as designated by different letters (Figs. 16.1–16.4). Although the categories are homologous among arthropods, the types are not identical. Consequently, a type wp-sw/A sensillum of *Periplaneta americana* is not necessarily structurally identical to the type wp-sw/A sensillum of *Amblyomma variegatum.* However both have features in common and both are olfactory. Among ticks, the different types can, however, be considered as structurally identical, but not necessarily have identical functional specificity. Thus a wp-sw/B sensillum have identical structural characteristics in both *A. variegatum* and *O. moubata*, and it contains olfactory units in both species, but the specificities of the receptor neurons in the two are not necessarily identical. The important question of functional specificity can only be evaluated by electrophysiological means and has not been resolved in most cases. Electrophysiological studies on some of the tarsal sensilla have confirmed functions hypothesized from structural studies. Np/C sensilla of *A. variegatum* contain mechanosensory and antagonistic thermosensitive receptor neurons (Fig. 16.5) (Hess & Loftus 1984), and the corresponding sensillum of *B. microplus* larvae also contains a cell sensitive to heat (Jorgensen 1985).

The olfactory function of wp-sw/A sensilla has been demonstrated unequivocally in several species. Hess & Vlimant (1980) reported cells reacting to short-chain fatty acids and 2,6-DCP (dichlorophenol) in *A. variegatum* sensory hairs of this type, and recently Schöni & Hess (in preparation) identified chemosensory neuron in the same sensilla which is specific for *o*-nitrophenol, a component of the aggregation–attachment

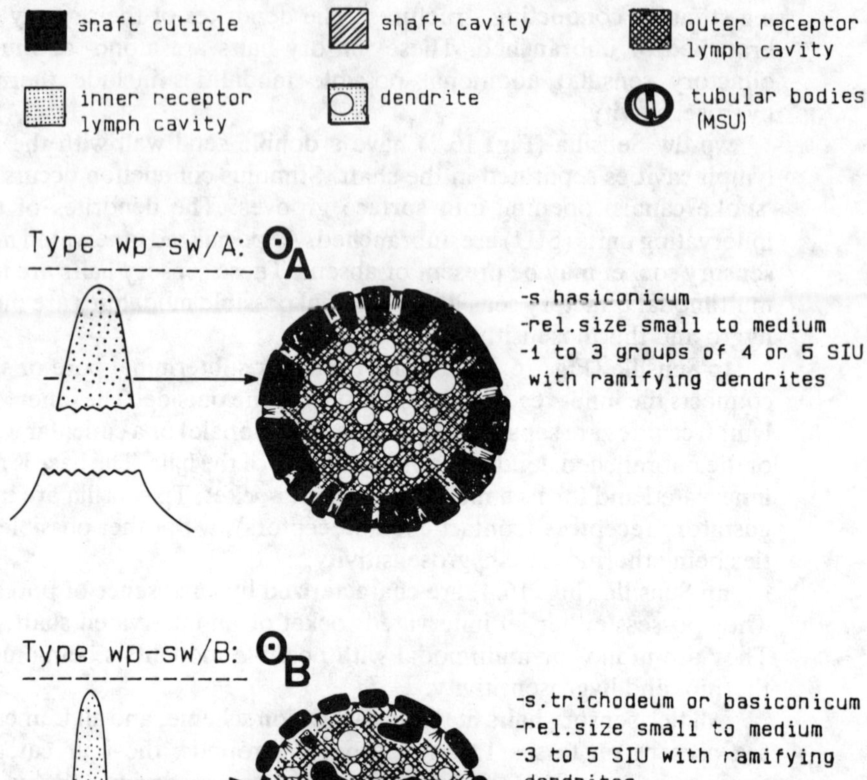

Fig. 16.1 — Characteristic structural features of tick leg sensory hairs. Wall pore single-walled (wp-sw) sensilla. SIUs shaft inervating units.

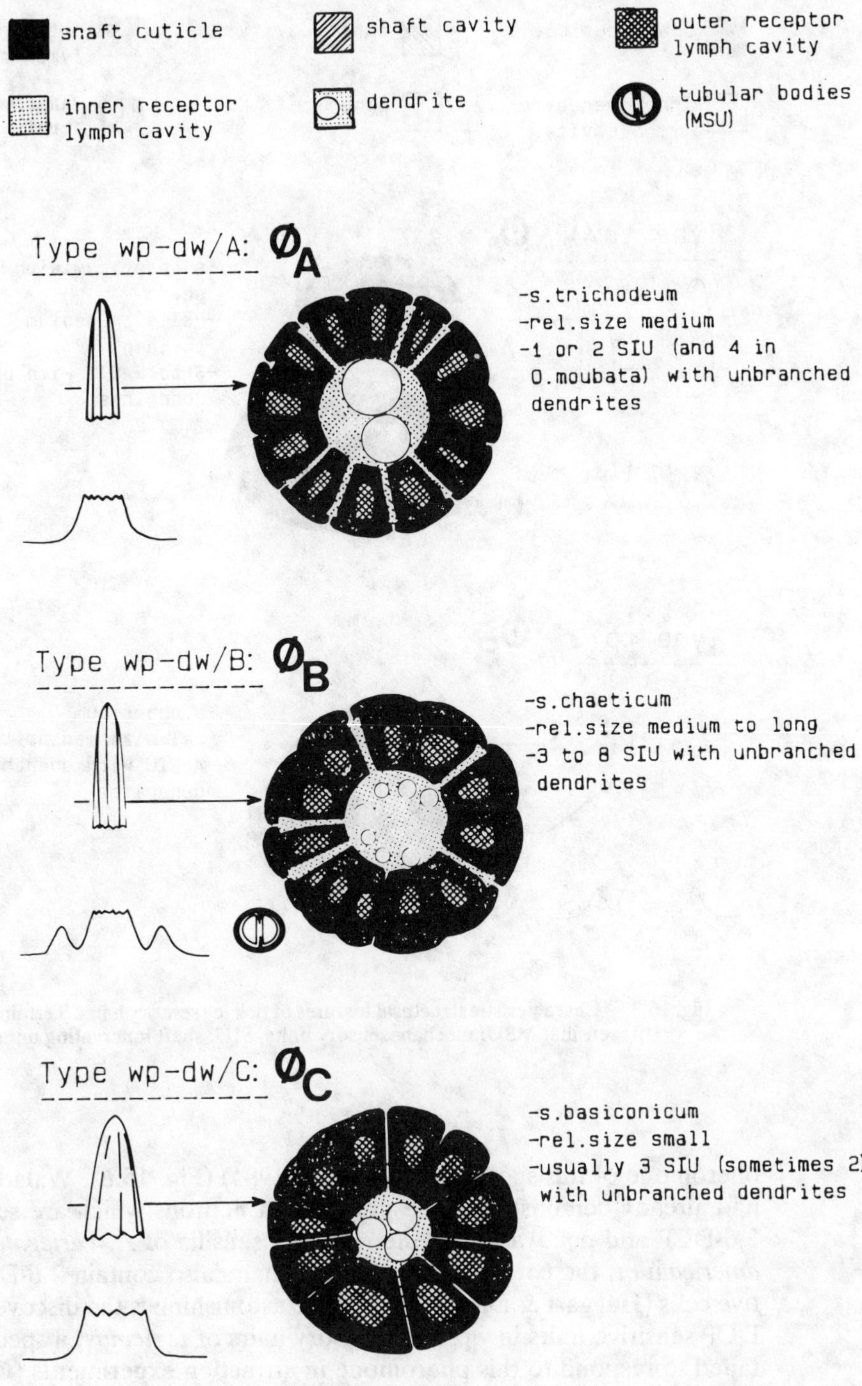

Fig. 16.2 — Characteristic structural features of tick leg sensory hairs. Wall pore double-walled (wp-dw) sensilla. MSU: mechanosensory units, SIU: shaft innervating units.

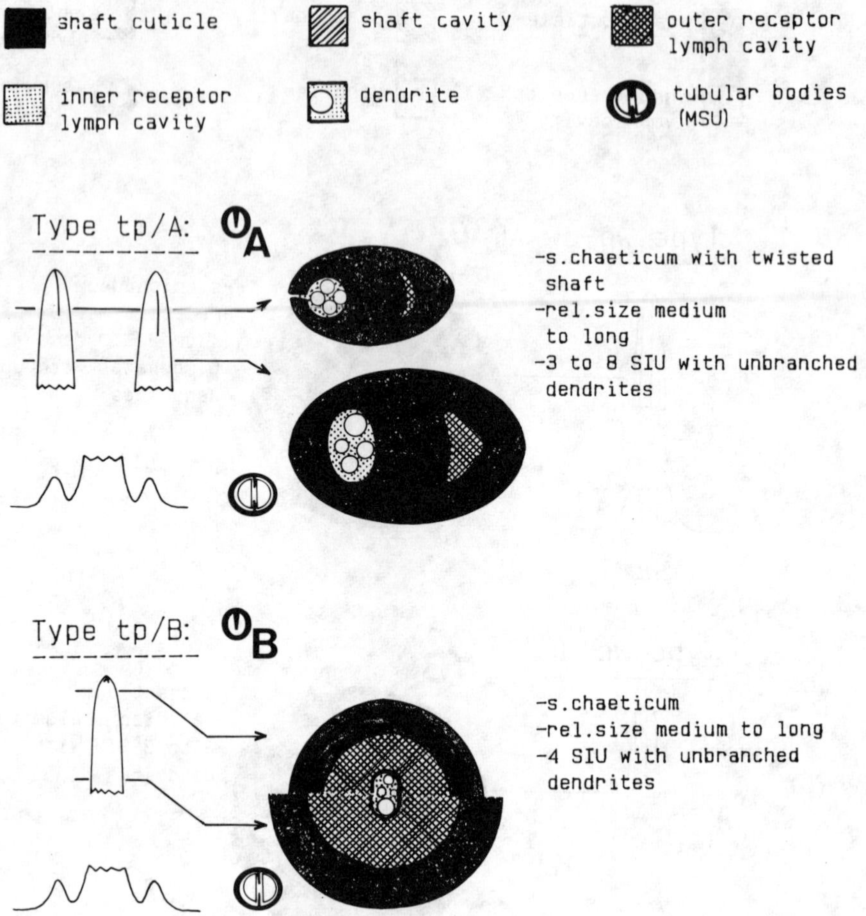

Fig. 16.3 — Characteristic structural features of tick leg sensory hairs. Terminal pore (tp) sensilla. MSU: mechanosensory units, SIU: shaft innervating units.

pheromone of this species (Schöni *et al.* 1984) (Fig. 16.6). Waladde (1982) had already demonstrated the presence of neurons which are sensitive to 2,6-DCP and tick washes in the wp-sw/A sensilla of *A. variegatum*. In *A. americanum*, the corresponding sensory hairs also contain 2,6-DCP sensitive cells (Haggart & Davis 1981). More astonishing is the discovery of 2,6-DCP sensitive units in wp-sw/C sensory hairs of *I. ricinus*, a species which failed to respond to this pheromone in attraction experiments (Graf 1975, Thonney 1986).

We recently identified wp-sw/B sensilla within the capsule of Haller's organ with temperature-sensitive units and receptor cells specific to methyl salicylate (Fig. 16.7) (Hess & Blendermann, in preparation), another

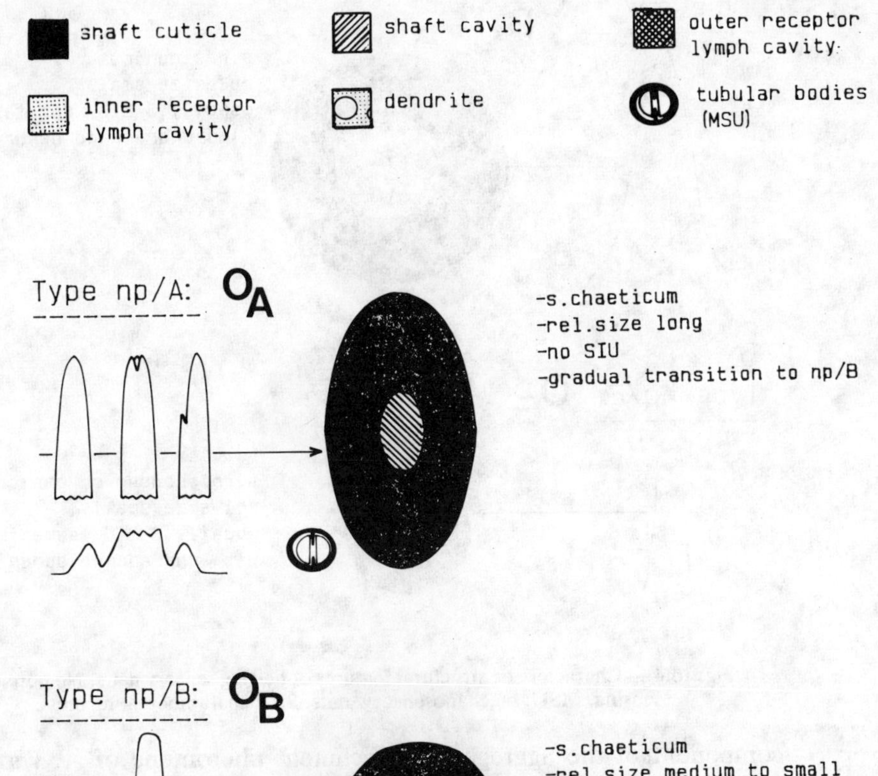

continued on p. 368.

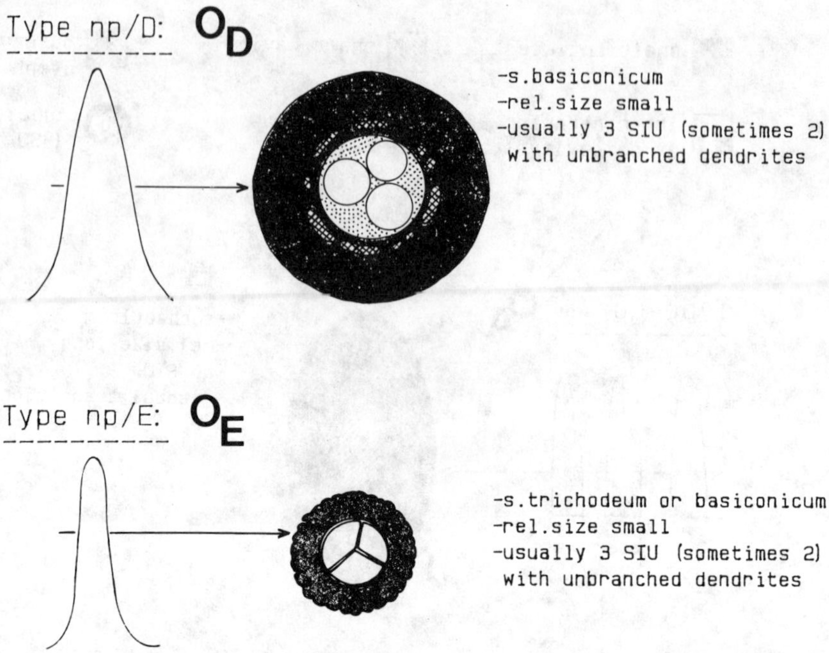

Fig. 16.4 — Characteristic structural features of tick leg sensory hairs. No pore (np) sensilla. MSU: mechanosensory units, SIU: shaft innervating units.

component of the aggregation-attachment pheromone of *A. variegatum* (Schöni *et al.* 1984). Wp-dw sensilla have been tested electrophysiologically only by Haggart & Davis (1980) who found ammonia sensitive cells within wp-dw/A sensilla of the 'medial group' of *R. sanguineus.*

Tp sensilla have been demonstrated to be gustatory by Elizarov (1963) and Balashov *et al.* (1976). Some contain receptor cells that react to NaCl and water as well, and Waladde *et al.* (1981) identified neurons in tp/A sensory hairs that react to a decrease in temperature.

Other electrophysiological studies have been published which unfortunately do not allow the precise determination of the sensilla from which the recordings were made. For example, Elizarov (1963) demonstrated repellent sensitive neurons in Haller's organ of *H. asiaticum* and *Ixodes persulcatus*, and Sinitzina (1974) identified cells in the anterior pit of the same species reacting to *n*-valerianic acid and odours from white mouse, blood, and the respiration of man. Haggart and Davis (1981) found ammonia sensitive units at the same location in *R. appendiculatus.* In a study of CO_2 perception, the source of the electrical impulses was undetermined (Holsher *et al.* 1980).

16.2.2 Distribution of sensory hairs

A superficial look at the comparative functional chaetotaxy of tick foreleg tarsi reveals that the distribution of sensory hairs is similar in all tick species. To what extent are they actually comparable? To help answer this

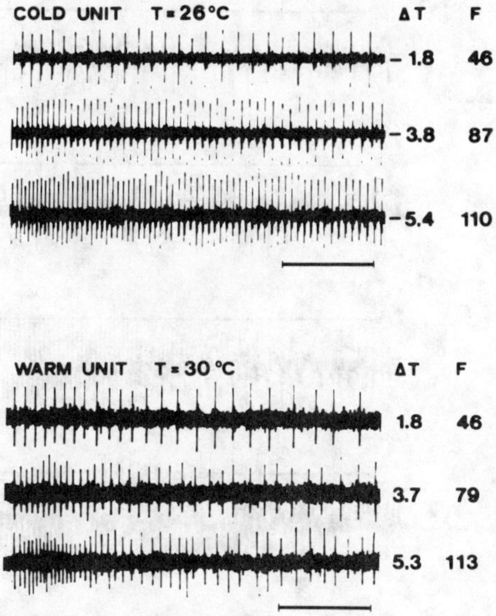

Fig. 16.5 — *A. variegatum* female: type np/C sensillum (dI.2). Impulse trains of a warm and a cold unit in response to abrupt temperature changes. Response magnitude for first 100 ms indicated under F. Time scale: 200 ms. Interstimulus conditioning time at T: 3 min. Exposure to △T: 1.3 s. (Hess & Loftus, (1984) by courtesy of Springer Verlag).

question, we need a system of designating sensory hairs which can easily be applied to all tick species and to all instars, to facilitate rapid vertical (stage) and horizontal (species) comparison. The system used in our laboratory assumes that the different types of sensory hairs are distrubuted in a comparable manner in all species, and takes into account the possibility of specific variations. It assumes homologous groups of setae in adults, but it is also relevant for nymphs and larvae. In most ixodids homologous groups of setae consist of the same types of sensilla and are arranged in a similar or identical manner . However, in some species or stages one or several groups may be modified or even missing. In others, two groups may fuse to form a combined group consisting of setae of the original groups. An example is the combined group dI/II in argasids which is composed of the sensilla of the original groups dI and dII, which are separated in ixodids.

Our alphanumeric code of designation of the setal groups includes letters to indicate the topographical location of the group (d=dorsal, v=ventral, la=lateral-anterior, lp=lateral-posterior, C=capsule of Haller's organ) and a Roman numeral to number the groups in the disto-proximal direction. Thus, the group dI is the most distal group of sensilla on the dorsal face of the tarsus. When a cypher is added to the group-code it is possible to designate individual sensilla from anterior to posterior and from distal to proximal. Thus, for example, the sensillum dI.1 (a type wp-sw/A in all metastriates) is

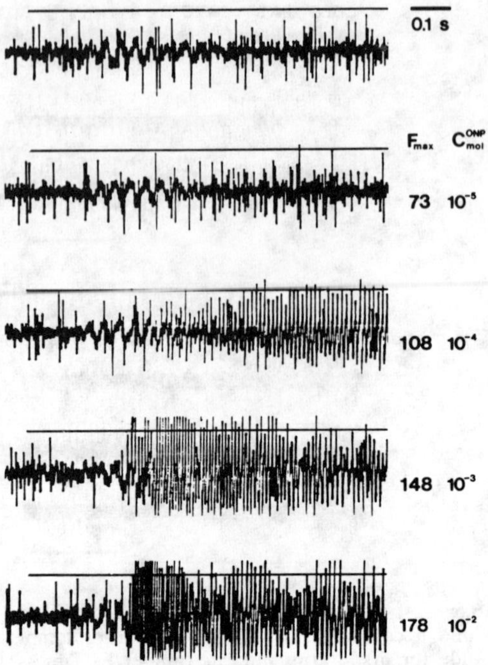

Fig. 16.6 — *A. variegatum* female: type wp-sw/A sensillum (group d II). Impulse trains of a unit in response to dry air control (upper trace) and to *o*-nitrophenol (0NP). Maximum response magnitude indicated under F_{max}. Molar concentration of the stimulating source indicated under C_{mol}. Horizontal bar represents stimulation (duration 1.5 s). Interstimulus time 3 min. (Schöni & Hess, in prep.).

the anterior sensillum of the group dI, and sensillum dI.2 (a type np/C sensillum in all metastriates) is posterior to dI.1.

Figs. 16.8–16.11 demonstrate the high degreee of homology of the functional chaetotaxy of tarsus 1 among tick spcies. A closer look at the figures shows that metastriate ixodids strongly resemble each other; some species such as *A. variegatum* and *A. nuttalli* are identical. Prostriates (*I. ricinus*) differ from metastriates by having a more homogeneous ventral side lacking the group vIV, a higher density of sensory hairs, and the group dI composed of 2 np/C and 2 wp-sw/C sensilla. This latter type is restricted to prostriates. We suppose that the anterior np/C sensillum of the group dI represents the original lateral np/C sensory hairs (laI.1 of metastriates) shifted to a dorsal position in *Ixodes*. This dorsal shift of the sensillum laI.1 has also been observed on the larval tarsus of *A. variegatum* (Hess & Vlimant 1983a).

The tarsus of the foreleg of *O. moubata* is characterized by the fused group dI/dII with 3 wp-sw/A sensilla; the fusion of groups dIV and dV, and a striking small number of setae as compared to hard ticks. As will be seen later, the lack of sensory hairs is partly compensated by intracuticular

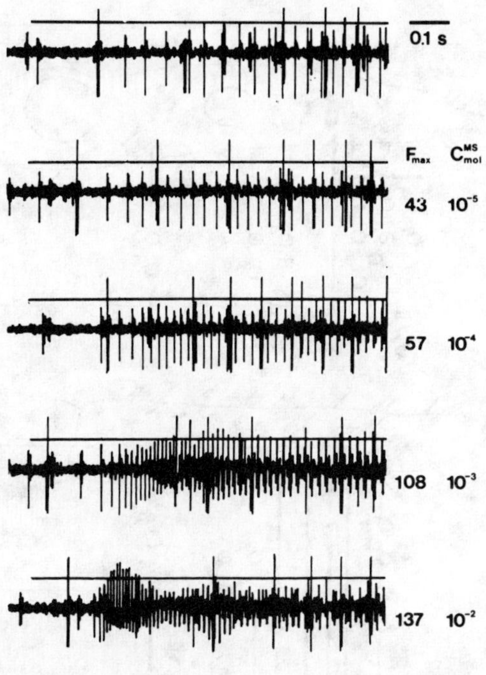

Fig. 16.7 — *A. variegatum* female: type wp-sw/B sensillum (capsule of Haller's organ). Impulse trains of a unit in response to dry air (upper trace) and to methyl salicylate (MS). Maximum response magnitude indicated under F_{max}. Molar concentration of the stimulating source indicated under C_{mol}. Horizontal bar represents stimulation (duration 1.5 s). Interstimulus time 3 min. The unit producing high-amplitude spikes does not respond to the component. (Hess & Blendermann, in preparation).

sensilla. This situation seems to be characteristic of soft ticks. It would be interesting to determine if this represents a primitive condition or a highly differentiated stage.

Even more interesting than the morphological comparison of the groups of sensilla is the comparison of total tarsal sensory input in terms of receptor cell types and sensory modalities (Table 16.1). This method allows assessment of the real sensory potential and it standardizes differences occuring in the number of sensory hairs. In fact, a lower number of setae observed in a given species can be compensated for by a higher number of neurons per sensillum as observed in the capsule of Haller's organ.

Before we discuss in detail the total foreleg tarsal sensory input of the different species, we must point out the identical structure of the tarsal sensory systems of *A. variegatum* and *A. nuttalli*, which share a common column in Table 16.1.

Comparing the total number of sensory hairs among the different species studied we note that *I. ricinus* has the largest number (79), followed by the two species of *Amblyomma* (74), *D. marginatus* (64), and *B. microplus* (56). *O. moubata* has only 30 sensory hairs. When we compare, however, the

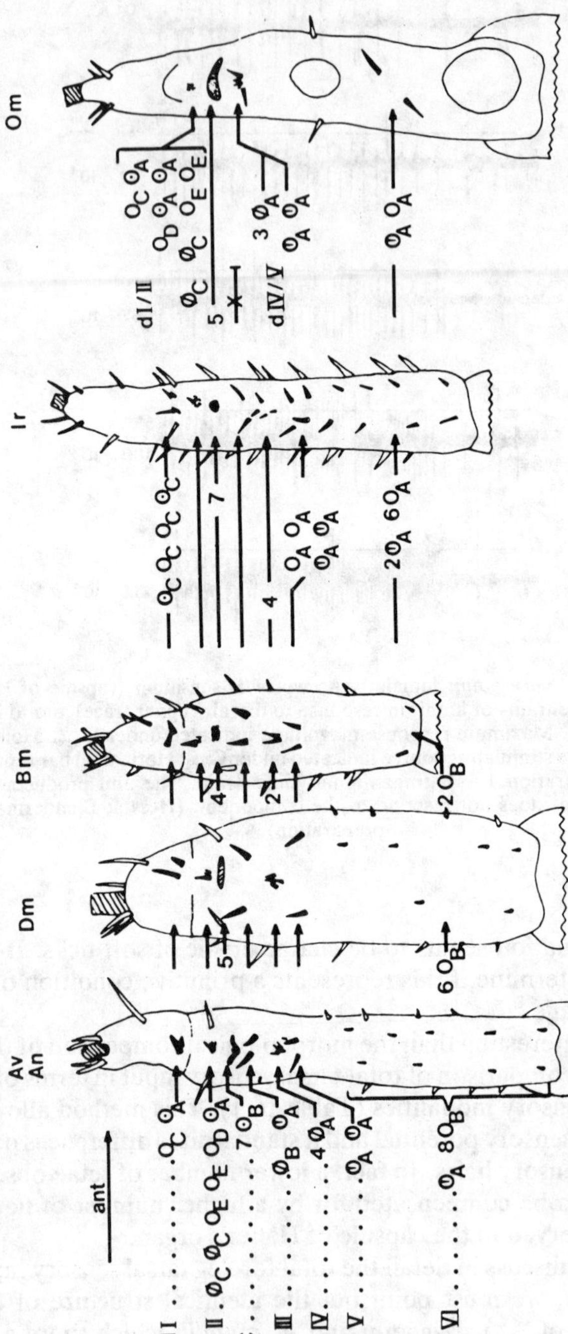

Fig. 16.8 — Comparative functional chaetotaxy of foreleg tarsi of adult *A. variegatum* (Av), *A. nuttalli* (An), *D. marginatus (Dm)*, *B. microplus* (Bm), *I. ricinus* (Ir), and *O. moubata (Om)*. Dorsal faces with groups dI to dVI (black sensilla). Not to scale. The symbols of the different types of sensory hairs are explained in Figs. 16.1–16.4. The figures should be read from left to the right beginning with the composition of the setal groups as found in *A. variegatum* and *A. nuttalli*. The arrows point to the corresponding groups in the other species, specific modifications in respect to the left species being indicated on the arrow.

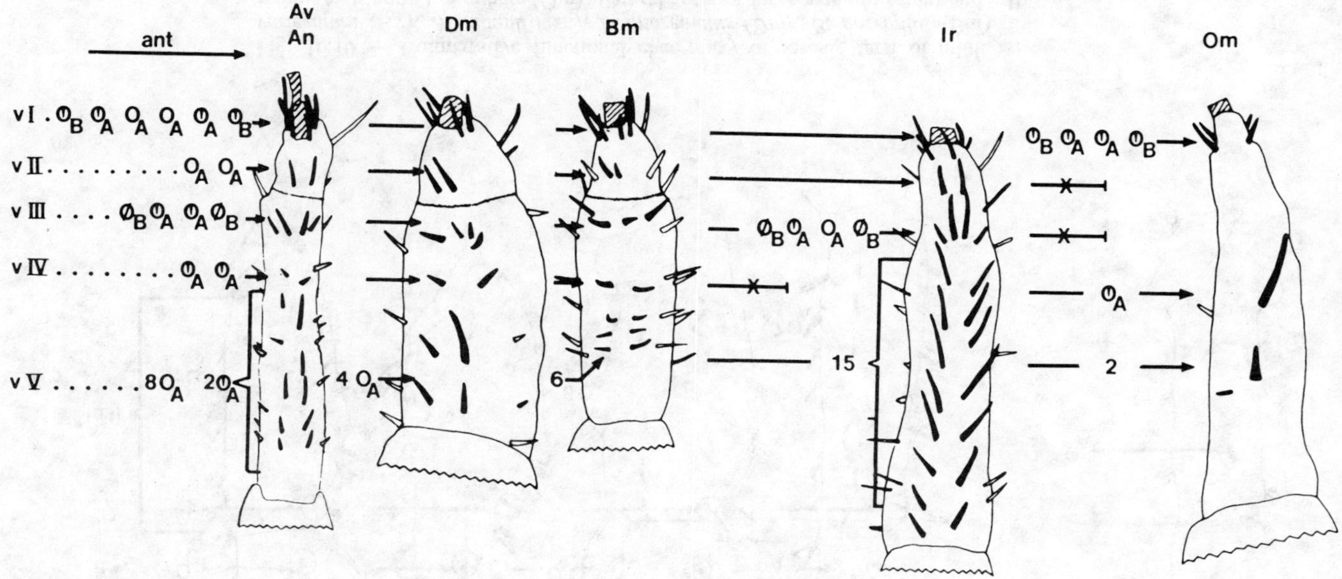

Fig. 16.9 — Comparative functional chaetotaxy of foreleg tarsi of adult *A. variegatum* (Av), *A. nuttalli* (An), *D. marginatus (Dm)*, *B. microplus* (Bm), *I. ricinus* (Ir), and *O. moubata* (Om). Ventral faces with groups vI to vV (black sensilla). Not to scale. The symbols of the different types of sensory hairs are explained in Figs. 16.1–16.4. The figures should be read from left to the right beginning with the composition of the setal groups as found in *A. variegatum* and *A. nuttalli*. The arrows point to the corresponding groups in the other species, specific modifications in respect to the left species being indicated on the arrow.

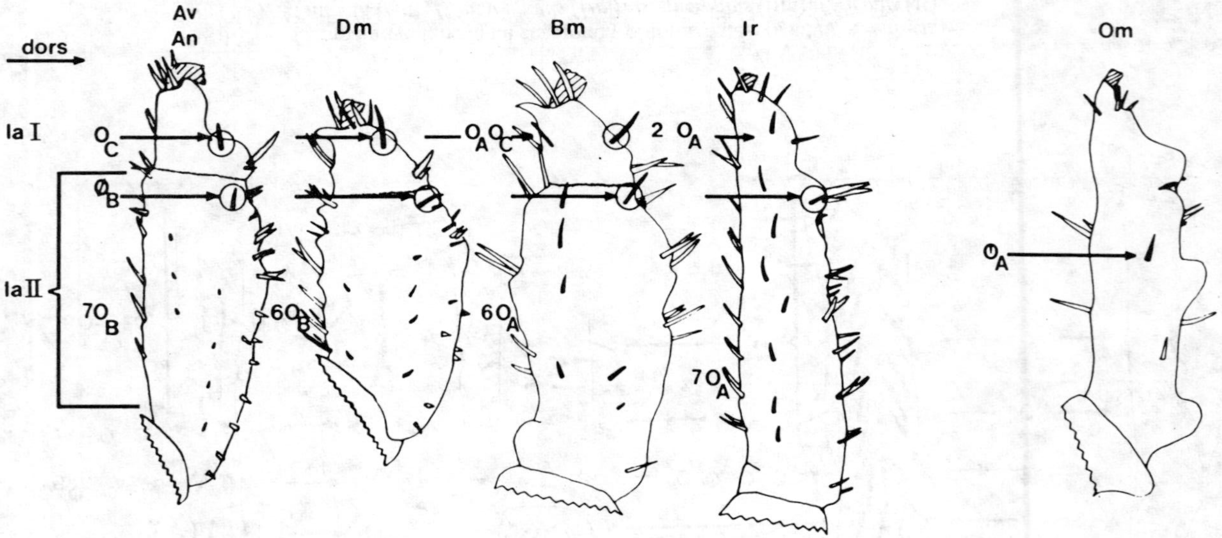

Fig. 16.10 — Comparative functional chaetotaxy of foreleg tarsi of adult *A. variegatum* (Av), *A. nuttalli* (An), *D. marginatus (Dm), B. microplus* (Bm), *I. ricinus* (Ir), and *O. moubata* (Om). Lateral anterior faces with groups laI and laII (black sensilla). Not to scale. The symbols of the different types of sensory hairs are explained in Figs. 16.1–16.4. The figures should be read from left to the right beginning with the composition of the setal groups as found in *A. variegatum* and *A. nuttalli.* The arrows point to the corresponding groups in the other species, specific modifications in respect to the left species being indicated on the arrow.

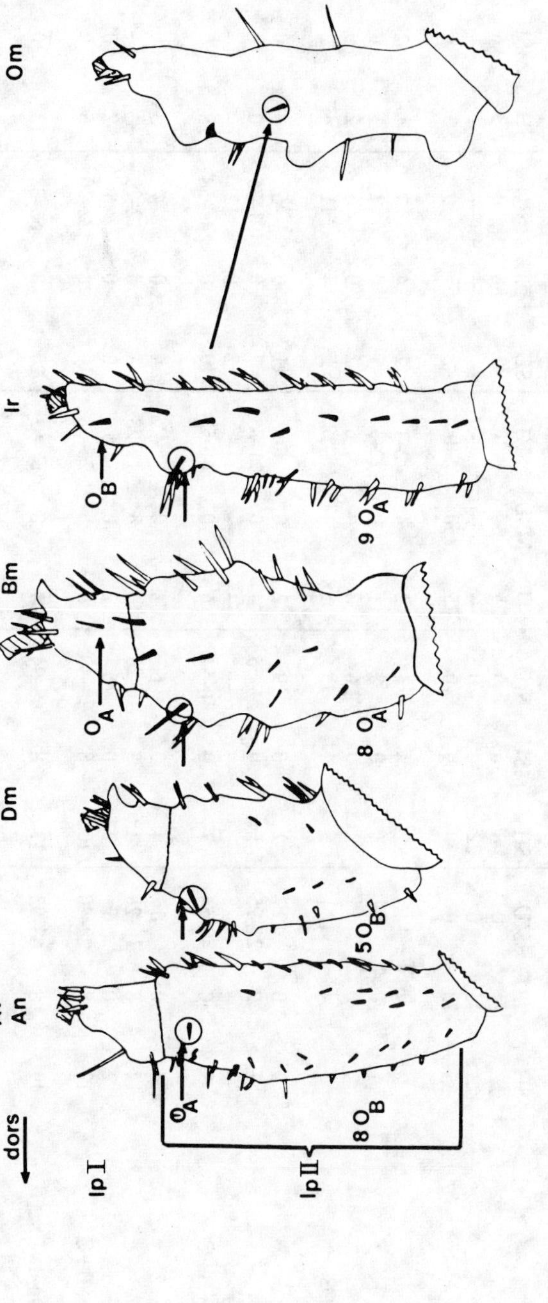

Fig 16.11 — Comparative functional chaetotaxy of foreleg tarsi of adult *A. variegatum* (Av), *A. nuttalli* (An), *D. marginatus (Dm)*, *B. microplus* (Bm), *I. ricinus* (Ir), and *O. moubata* (Om). Lateral posterior faces with groups lpI and lpII (black sensilla). Not to scale. The symbols of the different types of sensory hairs are explained in Figs. 16.1–16.4. The figures should be read from left to right beginning with the composition of the setal groups as found in *A. variegatum* and *A. nuttalli*. The arrows point to the corresponding groups in the other species, specific modifications in respect to the left species being indicated on the arrow.

Table 16.1 — Numbers of sensory hairs (SH) of the different types and corresponding numbers of mechanosensory units (MSU) and shaft innervating units (SIU) occuring on one foreleg tarsus of 6 species of ticks. For information about the functions of the different types refer to the text.

Types of SH	*A. variegatum* *A. nuttalli*			*D. marginatus*			*B. microplus*			*I. ricinus*			*O. moubata*		
	SH	MSU	SIU	SH	MSU	SIU	SH	MSU	SIU	SH	MSU	SIU	SH	MSU	SIU
Np/A	12	24	0	8	16	0	25	50	0	43	86	0	3	6	0
Np/B	23	46	0	22	44	0	—	—	—	—	—	—	—	—	—
Np/A+B	35	70	0	30	60	0	25	50	0	43	86	0	3	6	0
NP/C	2	4	4	2	4	4	2	4	2–3	2	4	3	1	2	1
Np/D	1	0	3	1	0	3	1	0	2–3	1	0	2	1	0	2
Np/E	2	0	6	2	0	6	2	4	4–6	2	0	6	2	0	6
Np/C+D+E	5	4	13	5	4	13	5	4	8–12	5	4	11	4	2	9
Wp-sw/A	2	0	19	2	0	19	2	0	20	1	0	6	3	0	16
Wp-sw/B	7	0	21–25	5	0	21–35	4	0	21–35	7	0	21–35	0	0	18
Wp-sw/C	—	—	—	—	—	—	—	—	—	2	0	17	—	—	—
Wp-sw/A+B+C	9	0	40–54	7	0	40–54	6	0	41–55	10	0	44–58	8	0	34
Wp-dw/A	4	0	6	4	0	6	2	0	4	4	0	4	3	0	7
Wp-dw/B	4	8	16–28	4	8	16–28	4	8	16–28	4	8	12–20	—	—	—
Wp-dw/C	2	0	6	2	0	6	2	0	6	2	0	6	2	0	4
Wp-dw/A+B+C	10	8	28–40	10	8	28–40	8	8	26–38	10	8	22–30	5	0	11–13
Wp-sw+wp-dw	19	8	68–93	17	8	68–94	14	8	67–93	20	8	66–88	13	0	45–47
Tp/A	13	26	39–78	10	20	40–70	10	20	40–70	9	18	27–45	8	16	32–64
Tp/B	2	4	8	2	4	8	2	4	8	2	4	8	2	4	8
Tp/A+B	15	30	47–86	12	24	48–78	12	24	48–78	11	22	35–53	10	20	40–72
Grand totals	74	112	128–193	64	96	129–185	56	86	123–183	79	120	112–152	30	28	94–128
			112			96			86			120			28
Total number of sensory cells			240–305			225–281			209–269			232–272			122–156

total number of the sensory cells innervating the sensilla instead of the number of sensilla, the two species of *Amblyomma* surpass *I. ricinus*. This reversal is due mainly to the *Amblyomma* species having more contact chemoreceptors containing multiple shaft innervating units (SIU) while *I. ricinus* possesses a higher number of monomodal mechanosensory sensilla (type np/A with 2 mechanosensory cells each).

In the metastriates we observe a decrease in the number of sensory hairs in *B. microplus* and *D. marginatus* as compared to the *Amblyomma* species. This seems to be related to the size of the segments. The smaller the tarsus of a metastriate ixodid, the lower the number of sensory hairs in the proximal groups dVI, vV, laII and lpII. Unlike metastriates, the small prostriate ticks such as *I. ricinus*, have well developed proximal groups providing them with a larger number of monomodal mechanosensory sensilla. In *I. ricinus* as well as in *B. microplus*, the typical short and blunt-tipped np/B sensilla are replaced by the longer np/A setae. This probably represents an adaptation towards a higher sensibility to mechanical stimulation and air currents. *O. moubata* has only 3 monomodal mechanosensory hairs, indicating a poorly developed sense of touch.

When we examined the putative hygro-thermosensitive sensilla np/C, np/D and np/E, we found 5 in hard ticks and 4 in soft ticks. The latter lack the np/C sensillum laI.1. The innervation of *B. microplus*, *I. ricinus*, and *O. moubata* np/C sensilla differs from other ticks in that one or both of their np/C setae possess only one SIU, suggesting that either the warm or the cold temperature sensitive unit is lacking. This has also been observed in immatures of *A. variegatum* (Hess & Vlimant 1983a). In *B. microplus*, the number of neurons associated with np/C, np/D, and np/E sensory hairs seems to be more variable than in the other species. In general, the total thermo-hygrosensitive potential is difficult to assess as tp and wp sensilla possess an unknown number of neurons reacting to temperature and humidity (Waladde *et al.* 1981, Hess & Blendermann, in preparation). We believe that thermo-hygrosensitivity in argasids is reduced as compared to ixodids because soft ticks have fewer neurons related to np-thermoreceptors and olfactory chemosensory hairs.

Analysis of the olfactory wp-sw sensilla of the different species is a good example of why it is insufficient to compare only the number of setae among the species. Such differences in no way inform us about the real sensory potential provided by this category of sensory hairs. The best example is the innervation of the capsule of Haller's organ (wp-sw/B sensilla) of hard ticks. The number of sensory hairs varies between 4 and 7, but the total number of neurons innervating these sensilla is the same in all species. This is explained by the fact, that in the smaller metastriates, some of the capsule-sensilla have apparently fused. This becomes evident when we examine the fine structure of the sensory hairs. Fused sensory hairs are equipped with 2 separated bundles of 3–5 dendrites, but the total number of bundles in the capsule is 7 in all species. Fusion of olfactory sensory hairs is also observed in the type wp-sw/A of the group dII. The ticks we studied have only one wp-sw/A sensillum there, while in *Amblyomma americanum* (Foelix & Axtell

1972), some representatives of the genus *Dermacentor* (Arthur 1962), and *Haemaphysalis* (Roshdy *et al.* 1972) there are 2. Fine structural analyses of these wp-sw/A sensilla reveal the presence of 3 bundles of neurons with a total of 14 neurons except *B. microplus* with 15. These 3 bundles are distributed on two sensilla as in *A. americanum* or they share a common shaft as in *A. variegatum.* We believe that the presence of 3 bundles indicates a primitive state of 3 wp-sw/A sensilla in the group dII.

I. ricinus has only one wp-sw/A sensillum but possesses 22 additional wp-sw/C sensory hairs (restricted to this genus) with a slightly higher total of wp-sw-neurons as compared to the metastriates.

In conclusion, of all ticks studied, prostriates have the highest number of neurons associated with wp-sw sensilla, while argasids have the least.

Concerning the wp-dw sensory hairs, *B. microplus* differs from the other hard ticks by the absence of the 2 single-innervated sensilla of the group dIV (Hess & Vlimant 1982), and in *I. ricinus* the innervation of wp-dw/B sensilla is slightly reduced as compared to the other hard ticks. In *O. moubata* and other argasids this type is apparently lacking.

Considering wp-sw and wp-dw sensory hairs together, it can be concluded with some certainty, that ixodids possess a better olfactory sense than argasids if we assume identical proportions of olfactory and thermo-hygrosensitive neurons in all species.

Concerning contact chemoreception (tp sensilla) the metastriate hard ticks appear best equipped, followed by *O. moubata* and *I. ricinus.* Compared to species of *Amblyomma* the smaller metastriates lack some tp/A sensilla related to the reduction of the proximal setal groups, while *I. ricinus* lacks tp/A sensilla on the ventral face of the tarsus. There is no difference in tp/B sensilla in all of the 6 species.

We hypothesize that taste is better developed in metastriates, than in postriates and argasids.

Thus far we have only compared the absolute numbers of the different types of foreleg tarsal sensilla and their innervation among species. We can also determine the relative importance of the different sensory modalities for each of them (Table 16.2). We observe that the ixodids differ greatly from the argasids in this respect. The latter have relatively few seta-associated mechanosensory units, and a higher precentage of thermo-hygrosensitive and gustatory neurons.

Among ixodids, Metastriata and Prostriata differ somewhat in the relative importance of taste and mechanosensitivity.

The relative importance of olfaction is similar in all observed ticks. It is difficult to estimate the biological value of this comparison, but it could indicate that hard ticks and soft ticks have different modes of host seeking.

Comparison of larval ticks revealed identical functional chaetotaxy in all the ixodid species studied (Fig. 16.12) and showed that *O. moubata* larvae are very similar to hard ticks in this respect. Interestingly, almost all of the sensory hairs are present in the latter species, but not all of them are differentiated. This is explained by the fact that larval *O. moubata* and other larval aragasids do not feed.

Table 16.2 — Estimation of the relative importance of sensory modalities of six species of ticks expressed in % of the total foreleg tarsal sensory hair input.

	Mechano-sensory neurons (%)	Thermo-hygro sensitive neurons (%)	Olfactory neurons (%)	Gustatory neurons (%)
A. variegatum *A. nuttalli*	41	5	30	24
D. marginatus	40	5	30	25
B. microplus	36	5	34	26
I. ricinus	47	4	31	18
O. moubata	21	7	34	38

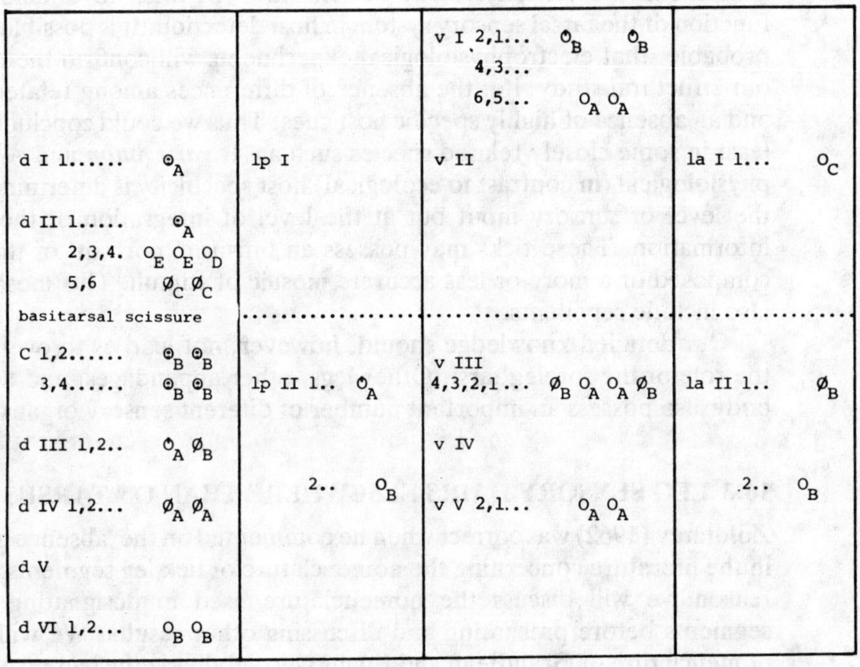

Fig. 16.12 — Map of larval functional chaetotaxy of foreleg tarsi of *A. variegatum*, *A. nuttalli*, *D. marginatus*, *B. microplus* and *I. ricinus* with dorsal (d), lateral posterior (lp), ventral (v), and lateral anterior (la) setal groups. Compare also with Figs 16.8–16.11. The symbols are explained in Figs. 16.1–16.4.

We can now attempt to answer the questions asked at the beginning of this chapter, i.e. do systematic position, life cycle, and ecology affect the structure of the foreleg tarsal sensory system? Comparison of the six species which differ greatly in these respects, and the integration of observations already published by other authors demonstrate, that important modality-related differences in the tarsal sensory components are not expressed in the species but at a higher systematic level. We observe characteristic differences between argasids and ixodids and between the Metastriata and Prostriata, but there is no difference between *A. variegatum* and *A. nuttalli* parasitizing warm-blooded vertebrates or reptiles respectively. On the other hand, it is true that adult *B. microplus* have a reduced tarsal sensory system when compared to species of *Amblyomma.* It is not easy to determine if this is due to *B. microplus* not needing to seek a host for feeding, or to the general trend of reduction in the tarsal sensory system observed in the smaller metastriates. Because most of the missing sensilla in *B. microplus* belong to the proximal groups and because a reduction in sensory hairs occurs in *D. marginatus* as well (a small three-host tick), we strongly suggest the second hypothesis.

The lack of structural and morphological differences among related tick species with completely different life cycles demonstrates that we have to concentrate on the physiology of sensilla if we wish to understand the function of the tarsal sensory system in host detection. It is possible, or even probable, that electrophysiological experiments will confirm the results of our structural study, i.e. the absence of differences among related species and an absence of highly specific host cues. Thus we could conclude that, at least in some closely related species such as *A. variegatum* and *A. nuttalli*, physiological (in contrast to ecological) host specificity is determined not at the level of sensory input but at the level of integration of the sensory information. These ticks may possess an intrinsic 'portrait' of their hosts composed of a more or less accurate mosaic of stimuli. This mosaic could also include repellents.

Our detailed knowledge should, however, not lead us to overestimate the role of the foreleg tarsi. Other legs, other appendages, and the whole body also possess an important number of diferent sensory organs.

16.3 LEG SENSORY HAIRS ELSEWHERE THAN ON TARSUS 1

Zolotarev (1962) was correct when he commented on the 'absence of clarity' in the literature concerning the nomenclature of tick leg segments. For this reason we will discuss the nomenclature used in designating tick leg segments before presenting and discussing other results. We will use the nomenclature of Grandjean and Evans who subdivide the legs of Anactinotrichida (to which the Ixodida belong) into coxa, trochanter (Tr), femur (Fe), genu (Ge), tibia (Ti), tarsus (Ta), and an apotele (ambulacrum) (Grandjean 1954, Evans *et al.* 1961, Van der Hammen 1976, Krantz 1978).

The femora and tarsi of ixodid adults are subdivided into a basi- and telofemur, a basi- and telotarsus by the adesmatic basifemoral (Fig. 16.14) and basitarsal scissures (Fig. 16.15) respectively. In metastriate ticks, the basitarsal scissure is less visible on the forelegs than on legs 2–4, and in the Prostriata and argasids it appears to be absent on tarsus 1. In immatures, distinct secondary segmentation seems to be lacking.

Near the proximal end of the foreleg tarsi and of each of the tibiae and genua of ixodid adults there are incomplete ventral fissures designated Foreleg Tarsal Fissure, Tibial Fissure, and Genual Fissure respectively. These scissures and fissures are mechanosensory organs which will be discussed in the next section.

All segments of the four pairs of legs bear sensory hairs. We studied these sensilla in *A. variegatum* by SEM and some by TEM in order to determine their modality.

The respective numbers of sensory hairs are approximately the same on the homologous leg segments of the four pairs of legs except tarsus 1, and no differences were observed between males and females. Moderate individual variations will not be discussed here.

On tarsi 2–4 the dorsal face is occupied by 2 distal np/A and 5–6 proximal np/B sensilla. The lateral anterior and lateral posterior faces bear 5–6 np/B sensory hairs. The most distal of these are longer and therefore could be considered as np/A as well. On the ventral face we found about 13 sensory hairs, 7 of which are arranged around the apotele. One 'vI'-sensillum is of the tp/A type; all the other ventral setae belong to the type np/A. A total of 31 np/A and B and 1 tp/A were observed on each of the tarsi 2–4.

The dorsal tibial face bears 10 np/B and a distal pair of gustatory tp/A sensilla (Fig. 16.13). The lateral anterior and posterior faces bear about 7 np/B sensilla each, the most distal being longer than the others. The ventral face bears 8 np/A sensory hairs. There are a total of 32 np/A and B and 2 tp/A sensilla for each tibia.

The genu, which is a relatively long segment, has a dorsal face with about 13 np/B and 2 distal tp/A sensilla, 10–11 np/B sensory hairs on each of the lateral faces and 6 ventral np/A setae. A total of 40 np/A and B, and 2 tp/A were found for each genu.

The femur has 11 dorsal np/B sensilla arranged in pairs. In the second pair we find a single tp/A sensory hair in the anterior position. The lateral faces bear 8–11 np/B sensory hairs on the telofemur and 1–2 on the basifemur. The ventral face has about 7 np/A sensilla of which 2 are located on the basifemur. A total of 40 np/A and B and 1 tp/A were found for each femur.

In nymphs, 1 tp/A sensillum is located on the dorsal faces of tibia, genu, and femur; the other sensory hairs are of the np/A and B types.

This study confirms that tarsus 1 is a highly specialized leg segment bearing 12 different types of sensory hairs. The other segments are equipped with np/A and B, and tp/A sensilla only, the former representing the largest number. Investigation of the total number of neurons associated with leg sensory hairs clearly indicate, however, that the 'other' leg segments play an important role in the sensory physiology of the tick. In adult *A. variegatum*, the total mechanosensory input provided by setae located on the legs is 2480 MSU (224 on the foreleg tarsi, 2256 on the other leg segments). The number of putative thermo-gustatory SIU associated with tp/A sensilla is 216–432 (78–156 on the foreleg tarsi and 138–276 on the other segments). Thus 91% of the total seta-related mechanosensory input and 63% of the neurons associated with putative thermohygro-gustatory tp/A neurons are located on leg segments other than the foreleg tarsi.

We conclude, that the sensory hairs of the 'other' leg segments play an important role in taste and possibly in the detection of temperature differences, sense of touch and of substrate vibrations. An important role for np sensilla in the detection of air currents in *A. variegatum* seems improbable because np/A sensilla are too rigid and np/B sensilla too short. In *I. ricinus* these sensory hairs are much finer.

16.4 INTRACUTICULAR LEG SENSILLA

As in other chelicerates, the legs of ticks bear intracuticular sensilla distributed on all segments in addition to sensory hairs. During the last 3 years we have discovered and analyzed ten different structural types of such organs. They are mechanoreceptors that act as strain detectors since they possess tubular bodies. Some of them represent structural intermediates between arachnid slit sense organs and insect campaniform sensilla. One of these sensilla, called the *Distal Tarsal Slit Sense Organ* (DTSSO), has been described in detail (Hess & Vlimant 1984). In this chapter, we will not present and discuss the structural details of each of the ten types of sensilla (the corresponding papers are in preparation), but we will confine ourselves to describing their frequency and distribution on the legs of *A. variegatum*, *I. ricinus* and *O. moubata.*

The intracuticular sensilla are located in the cuticle or are associated with moving cuticular elements joined by membranes. All of them consist of the basic elements described for the DTSSO (Hess & Vlimant 1984), i.e. inner and outer receptor canals separated from each other by a fibrous cup (Fig. 16.19, 16.20). Either of the receptor canals may be reduced, depending on the type of sensillum. Externally, the outer receptor canal is closed by the covering membrane epiostracum which represents a specialized part of epicuticle (Figs. 16.15, 16.16, 16.18–20). The outer canal of some types of intracuticular mechanoreceptors contains globular cuticular particles (Figs 16.19, 16.20). The fibrous cup which acts as a stimulus transducer is penetrated by two structurally different tubular bodies protected by a scolopale (Fig. 16.20). The tubular bodies are specialized mechanosensory dendritic endings that respond to compressional forces. In most sensilla, the receptor elements are associated with slits and membranes that represent mechanical filters (Barth & Blickhan, 1984). Sensilla integrated in scissures and fissures probably represent composed organs. The ten types of mechanosensory organs differ in their arrangement of these components and their location on the segments. When normal preparative techniques are used the sensilla are invisible even when using SEM, because most are covered by cuticular lipids and dirt (Hess & Vlimant 1984).

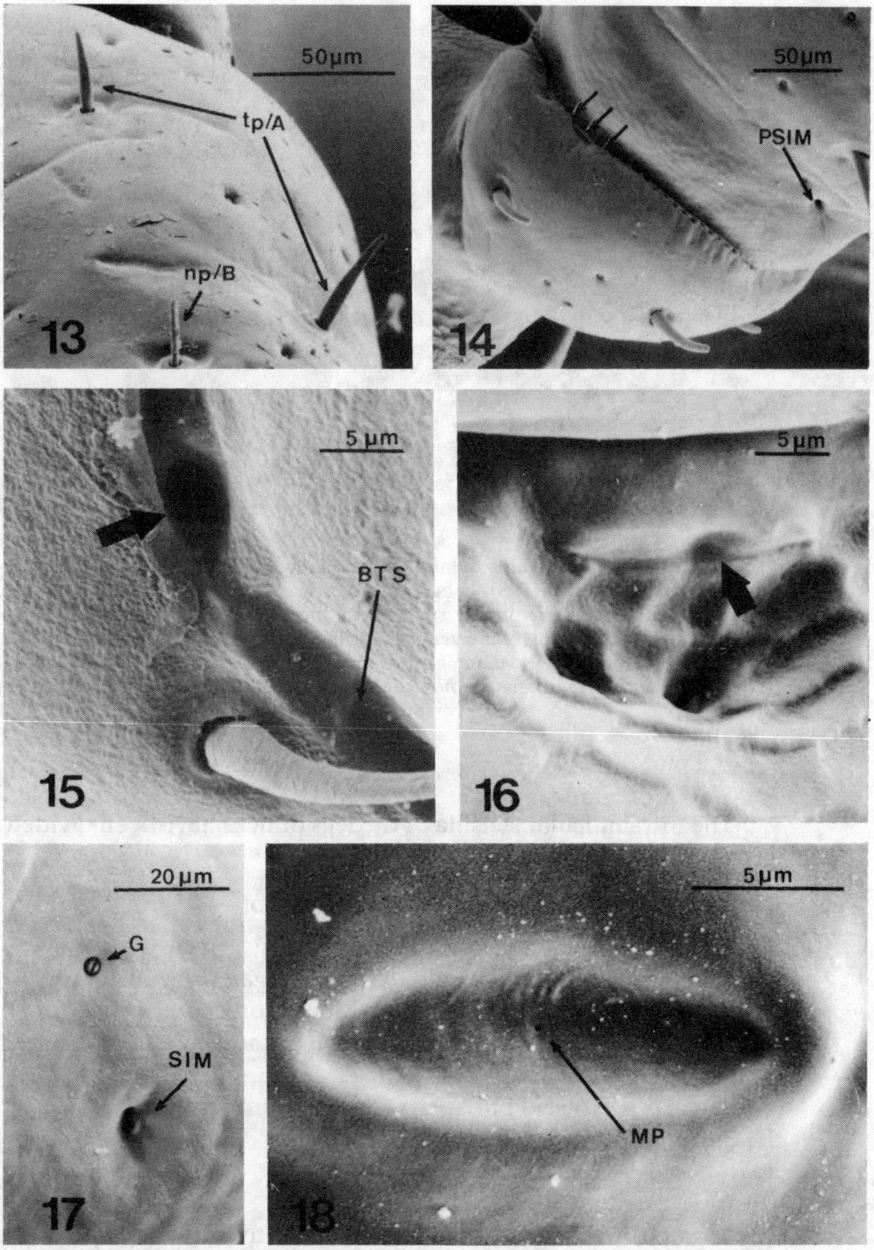

Fig. 16.13 — *A. variegatum* adult: Distal end of tibia with one np/B and two tp/A sensilla. The latter are easily recognizable by their twisted shafts. Fig. 16.14 — *A. veriegatum* adult: Basifemur and telofemur are separated by the Basifemoral Scissure (BFS) which is equipped with numerous mechanoreceptors (arrows); PSIM: simple intracuticular mechanoreceptor associated with a pore; see also Fig. 16.21.

Fig. 16.15 — *A. variegatum* adult: The Basitarsal Scissure (BTS) of the legs 2–4 is equipped with two lateral and one ventral mechanosensory sensilla. Arrow: depression (covering membrane) of the anterior lateral mechanoreceptor; see also Fig. 16.21. Fig. 16.16 — *I. ricinus* adult: Proximal Tarsal Slit Sense Organ (PTSSO). The arrow indicates the central depression of the covering membrane invaginated into the fibrous cup; see also Fig. 16.22. Fig. 16.17 — *A. variegatum* adult: Simple Intracuticular Machanoreceptor (SIM) and gland opening (G); see also Fig. 16.21. Fig. 16.18 — *O. moubata* adult: Argasid Slit Sense Organ (ASSO) with a central moulting pore (MP); see also Fig. 16.23.

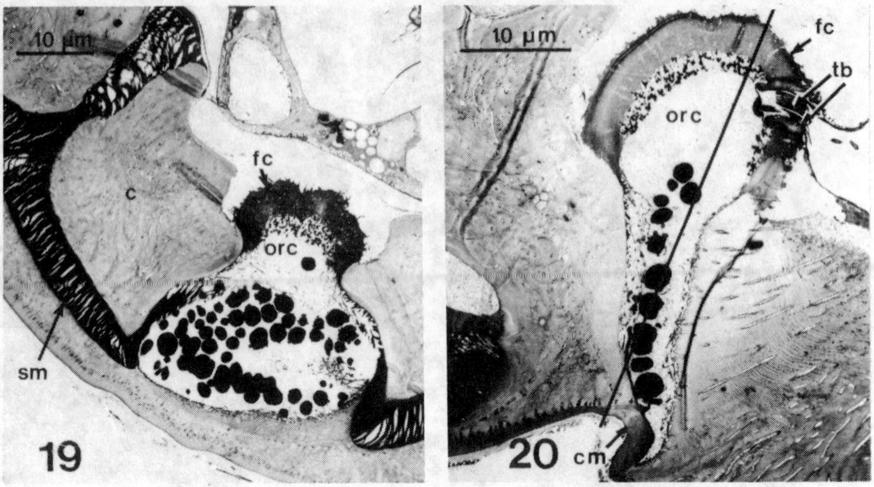

Fig. 6.19 — *A. variegatum* adult: Transverse section of the ventral sensillum of the Basitarsal Scissure (BTS). c: basitarsal cuticle; Fc: fibrous cup; orc: outer receptor canal with cuticular globules; sm: scissure membrane. Fig. 16.20 — *A. variegatum* adult: Sagittal section of the ventral sensillum of the Basitarsal Scissure (BTS). The black line indicates the plane of section of Fig. 16.19. cm: covering membrane; fc: fibrous cup; orc: outer receptor canal with cuticular globules; tb: tubular bodies piercing the fibrous cup.

The intracuticular sensilla of the legs of ticks have been divided into two groups:

(1) single intracuticular organs that consist of:

DTSSO:	Distal Tarsal Slit Sense Organ;
PTSSO:	Proximal Tarsal Slit Sense Organ (Fig. 16.16);
SIM:	Simple Intracuticular Mechanoreceptor (Fig. 16.17);
PSIM:	Simple Intracuticular Mechanoreceptor associated with a Pore (Fig. 16.14);
ASSO:	Argasid Slit Sense Organ (Fig. 16.19);
FTF:	Foreleg Tarsal Fissure;
GF:	Genual Fissure;

(2) Composite intracuticular mechanosensitive organs consisting of:

BFS:	Basifemoral Scissure composed of one ventral (asymmetrical) and several lateral (asymmetrical) sensilla (Fig. 16.14);
BTS:	Basitarsal Scissure composed of one ventral (symmetrical) and a pair o lateral (asymetrical) sensilla (Fig. 16.15);
TF:	The Tibial Fissure consisting of one ventral (symmetrical) and a pair of lateral (asymmetrical) sensilla.

The distribution of the mechanosensory organs on the legs of *A.*

variegatum, *I. ricinus* and *O. moubata* are represented in Figs. 16.21–16.23. The following details should be noted:

DTSSO, PTSSO, SIM and FTF are restricted to ixodids and PSIMs have been observed only in *A. variegatum*. FTFs instead of BTSs occur on ixodid foreleg tarsi. In *A. variegatum*, the SIMs are located on the anterior face and in *I. ricinus* on the posterior face of the segments. In *O. moubata* we find ASSOs instead of SIMs, but the total number of ASSOs in *O. moubata* is much higher than the number of SIMs in the ixodids. *O. moubata* females have a slightly higher number of ASSOs than males. Except for the foreleg tarsi, the homologous segments of the four pairs of legs seem to have a identical sensillar apparatus. Compared to adults, immatures possess a lower number of intracuticular mechanoreceptor organs (Figs. 16.21–23).

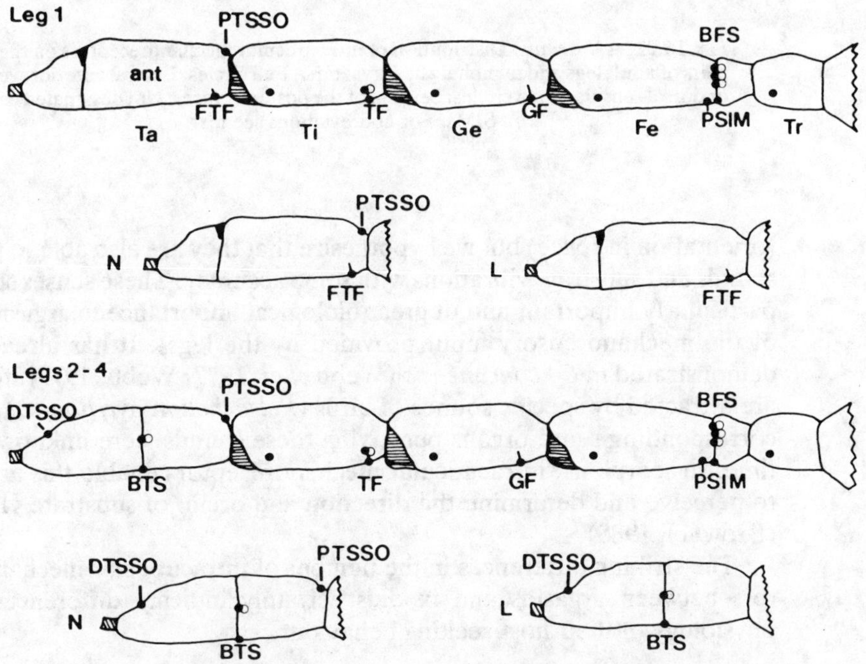

Fig. 16.21 — *A. variegatum*: Distribution of intracuticular mechanoreceptors on the four pairs of adult legs and nymphal and larval tarsi. Full circles: Lateral anterior, ventral or dorsal sensilla, open circles: Sensilla of the posterior faces. Not designated circles: SIMs. For abbreviations see text.

What is the contribution of the intracuticular mechanosensory sensilla to the total sensory input related to the legs? In *A. variegatum* it represents 324 neurons (mechanosensory units, MSU) i.e. 12% of the total population of leg MSU, in *I. ricinus* it represents 308 MSU i.e. 13%, and in *O. moubata* it represents about 1340 MSU i.e. 70% of the mechanosensory neurons of the legs. This means that ticks not only have a well developed sense of gravity

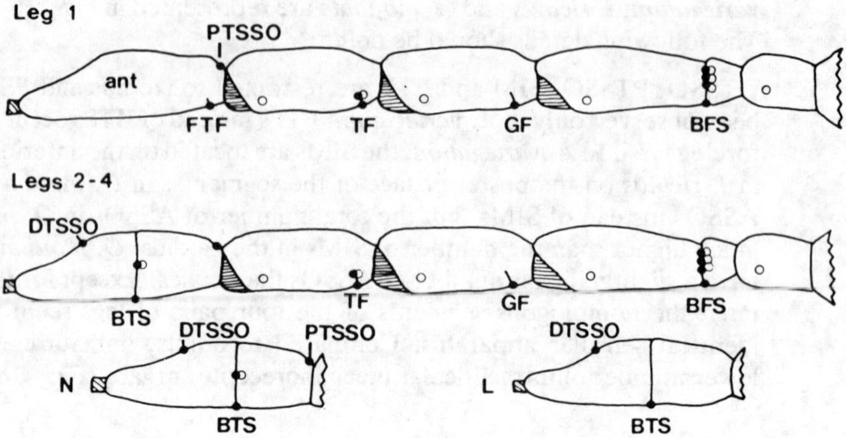

Fig. 16.22 — *I. ricinus*: Distribution of intracuticular mechanoreceptors on the four pairs of adult legs and nymphal and larval tarsi. Full circles: Lateral anterior, ventral or dorsal sensilla, open circles: Sensilla of the posterior faces. Not designated circles: SIMs. For abbreviations see text.

(orientation in space) but we hypothesize that they are also able to perceive sounds and substrate vibrations with some accuracy. These senses should be particularly important and of great biological importance in argasids (70% of the mechanosensory input provided by the legs). It has already been demonstrated in *O. concanensis* (Webb *et al.* 1977, Webb, 1979), that ticks are attracted by specific sounds of birds (*Petrochelion pyrrhonota*), but the corresponding sense organs perceiving these sounds were unknown at the time. In scorpions, intracuticular mechanoreceptors enable this arthropod to perceive and determine the direction and origin of substrate vibrations (Brownell 1985).

The striking differences in the number of intracuticular mechanoreceptors between argasids and ixodids certainly indicate differences in the physiology of their host-seeking behaviour.

16.5 CONCLUSIONS

The foreleg tarsi of ticks are highly specialized leg segments bearing different types of mechanosensitive, thermo-hygrosensitive, gustatory, and olfactory sensilla. Characteristic differences in the number and distribution of these structure occur among metastriate, prostriate, and argasid ticks, but not among related species even if the latter differ greatly in their life cycle and ecology.

The prominent role of the foreleg tarsi in the sensory physiology of ticks is qualified somewhat by the presence of mechanosensitive and numerous thermohygro-gustatory sensory hairs on other leg segments.

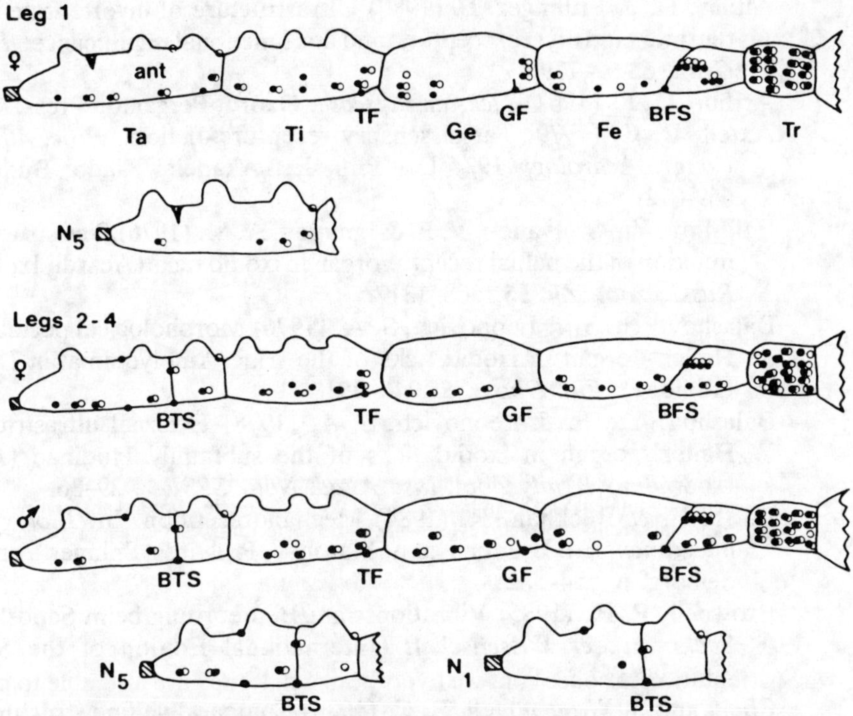

Fig. 16.23 — *O. moubata*: Distribution of intracuticular mechanoreceptors on the four pairs of adult legs and nymphal 5 and nymphal 1 tarsi. Full circles: Lateral anterior, ventral or dorsal sensilla, open circles: Sensilla of the posterior faces. Not designated circles: ASSOs. For abbreviations see text.

The presence of ten different types of intracuticular mechanosensory leg sensilla suggests that ticks most probably possess a well developed sense of gravity and the capacity to detect and to analyze substrate vibrations and sounds. This is especially true for argasids which are equipped with a great number of such sensilla which are probably used for the detection and identification of hosts.

Comparative functional chaetotaxy is a useful tool for approaching the physiology of host-seeking behaviour in ticks, but it could also be useful in studying the evolution of ticks.

ACKNOWLEDGEMENTS

Supported by the Swiss National Foundation for Scientific Research grant No. 3.070.81 and by the Stanley Thomas Johnson Foundation. We are grateful to Miss Ellen Dotson for her help with the English translation.

REFERENCES

Altner, H. (1977) Insektensensillen; Bau und Funktionsprinzipien, *Verh. dtsch. zool. Ges.* 1977 139–153.

Altner, H. & Prillinger, L. (1980) Ultrastructure of invertebrate chemo-, thermo- and hygroreceptors and its functional significance, *Int. Rev. Cytol.* **65** 69–139.

Arthur, D. R. (1962) *Ticks and disease.* Oxord, Pergamon Press, 445 pp.

Axtell, R. C. (1979) Tarsal sensory receptors of ticks. *Proc. 4th Intern. Congr. Acarology 1974* Ed. Piffl, E., Akadeiai Kiado, Budapest p. 669-672.

Balashov, Yu. S., Ivanov, V. P. & Ignatiev, A. M. (1976) Fine structure and function of the palpal receptor organ in ixodid ticks (Acardi, Ixodoidea), *Russk. zool. Zh.* **55** 1308–1317.

Balashov, Yu. S. & Leonovich, S. A. (1976) Morpholocical pecularities of Haller's organ in ixodid ticks of the tribe Amblyommatini (Acarina, Ixodidae), *Ent. Obozr.* **55** 946–952.

Balashov, Yu. S. & Leonovich, S. A. (1978) External ultrastructure of Haller's organ in ixodid ticks of the subfamily Ixodinae (Acarina, Ixodoidea), *Trudy Zool. Inst. Akad. Nauk SSSR* **77** 29–36.

Barth, F. & Blickhan, R. (1984) Mechanoreception. In: *Biology of the integument* Ed. Breiter-Hahn, Matoltsy, Richards, Spinger Verlag Heidelberg. p. 554–582.

Brownell, P. H. (1985) Vibrationssinn: Beuteortung beim Sandskorpion, *Spektrum der Wissenschaft* (International Edition of the Scientific American) **2** 84–92.

Chu-wang, I. W. & Axtell, R. C. (1973) Comparative fine structure of the claw sensilla of a soft tick, *Argas* (*P.*) *arboreus* Kaiser, Hoogstraal and Kohls, and a hard tick, *Amblyomma americanum* L., *J. Parasitol.* **59** 545–555.

Clifford, C. M. & Anastos G. (1960) The use of chaetotaxy in the identification of larval ticks (Acarina: Ixodidae), *J. Parasitol.* **46** 567–578.

Elizarov, Yu. A. (1963) Insect and tick chemoreception: electrophysiologic study of the chemoreception of Ixodidae, *Vestn. Moskov. Univ. Ser VI Biol. Pochvovded* **18** 16–24.

Evans, G. O., Sheals, J. G., & Macfarlane, D. (1961) *The terrestrial acari of the British Isles, an introduction to their morphology, biology and classification,* Printed by order of the Trustees of the British Museum.

Foelix R. F. (1972) Permeability of tarsal sensilla in the tick *Amblyomma americanum L. (Acarina, Ixodidae), Tissue and Cell* **4** 130-135.

Foelix, R. F & Axtell R. C. (1971) Fine structure of tarsal sensilla in the tick *Amblyomma americanum L., Z. Zellforsch.* **114** 22–37.

Foelix, R. F. & Axtell, R. C. (1972) Ultrastructure of Haller's organ in the tick *Amblyomma americanum*, L. *Z. Zellforsch* **124** 275–292.

Graf, J. F. (1975) Ecologie et ethologie d'*Ixodes ricinus* L. en Suisse (Ixodoidea; Ixodidae). Ciquieme note: Mise en evidence d'une pheromone sexuelle chez *Ixodes ricinus*, *Acarologia* **17** 436–441.

Grandjean, F. (1954) Sur les nombres d'articles aux appendices des acariens actinochitineux, *Arch. Sci. Geneve* **7** 335–362.

Haggart, D. H. & Davis, E. E. (1980) Ammonia sensitive neurones on the first tarsi of the tick, *Rhipicephalus sanguineus*, *J. Insect Physiol.* **26**

517–523.

Haggart, D. H. & Davis, E. E. (1981) Neurons sensitive to 2,6-dichlorophenol on the tarsi of the tick *Amblyomma americanum* (Acari: Ixodidae) *J. Med. Entomol.* **18** 187–193.

Hess, E. & Loftus, R. (1984) Warm and cold receptors of two sensilla on the foreleg tarsi of the tropical bont tink *Amblyomma variegatum*, *J. Comp. Physiol. A* **155** 187–195.

Hess, E. & Vlimant M. (1980) Morphology and fine structure of tarsal chemosensory hairs of the tick *Amblyomma variegatum* (Acarina, Ixodidae, Metastriata), including preliminary electrophysiological results, *Proc. Olfaction and Taste* **7** 190.

Hess, E. & Vlimant, M. (1982) The tarsal sensory system of *Amblyomma variegatum* Fabricius (Ixodidae, Metastriata) I. Wall pore and terminal pore sensilla, *Revue suisse Zool* **89** 713–729.

Hess, E. & Vlimant, M. (1983a) The tarsal sensory system of *Amblyomma variegatum* Fabricius (Ixodidae, Metastriata) II. No pore sensilla, *Revue suisse Zool* **90** 157–167.

Hess, E. & Vlimant, M. (1983b) The tarsal sensory system of *Amblyomma variegatum* Fabricus (Ixodidae, Metastriata) III. Mapping of sensory hairs and evolution of the relative importance of sensory modalities during post-embryonic development, *Revue suisse Zool* **90** 887–897.

Hess, E. & Vlimant, M. (1984) The distal tarsal slit sense organ (DTSSO). A new type of mechanoreceptor on the walking legs of the ixodid tick *Amblyomma variegatum* Fabricius 1794 (Ixodidae: Metastriata), In: *Acarology VI* Vol 1, Ed. Griffiths, D. A. & Bowmann, C. E., Ellis Horwood Limited, Chichester, p. 253–260.

Holsher, K. H., Gearhart, H. L., & Barker, R. W. (1980) Electrophysiological responses of three tick species to carbon dioxide in the laboratory and field, *Ann. Entomol. Soc. Am.* **73** 288–292.

Homsher, P. J. & Sonenshine, D. E. (1975) Scanning electron microscopy of ticks for systematic studies: Fine structure of Haller's organs in ten species of Ixodes, *Trans. Amer. Micros. Soc.* **94** 368–374.

Homscher, P. J. & Sonenshine, D. E. (1977) Scanning electron microscopy of ticks for systematic studies. 2. Structure of Haller's organ in *Ixodes brunneus* and *Ixodes frontalis*, *J. Med. Entomol.* **14** 93–97.

Hoogstraal, H., Clifford, C. M., Roshby, M. A., & Keirans, J. E. (1983) The subgenus *Persicargas* (Ixodoidea: Argasidae: *Argas*). 38. Haller's organ roof and anterior pit setae in *Persicargas* and *Microargas*, *J. Parasitol.* **69** 977–985.

Jorgensen, W. K. (1985) The ultrastructure of the sense organs of tarsus 1, the palps and chelicerae of the larval cattle tick *Boophilus microplus* (Canestrini) Ixodidae including a study of the ultrastructure and function of the tick podium, PhD thesis, University of Queensland, Australia.

Krantz, G. W. (1978) *A manual of acarology* 2nd edition, Ed. Oregon State University Book Stores Inc., Corvallis.

Leonovich, S. A. (1977) Electron microscopic studies of Haller's organ of the tick *Ixodes persulcatus* (Ixodidae), Parazitologija **11** 340–347.

Leonovich, S. A. (1978) Fine structure of Haller's organ in the tick *Hyalomma asiaticum* P.Sch. & E. Schl. (Parasitiformes, Ixodidae, Amblyomminae), *Revue Ent. URSS* **57** 221–225.

Leonovich, S. A. (1979) Ultrastructural investigations of Haller's organ in the argasid tick, *Argas tridentatus* (Argasinae), *Parazitologija* **13** 483–487.

Leonovich, S. A. (1980) Fine structural investigations of Haller's organ in the argasid ticks, *Ornithodoros moubata* and *Alveonasus lahorensis*, *Parazitologija* **14** 376–385.

Roshdy, M. A., Foelix, R. F., & Axtell, R. C. (1972) The subgenus *Persicargas* (Ixodidae: Argasidae: *Argas*). 16. Fine structure of Haller's organ and associated tarsal setae of adult *A. (P.) arboreus* Kaiser, Hoogstraal & Kohls, *J. Parasitol.* **58** 805–816.

Schöni, R., Hess, E., Blum, E., & Ramstein, K. (1984) The aggregation-attachment pheromone of the tropical bont tick *Amblyomma variegatum* Fabricius (Acari, Ixodidae): isolation, identification and action of its components, *J. Insect. Physiol.* **30** 613–618.

Sinitzina, E. E. (1974) Electrophysiological reactions of the neurons of the Haller's organ to odour stimuli in the tick *Hyalomma asiaticum*, *Parazitologija* **8** 223–226.

Thonney, F. (1980) Morphology and structure of presumed tarsal chemoreceptors of the first pair of legs of *Ixodes ricinus* L. (Acarina, Ixodidae, Prostriata), *Proc. Olfaction and Taste* **7** 191.

Thonney, F. (1986) Etude morphologique, structurale et physiologique des recepteurs sensoriels du tarse 1 de la tique *Ixodes ricinus* L. These de doctorates sciences, Université de Neuchâtel, Suisse.

Van der Hammen, L. (1976) Glossaire de la terminologie acarologique (glossary of acarological terminology) Ed. Dr. W. Junk B.V. The Hague.

Waladde, S. M. (1979) A study of the structure and physiology of certain sensory receptors of the adult cattle tick *Boophilus microplus* (Canestrini) Ixodidae. PhD thesis, University of Queensland, Australia.

Waladde, S. M. (1982) Tip-recording from ixodid tick olfactory sensilla: responses to tick odours, *J. Comp. Physiol.* **148** 411–418.

Waladde, S. M., Kokwaro, E. D., & Chimtavi, M. (1981) A cold receptor on the tick, *Rhipicephalus appendiculatus:* electrophysiological and ultrastructural observations, *Insect Sci. Application* **1** 191–196.

Webb, J. P. (1979) Host-locating behavior of nymphal *Ornithodoros concanensis* (Acarina: Argasidae) *J. med. Entomol.* **16** 437–447.

Webb, J. P. jr., George, J. E., & Cook, B. (1977) Sounds as a host detection cue for the soft tick *Ornithodoros concanensis, Nature* **265** 443–444.

Zolotarev, E. K. (1962) The leg and its terminology in parasitiform ticks, *Zool. Zh.* **41** 1739–1741.

Zumpt, F., (1949) Das Hallersche Organ der Zecken und seine systematische Bedeutung, *Verh. dtsch. zool. Ges.* 1948 427–431.

17

Ticks on reptiles, a difference of scale

C. M. Bull

17.1 INTRODUCTION

Tick research has been concentrated largely on those tick species which infest man or his domesic livestock and which are of direct economic or medical interest. Other phylogenetic groups possess a diversity of adaptations reflecting the separate ecological circumstances of different species. If this is so for ticks, then the present concentration of effort on a relatively small proportion of all tick species, mostly those infesting large mammals, will seriously under-represent the biological variability present within the group and will lead to inaccurate generalisations.

For instance, there are many tick species which infest only reptiles, but comparatively little work has been done on them. Ticks of two genera, *Aponomma* and *Amblyomma*, infest large snake and lizard species in Australia. Roberts (1970) lists eleven species, and there are at least three currently undescribed. The sleepy lizard (*Trachydosaurus rugosus*), a large (500–800 gm) scincid lizard whose distribution covers much of southern Australia, is the host species for three tick species, *Aponomma hydrosauri, Amblyomma albolimbatum*, and *Amb. limbatum*.

At Flinders University, South Australia, we have been investigating the biology of these three ectoparasite species and of their host. Our work has led us to conclude that the three tick species, from two phylogenetically distinct genera, have developed patterns of behaviour, ecology, and population structure which are similar to each other, but are distinct from other tick species adapted to infest large mammals. Many aspects of the difference can be explained as adaptations to their host species, the sleepy lizard.

The work reported is not only mine. Ross Andrews, Ingrid Belan, Dale Burzacott, Neil Chilton, Trevor Petney, Rojchai Satrawaha, and Robert Sharrad have been involved in various aspects of the project, and all have contributed ideas and results.

17.2 GENERAL BIOLOGY OF THE TICKS

Ap. hydrosauri, Amb. albolimbatum, and *Amb. limbatum* are three-host ticks. Larvae, nymphs, and adults infest the same range of host species. All three life history stages are often found together on the same host individuals (Bull 1978a). The sleepy lizard is the most commonly infested host. Approximately 85% of individuals of this species are infested with one of the three tick species (Sharrad 1980, Bull *et al*. 1981), although infestation rates vary geographically and temporally (Bull 1978a). Other tick species are rarely found on sleepy lizards.

Each of the three species is also found infesting other reptile species (Smyth 1973, Bull *et al*. 1981, Sharrad & King 1981), including snakes and lizards from the varanid and agamid families. Alternative hosts are least common for *Amb. albolimbatum,* whose range in south-western Australia is almost entirely within that of the sleepy lizard, and most common in *Amb. limbatum* whose distribution extends beyond the northern limit of the sleepy lizard.

The three ticks have rarely been reported from non-reptile hosts, with the exception of records of *Ap. hydrosauri* infesting the monotreme mammal *Tachyglossus aculeatus* (the echidna) near Rockhampton in central Queensland (Roberts 1970). This population is over 1000 km further north than any other records of *Ap. hydrosauri* (Smyth 1973). I have examined specimens from Rockhampton echidnas. They are morphologically similar to the southern *Ap. hydrosauri* which only infest reptiles, but investigation of the specific status of the Rockhampton population using other techniques (see later in this chapter) would be useful. Apart from this one small population, *Ap. hydrosauri, Amb. albolimbatum,* and *Amb. limbatum* are exclusively parasites of reptiles. I shall describe some of the unique properties of these tick species, and propose that they derive, at least in part, as adaptations to properties of their hosts.

17.3 GENERAL BIOLOGY OF THE HOST SPECIES

All of the laboratory ticks we have studied were obtained from populations where the sleepy lizard, *T. rugosus*, was the common host. Sleepy lizards are large skinks. Adult females attain 800 gm in weight with snout–vent lengths of 32 cm. Their bodies are covered with thick overlapping scales. Ticks can easily crawl under the posterior edge of a scale and attach to the skin beneath the scale. The lizards are omnivorous and long-lived (some lizards in field populations I have studied must be at least 10 years old (Bull, unpublished data)). Females usually bear two live young in the autumn, but there is no evidence of any subsequent maternal care. They have low dispersal rates with 85% of recaptures after 6 years being within 100 m of the original capture site (Bull, unpublished data).

There are a number of properties of sleepy lizards, and reptiles in general, which make them different from mammalian hosts. One obvious difference, that reptiles are ectothermic, has two major implications for the ticks which infest them. Firstly, host body temperature will fluctuate diurnally and seasonally, and will be generally low during winter, and at night over the entire year. This will influence tick development rates, which are generally slower on reptile than mammal hosts (Bull *et al.* 1978, Sharrad 1980). Attached ticks cease development altogether during the winter (Bull & Sharrad 1980). Secondly, lizard activity is less frequent than that of mammals. Even during their active season sleepy lizards will not move during cool days (Satrawaha & Bull 1981). When they do emerge they spend much time basking, and their movements are slow and cover relatively short distances (Bull 1978b). There is less chance that a lizard will move past the place where a tick is waiting for a host, than a more active, and more regularly active mammal.

Another distinctive property of reptiles is an absence of legs, or the possession of legs that project laterally, with the consequence that much of the body is in contact with the substrate during both walking and resting. In contrast, only a small part of a mammal body surface touches the substrate and when it does so, more briefly. Thus reptile ticks have a greater opportunity to detect and attach to passing hosts.

Sleepy lizards are solitary much of the year, although they occupy overlapping home ranges (Bull 1978b, Satrawaha & Bull 1981). Within each home range a lizard has a variety of refuge sites that it uses during non-active periods (e.g. under bushes, down rabbit burrows, buried under litter) (Satrawaha 1980), but they do not consistently return to the same refuges (Bull 1978b, Satrawaha & Bull 1981). Sleepy lizards do not have large social groups, or regularly-visited resting sites. In this way they differ from many mammal species, and their ticks have a lower chance of reliably locating hosts.

17.4 TICK DISTRIBUTION PATTERNS

The first striking property of the three sleepy lizard ticks is their parapatric distributions. This means their ranges are mutually exclusive (allopatric), abutting along common boundaries with little or no overlap. This was first observed by Smyth (1973) in the state of South Australia (Fig. 19.1). He found ticks distributed throughout the state, but only one species was found in any one area. *Ap. hydrosauri* occupied the south-eastern section and the southern part of Eyre Peninsula, *Amb. albolimbatum* were found in the western part of the state and to the north of *Ap. hydrosauri*, and *Amb. limbatum* occupied an area to the north of the other two.

Detailed collections showed that wherever the ranges of two of the three species made contact, boundaries for their distribution were abrupt, with one species completely replacing the other over distances as small as 100 m (Smyth 1973, Bull *et al.* 1981). Boundaries have remained in a stable position for more than 16 years at two sites (the boundary between *Ap. hydrosauri* and *Amb. limbatum* near Mt Mary (Bull *et al.* 1981), and that between *Ap. hydrosauri* and *Amb. albolimbatum* near Cleve on Eyre Peninsula (Petney *et al.* 1982)).

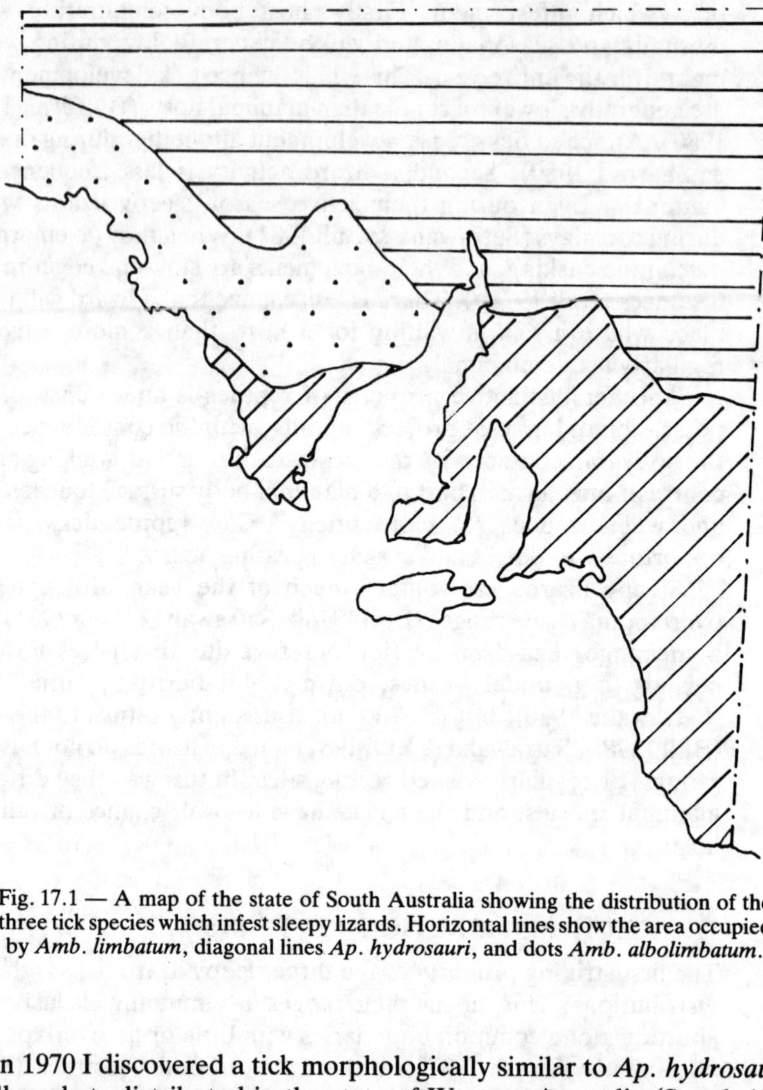

Fig. 17.1 — A map of the state of South Australia showing the distribution of the three tick species which infest sleepy lizards. Horizontal lines show the area occupied by *Amb. limbatum*, diagonal lines *Ap. hydrosauri*, and dots *Amb. albolimbatum*.

In 1970 I discovered a tick morphologically similar to *Ap. hydrosauri* in small pockets distributed in the state of Western Australia (Smyth 1973). Subsequently all three species were found to have similar non-overlapping distributions (Sharrad & King 1981). A survey of the boundary near Albany (W.A.) between the *Ap. hydrosauri*-like tick and *Amb. albolimbatum* revealed the same abrupt transition between species (Bull & King 1981).

Detailed distribution maps have not yet been published for eastern Australia, although a similar pattern has been predicted (Smyth 1973). A fourth species, *Amb. moreliae*, replaces *Amb. limbatum* along the east coast of Australia, but the distributional interaction of these two species is not precisely known.

Several generalities can be made about the parapatric boundaries. They do not coincide with any discontinuity of distribution of any major host species (Sharrad 1980, Bull *et al.* 1981). In particular, the sleepy lizard is common on both sides of all boundaries mapped thus far. However, boundaries sometimes coincide with environmental changes. For instance, the boundary between *Ap. hydrosauri* and *Amb. limbatum* broadly follows the 250 mm rainfall isohyet (Smyth 1973), and, near Mt Mary, it coincides with a sharp vegetational transition (Sharrad 1980, Bull *et al.* 1981, Petney & Bull 1984). We have argued elsewhere (Bull *et al.* 1981) that this specific environmental ecotone could not alone maintain the tick boundary, since this does not explain the generality of the interaction wherever any two of the three species are in contact in a variety of environmental circumstances (Bull & King 1981, Petney *et al.* 1982). Rather, we believe that parapatry is an intrinsic property of the interaction between the species, and that environmental ecotones simply set the position of the boundary (Bull *et al.* 1981). We still do not understand the processes which maintain the boundaries and prevent co-occurrence of two species (Bull *et al.* 1981).

In this chapter I emphasise the uniqueness of this distribution pattern. Although parapatric boundaries are known from other taxa (e.g. Odendaal & Bull 1982), I am not aware of any comparable case described for ticks. Generally, where a species can serve as host for several different ticks, those ticks overlap in their ranges and often simultaneously infest individuals of the same host. The case of the sleepy lizard whose three tick species are distributed parapatrically, appears to be unique. Even other reptile ticks in Australia overlap in distribution and can parasitize the same host. For instance, *Ap. fimbriatum* is a tick which infests snakes and varanid lizards. In northern Australia it is commonly found in the same area, and on the same host as *Amb. limbatum* (Sharrad & King 1981, Bull unpublished data). Possibly, sleepy lizards have some unique feature of their biology which predisposes them to cause this unusual distribution pattern of their ectoparasites. For instance, slow host movement may be important in restricting dispersal of ticks across the boundary, and overlapping host home ranges may ensure that any tick colonists will always make contact with the resident tick species.

17.5 GENETIC VARIATION

One tool we use to study the dynamics of the boundary is enzyme electrophoresis. Our initial intention was to detect polymorphic enzyme loci, and record microgeographic variation in allele frequencies to allow assessment of gene flow and dispersal rates (Selander 1970), important parameters in all of our models of boundary maintenance in reptile ticks.

We anticipated finding genetic variation in reptile ticks for three reasons. First, higher isoenzyme variability within populations in invertebrate taxa than in vertebrate taxa have been reported (Nevo 1978). Second, from various properties of the adaptations of parasites to hosts, Price (1977) deduced that there should be greater genetical variation between populations in parasite taxa than in non-parasite taxa. Third, Healy (1979a,b) found high levels of variation, for at least two enzyme loci, both within and between populations of *Ixodes ricinus* in Ireland.

We examined the variability at up to 25 enzyme loci in six reptile tick species, the three sleepy lizard ticks plus *Ap. fimbriatum, Ap. undatum,* and an undescribed *Amblyomma* species endemic to the Flinders Ranges of South Australia (Bull *et al.* 1984). The results from 67 tick populations were quite different from our expectations. There was a universally low level of variation within individual populations. The average heterozygosity per locus was 2.35% in *Ap. hydrosauri*, and lower for the other five tick species (Bull *et al.* 1984). The average heterozygosity for 93 other invertebrate species was 11.23% (Nevo 1978). None of the loci examined in reptile ticks approached the variation reported for two loci in *I. ricinus* (Healy 1979a,b). There were few geographical clines in the reptile ticks. Populations of *Amb. albolimbatum*, separated by over 2500 km, were electrophoretically indistinguishable (Bull *et al.* 1984).

Enzyme electrophoresis has not often been used to study ticks, so comparative statements must be speculative until more data are produced. *Ixodes ricinus*, a mammalian tick, and the Australian reptile ticks, are clearly different in the genetical structure of their populations. A possible explanation lies in the properties of their host species. Price (1977) predicted that if the host represented a resource of extreme patchiness in space and time, parasites could adapt by increasing clutch size and decreasing dispersal and outbreeding. There would be strong selection for adaptations to local host populations. Price (1977) proposed that all of these factors should lead to reduced genetical variation in parasite populations. We argued (Bull *et al.* 1984) that lizards, with their slow dispersal, small home ranges, and solitary nature do represent patchy resources for their ticks, so that the low variability we discovered is in agreement with the predictions. On the other hand, *I. ricinus*, a tick of domestic mammalian stock, would often have a variety of host individuals to choose from, and there could be frequent host movement and gene mixing as a result of agricultural practice. Price's predictions are unlikely to apply in these circumstances. Thus the difference between the ticks can be explained through differences in the way the hosts are presented as resources.

Although the normal pattern in Australian reptile ticks showed little geographical variation in isoenzyme allele frequencies, three cases were detected (Bull *et al.* 1984) where peripheral populations (two in *Ap. hydrosauri* and one in *Amb. albolimbatum*) were quite different from all other populations. These populations were monomorphic for unique alleles at one or more isoenzyme loci, implying genetic differentiation had occured, possibly during a period of geographic separation. In each case the divergent population was morphologically similar to populations with the more common genetic structure. The implication is that genetical divergence, and perhaps speciation, can occur without morphological change. Sibling species are recognized in many other invertebrate taxa but have not yet, to my knowledge, been reported in ticks where morphological criteria are the main taxonomic tool. It seemed possible from our electrophoretic survey that the Australian reptile tick fauna contains some sibling species. (A test of this hypothesis appears in the next section).

Whether or not this is a unique property of reptile ticks remains untested because biochemical systematics have not been used in other tick groups. I believe this should be a priority in tick taxa which are suspected disease vectors. Intrinsically, however, I suspect that ticks with slow dispersing lizards as hosts might have more opportunities for genetical divergence than ticks of more mobile mammalian hosts.

17.6 MATING BEHAVIOUR

The tests which we used to detect sibling species were interpopulation mating trials. I must first explain some details of our investigations of the mating behaviour of Australian reptile ticks. Our interest was aroused by the widely reported phenomenon of a single pheromone, used during tick mating, which attracted both conspecific and non-conspecific males (e.g. see Chapter 15, Sonenshine *et al*. 1976, Kellum & Berger 1977). We asked whether reptile ticks had a similar common pheromone? and if so: (i) what happens at parapatric boundaries where adults of two species meet? and (ii) how are genetical differences, as detected by electrophoresis, maintained? because if two populations have the same sexual signals there should be free interbreeding and introgression.

Ross Andrews studied the mating behaviour of the three sleepy lizard ticks (Andrews 1982), describing in most detail the courtship of *Ap. hydrosauri* (Andrews & Bull 1980). In some aspects their behaviour is similar to that described for other hard tick species. Mating takes place on the host. It ends with the male and female ventral surfaces together (Andrews & Bull 1980). Females must feed from their host before they are sexually attractive (Andrews & Bull 1981). Fed females emit an air-borne pheromone which lures mobile males to the proximity (Andrews & Bull 1982a). Males of all three species are attracted to females of each species (Andrews & Bull 1982a). We do not yet know the composition of the attractant signal, but it induces behaviour similar to that reported for 2,6-dichlorophenol, the attractant pheromone in other tick species (Sonenshine *et al*. 1976, Kellum & Berger 1977).

At this point, our similarity with previously reported studies ends. Most importantly, early in the courtship of reptile ticks there is a stage which precedes the attraction of mobile males to females. Males must first be stimulated to detach from their position on the host to search for receptive females. Males attach to lizards and remain there for many months waiting for new females to attach. Several weeks may pass without a female attaching. The males conserve energy and reduce the risk of dislodgement by rarely moving from one attachment site when no females are present. Females attach to sites independent of male position (Andrews & Bull 1980). Experimental observations (Andrews & Bull 1980, Andrews 1982) showed that male ticks moved significantly more often on lizards after females had attached than before, and more than on lizards with only males attached. The male moves were not directed specifically towards females, and did not always result in a male finding a female (Andrews & Bull 1980).

We attributed this increased male activity to an excitant signal released by females in order to stimulate the males to detach and begin searching for the females. We suggested (Bull & Andrews 1984) that this was a different signal from the attractant signal for three reasons. Firstly, the attractant was non-specific but the excitant response was species specific (Andrews 1982). Males only started searching actively in the presence of conspecific females. Secondly, the excitant only worked when the males had fed, whereas the attractant acted equally well on fed and unfed males (Andrews & Bull 1981). Thirdly, the searching response to the excitant was non-directional, while the attractant directs males towards females (Andrews & Bull 1982a).

Although we have not identified the structures of the two signals, we have demonstrated that the excitant is also an air-borne pheromone, because males can be induced to move when alone on a lizard, if that lizard is placed next to a lizard with only female ticks attached (Andrews *et al.* 1986). The alternative we rejected was that the excitant signal was passed from female tick to male tick through the host circulation.

In summary, we believe that the early stages of courtship in reptile ticks involves the male attaching to a host in wait of a female. The female attaches, feeds, and then produces an excitant signal which induces males to detach and move randomly, about on the surface of their lizard host, searching for females. If males come within the range of the female attractant signal they orientate and move towards their potential mate.

It has recently been recognized that tick courtship involves more than one sexual attraction signal (e.g. Sonenshine *et al.* 1982). However, I know of no other reports where female ticks use an excitant signal to stimulate males to start searching. One explanation is that this behaviour is present in other tick species but has not been detected. In many studies of tick mating behaviour, experimental observations start when active males and females are confined together, so the necessity of initial male excitation has been bypassed before observations begin.

Alternatively, this system may be unique to reptile ticks. In other ticks, which infest mobile or communally dwelling host species, male ticks could actively search for females by moving from host to host. A male on a solitary lizard host may increase its reproductive chances by staying for long periods on that host and waiting for females, because of the low chance of finding a second host if it detaches from the first. In these circumstances males must become quiescent and a female, newly attached to a host, may need to indicate her presence to waiting males. Again the behaviour of reptile ticks can be explained in terms of the biology of the host.

There are generally many more males than females on a host (Bull 1978a). What is unclear to me, in the light of current concepts of sexual selection, is why males do not signal to attract females in some way, or compete amongst themselves for mates. These questions require further investigation.

The species-specificity of the excitant signal allowed us to answer our two initial questions. The first concerned the reproductive interactions of two different species where their distributions contact at parapatric boundaries.

Since males of one species do not respond to females of another, interspecific mating attempts would be unlikely. We made an unexpected discovery, however, of another form or reproductive interaction, i.e. interspecific signal jamming (Andrews *et al.* 1982a). Males, which responded to conspecific females, did not respond when conspecific and non-conspecific females were attached to the same host. We suggested that the signal from one species interferes with the reception of signals from conspecific females.

This interspecific interaction may be an important mechanism in preventing co-existence. It is discussed in more detail later in this chapter. If the excitant phase of courtship is unique to Australian reptile ticks, this form of interaction could also be unique and explain why parapatric boundaries are found in no other groups of tick species.

The second initial question concerned the maintenance of genetical divergence between populations. We found that genetically similar populations within a species had compatible excitant signals even though geographically separated (Andrews & Bull 1982b). For instance, there was successful interpopulation mating between *Amb. albolimbatum* populations from 2000 km apart. On the other hand, ticks from Western Australia and morphologically similar *Ap. hydrosauri* from South Australia, had fixed genetical differences and incompatible excitant signals so that no interpopulation mating took place in experimental trials (Andrews & Bull 1983). Genetical differences between the populations would be preserved. This behavioural assay together with enzyme electrophoresis have been useful techniques for studying evolutionary divergence within the species of Australian reptile ticks.

Up until now I have concentrated on behavioural events leading up to contact between male and female ticks. In reptile ticks two further differences between the species were observed in the behaviour following contact.

The first was in the female response to male courtship. A female attached under a lizard scale can do little to discourage the attention of an unwanted courting male. The simple strategy of moving away, commonly adopted by reluctant females of other taxa, is not available to an attached female. But to complete courtship a male must have access to the female ventral surface, which normally lies against the lizard surface. Receptive females will lift their body at an angle to the host surface by 'standing up' with their last pair of legs, and will allow conspecific males to crawl under (Andrews & Bull 1980). However, in experimentally induced interspecific encounters, females did not respond, and kept their bodies pressed to the substrate (Andrews 1982). Presumably there is a signal transmitted by the male during the early stages of contact which allows the females to recognize conspecific males and respond appropriately. Such a species specific response by females to male courtship has not been previously reported in ticks.

A second interspecific difference is in leg arrangements. When the male crawls over the female ventral surface, he clasps her sides with his legs. In the final courtship position each species has a fixed and unique arrangement of male and female legs, such that the male mouth parts end up opposite to the female genital pore (Andrews 1982). In experimentally induced interspecific couplings, males adopted their usual leg arrangement which led to the non-alignment of the appropriate organs because the species differed in body dimensions (Andrews 1982). Thus even when males achieved the final stage of courtship with non-specific females, copulation was prevented.

In other taxa, differences have been described at several stages of the courtship behaviour of closely related species. It is not surprising that there should be several differences between the reptile tick species. It is more surprising that tick biologists have held the view for so long of little variation between tick species in mating behaviour. This may be because there have been few accounts of the intimate details of tick courtship until recently. The new discovery of species specific genital pheromones in *Dermacentor* species (Sonenshine *et al.* 1982) reinforces the view that differences in mating behaviour should be as widespread amongst tick species as it is in species of other taxa.

17.7 ECOLOGICAL INTERACTIONS

Ecology is another area in which our research on reptile ticks has led us in directions different from most other tick research. A major current controversy in ecology concerns the extent of interaction between species in a community and the role of interspecific competition in community structure. Most ecological studies of ticks have ignored community interactions and concentrated only on the dynamics of the tick population and its host.

There have been occasional references to tick predators (e.g. Ault & Elliott 1979) and parasites (Doube & Heath 1975), but few studies have considered interspecific interactions between tick species. An exception is the study by Norval & Short (1984) on the competitive replacement of *Boophilus decoloratus* by *B. microplus* in Zimbabwe.

In our studies of the reptile ticks in Australia we have been examining both interspecific competition and predation. An initial hypothesis to explain the dramatic tick boundaries in South Australia was that species competed with each other for limited resources. However, laboratory experiments (Sharrad 1980, Bull *et al.* 1981) failed to demostrate any major disadvantage either in attachment, feeding, or subsequent moulting for ticks on a host with another species. Furthermore, hosts in the field rarely carry the heavy infestation loads they can tolerate in the laboratory (Bull 1978a). Those lizards which do have heavy infestations show no evidence of greater resistance to subsequent infestations (Bull *et al.* 1981, and unpublished data). In natural tick populations, hosts are not a limiting resource. Whatever the interaction between the species which results in sharp parapatric boundaries, we do not consider it as the result of interspecific competition for food resources.

Interference competition can be effective between species even when resource levels are adequate. The tick species may compete through reproductive interference. I described earlier an experiment which showed that females of two species, when attached on the same host, appeared to prevent males from responding to conspecific excitant signals. A model to explain the importance of interference competition at the boundary can now be devised. Since lizard dispersal is slow, colonization by ticks across the boundary would be relatively rare. A female tick which has moved across the boundary would have a high probability of attaching to a host with females of the resident species also attached. Even if colonist males found the same host they would be prevented from recognizing the colonist female because of signal interference from residents. Thus any colonization across the boundary would rarely be reproductively successful.

So far we have not gathered evidence from the field to test this hypothesis (Andrews & Petney 1982), but more detailed studies are now in progress. However, one apparent consequence of the proposed signal interference, has resuted in a niche shift by ticks (Andrews *et al*. 1982b). *Ambloymma albolimbatum* males and females have a marked preference for attaching inside the ears of their host lizards. *Ap. hydrosauri*, over most of its distribution, attaches in the ears quite commonly, but also under the scales on the flanks and neck of the host. However, in two geographic regions immediately adjacent to the *Amb. albolimbatum* distribution, *Ap. hydrosauri* males have a significantly reduced frequency of attachment in host ears. This niche shift was retained by isolated laboratory stocks, implying a genetical basis for the behavioural change (Andrews *et al*. 1982b). We postulated that the move away from host ears evolved when those male *Ap. hydrosauri* which attached frequently encountered female *Amb. albolimbatum*. The subsequent signal interference reduced the probability of these males detecting conspecific females, providing a selective advantage to males which attached elsewhere.

A second area of tick ecology which we have started to investigate is the role of predation on population dynamics. We have been working at a site near Mt Mary in South Australia where the distributions of *Ap. hydrosauri* and *Amb. limbatum* meet. The site has a high density and diversity of ants (Chilton, unpublished data). Also we know that, at the site, experimentally released ticks in cages with predator access have higher mortality than ticks where predators are excluded (Bull, unpublished data), so the risk of predation is high.

Another risk for ticks while waiting for hosts is desiccation (Bull & Smyth 1973), particularly since most lizard activity (Bull 1978b) and most tick population growth (Bull & Sharrad 1980) takes place during the hot, dry summer months.

We believe that predation and desiccation have a major influence on reptile tick behaviour between bloodmeals. They bury themselves deep into the litter and remain quiescent (Petney & Bull 1984) and cannot be induced to quest in response to lizard odour, heat, or shadows (Downes 1984, Belan, personal comm.). Only physical disturbance has a stimulating effect (Downes 1984). They do not have a diurnal rhythm of regular searching activity which some mammalian ticks possess (e.g. Milne 1947), and under field conditions cannot be induced to attach to lizards as near as 50 cm (Petney *et al*. 1983). We proposed that sleepy lizard ticks are 'sit-and-wait' parasites, in contrast to many others which actively forage for hosts.

This ecological difference can again be explained by differences between the host species. Since sleepy lizards are slow-moving, solitary, and inactive for much of the year, ticks may have to wait long periods before encountering a host. Active questing would reap few rewards. Ticks which remain quiescent between meals will avoid both loss of energy and exposure to dehydration and predation. A property of lizards which makes this strategy favourable is their lowness to the ground. Unlike mammals, movement of lizards through litter will expose their ventral surface to the substrate, physically disturb the resting ticks and provide direct opportunities for attachment. A mammal tick does not have such immediate access and they cannot afford the luxury of concealment from predators and dehydration.

17.8 SUMMARY

In many respects the biology of the three ticks of sleepy lizards is similar to most other tick species. There are certain phylogenetic constraints on the range of adaptations possible in any taxa. However, I have tried to show some adaptations the ticks have made to their specific host species which have led to apparently unique characteristics.

Sleepy lizards are metabolically and physiologically slower than the endothermic hosts of many of the other ticks which have been studied in detail. Their body temperature drops at night and during cool seasons of the year. They move slowly and infrequently, have low dispersal patterns, and are generally solitary. They do not have permanent burrows or resting sites, but move from place to place. One consequence is that their development rates are generally slower than for other species (Bull *et al.* 1978, Sharrad 1980).

I have used host biology to explain unusual host-finding behaviour (sit-and-wait for host), and unusual mate-seeking behaviour by males (sit-and-wait on host until female attaches and advertises her receptiveness). I have used the host behaviour to explain the apparent anomolous genetic structure of the reptile tick population (extremely low genetic variation both between and within populations), and the unique distribution pattern (parapatric boundaries). I hypothesize that behavioural and ecological properties of tick species reflect the hosts they infest. I suggest that unusual properties found in Australian reptile ticks appear unusual only because little attention has been paid in the past to ticks of non-mammalian hosts.

ACKNOWLEDGEMENTS

This research has been supported by grants from the Australian Research Grants Scheme and from the Flinders University Research Budget. Many people have assisted in various stages of the work. They are too numerous to mention here, but their contribution has been acknowledged in the individual papers. Kathy White has typed this manuscript and many others on the subject of reptile ticks.

REFERENCES

Andrews, R. H. (1982) Mating behaviour and reproductive isolation of three species of reptile tick. *Anim. Behav.* **30** 514–524.

Andrews, R. H. & Bull, C. M. (1980) Mating behaviour in the Australian reptile tick *Aponomma hydrosauri. Anim. Behav.* **28** 1280–1286.

Andrews, R. H. & Bull, C. M. (1981) Inhibition of mating behaviour before feeding in the tick *Aponomma hydrosauri. Anim. Behav.* **29** 518–522.

Andrews, R. H. & Bull, C. M. (1982a) An attractant pheromone with common properties in three reptile ticks. *Experientia* **38** 99–100.

Andrews, R. H. & Bull, C. M. (1982b) A comparison of mating behaviour between populations of the reptile tick, *Aponomma hydrosauri* in South Australia. *Aust. J. Zool.* **30** 635–642.

Andrews, R. H. & Bull, C. M. (1983) Premating reproductive isolation between geographically isolated populations of an Australian reptile tick. *J. Parasitol.* **69** 1125–1130.

Andrews, R. H. & Petney, T. N. (1982) Competition for sites of attachment to hosts in three parapatric species of reptile ticks. *Oecologia* **51** 227–232.

Andrews, R. H., Petney, T. N. & Bull, C. M. (1982a) Reproductive interference between three parapatric species of reptile tick. *Oecologia* **52** 281–286.

Andrews, R. H., Petney, T. N. & Bull, C. M. (1982b) Niche changes between parasites populations: an example from reptile ticks. *Oecologia* **55** 71–80.

Andrews, R. H., Bull, C. M. & Orbach, J. (1986) The mode of transmission of the excitant sex pheromone in Australian reptile ticks. *J. Parasitol.* **72** (in press).

Ault, S. K. & Elliott, K. D. (1979) Spider predation (Araneae, Salticidae) on *Ornithodoros coriaceus* (Acarina, Argasidae), with a survey of the predators of the genus *Ornithodoros. J. Med. Entomol.* **15** 570–571.

Bull, C. M. (1978a) Heterogeneity of resource utilization in a population of the Australian reptile tick, *Aponomma hydrosauri* (Denny). *Ecol. Entomol.* **3** 171–179.

Bull, C. M. (1978b) Dispersal of the Australian reptile tick *Aponomma hydrosauri* by host movement. *Aust. J. Zool.* **26** 689–697.

Bull, C. M. & Andrews, R. H. (1984) Two different mating signals used by female reptile ticks. In: Griffiths, D. A. & Bowman, C. E. (eds.) *Acarology* **6** Vol. I Ellis Horwood Ltd, Chichester p. 427–430.

Bull, C. M. & King, D. R. (1981) A parapatric boundary between two species of reptile ticks in the Albany area, Western Australia. *Trans. Roy. Soc. S.A.* **105** 205–208.

Bull, C. M. & Sharrad, R. D. (1980) Seasonal activity of the reptile tick,

Aponomma hydrosauri (Denny) (Acari: Ixodidae) in experimental enclosures. *J. Aust. Ent. Soc.* **19** 47–52.

Bull, C. M. & Smyth, M. (1973) The distribution of three species of reptile ticks, *Aponomma hydrosauri* (Denny), *Amblyomma albolimbatum* Neumann, and *Amb. limbatum* Neumann. II. Water balance of nymphs and adults in relation to distribution *Aust. J. Zool.* **21** 103–110.

Bull, C. M., Sharrad, R. D. & Smyth,M. (1978) The pre-molt period of larvae and nymphs of the Australian reptile tick *Aponomma hydrosauri*. *Acarologia* **19** 593–600.

Bull, C. M., Sharrad, R. D. & Petney, T. N. (1981) Parapatric boundaries between Australian reptile ticks. *Proc. Ecol. Soc. Aust.* **11** 95–107.

Bull, C. M., Andrews, R. H. & Adams, M. (1984) Patterns of genetic variation in a group of parasites, the Australian reptile ticks. *Heredity* **53** 509–525.

Doube, B. M. & Heath, A. C. G. (1975) Observations on the biology and seasonal abundance of an encyrtid wasp, a parasite of ticks in Queensland. *J. Med. Entomol.* **12**, 443–447.

Downes, B. J. (1984) Host-location by the reptile tick *Aponomma hydrossauri*: the roles of physical disturbance and odor. *J. Parasitol.* **70** 164–166.

Healy, J. A. (1979a) Phosphoglucomutase polymorphism in the tick *Ixodes ricinus*. *Parasitology* **78** 7–18.

Healy, J. A. (1979b) Analysis of the α-glycerophosphate dehydrogenase variability in the tick *Ixodes ricinus* (Acari: Ixodidae). *Genetica* **50** 19–30.

Kellum, D. & Berger, R. S. (1977) Relationship of the occurrence and function of 2,6-dichlorophenol in two species of *Amblyomma* (Acari: Ixodidae). *J. Med. Entomol.* **13** 701– 705.

Milne, A. (1947) The ecology of the sheep tick, *Ixodes ricinus* L. Some further aspects to activity, seasonal and diurnal. *Parasitology* **30** 27–33.

Nevo, E. (1978) Genetic variation in natural populations: pattern and theory. *Theor. Pop. Biol.* **31** 121–177.

Norval, R. A. I. & Short, N. J. (1984) Interspecific competition between *Boophilus decoloratus* and *Boophilus microplus* in southern Africa. In: Griffiths, D. A. & Bowman, C. E. (eds.) *Acarology* **6** Vol. II. Ellis Horwood Ltd, Chichester, p. 1242–1246.

Odendaal, F. J. & Bull, C. M. (1982) A parapatric boundary between *Ranidella signifera* and *R. riparia* (Anura: Leptodactylidae) in South Australia. *Aust. J. Zool.* **30** 49–57.

Petney, T. N. & Bull, C. M. (1984) Microhabitat selection by two reptile ticks at their parapatric boundary. Aust. J. Ecol. **9** 233–239.

Petney, T. N., Bull, C. M. & Andrews, R. H. (1982). A stable boundary between two species of reptile ticks on Eyre Peninsula, South Australia. *Trans, Roy. Soc. S. Aust.* **106** 159–161.

Petney, T. N., Andrews, R. H. & Bull, C. M. (1983) Movement and host finding by unfed nymphs of two Australian reptile ticks. *Aust. J. Zool.* **31** 717–721.

Price, P. W. (1977) General concepts on the evolutionary biology of parasites. *Evolution* **31** 405–420.

Roberts, F. H. S. (1970) *Australian ticks*. C.S.I.R.O.: Melbourne.

Satrawaha, R. (1980) Ecology and activity patterns of the lizard *Trachydosaurus rugosus*. MSc thesis: Flinders University.

Satrawaha, R. & Bull, C. M. (1981) The area occupied by an omnivorous lizard *Trachydosaurus rugosus*. *Aust. Wildlife Research* **8** 435–442.

Selander, R. K. (1970) Behaviour and genetic variation in natural populations. *Amer. Zool.* **10** 53–66.

Sharrad, R. D. (1980) Studies of factors which determine the parapatric distributions of three species of South Australian reptile ticks. PhD thesis: University of Adelaide.

Sharrad, R. D. & King, D. R. (1981) The geographical distribution of reptile ticks in Western Australia. *Aust. J. Zool.* **29** 861–873.

Smyth, M. (1973) The distribution of three species of reptile ticks, *Aponomma hydrosauri* (Denny), *Amblyomma albolimbatum* Neumann, and *Amb. limbatum* Neumann, I. Distribution and hosts. *Aust. J. Zool.* **21** 91–101.

Sonenshine, D. E., Silverstein, R. M., Plummer, E., West, J. R. & McCullough, T. (1976). 2,6-dichlorophenol, the sex pheromone of the Rocky Mountain wood tick, *Dermacentor andersoni* Stiles and the American dog tick *Dermacentor variabilis* (Say). *J. Chem. Ecol.* **2** 201–209.

Sonenshine, D. E., Khalil, G. M., Homsher, P. J. & Mason, S. N. (1982). *Dermacentor variabilis* and *Dermacentor andersoni*: genital sex pheromones. *Exp. Parasitol.* **54** 317–330.

18

Behavioral ecology of *Amblyomma americanum* (L.)

Jakie A. Hair and Jerry L. Bowman

18.1 INTRODUCTION

Amblyomma americanum (L.), commonly known as the lone star tick owing to the silvery spot on its dorsal surface, has the distinction of being the first tick described in the United States (Bequaret 1946). The description was given in 1754. In 1739, according to Bequaert, Thomas Salmon wrote the following on the presence and pestilence of the lone star tick in his description of Virginia (See Oudemans 1926):

> Seed-ticks and redworms are small insects that annoy the people by day as musqueto's and chinches do by night; but both these keep out of your way if you keep out of theirs; for seed ticks are no where to be met with but in the track of cattle, upon which the great ticks fasten and fill their skin's so full of blood that they drop off, and where ever they happen to fall they produce a kind of egg, which lies about a fort-night before the seedlings are hatch'd. These seedlings run in swarms up the next blade of grass that lies in their way, and then the first thing that brushes the blade of grass gathers off most of these vermine, which stick like burs upon anything that touches them...

Although two centuries have passed, Salmon's description as recorded above adequately introduces one of the earliest recognized serious arthropod pests of man and animals in the United States. In my opinion, only *Dermacentor variablis* (Say) approaches *A. americanum* in overall importance as a tick pest of man and animals in the United States.

The lone star tick is distributed from as far west as west-central Texas, north to Missouri and eastwards, infesting parts of most states where abundant hosts and 'scrub-brush' habitat exists (Fig. 18.1). Historically,

Fig. 18.1 — Probable distribution of lone star tick populations of economic importance in the United States.

epidemic-like populations have been associated with definable habitat-types and significant populations of large ungulates, specifically white-tailed deer (*Odocoileus virginanus* (Boddaert)) and/or domestic cattle (*Bos taurus* (L.)). All parasitic stages are non-specific as to host preference, but most adults infest medium to large mammals.

Although reference is frequently made to this species as a vector of disease, its greatest impact on man and animals results from its tenacious feeding behavior and the resulting inflammations. The latter invariably result in bacterially infected pus-filled lesions or physiological toxicosis.

Because of its importance as a pest of man and his possessions in the south-central portion of the United States, extensive work has been done on the ecology, behavior, and control in the lower Ozark region. Much of this work will be summarized in this chapter.

18.2 HOST SPECIFICITY

A survey of common mammals and birds by Clymer *et al.* (1970) showed this species to comprise 92% of all ticks removed from examined hosts. One or more stages of the lone star tick were found on nearly every species of animal sampled, and also one or more stages appeared throughout the year regardless of the season. Little host specificity was demonstrated, and the extent of parasitism depended on the size of the host and the host habitat. Clymer's data agree with those of Mohr & Stumpf (1962) and Worth (1950), which showed that larger animals were more heavily parasitized than smaller animals. Those ranging widely were also more heavily parasitized. The immature stages of *A. americanum* parasitized all sizes of mammals and birds, while the adults were found predominantly on medium to large animals. Large populations of ticks were often found in or near the bedding areas of animals.

18.3 HOST AND HABITAT AS REGULATORS OF LONE STAR TICKS

Several published works deal with *A. americanum* feeding behavior and fecundity on bovine and white-tailed deer (*Odocoileus virginanus* (Boddaert)). These are the two more important hosts for lone star tick adults. Reference is made by Garris *et al.* (1979) to differences in cattle breed susceptability to this tick. These workers noted that as compared to Hereford cattle, Brahman supported fewer larvae, nymphs, and females at various stages of engorgement. Lone star ticks removed from the Brahman cattle were lighter, females laid fewer eggs, and fewer immature ticks molted.

It is assummed that host range and feeding behavior, grooming ability, and natural and acquired immunity play important roles in host acquisition and the feeding success of lone star ticks. Although extensive use of crossbreeding of cattle is done in Oklahoma for the purpose of combating lone star ticks, epidemic level populations occur where cattle and deer cohabit woodlot range. Equally as abundant, it seems, are those number of *A. americanum* sometimes associated with white-tailed deer in wildlife refuges exempt from cattle use (Hair 1968, Partick & Hair, 1977). Patrick & Hair (1978) made a study of the utilization of different habitats by white-tailed deer and its influence on lone star tick populations. Their findings showed, however, that host utilization of specific habitats does not necessarily correlate with the abundance and distribution of the tick. Uncorrelated to usage patterns, habitat types of upland and bottom oak–hickory supported larger populations of all developmental stages of this tick than did an open, more exposed meadow habitat. High numbers of replete female ticks were found on deer when meadow habitat was receiving its highest use. This should have been reflected by the presence of large numbers of larvae in meadows in July, August, and September. However, the high temperatures and low humidities that replete females were subjected to in such meadows resulted in their death, no oviposition, low-egg hatch, or low survival of the larvae (Patrick & Hair 1978). That larvae of this species have such low survivorship under meadow conditions during the summer is well documented (Robertson 1975b). Habitat type was shown to be an important factor in lone star tick population build-up irrespective of host abundance and distribution (Patrick & Hair 1978). For example, these workers showed bottom oak–hickory habitat as receiving considerable less deer use than upland oak–hickory, yet tick abundance was similar. In this study tick population numbers in the bottom-land remained substantially constant. However, because of greater extremes in the upland environment, tick populations there were more erratic.

The summary given by Patrick & Hair (1978) details the important factors in lone star tick population maintenance and build-up. In an ideal habitat, heavy deer or cattle use can result in ‘epidemic’ populations. In many areas, if we assume adequate host abundance, lone star populations are regulated by (1) available host food supply, which in turn influences grazing and browsing activity, and (2) physiography of the area which provides environmental protection for the host and conserves moisture through the presence of canopy and ground cover. The above study suggested that, upland oak-hickory habitat would normally, support large numbers of ticks under average Ozark Mountain climatic conditions, and would be suitable for greater tick longevity. However, when climatic conditions produce dessicating conditions, even with continued high deer use, this habitat type would support very low populations of ticks as compared to similar areas receiving more rainfall or less desiccation. In contrast, bottom oak–hickory habitat was more capable of consistently supporting large populations of ticks even under less suitable climatic conditions. Since the bottom oak–hickory produced low mass for deer, the net result was not necessarily high populations in the bottom-land, but a longer lived population. A harsher habitat, the meadow, received high deer use, but climatic conditions were such that ticks did not survive. Therefore, any possible build-up of ticks was precluded. Although deer use introduced the lone star tick to various habitats, climatic factors ultimately controlled the ticks’ ability to survive, and consequently were of major importance in population build-up.

18.4 SEASONAL ABUNDANCE ON WHITE-TAILED DEER

The earliest significant activity of lone star ticks begins in mid-February when average ambient temperatures are 10–12 °C. Of course, soil surface temperatures may rise significantly above this. Peak populations occur in early June and then decline until almost none are observed in late October

(Fig. 18.2). Replete females peak in late June and decline until the end of August. Fig. 18.2 illustrates a seasonal repression of repletion for female lone star ticks. This phenomenon was also observed during other years. Even though large numbers of flat females (100 mg) were feeding early in the summer, repletion (300–900 mg) did not begin until May. Engorgement appeared to be a photoperiod response that resulted in a narrow replete drop-off period and subsequent larval hatch period. Male lone star ticks are found on deer throughout the year.

The period of nymphal infestation on deer is longer than any other stage and occurred from March to October with no distinct single peak in activity. This is probably due to the free-living nymphal activity patterns observed by Semtner & Hair (1973a). They indicated that nymphal activity occurred from March to October with no general peak as observed for adult ticks. They noted that peak nymphal activity varied by as much as a month from one habitat type to another, thus representing a continuous exposure to the wide-ranging white-tailed deer. The same repression of repletion was observed for the nymphal stage as reported for females, in that significant feeding did not begin until June of each year (Fig. 18.3).

Larvae first appear on deer in late June and peak in late July or early August. This activity may continue until mid-November (Fig. 18.4).

18.5 MACROHABITAT AND TICK ABUNDANCE

Certain plant associations support higher tick populations than others (Table 18.1). The largest numbers of adult and nymphal ticks generally occur in persimmon-sassafras groves or in winged elm woods. These two types of habitat are frequently found between meadows and the main forest body and are generally the result of secondary 'old field' succession (Bazzaz 1968). With the exception of persimmon–sassafras, small openings in woods and old fire trails within woodlots often have higher tick populations than other parts of a particular woodlot.

To a great extent nymphal and adult lone star ticks are concentrated in the ecotone (Fig. 18.5). Samples taken in prairies, woodlots, and woods indicated a direct correlation between tick abundance and proximity of the sample to the ecotone. For example, Semtner *et al.* (1971) showed only 12% of all adults collected were found outside the ecotone proper.

18.6 ACTIVITY AND SURVIVAL OF ADULT TICKS IN DIFFERENT ECOLOGICAL HABITATS

Semtner *et al.* (1971b) showed a correlation between the survival of mixed-aged adult lone star ticks and habitat type. The average longevity of caged, field collected adults was less than 22 days in a meadow habitat, 35 days in persimmon and upland oak–hickory, and more than 65 days in bottom-land

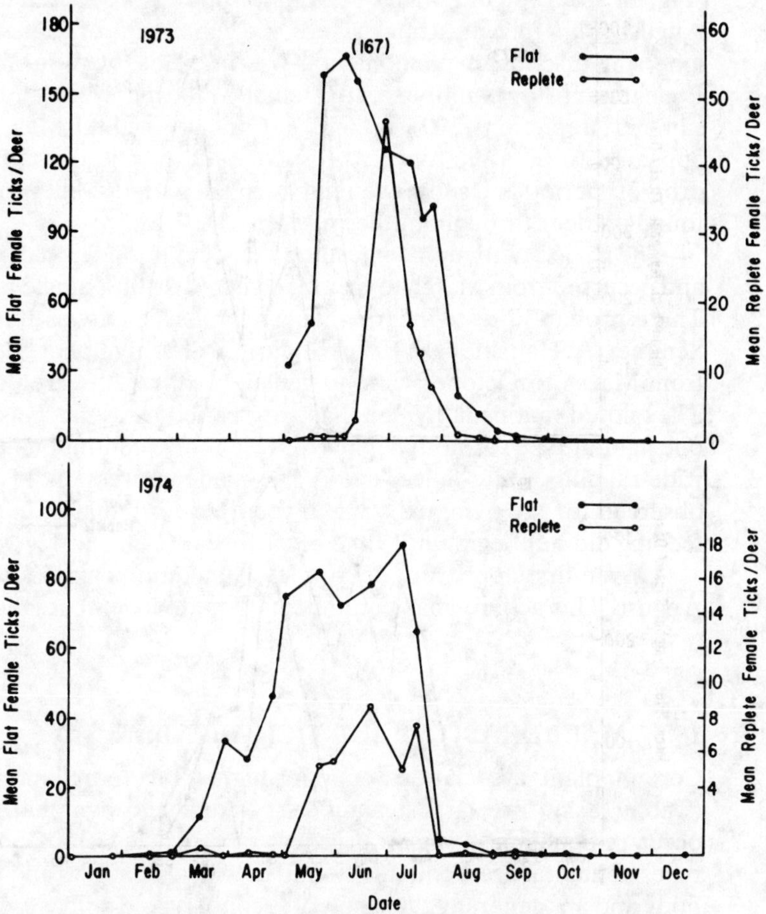

Fig. 18.2 — Mean number of flat and replete female ticks on the areas of deer in the Cookson Hills State Game Refuge in easter Oklahoma, 1973 and 1974.

forest (Fig. 18.6). Temperature and humidity measurements in the four habitats indicated very high day temperatures and low humidities in the meadow with considerable cooling and a rise in relative humidity during the night. Based on survivorship curves, this habitat was considered to be the most undesirable as a tick habitat. Bottom oak–hickory habitat generally showed the lowest day temperatures and highest humidities, and on a daily and seasonal basis this habitat showed less variability among the parameters measured.

High midsummer temperatures influence the behavior of adult ticks in all habitat types. This condition probably results in a shorter longevity in certain habitats. As a behavioral response to rising temperatures during mid- and late July, adults migrate down the vegetation to the soil and/or duff.

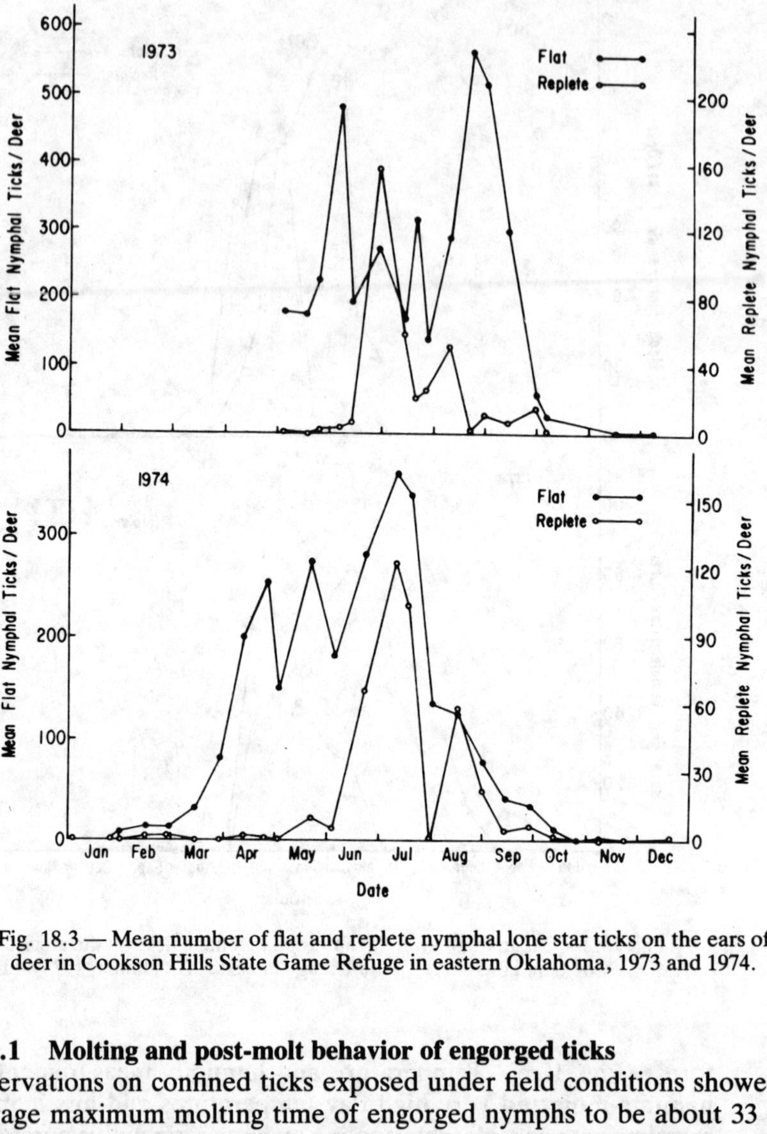

Fig. 18.3 — Mean number of flat and replete nymphal lone star ticks on the ears of deer in Cookson Hills State Game Refuge in eastern Oklahoma, 1973 and 1974.

18.6.1 Molting and post-molt behavior of engorged ticks

Observations on confined ticks exposed under field conditions showed an average maximum molting time of engorged nymphs to be about 33 days during most summer months (Fig. 18.7). Cooler temperatures either retarded or prevented molting. Adults from nymphs molting prior to mid-July were active for a very limited time that summer, but mid-to-late season molting produced quiescent adults. Bottom woods apparently provided the best environment for overwintering adults, and a greater percentage of ticks were active in this habitat type as compared to others studied (Fig. 18.8).

18.6.2 Activity of adults

Hair *et al.* (1975) demonstrated tht *A. americanum* is highly dependent on moist microenvironments because of its great susceptibility to percentage losses of total body water, and drop in hemolymph volume at low humidi-

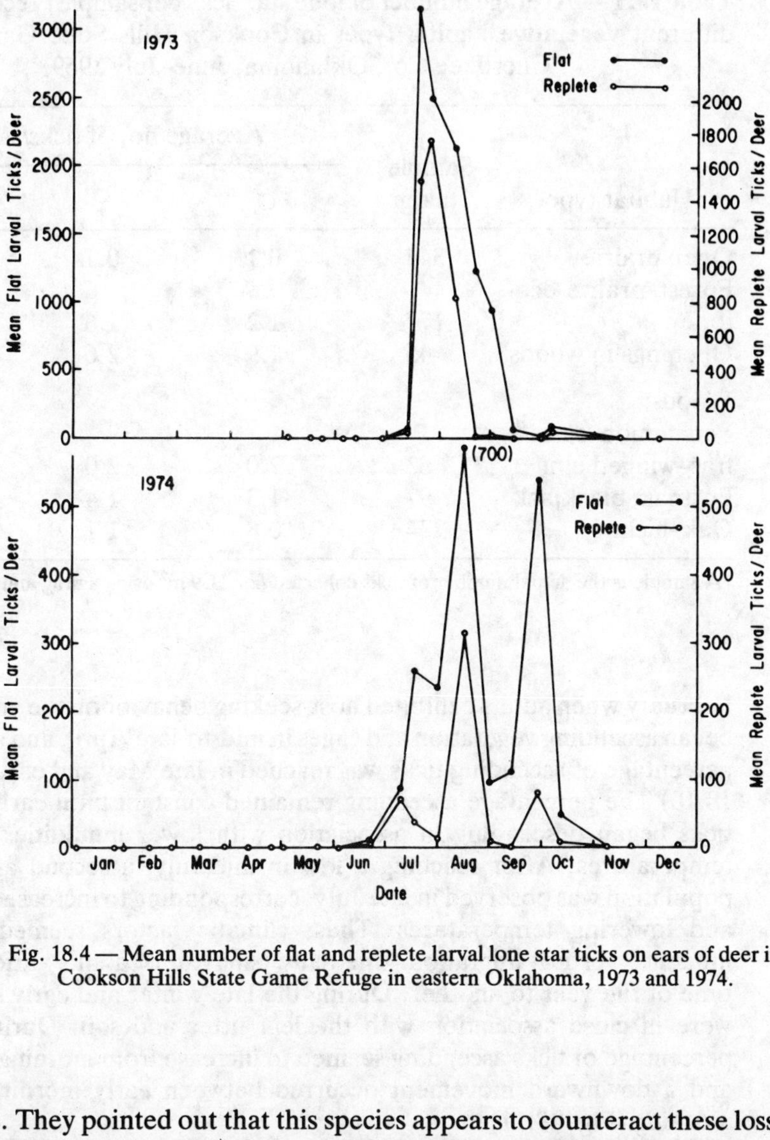

Fig. 18.4 — Mean number of flat and replete larval lone star ticks on ears of deer in Cookson Hills State Game Refuge in eastern Oklahoma, 1973 and 1974.

ties. They pointed out that this species appears to counteract these losses as adults by seeking environments with relative humidities above its critical equilibrium humidity (CEH) and that it sought these conditions very efficiently after water losses of ca. 10–15% of its hydrated weight (Fig. 18.9). Numerous works, including Hoch *et al.* (1971), illustrate that lone star ticks perish rapidly when their humid refuge is disrupted. The work of Semtner & Hair (1973b) illustrate some of the behavioral activity by adults within their microhabitat to avoid adverse daily and seasonal fluctuations of relative humidity and temperatures within several different habitats. When caged adults were exposed to four prominent vegetative associations within the study area, ticks were least active in the meadow habitat and most active in the more protected bottom woods (Fig. 18.8). Seasonal activity began in

Table 18.1 — Average number of lone star ticks per sample† recovered from different vegetative habitat types in Cookson Hills State Game Refuge, Cherokee Co., Oklahoma, June–July 1969.

Habitat type	Samples taken	Average no. of ticks/sample ♂	♀	Nymphs
Open prairie	344	0.1	0.1	0.9
Forest prairie ecotone	177	1.2	1.6	8.5
Openings in woods	45	1.9	2.6	24.1
Woods:				
Persimmon–sassafras–winged elm	83	2.0	2.0	68.4
Postoak–blackjack	97	1.3	1.4	16.8
Oak–hickory	234	0.8	1.1	8.9

†A sample is the actual number of ticks collected per 21.9 m^2 using a drag and sweep.

February when adults exhibited host-seeking behavior on the ground. Ticks began ascending vegetation and cages in mid-to-late April, and a peak in the percentage of ascending ticks was reached in late May and early June (Fig. 18.10) The percentage ascending remained constant until early July when ticks began descending in association with lower humidities and higher temperatures. After reaching a low in mid-July, a second ascent of the population was observed in late July, corresponding to increased humidities and lowering temperatures. Thus, climatic factors seemed to greatly influence vertical migration. The daily behavior of adults varied from one time of the year to another. During the late winter and early spring, ticks were in close association with the leaf litter and soil. During May the percentage of ticks ascending seemed to increase from morning to evening, and a downward movement occurred between early morning and mid-morning (Fig. 18.11).

Semtner & Hair (1973a, b) also calculated 'expected activity' for adults under different habitat conditions (Modified from Cox 1972). This period was calculated not only from the expected survival time, but also considered when they became inactive owing to high temperatures, photoperiodic changes, or other unfavorable factors. It is obvious that this is an important consideration in studying the population dynamics of this species. Little is known or has been published on the host-seeking behavior of this species. Semtner & Hair noted that the expected activity of those kicks in cages during early June, by habitat, was as follows: meadow, 30.1 days; persimmon, 36.7 days; upland oak–hickory, 38.7 days; and bottom woods, 41.2 days (Table 18.2).

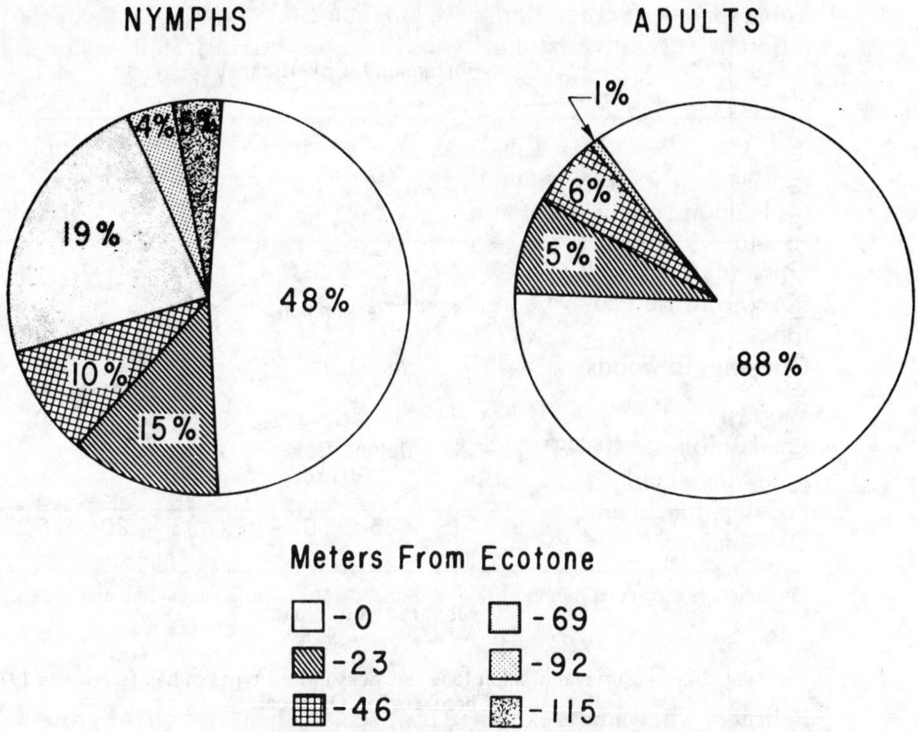

Fig. 18.5 — Percentages of nymphal and adult *Amblyomma americanum* occurring in prairies at various intervals from forest–prairie ecotones in Cherokee Co., Oklahoma, summer 1969.

When Semtner & Hair (1973b) surveyed different vegetative associations for the different development stages of *A. americanum*, distinct activity patterns were noted for each stage, depending on habitat type and thus environmental conditions (Figs. 18.12, 18.13 and 18.14). For example, they compared sassafras (11), black oak (07) and post oak (08), and noted drastic differences in nymphal activity during July and August in these habitats. Though not measured in their study these activity patterns were surely regulated by differing humidity and temperature regimes. They concluded that even though considerable differences occur in tick abundance in plots of similar plant association, high tick populations occur more frequently in certain associations than in others.

Robertson *et al.* (1975a) gave data that implied greater host-seeking activity by adults when placed in adverse habitats. Their comparison of active ticks before and after disturbance (Fig. 18.15) indicated that a higher percentage of adults were active in the more unfavorable meadow. They attributed the more active behavior which was distincly different from that observed in more favorable habitats, to extreme soil surface temperatures (>40 °C). They also thought that high numbers of adult lone star ticks in

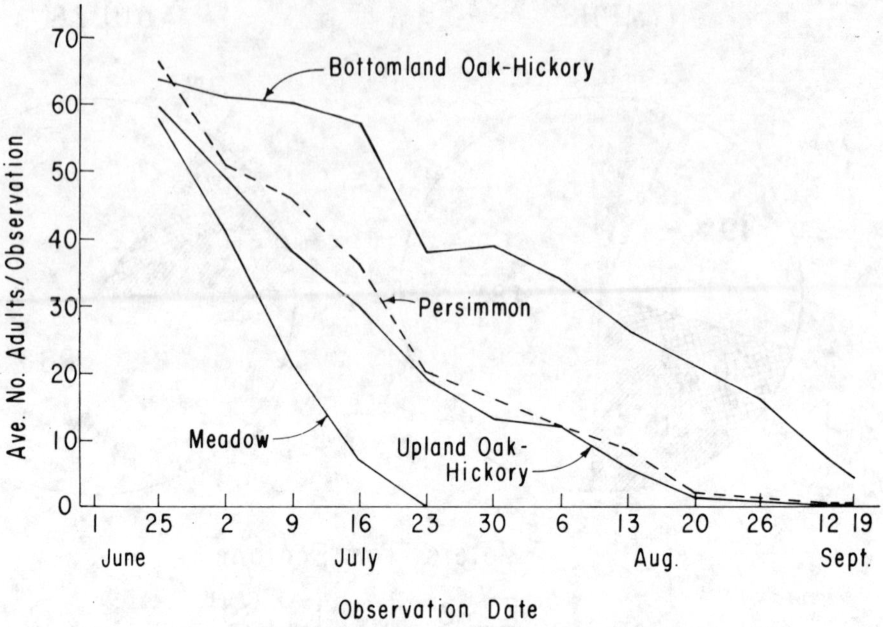

Fig. 18.6 — Survival of adult lone star ticks in four types of habitats during 1970, Cherokee Co., Oklahoma.

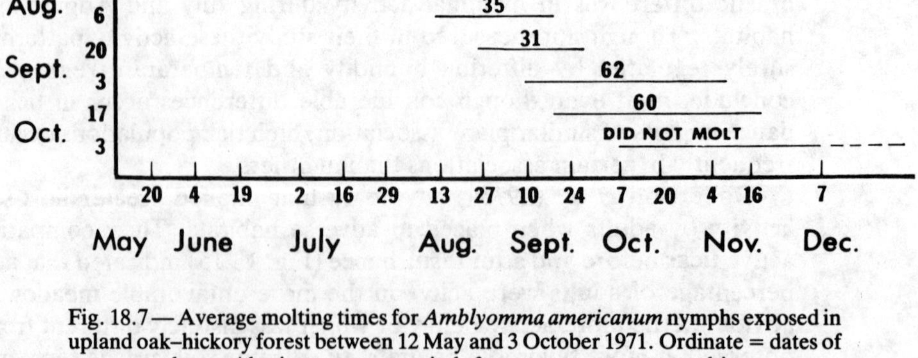

Fig. 18.7 — Average molting times for *Amblyomma americanum* nymphs exposed in upland oak–hickory forest between 12 May and 3 October 1971. Ordinate = dates of release. Abscissa = mean periods from engorgement to molting.

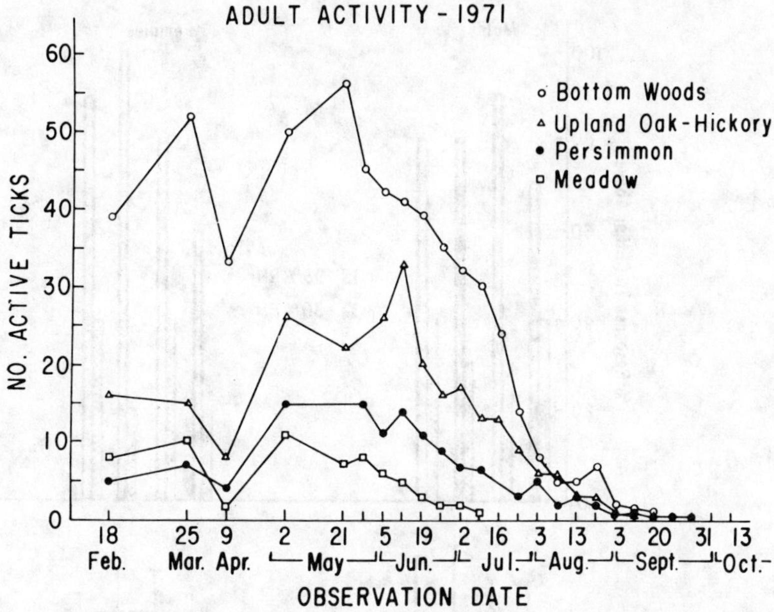

Fig. 18.8 — Average number of active adult lone star ticks per cage during 1971 of ticks released in late June of 1970 in four different habitats.

meadows which were observed at the vegetation tips from mid-June to mid-July were selecting a zone that was characterized by lower temperatures and higher humidities than could be found at the plant bases. Their conjecture was that the high mortality experienced by ticks in meadows may have been a result of high temperatures amplifying the effects of low humidities. This was previously pointed out by MacLeod (1935) while working with *Ixodes ricinus,* in that he noted that as temperature increased, the effect of humidity in prolonging the survival period steadily diminished.

Although photoperiod appears to have a bearing on the initiation of questing by lone stars ticks (Robertson *et al.* 1975a), subsequent host seeking or activity is regulated predominantly by temperature and humidity fluctuations. These workers also made observations on the daily activity patterns of individual, marked ticks. The duration of activity is normally considerably shorter than the length of the whole life span. In their study male adult ticks were active an average of 36 days, while females were active an average 52 days (Table 18.3). There appeared to be no regularity to successive active phases (Fig. 18.16). The quiescent phase of 9.4 days for both males and females was usually shorter than the active phases of 13.1 days for males and 15.1 days for females. Robertson and his colleagues thought that these behavioral changes were probably related to body water reserves. Body water loss of 10–15% of the original weight is enough to elicit a response to high humidities in a humidity gradient chamber (Hair *et al.*

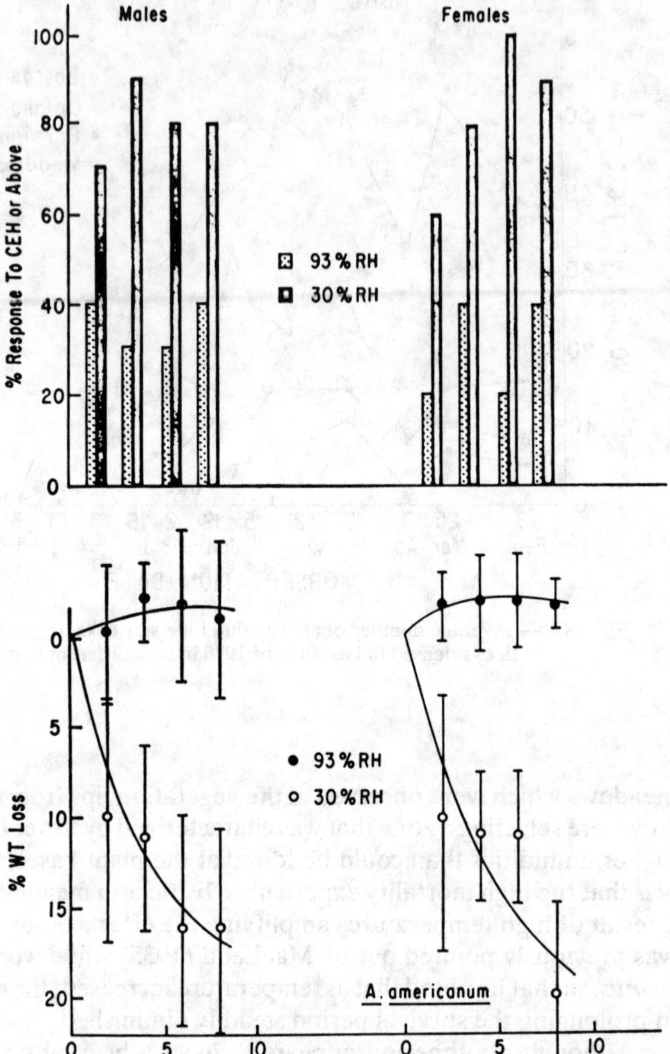

Fig. 18.9 — Percentage response of *A. americanum* to humidities equal to, or above, the critical equilibrium humidity (CEH) when predesiccated at 30% RH or held at 93% RH to achieve varying weight loss percentages.

1975). This behavior prevents excessive desiccation during prolonged periods of activity, enables the lone star tick to remain at the tips of vegetation for more total days than would otherwise be possible, and exposes the tick to the host for a much longer period.

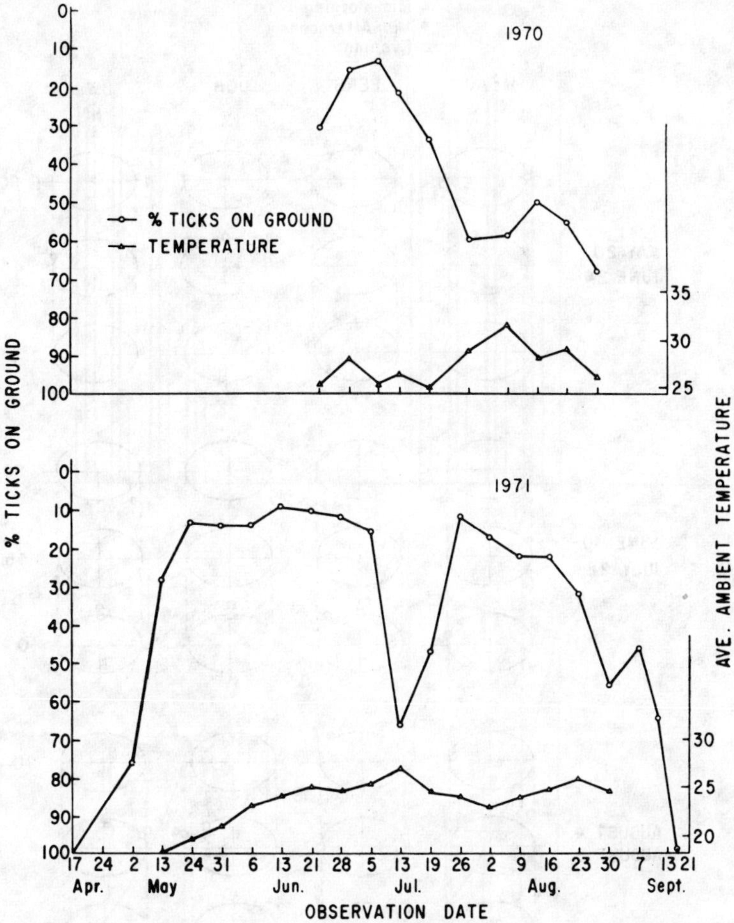

Fig. 18.10 — Seasonal vertical migration of adult lone star ticks in all habitats during 1970 (abstracted from Semtner *et al.* 1971a) and 1971 in relation to the weekly mean temperature.

An excellent review of pre- and post-molt behavior of the immature stages of this tick in different ecological associations is given by Robertson *et al.* (1975b) (Figs 18.17, 18.18). They noted that engorged nymphs and engorged larvae had limited mobility after release and sought shelter in plant litter and debris near the point of release. The molting times of nymphs ranged from 16 to 66 days. The post-molt developmental phase was from 8 to 15 days, depending on engorgement season. Those nymphs after mid-July remained inactive until the following spring. In general, habitat type affected molting time. Of those nymphs molting to adults prior to 15 July,

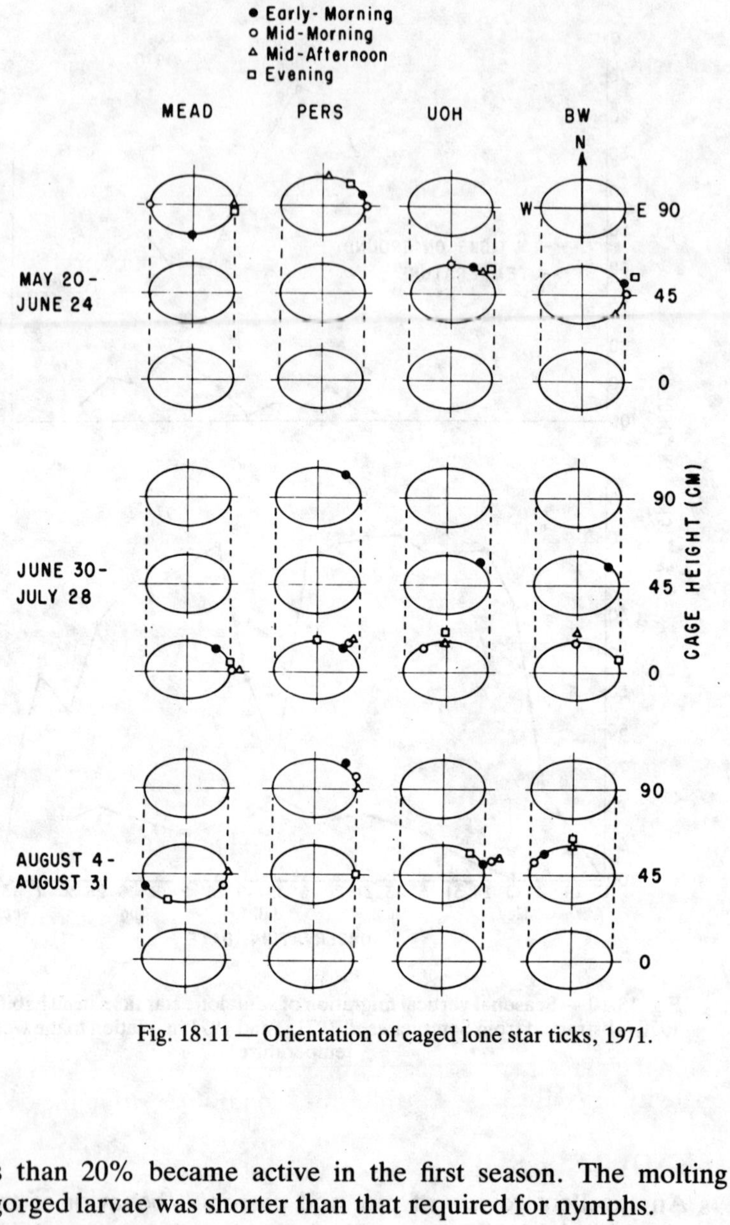

Fig. 18.11 — Orientation of caged lone star ticks, 1971.

less than 20% became active in the first season. The molting time of engorged larvae was shorter than that required for nymphs.

18.7 SUMMARY

The lone star tick is a mult-host tick that attacks a wide range of avian and mammalian hosts. Its rise to prominence as a severe pest is generally dictated by an abundance of large mammals such as cattle and/or deer inhabiting brushy-type vegetative associations. Its abundance in these associations, of course depends firstly on hosts such as deer or cattle, but equally

Table 18.2 — Expected activity† time (in days) of field-collected adult lone star ticks in 4 habitats at weekly intervals during late spring and summer of 1971.

	Habitat			
Date	Meadow	Persimmon	Upland oak–hickory	Bottom woods
4 June	30	37	39	41
10	25	32	32	38
17	19	27	29	33
24	16	25	25	31
30	13	23	21	28
7July	11	17	17	26
14	10	19	20	26
21	15	19	22	31
28	9	17	20	24
4 August	10	15	19	21
10	13	12	18	16
17	18	13	16	12
24	11	10	12	11
31	4	10	11	12
7 Sept.		5	12	9
14		2	7	9
21			4	6
28				4

†Expected activity = an estimate of the average number of days of tick activity remaining in specific habitats. (Modified from Cox 1972).

as important are the climatological conditions, especially relative humdity, within the plant associations. Ultimately egg-laying, molting, behavior, and longevity are determined by moisture conditions within the ticks' habitat.

ACKNOWLEDGEMENTS

Data referenced herein were abstracted from the works of the outstanding young men who have worked with the Oklahoma Agricultural Experiment Station on this serious arthropod pest. We are particularly appreciative of their labors and contributions to management strategies for this pest.

REFERENCES

Bazzaz, F. A. (1968) Succession on abandoned fields in the Shawnee Hills, Southern Illinois. *Ecology* **49:** 924–36.

Bequaret, J. C. (1946) The ticks, or Ixodoidea, of the northeastern United States and eastern Canada. *Entomol. Amer. XXV* (2): 73–120.

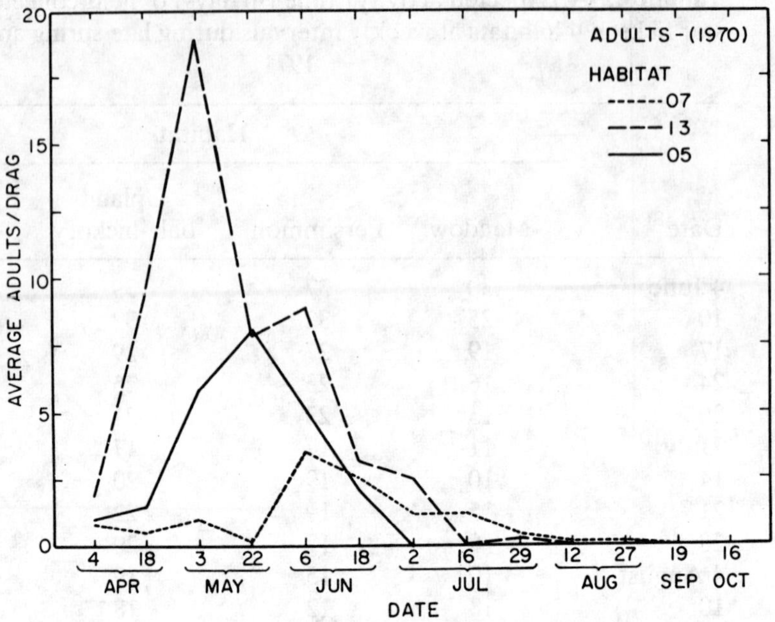

Fig. 18.12 — Seasonal activity of adult lone star ticks in different habitats of Cherokee Co., Oklahoma, 1970.

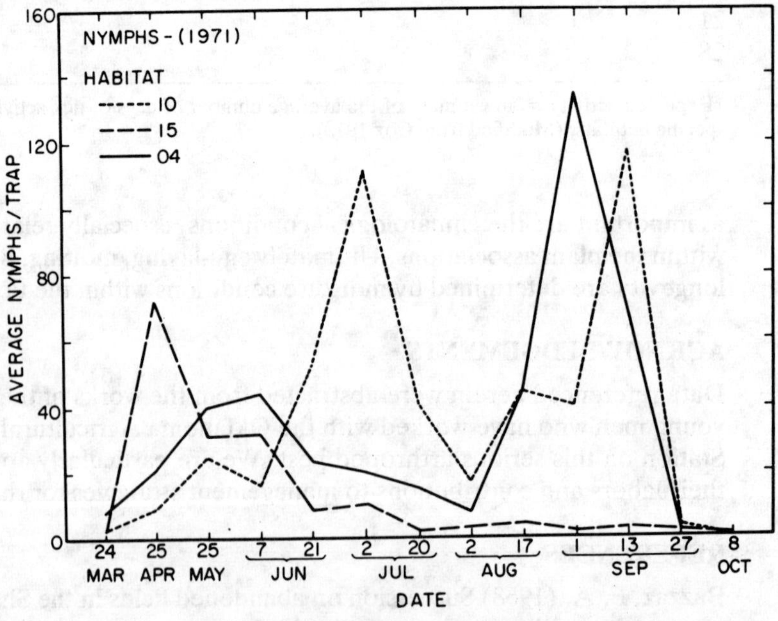

Fig. 18.13 — Seasonal activity of nymphal lone star ticks in different habitats of Cherokee Co., Oklahoma, 1971.

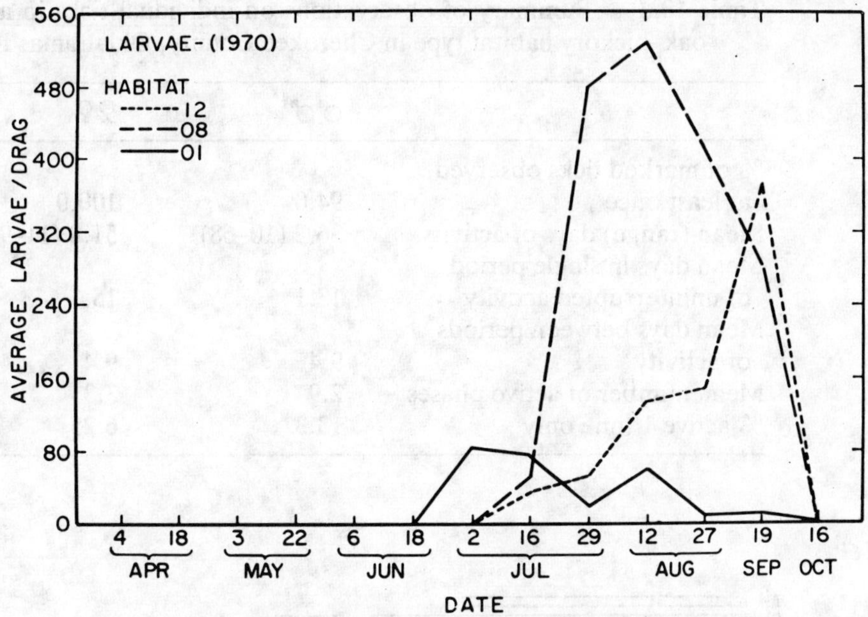

Fig. 18.14 — Seasonal activity of larval lone star ticks in different habitats of Cherokee Co., Oklahoma, 1970.

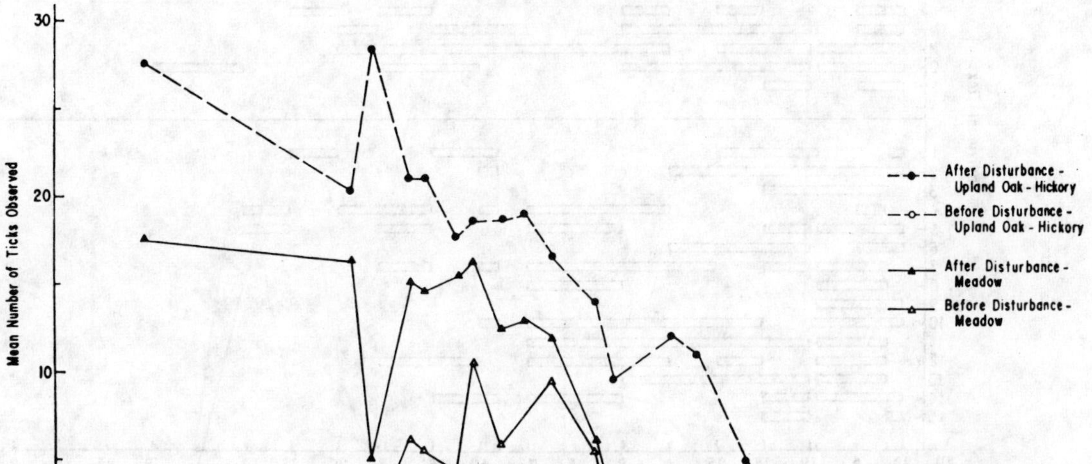

Fig. 18.15 — Activity of adult lone star ticks in two habitat types in Cherokee County, Oklahoma, 1973.

Table 18.3 — Summary of observations on individual ticks in an upland oak–hickory habitat type in Cherokee County, Oklahoma, 1972.

	♂♂	♀♀
% of marked ticks observed at least once	94.0	100.0
Mean (range) days of activity	36.2 (10–68)	51.8 (11–99)
Mean days in single period of uninterrupted activity	13.1	15.1
Mean days between periods of activity	9.4	9.4
Mean number of active phases	2.9	3.3
% active 1 time only	13.3	6.25

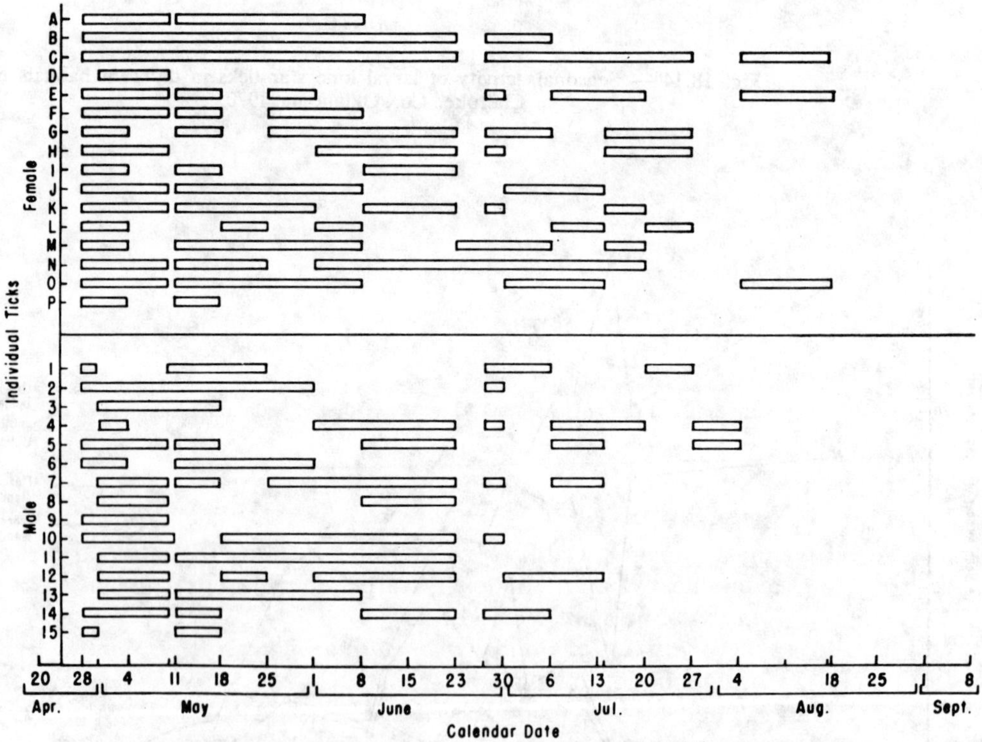

Fig. 18.16 — Activity records of individual ticks observed in an upland oak–hickory habitat type in Cherokee County, Oklahoma, 1972.

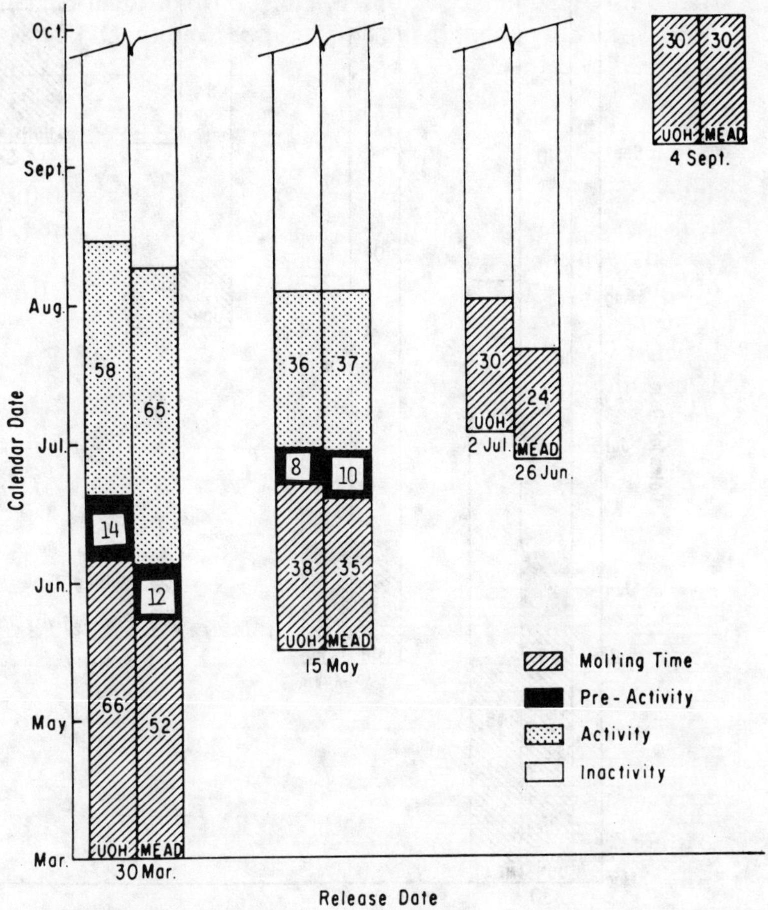

Fig. 18.17 — Behavior of engorged nymphal lone star ticks released in meadow and upland oak–hickory habitats, Cherokee County, Oklahoma, summer 1973.

Clymer, B. C., Howell, D. E. & Hair, J. A. (1970) Animal hosts of economically important ticks (Acarina) in east central Oklahoma. *Ann. Entomol. Soc. Amer.* **63:** 612–4.

Cox, G. W. (1972) *Laboratory manual of general ecology*. Wm. C. Brown Co., Dubuque, Iowa. p. 103–09.

Garris, Glen I., Stacey, B. R., Hair, J. A. & McNew, R. W. (1979) A comparison of lone star ticks on Brahman and Hereford cattle. *J. Econ. Entomol.* **72:** 869–872.

Hair, J. A. (1968) *Ticks can kill. Outdoor Oklahoma* **24:** 2–7.

Hair, J. A., Sauer, J. R. & Durham, K. A. (1975) Water balance and humidity preference in three species of ticks. *J. Med. Entomol.* **12:** 37–47.

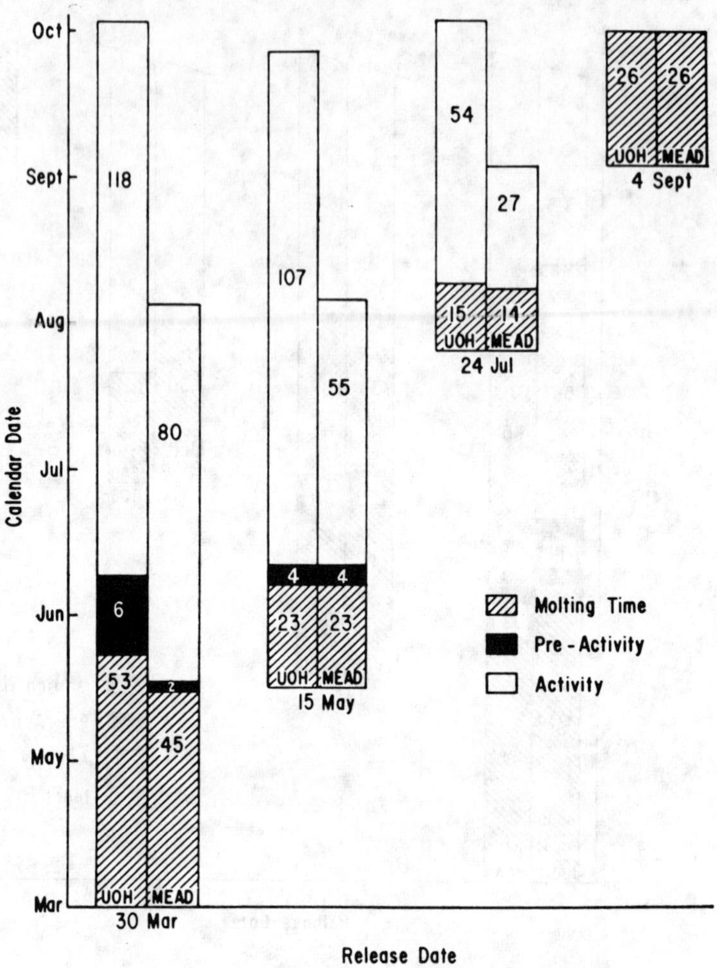

Fig. 18.18 — Behavior of engorged larval lone star ticks released in meadow and upland oak–hickory habitats, Cherokee County, Oklahoma, summer 1973.

Hoch, A. L., Barker, R. W. & Hair, J. A. (1971) Measurement of physical parameters to determine the suitability of modified woodlots as lone star tick habitat. *J. Med. Entomol.* **8:** 725–30.

MacLeod, J. (1935) *Ixodes ricinus* L. in relation to its physical environment. II. The factors governing survival and activity. *Parasitology* **27:** 123–44.

Mohr, C. O. & Stumpf, W. A. (1962) Relation of ectoparasitic load to host size and home area in small mammals and birds. *N. Amer. Wildlife Natur. Resour. Conf. Trans.* **27:** 174–83.

Oudemans, A. C. (1926) Kritisch historisch overzicht der Acarologie. I. 850 v. Chr.-1758. *Tijdschr. v. Entomol.* **69** Suppl. VIII 500 pp.

Patrick, C. D. & Hair, J. A. (1977) Seasonal abundance of lone star ticks on white-tailed deer. *Environ. Entomol.* **6:** 263–9.

Patrick, C. D. & Hair, J. A. (1978) White-tailed deer utilization of different habitats and its influence on lone star tick populations. *J. Parasitol.* **64:** 1100–06.

Robertson, A. S., Patrick, C. D., Semtner, P. J. & Hair, J. A. (1975a) The ecology and behaviour of the lone star tick (Acarina: Ixodidae). VI. Response of unfed adults to certain environmental parameters. *J. Med. Entomol.* **12:** 525–9.

Robertson, A. S., Patrick, C. D., Semtner, P. J. & Hair, J. A. (1975b) The ecology and behavior of the lone star tick (Acarina: Ixodidae). VII. Pre- and post-molt behavior of engorged nymphs and larvae. *J. Med. Entomol.* **12:** 530–4.

Semtner, P. J. & Hair, J. A. (1973a) The ecology and behavior of the lone star tick (Acarina: Ixodidae). IV. The daily and seasonal activity patterns of adults in different habitat types. *J. Med. Entomol.* **10:** 337–44.

Semtner, P. J. & Hair, J. A. (1973b) The ecology and behavior of the lone star tick (Acarina: Ixodidae). V. Abundance and seasonal distribution in different habitat types. *J. Med. Entomol.* **10:** 618–28.

Semtner, P. J., Barker, R. W. & Hair, J. A. (1971a) The ecology and behavior of the lone star tick. II. Activity and survival in different ecological habitats. *J. Med. Entomol.* **8:** 719–25.

Semtner, P. J., Howell, D. E. & Hair, J. A. (1971b) The ecology and behavior of the lone star tick (Acarina: Ixodidae) I. The Relationship between vegetative habitat type and tick abundance and distribution in Cherokee Co., Oklahoma. *J. Med. Entomol.* **8:** 329–35.

Worth, B. C. (1950) Observations on ectoparasites of some small mammals in the Everglades National Park and Hillsborough Co., Florida. *J. Parasitol.* **36:** 326–35.

19

Aspects of the bovine host–lone star tick interaction process in forage areas

Donald R. Barnard†

19.1 INTRODUCTION

The meeting of a bovine host and an ixodid tick co-inhabiting a pasture seems a fortuitous event. In fact this process is the culmination of a sequence of events which occur simultaneously but separately in each organism. When undisturbed, the host–tick contact process evolves to one of low risk and positive growth potential for the tick population. Extraordinary densities of ticks often result. In the south-eastern United States where climatic factors, vegetation, and an abundance of host animals favour growth in populations of the lone star tick, *Amblyomma americanum* (L.), such high densities are commonplace (Hair & Howell 1970).

Amblyomma americanum attacks a wide spectrum of host animals including domestic cattle, *Bos taurus* (L.). In the Ozark region, which includes parts of Oklahoma, Arkansas, and Missouri, beef producers utilize forested pastures as forage areas for cattle. Indigenous populations of the lone star tick in these situations grow quickly by exploiting the increased host availability. Beef producers, in turn, are confronted with recurring and high levels of tick infestation on their cattle. Management of these tick populations is critical from the standpoint of profit. Recent studies (Barnard 1985, Barnard *et al.* 1986) indicate that average daily gain in tick-infested nursing calves, when cows are also infested, begins to decline with as few as 15 feeding female *A. americanum*. The economic threshold (where the cost of control equals the value of increased production) is on average from 26 to 38 female ticks per calf per day during the tick season.

But how can knowledge of the bovine host–tick interaction be used to manage populations of ticks in forage areas? One way is to prevent host–tick contact. Host exclusion and pasture rotation are two examples. Strategic application of acaricides to cattle for tick control (Barnard *et al.* 1983) is another example. In the latter case, an ecological event (the appearance of females of a certain size on cattle (Barnard 1981) determines when treatment commences, whereas the length of time females feed on cattle defines the interval between treatments. This procedure exploits elements of the natural history of the tick and the host–tick interaction, thereby enhancing treatment efficacy.

Nor is it difficult to visualize other, perhaps more sophisticated, uses of such knowledge. Temporal and spatial congruencies in the activity patterns for ticks and bovine hosts, for example, may suggest circumstances where manipulation of the host population would reduce the probability of host–tick contact, would lower survivorship in parasitic and post-parasitic ticks, or would lower natality in the tick population.

19.1.1 Host and tick activity functions in time and space: a model of the host–tick contact process

How do cattle and ticks come together in a pasture? Obviously, physical contact must be made, but does it occur by chance or is it the result of programmed behavior in the host, the tick, or both? Does the likelihood of contact remain constant, or does it change through time and space and can we calculate the probability of host–tick contact? Fragments of the answers to these questions exist in the domestic animal ethology, acarology, and entomology literature. However, the relationship between cycles of activity in the bovine host and *A. americanum* has yet to be examined in a systematic manner. For this purpose, a model of the system comprising factors that influence bovine host–tick contact is presented (Fig. 19.1). Simply stated, the probability of host–tick contact is high, according to this model, when both the host and the tick are in the same part of the pasture and when the tick is responsive to the presence of the host. The probability of contact is low when activity patterns in either organism do not coincide or when the tick does not respond to the presence of a host.

19.2 UTILIZATION OF THE HABITAT BY THE BOVINE HOST

19.2.1 Spatial and temporal patterns of host activity

Cattle do not utilize forage areas in a random manner (Hafez 1975). Such use, in fact, varies with breed, age, sex, the physical and biological structure of the environment, weather, and patterns of maintenance and social behavior (Arnold & Dudzinski 1978). Few data exist that describe such variation. None consider it in relation to the dispersion of ticks in forage areas. To eventually calculate the probability of bovine host–tick contact we must be able to specify, with regard to location and time, the use of discreet units of the pasture by cattle as a function of some key factor(s) such as vegetation type. Then, dispersion and density, or both, of bovine hosts

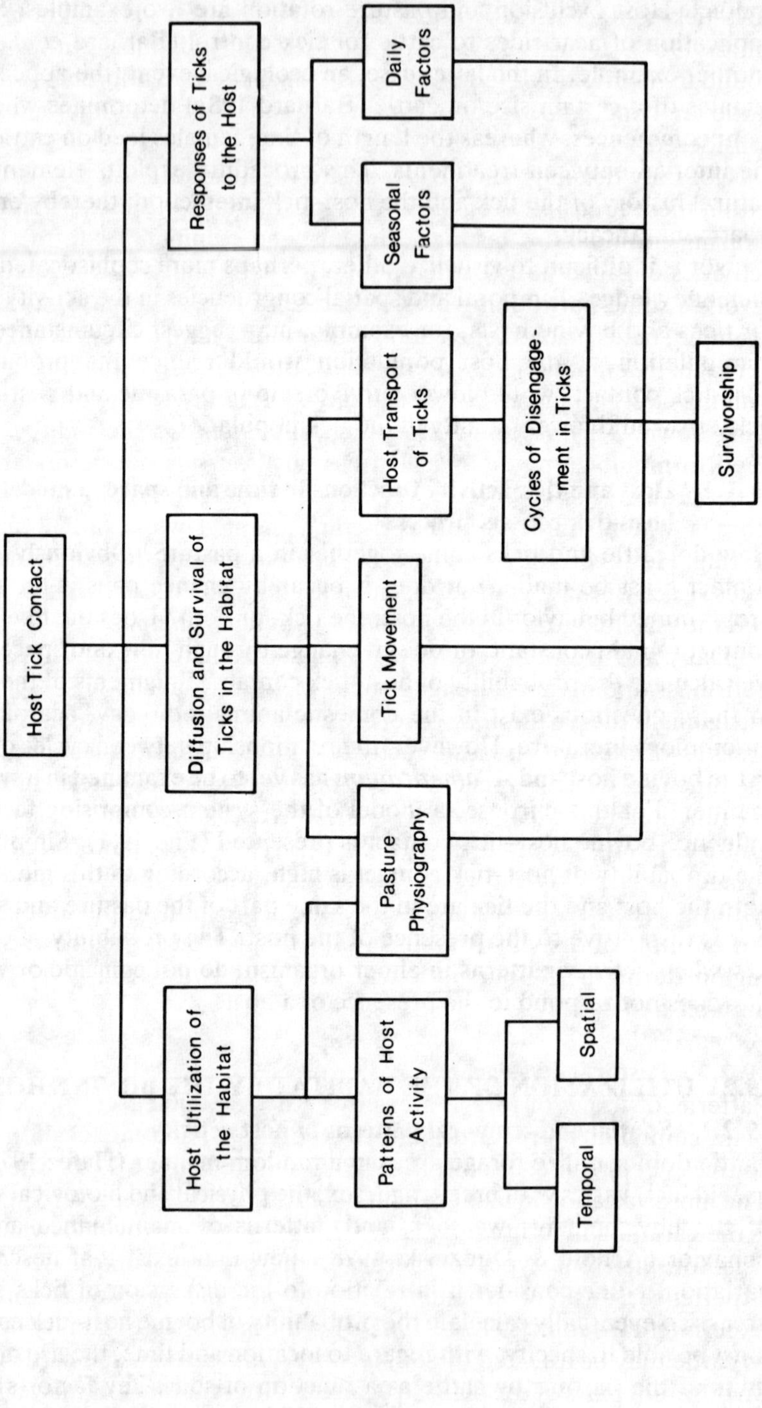

Fig. 19.1 — Model of the system comprising factors affecting the bovine host–lone star tick contact process.

within the units so defined needs to be elucidated. A measure of the dispersion of cattle that may be useful for defining the probability of host–tick contact is the average distance to nearest first neighbor (Greig-Smith 1964). Density is the average area per individual within the herd boundary (Arnold & Dudzinski 1978). That these relationships vary when cattle are lying, grazing, or in transit from one spot to another must also be considered.

Not only do some parts of a pasture receive more use than others, but within these areas different activities in cattle take place at different times. Grazing, rumination, resting, drinking, and elimination are the principal activities engaged in by cattle on a daily basis (Hafez 1975). Each process involves some partitioning of the pasture in time and space. Grazing, for example, takes place in open areas where cattle have unobstructed access to herbage. Rumination can occur while the animal is standing quietly or walking slowly, but cattle show a marked preference for ruminating while lying down. This process and resting, depending on the weather and the need to minimize physiological stress, may or may not be undertaken in areas protected by overstory vegetation. Drinking involves travel to and from a water source, whereas cattle eliminate while both grazing and resting. However, they avoid grazing in areas previously contaminated with feces. Each of these activities is influenced by breed, sex, season, weather, and the physiological requirements of individual animals.

Daily and seasonal activity patterns in cattle are relevant to patterns of both host-finding and disengagement of ticks from the host. Particularly intriguing is the partitioning of forage areas in time and space, by cattle, into areas associated with a specific activity. Thus a pertinent question is whether the functions describing categories of bovine host activity are congruent with the functions describing attack rates or with cycles of disengagement in ticks. Unfortunately, little data exist that describe spatial patterns of usage of forage areas by bovine hosts. We do know something of the temporal patterns of activity by pastured beef cattle in the spring in our area, however. An example of daily activity in *B. taurus* on a typical pasture in the Ozark region during a 24-hour period in May is shown in Fig. 19.2 (Barnard, unpublished data).

19.2.2 Pasture physiography

Patterns of usage of forage areas by cattle are influenced by pasture age, plant community structure, pasture typography, and the location of water sources (Peterson & Woolfolk 1955, Quinn & Harvey 1970, Sneva 1970). The more variation in plant communities that exists on a pasture, the greater is the variation in its usage by cattle. Little insight is provided by preference lists for different types of herbage, these being confounded by early experience, local environmental variation, and individual variation in experimental subjects (Kare 1961). Furthermore, acceptability of herbage to cattle may be affected by soil, fertilizer, and season, and cattle may feed on different parts of the plant structure at different times of the year (Stobbs 1974). Plant species with objectionable surface vestitures are avoided or

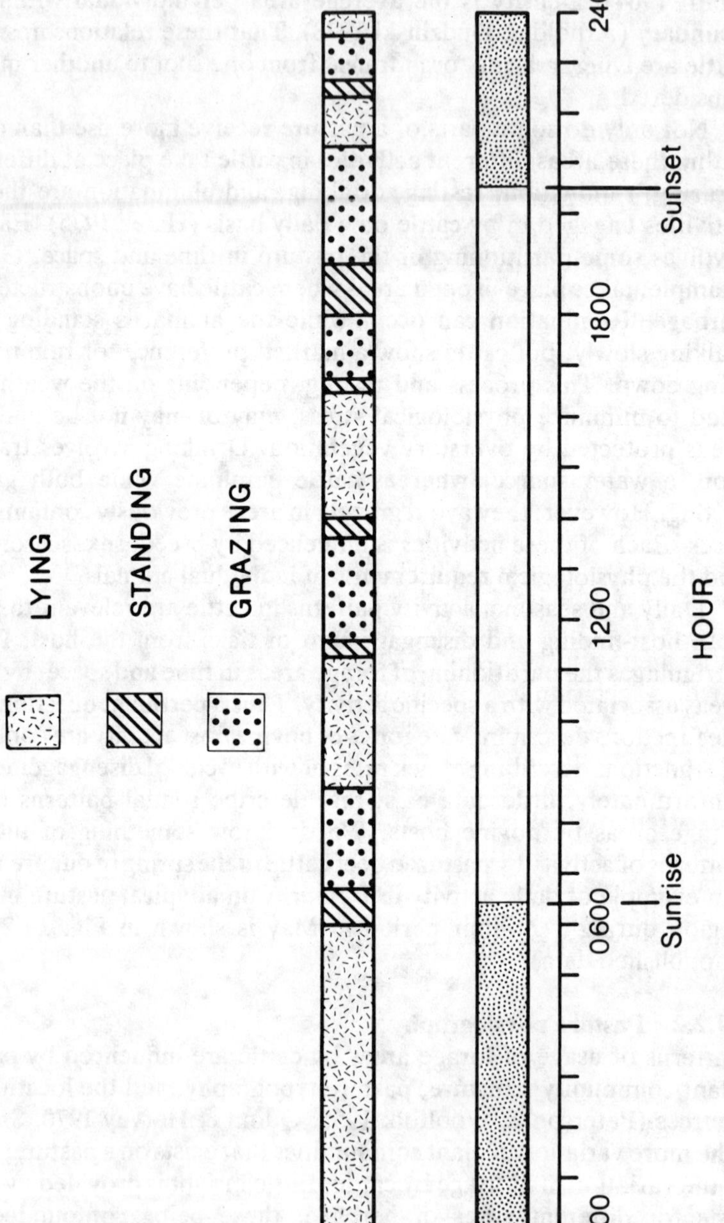

Fig. 19.2 — Diel activity pattern (May) of pastured *Bos taurus* in the Ozark region.

eaten only in the absence of other herbage (Brownlee 1951). Shade from vegetation provides a refuge for resting *B. taurus* during the day, but is not used by *B. indicus* (Rhoad 1938). In the absence of shade, cattle camp near water during the day (Schmidt 1969). Interestingly, artificial shade, to be competitive with natural shade, must provide enough space for animals to maintain social distance and to allow adequate air movement (McDaniel & Roark 1956). Pasture typography, including location of water sources, likewise influences utilization of the environment by cattle. Steep slopes (>10%) inhibit grazing, and in hilly rangeland cattle develop grazing routes which are the most direct path between grazing or camping sites and water sources (Weaver & Tomanek 1951). The location of water is a key element in the exploitation of forage areas by cattle. When water is freely available, cattle range far for feed, but when water sources are few they range repeatedly over relatively fixed distances (Schmidt 1969).

19.3 DIFFUSION AND SURVIVAL OF TICKS IN THE HABITAT

19.3.1 Disengagement of ticks from the host

What factors contribute to the infestation of forage areas with ticks? As these organisms are obligate ectoparasites, it is unlikely that they existed on pastures before the presence of host animals. In the case of *A. americanum*, cattle are the principle host for adult ticks and are an essential component of the host spectrum for nymphs in forage areas (Barnard 1981, Barnard *et al.* 1983). This means that bovine hosts play a role in the diffusion of engorged nymphs and females in pastures, and consequently influence the dispersion of larvae and adult ticks in these situations. At what time of day does *A. americanum* detach and disengage from the bovine host? Profiles of this activity in the field during May, for female and nymphal lone star ticks, in the Ozark region are shown in Fig. 19.3 (Barnard, unpublished data). Both life stages disengage primarily during daylight hours with a preponderance of activity before midday. During periods of peak disengagement by ticks, cattle are grazing or lying in the pasture (Fig. 19.2).

19.3.2 Movement of ticks

The diffusion of ticks in forage areas is further influenced by the movement of free-living individuals and by survivorship. Movement by replete female lone star ticks is influenced by ground cover and temperature, but females actively seek a site suitable for oviposition. In May (17.6°C), for example, in an area with light ground cover, replete females moved an average of 8.6 cm to the oviposition site, whereas in June (23.7°C), with the same ground cover, females moved 61.0 cm to the oviposition site (Patrick & Hair 1979). After eclosion from the egg, larvae travel only short distances — to locate and ascend vegetation on which they await a passing host. In contrast, adult

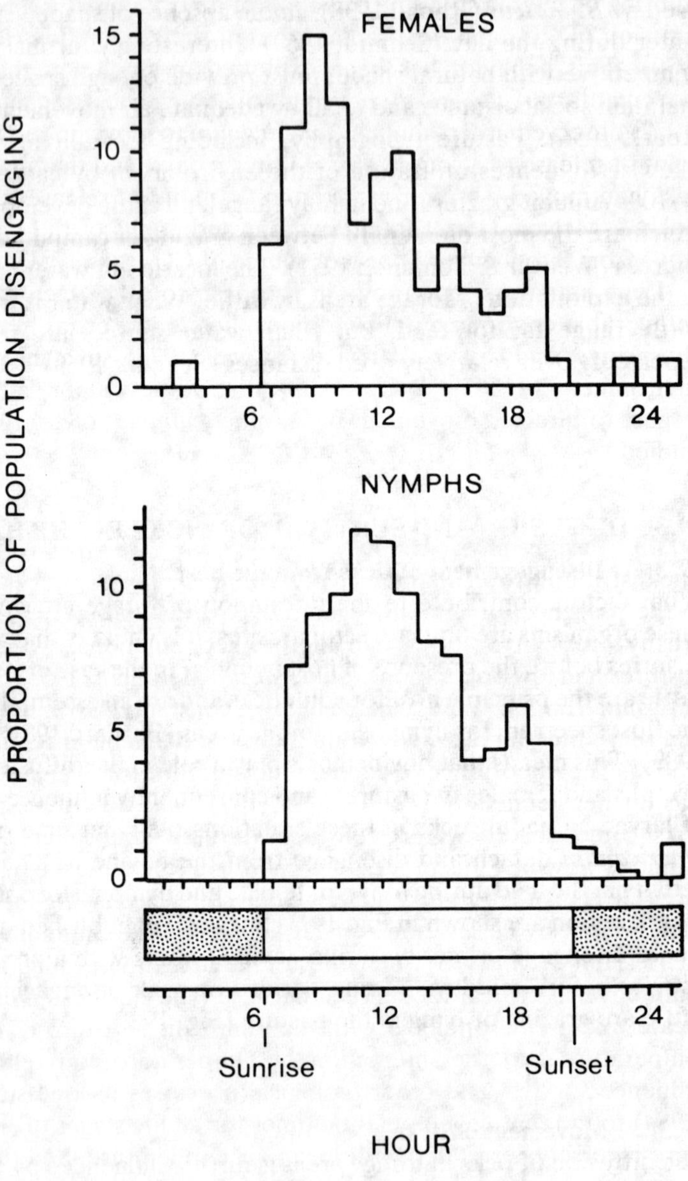

Fig. 19.3—Diel pattern (May) of disengagement of lone star ticks from pastured *Bos taurus*.

lone star ticks migrate from open areas to nearby woodlots, moving as much as 22.8 m from the point of release to recapture (Smittle *et al.* 1967). Hunting or migration in the direction of a host animal is a response to the emission of CO_2 and heat radiating from the host's body (Hair & Howell 1970). This behavior has been exploited in studies of the host-seeking movements of *A. americanum* using Co_2 as a chemoattractant. Wilson *et al.* (1972) observed unfed adult *A. americanum* to move up to 21.3 m from the point of release to a source of CO_2 after 3 days, and that after 7 days most ticks had moved 9.4 m or more from the point of release to the trap site. Koch & McNew (1982) found nymphs and adults of the lone star tick to move from 3 to 9 m per hour to a source of CO_2.

Lone star ticks may move in response to other factors, for example unfavorable environmental conditions, host odor, or, as in the case of the klipspringer and *Ixodes neitzi* (Rechav *et al.* 1978), substances placed in the environment by the host. The significance of these and related factors with respect to directed movement by *A. americanum,* however, remains to be studied.

19.3.3 Pasture physiography

The distribution and abundance of *A. americanum* in forage areas is related to the presence of suitable vegetative cover. Lancaster (1957) noted that brushy areas in pastures had high tick populations, whereas improved pastures had low populations. Lancaster & McMillan (1955) noted that eggs of *A. americanum* failed to hatch below a constant relative humidity of 73%, and that unfed larvae died within 2 days of hatching at relative humidities <60%. The importance of ambient moisture for survival in *A. americanum* was further defined by Sauer & Hair (1971) who determined that lone star ticks could absorb water from subsaturated atmospheres. The relative humidity at which this process commences is the critical equilibrium humidity (CEH). In laboratory studies, lone star ticks having lost 10% of their body weight by desiccation actively sought relative humidities above their CEH (Hair *et al.* 1975). Thus, in forage areas the availability of suitable plant cover is critical for survival because lone star ticks must periodically locate micro-environments with high ambient moisture levels. Similarly, temperature and the interaction of temperature and relative humidity influence survival and other biological processes in lone star ticks. Koch (1984) found that developmental times for all life stages of *A. americanum* were inversely correlated with mean air temperature, and that the habitat most favorable for ticks was also most moderate in terms of temperature and relative humidity. Furthermore, survival of lone star ticks in marginal habitats, i.e., those in which heat and desiccation normally preclude survival, was correlated with below-normal temperature and above-normal relative humidity. Consequently, the environmental conditions created within a plant community, specifically temperature and relative humdity, are more important for the survival of lone star ticks than the plant species comprising that community.

19.4 RESPONSES OF TICKS TO THE BOVINE HOST

19.4.1 Daily factors

Like many other ixodid species, lone star ticks can passively await a host on vegetation. With the approach of a host, perceived by mechano- and photoreceptors in the tick, a 'questing' posture, with forelegs extended, is assumed. If tick–host contact is not made the tick returns to a 'passive' posture with its forelegs folded under the body (Balashov 1968). At other times lone star ticks may undertake immediate and directed movement toward a host in response to a chemical stimulus or stimuli such as host odor. Body water content directly influences daily activity and determines if an individual lone star tick will be located in the environment such that it can respond to an approaching host. Body water content, in turn, is governed by temperature and relative humidity. Thus, when temperature is low and relative humidity high, for example in April and May, populations of adult *A. americanum* ascend vegetation and await a host. When temperature is high and relative humidity low, for example in July and August, adult ticks descend the vegetation and seek shelter in the duff and leaf litter (Semtner & Hair 1973). Daily activity in *A. americanum* in response to exogenous cycles in the environment is limited, but lone star ticks do orient their bodies to minimize exposure to direct sunlight (Semtner & Hair 1973). Other daily activities affecting the response of lone star ticks to a bovine host vary during the year but are generally keyed to times in a given 24–hour period in which temperature and relative humidity favor maintenance of water balance and, therefore, host seeking activity.

19.4.2 Seasonal factors

Patterns of seasonal activity for *A. americanum* on cattle in the Ozark region are shown in Fig. 19.4 (Barnard 1981). We know that lone star ticks seek hosts by moving across the ground in March and April and by ascending vegetation in May and June. But why do we find very few or no adults on cattle after mid July? At present we have not basis for proposing an answer to this question. Feeding is a process that begins with hunger, followed by response to and location of a host, and which ends with satiation. Waladde & Rice (1982) divide feeding into a sequence of nine events: appetence, engagement, exploration, penetration, attachment, ingestion, engorgement, detachment, and disengagement. Are some or all of these events governed seasonally by exogenous factors? Exploratory studies of the responses of lone star ticks to constant light and temperature (Barnard *et al.* 1986a) provide some insight into the importance of these two factors with respect to penetration, attachment, ingestion, engorgement, detachment, and disengagement. The results of these studies suggest that feeding activity in females (Fig. 19.5) and nymphs (Fig. 19.6) is in response to both photoperiod and temperature. In short-day compared to long-day conditions, fewer ticks feed: those that do, take longer to do so, ingest less blood, and the females lay fewer or no eggs (Fig. 19.7). Taken together, these observations suggest that exposure to a short light phase and perhaps high temperatures inhibits feeding and feeding related activity in *A. americanum* and sets the stage for a low biotic potential in the population (Fig. 19.8). Additional studies using field populations of lone star ticks are needed to elucidate the impact of photoperiod and temperature on the responses of ticks to bovine hosts over time. Moreover, to fully elucidate the mechanism(s) that controls the host–tick interaction process on a seasonal basis, we need to understand the influence of environmental factors on appetence, engagement, and exploration.

19.5 IMPLICATIONS FOR MANAGEMENT OF TICKS IN FORAGE AREAS

To calculate the probability of bovine host–tick contact in beef cattle forage areas we must know the activity status of each organism in time and space. We need to understand as well the relationship between plant species composition, feeding preferences of cattle for various plant species, and the favorability of the micro-environment of each plant community for the survival of *A. americanum*. Further, we must elucidate the bases of seasonal activity in lone star ticks on cattle with respect to exogenous cycles in the environment and endogenous cycles in the tick. This information will allow us to organize our understanding of the bovine host–tick interaction system so that we can predict, on the basis of carefully selected response variables, the probability of host–tick contact. With this capability we can identify the components of the host–tick contact model suitable for further study, or to which we can apply conventional tick management technology. Some of the possibilities are already well known. Modification of pasture physiography to reduce survivorship in free-living ticks and the introduction of Zebu characteristics into herds of cattle are two examples. With a better understanding of the host–tick interaction process we can manage herds of cattle to minimize host–tick contact or to cause the deposition of fed ticks in locations unsuitable for their survival or in which they are subject to destruction. Eventually, we may also discover circumstances in which populations of lone star ticks are vulnerable to management by the manipulation of genetic or physiological mechanisms.

REFERENCES

Arnold, G. W., & Dudzinski, M. L. (1978). *Ethology of free-ranging domestic animals*. Elsevier, Amsterdam.

Balashov, Y. S. (1968). Blood-sucking ticks (Ixodidae). Vectors of diseases of man and animals. Nauka Publishers, Leningrad Department, Leningrad. (Translated 1972). *Misc. Publ. Entomol. Soc. Am.* **8:** 161–376.

Barnard, D. R. (1981). *Amblyomma americanum:* comparison of populations of ticks free-living on pasture and parasitic on cattle. *Ann. Entomol. Am.* **74:** 507–511.

Barnard, D. R. (1985). Injury thresholds and production loss functions for

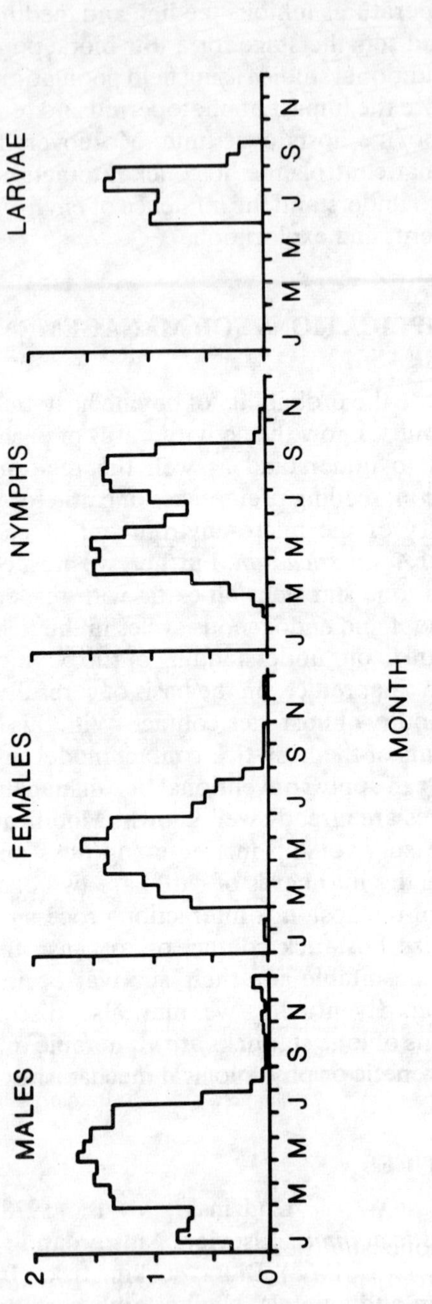

Fig. 19.4 — Seasonal abundance of lone star ticks on pastured *Bos taurus* beef cattle (Barnard 1981).

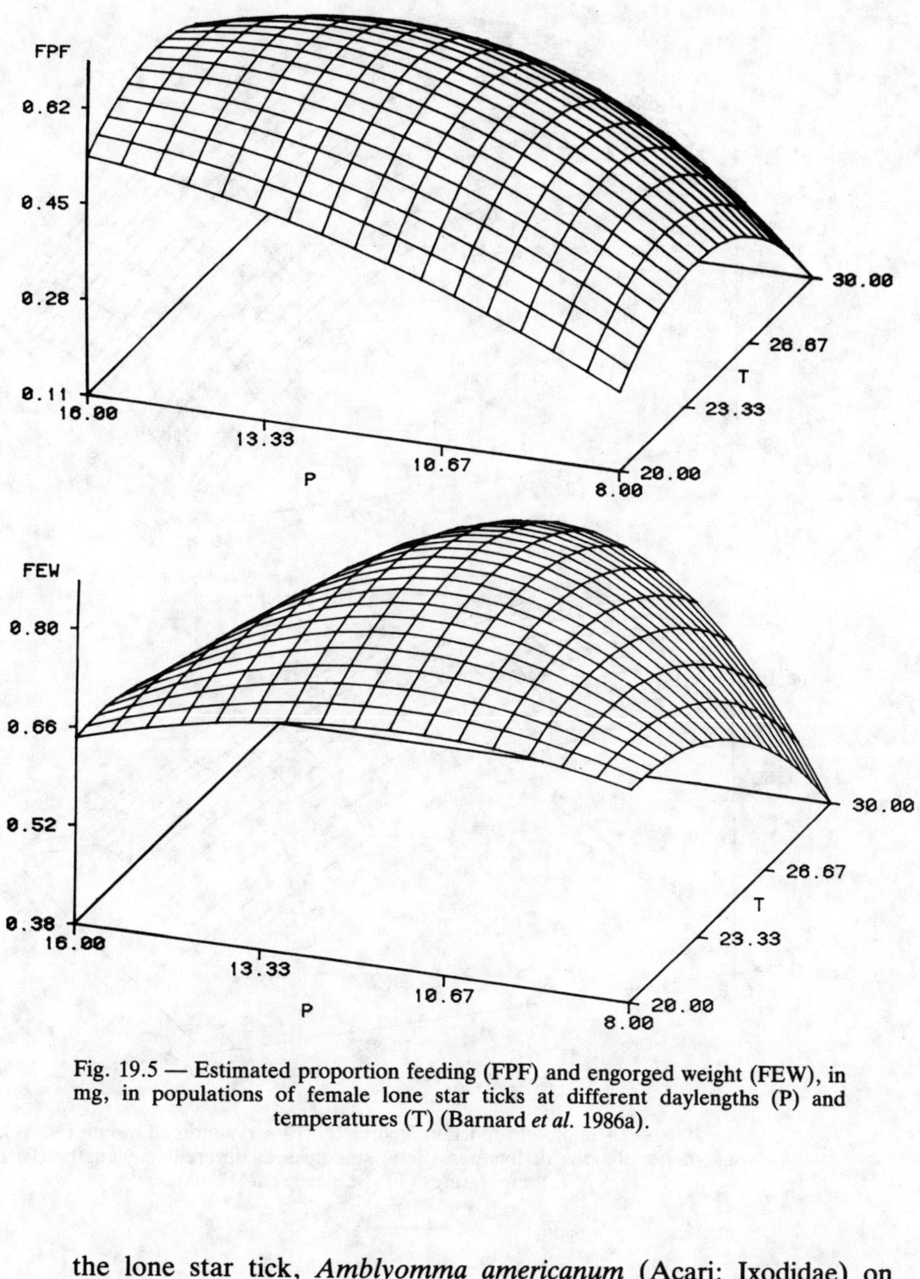

Fig. 19.5 — Estimated proportion feeding (FPF) and engorged weight (FEW), in mg, in populations of female lone star ticks at different daylengths (P) and temperatures (T) (Barnard *et al.* 1986a).

the lone star tick, *Amblyomma americanum* (Acari: Ixodidae) on pastured, preweaner beef cattle, *Bos taurus*. *J. Econ. Entomol.* **78:** 852–855.

Barnard, D. R., Rogers, G. D., & Jones, B. G. (1986). Strategic treatment of pastured cattle with acaricides: effect on populations of the lone star tick. *J. Econ. Entomol.* **76:** 99–102.

Barnard, D. R., Ervin, R. T., & Eplin, F. M. (1986). A production system based model for defining economic thresholds in preweaner beef cattle,

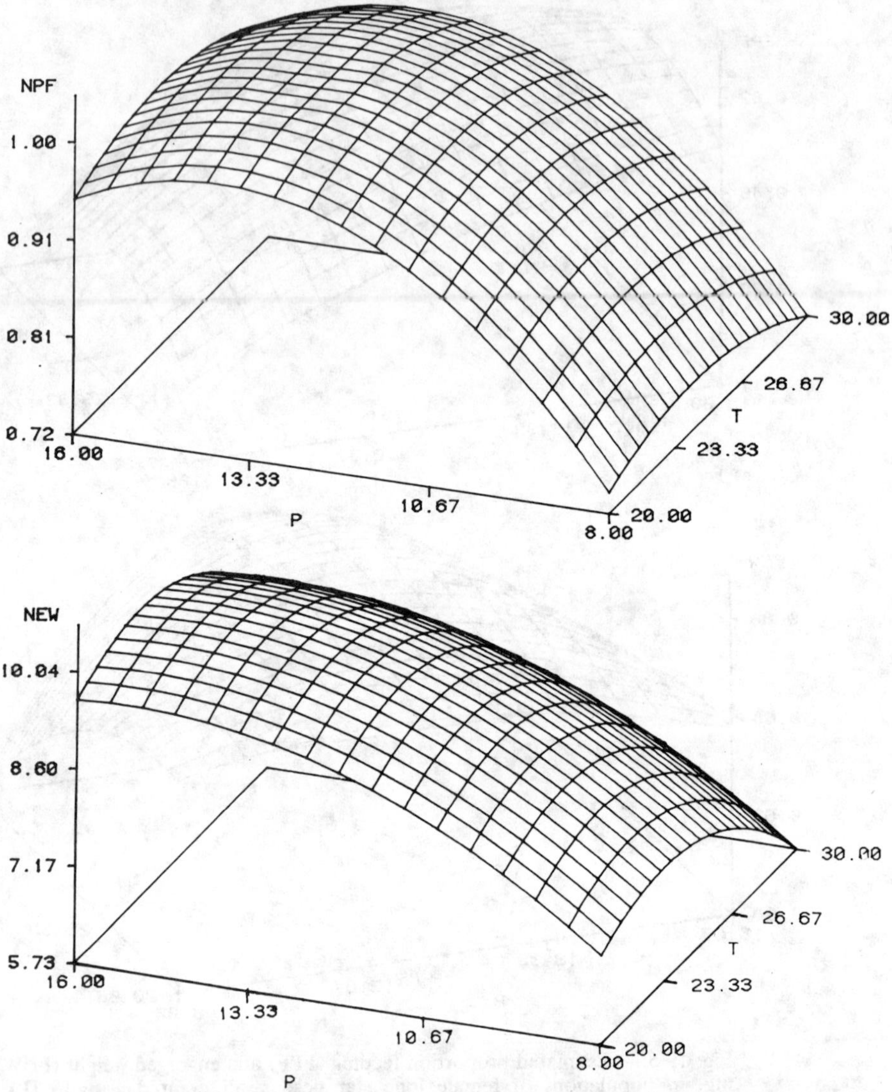

Fig. 19.6 — Estimated proportion feeding (NPF) and engorged weight (NEW), in mg, in populations of nymphasl lone star ticks at different daylengths (P) and temperatures (T) (Barnard *et al.* 1986a).

Bos taurus, infested with the lone star tick, *Amblyomma americanum* (Acari: Ixodidae). *J. Econ. Entomol.* **79:** 141–143.

Barnard, D. R., Morrison, R. D., & Popham, T. W. (1985b). Light and temperature sensitivity of feeding-related and reproductive processes in *Amblyomma americanum* (Acari: Ixodidae) on cattle. *Environ. Entomol.* **14:** 479–485.

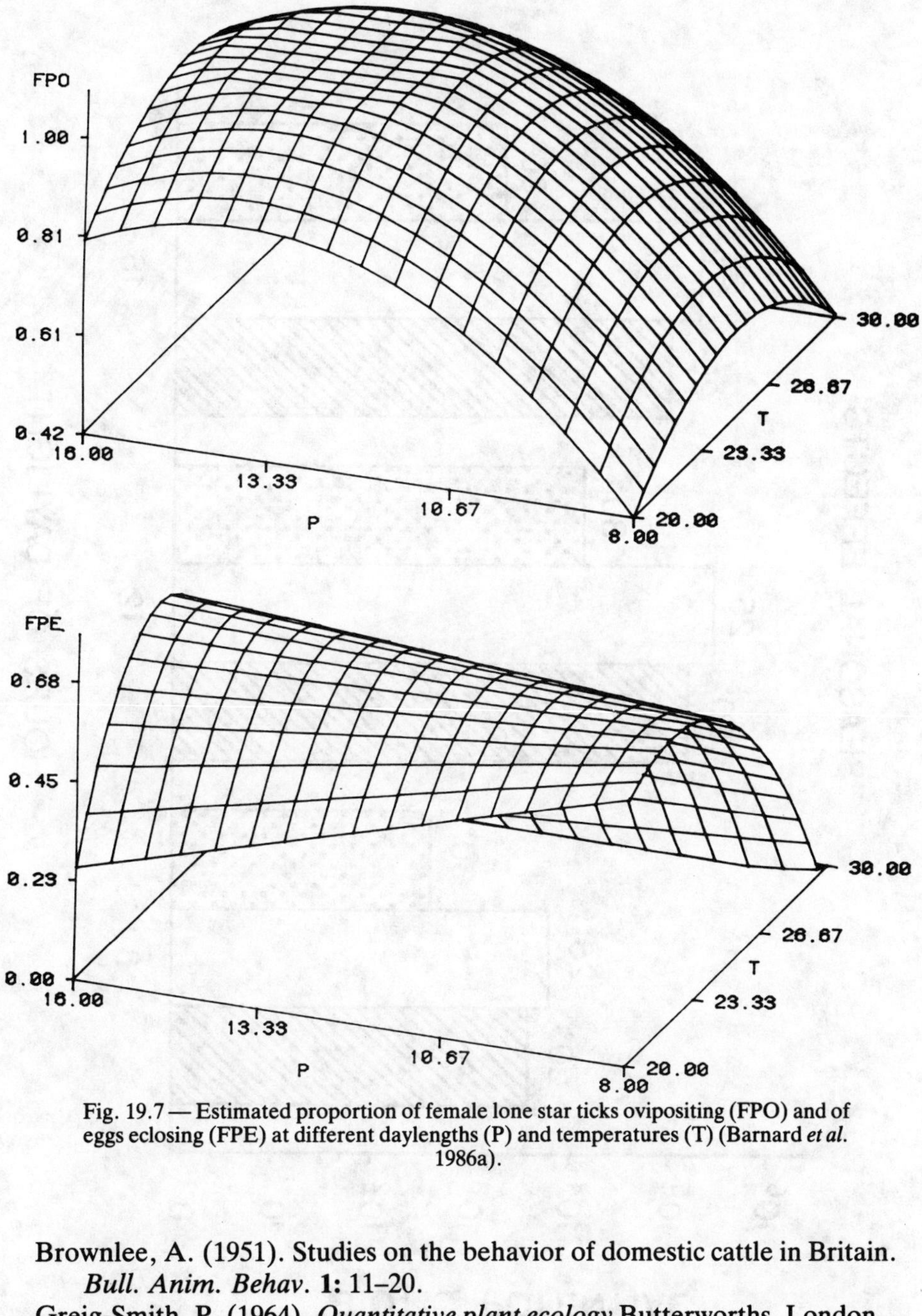

Fig. 19.7 — Estimated proportion of female lone star ticks ovipositing (FPO) and of eggs eclosing (FPE) at different daylengths (P) and temperatures (T) (Barnard *et al.* 1986a).

Brownlee, A. (1951). Studies on the behavior of domestic cattle in Britain. *Bull. Anim. Behav.* **1:** 11–20.

Greig-Smith, P. (1964). *Quantitative plant ecology* Butterworths, London.

Hafez, E. S. E. (1975). *The behavior of domestic animals.* The Williams and Wilkins Comp. Baltimore.

Hair, J. A., & Howell, D. E. (1970). Lone star ticks. Their biology and control in Ozark recreation areas. *Okla. Agric. Exp. Stn. Bull.* 679.

Hair, J. A., Sauer, J. R., & Durham, K. A. (1975). Water balance and

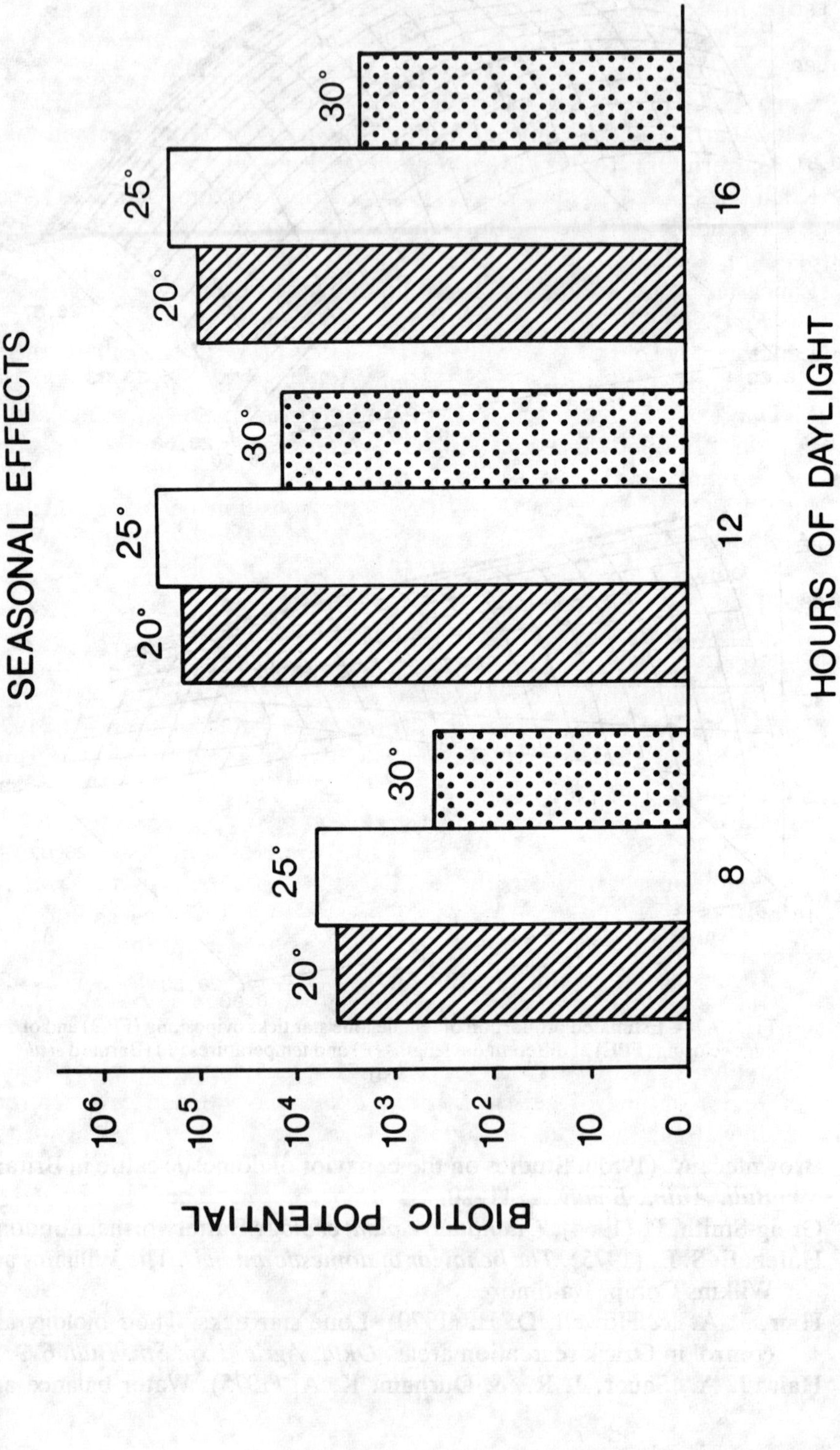

Fig. 19.8 — Estimated biotic potential of populations of lone star ticks at different daylengths and temperatures (Barnard *et al.* 1986a).

humidity preference in three species of ticks. *J. Med. Entomol.* **12:** 37–47.

Kare, M. R. (1961). Comparative aspects of the sense of taste. In: Kare, M. R., & Halpern, B. P. (eds.), *The physiological and behavioral aspects of taste,* University of Chicago Press, Chicago.

Koch, H. G. (1984). Survival of the lone star tick, *Amblyomma americanum* (Acari: Ixodidae), in contrasting habitats and different years in southeastern Oklahoma. *J. Med. Entomol.* **21:** 69–79.

Koch, H. G., & McNew, R. W. (1982). Sampling of lone star ticks (Acari: Ixodidae): Dry ice quantity and capture success. *Ann. Entomol. Soc. Am.* **75:** 579–582.

Lancaster, J. L., Jr. (1957). Control of the lone star tick. *Ark. Agric. Exp. Rep. Ser.* 67.

Lancaster, J. L., Jr., & McMillan, H. L. (1955). The effects of humidity on the lone star tick. *J. Econ. Entomol.* **48:** 338–339.

McDaniel, A. H., & Roark, C. B. (1956). Performance and grazing habits of Hereford and Aberdeen-Angus cows and calves on improved pastures as related to types of shade. *J. Anim. Sci.* **11:** 59–63.

Patrick, C. D., & Hair, J. A. (1979). Oviposition behavior and larval longevity of the lone star tick, *Amblyomma americanum* (Acarina: Ixodidae), in different habitats. *Ann. Entomol. Soc. Am.* **72:** 308–312.

Peterson, R. A., & Woolfolk, E. J. (1955). Behavior of Hereford cows and calves on short grass range. *J. Range Mgmt.* **8:** 51–57.

Quinn, J. A., & Harvey, D. F. (1970). Trampling losses and travel by cattle on sandhills range. *J. Range Mgmt* **23:** 50–55.

Rechav, Y., Norval, R. A. I., Tannock, J., & Colborne, J. (1978). Attraction of the tick *Ixodes neitzi* to twigs marked by the klipspringer antelope. *Nature* **275:** 310–311.

Rhoad, A. O. (1938). Some observations on the response of purebred *Bos taurus* and *Bos indicus* cattle and their crossbred types to certain conditions of the environment. *Proc. Am. Soc. Anim. Prod.* **38:** 284–295.

Sauer, J. R., & Hair, J. A. (1971). Water balance in the lone star tick (Acarina: Ixodidae): the effects of relative humidity and temperature on weight changes and total water content. *J. Med. Entomol.* **8:** 479–485.

Schmidt, P. J. (1969). Observations on the behavior of cattle in a hot dry region of the Northern Territory of Australia, with particular reference to walking, watering, and grazing. MSc. thesis, The University of New England, Armidale, N.S.W.

Semtner, P. J., & Hair, J. A. (1973). The ecology and behavior of the lone star tick (Acarina: Ixodidae). IV. The daily and seasonal activity patterns of adults in different habitat types. *J. Med. Entomol.* **10:** 337–344.

Smittle, B. J., Hill, S. O., & Phillips, F. M. (1967). Migration and dispersal patterns of Fe^{59} labeled lone star ticks. *J. Econ. Entomol.* **60:** 1029–1031.

Sneva, F. A. (1970). Behavior of yearling cattle on eastern Oregon range. *J. Range. Mgmt.* **23:** 155–158.

Stobbs, T. H. (1974). Components of grazing behavior of dairy cows on some tropical and temperae pastures. *Proc. Aust. Soc. Anim. Prod.* **10:** 299–302.

Waladde, S. M., & Rice, M. J. (1982). The sensory basis of tick feeding behavior. In: Obenchain, F. D., & Galun, R. (eds.), *Current themes in tropical science. Physiology of ticks.* Pergamon Press, New York.

Weaver, J. E., & Tomanek, G. W. (1951). Ecological studies in a midwestern range: the vegetation and effects of cattlc on its composition and distribution. *Nebr. Conserv. Serv. Bull.* 31.

Wilson, J. G., Kinzer, D. R., Sauer, J. R., & Hair, J. A. (1972). Chemo-attraction in the lone star tick (Acarina: Ixodidae). I. Response of different developmental stages to carbon dioxide administered via traps. *J. Med. Entomol.* **3:** 245–252.

20

Tick populations of two breeds of cattle under field conditions, with a note on blood components related to host resistance.

Y. Rechav and M. E. Zeederberg

20.1 INTRODUCTION

It is well known that ticks and tick-borne diseases cause high losses in livestock production. Steelman (1976) cited reports of the USA Department of Agriculture, estimating that in 1965 'tick-caused' losses were \$60 000 000 annually to cattle production, and that these figures were much higher at the beginning of the century. Similar high loss estimates were reported from Australia (Johnson & Haydock 1969, Waters *et al.* 1982) and Mexico (Woodham *et al.* 1983).

To avoid such a financial burden and allow the livestock industry to develop, control procedures must be applied. However, development of resistance by many tick species to some acaricides (Whithead 1965, 1973, Drummond 1970), the cost of dipping and the high cost of developing new acaricides in relation to the economic return expected from their use before resistance occurs (Durand 1976, Wharton 1976) raised the need to develop new management approaches for tick control.

Two of the most promising alternative measures are: the use of tick pheromones combined with acaricides (Gladney *et al.* 1974, Rechav & Whithead 1978) and host resistance to tick infestation (Wikel & Allen 1982).

Bovine resistance to *Boophilus microplus* (Canestrini) has been reported by many Australian workers and recently reviewed by Wikel &

Allen (1982). *Bos indicus* cattle have been reported to be more resistant to tick infestation than *Bos taurus*. Furthermore, some of the Australian authors found 'that cattle with more Zebu blood were more resistant than the European cattle'. As a result of the extensive studies done in Australia, a technique for the control of *B. microplus* by breeding resistant cattle has been developed (Waters *et al.* 1982). Unfortunately, *B. microplus*, as a one-host tick, cannot serve as a model for two- and three-host ticks.

Very little has been done in the USA with host resistance to three-host ticks such as *Amblyomma americanum* (L). However, it has been demonstrated that Brahman and Brahman X Hereford crossbreeds were more resistant than Hereford (Strother *et al.* 1974, Garris *et al.* 1979).

The situation in Africa is more complicated than in other continents: not only that more three-host species of ticks are abundant, but our knowledge of the indigenous cattle of southern africa and their resistance to the many tick species is limited. The pioneering work of Bonsma (1944) and the recent studies of Kaiser *et al.* (1982) and Latif (1984) indicated that European breeds (*B. taurus)* carry more ticks than *B. indicus* types. However, the existing information is not satisfactory to delineate the role of cattle breed in tick population maintenance.

The mechanisms of host immunity to ticks have been studied intensively during the last decade. These experiments were conducted with laboratory animals as well as with cattle, and were recently reviewed by Willadsen (1980), and Brown & Askenase (Chapter 13 of the present book). However, the mechanism of resistance to ticks in host animals is not fully understood. Furthermore, some researchers believe that *B. indicus* cattle have innate resistance, while others claim that the resistance has been aquired (Wikel & Allen 1982).

This chapter focuses primarily on field collections of ticks from two breeds of cattle in Transvaal, the Northern Province of South Africa. Practically no information is available concerning the tick populations on various types of cattle, and this information is essential prior to any introduction of control methods based on resistance of cattle. To determine the relationship between the tick burden on the different breeds of cattle and resistance to ticks, blood samples from each animal were analysed and various components, including globulins, have been monitored.

20.2 STUDY AREA AND SAMPLING TECHNIQUES

20.2.1 Study area and cattle used

The selected farm was situated in the northern parts of South Africa (24°18′S and 19°07′E). The vegetation according to Acocks (1975) is a red turfveld variation of the Springbok Flats Turf Thornveld, i.e., closed thronveld with a sparse grass cover. Temperature, relative humidity, and rainfall were recorded daily. The stocking density was 1 heifer per 8 ha. The animals which were used during this survey were part of a herd, of which ten animals (5 Brahman and 5 Hereford) were sampled at monthly intervals.

To minimize the differences between individuals, the same animals were used throughout the study, and these were kept under natural conditions without being dipped or sprayed with acaricides.

20.2.2 Sampling techniques

Removal of feeding ticks from the heifer hosts were carried out by an experienced team led by the senior author. For the removal of ticks from the hosts' heads, the animals were tethered by ropes to a wooden fence and their legs restrained; collections from the rest of the body were made after the animals had been restrained on the ground and tied (Rechav & Whithead 1978).

The ticks from each animal were placed in four separate bottles containing 70% ethyl alcohol: for the ticks which were collected from the head, from the perianal area and the tail, from the body, and from the abdomen and hooves.

A similar procedure was used for removing ticks from newly born calves. The objective of this was to collect preliminary data on tick numbers removed from young hosts which had not been previously infested with ticks.

20.3 TICK POPULATIONS ON CATTLE

Collections of ticks from cattle started in the winter of 1983 with the objective of recording abundance and seasonal occurrence of ticks in the Northern Transvaal. These data, the first to be collected in South Africa, will serve as basic information for any future studies dealing with the mechanism of resistance in hosts to ticks, and it will also provide facts for any selection of methods for controlling ticks by using immune breeds.

The information provided in this chapter is based on data collected during one season only (June 1983–June 1984), and thus the conclusions, although based on ten heifers, should be regarded as preliminary. The results obtained from each individual heifer of the same breed are grouped together, and each tick species is dealt with separately.

20.3.1 *Amblyomma hebraeum* (the bont tick)

Adults of this tick species were found on cattle throughout the year, but were very abundant during December, January, and February.

The numbers of males and females of *A. hebraeum* on Hereford heifers were significantly higher than those found on Brahman heifers (Figs. 20.1, 20.2). Unfortunately, very little is known about the resistance of different hosts to *A. hebraeum*. However, it appears that the abovementioned information does not agree with Norval's (1978) claim that '.... rabbits and sheep are unable to acquire resistance to immature stages', but it supports the preliminary results reported by Heller-Haupt *et al.* (1981). Furthermore, Hereford heifers carried more bont ticks than Brahman during the summer when the density of the bont tick population was high, but the burden on cattle was very similar during the winter, when the animals carry very few ticks (Figs. 20.1 & 20.2).

The reason for the abovementioned phenomenon is unclear. It might be linked to the deterioration in the nutritional value of the grass during winter, or to any other physiological factor which might affect the immune state of the heifers. However, a similar decrease in the resistance of cattle to *Boophilus microplus* towards the end of the winter has been described in Australia (Dicker & Sutherst 1981).

20.3.2 *Rhipicephalus appendiculatus* (the brown-ear tick)

The brown-ear tick showed a definite seasonal pattern of abundance on their hosts (Figs 20.3 & 20.4). The number of adults on heifers of the two studied breeds increased very sharply from December, reached a peak in January, then declined sharply in March/April, and finally reached a very low number in May. Similar seasonal activity was found in Natal (Baker & Ducasse 1967) and the Eastern Cape Province of South Africa (Rechav 1982). However, from the time of their appearance in December, throughout the summer, the adult brown-ear ticks are significantly more common on Hereford than on Brahman heifers. This situation was found for both males (Fig. 20.3) and for females (Fig. 20.4).

Although Kaiser *et al.* (1982) dealt with this species in Uganda, there is not enough evidence to prove that Zebu cattle carried fewer brown ear-ticks than any other breeds. A recent survey (Spickett, unpublished data) conducted near Loskop dam, in South Africa, did not show significant results because of the low numbers of adults found on the animals.

20.3.3 *Rhipicephalus evertsi evertsi* (the red-legged tick)

Adults of the red-legged ticks were removed from Hereford and Brahman heifers throughout the year, but the tick was more common in summer.

Adults were more abundant on Hereford than on Brahman, but these differences were not as clear as with *A. hebraeum* or *R. appendiculatus*. The burden of *R. evertsi evertsi* males on Hereford was greater than on Brahman only from the beginning of the summer (Fig. 20.5). A similar situation was found during the summer for females (Fig. 20.6). However, no differences in the number of adults on the two breeds were observed during winter.

The reasons for the decrease in host immunity during winter are unknown, and more reasearch should be conducted. This phenomenon was found also in *A. hebraeum* (Rechav, unpublished) and in *B. microplus* (Dicker & Sutherst 1981), and has been discussed above.

20.3.4 Other species of ticks

Hyalomma marginatum rufipes, *Hyalomma truncatum*, *Boophilus decoloratus* and *Rhipicephalus simus simus* were found frequently, but in low numbers, thus few conclusions could be drawn. However, results indicated that the Hereford breed appeared to carry more adult ticks of the abovementioned species than Brahman.

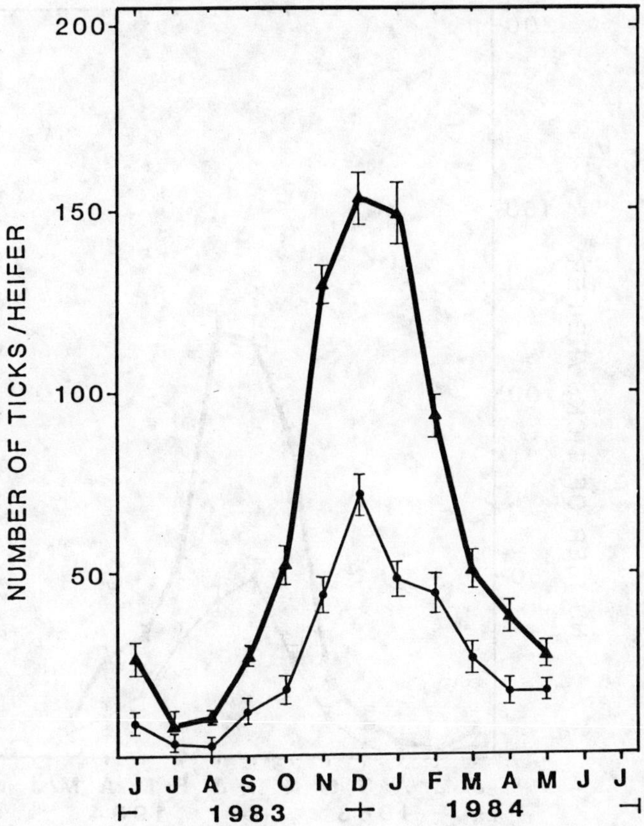

Fig. 20.1 — Number of *Amblyomma hebraeum* males (mean±SE) removed from Brahman (●—●) and Hereford (▲—▲) heifers at monthly intervals.

20.4 TICK POPULATIONS AND GAMMA GLOBULINS

Although this chapter does not deal with the mechanism of host immunity to ticks, the amount of gamma globulins in serum taken from Hereford and Brahman (Fig. 20.7) was compared with the number of ticks removed from the heifers.

It appears that there is a correlation between the number of ticks and the amount of gamma globulins. During the summer, when the tick species were abundant on the heifers (Figs. 20.1–20.6), the concentration of gamma globulins increased. Furthermore, Hereford heifers had higher amounts of gamma globulins then Brahman, and also they carried a higher number of ticks. This indicates that the production of antibodies in Hereford heifers might be higher than in Brahman, although the breed is less resistant. The presence of more ticks on Hereford heifers probably introduced more antigens to the host.

However, it should be emphasized that the heifers of the two breeds

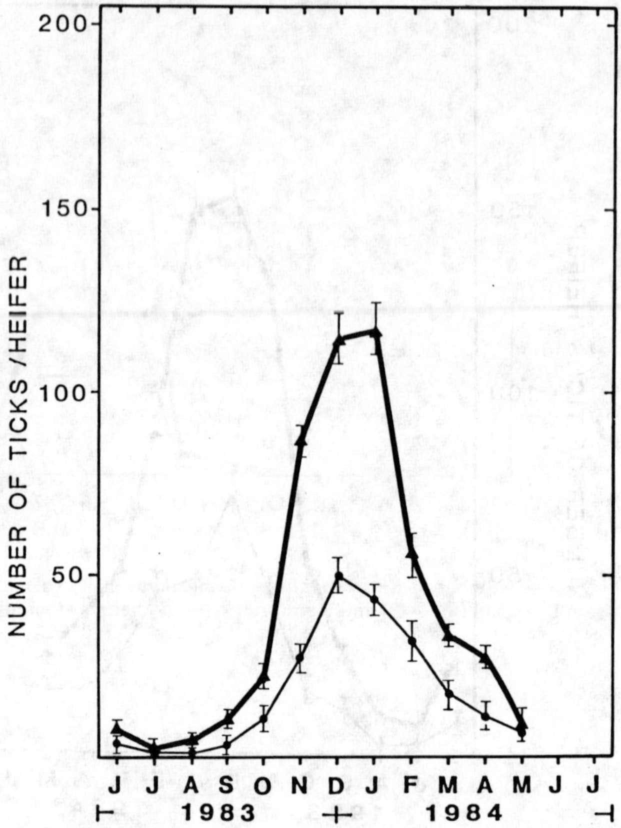

Fig. 20.2 — Number of *Amblyomma hebraeum* females (mean±SE) removed from Brahman (●——●) and Hereford (▲——▲) heifers at monthly intervals.

carried not only few species of ticks at the same time but also more than one stage (larvae, nymphs, and adults). Therefore, we do not know whether the increase in the amount of gamma globulins during the summer was caused by one species of tick or by more than one species. Furthermore, if more than one species contributed to the increase in the gamma globulin production, is their contribution equal to that of other species?

20.5 TICK POPULATIONS AND BLOOD

Analysis of the blood and its correlation with the number of ticks on Hereford and Brahman heifers showed that during the summer, when more ticks were found on Hereford, the haemoglobin concentration dropped from 15.0% in winter to 9.5% in summer. In Brahman, which carried fewer ticks, the decrease was more moderate, to a value of 12.3%. A similar situation was found for red blood corpuscles.

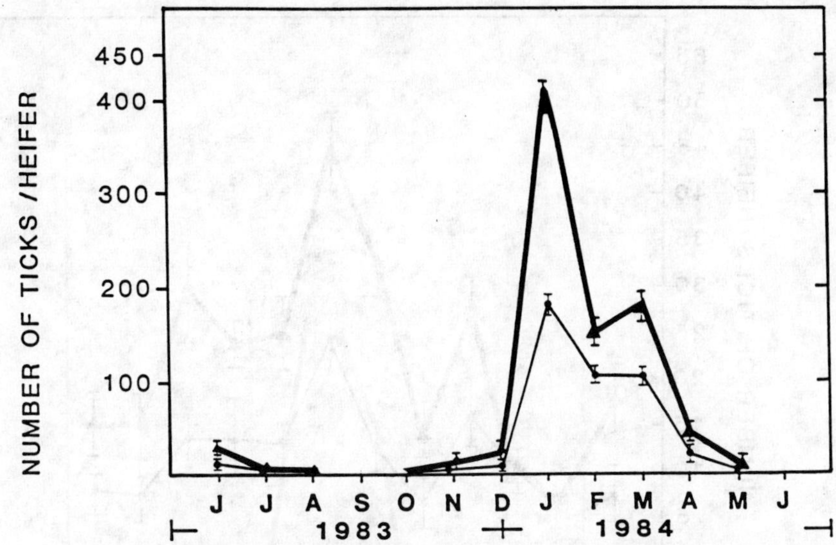

Fig. 20.3 — Number of *Rhipicephalus appendiculatus* males (mean±SE) removed from Brahman (●—●) and Hereford (▲—▲) heifers at monthly intervals.

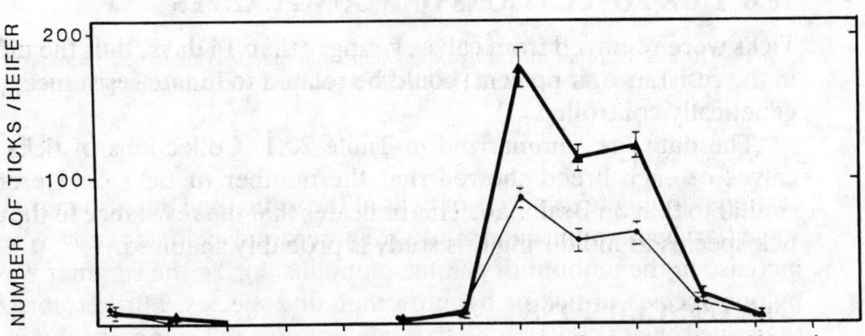

Fig. 20.4 — Number of *Rhipicephalus appendiculatus* females (mean±SE) removed from Brahman (●—●) and Hereford (▲—▲) heifers at monthly intervals.

These values clearly indicated that the heifers were suffering from anaemia, as was described by Rechav *et al.* (1980). No significant differences were found for the number of white blood corpuscles.

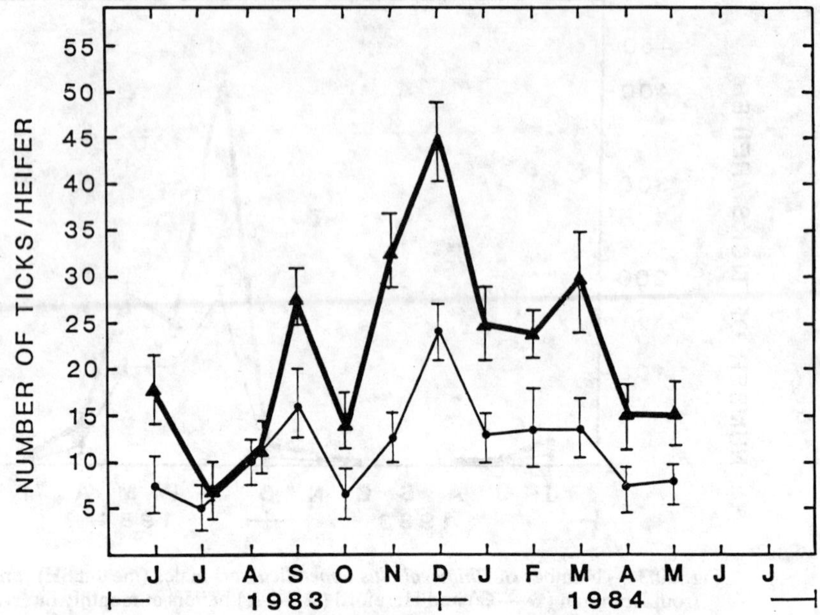

Fig. 20.5 — Number of *Rhipicephalus evertsi evertsi* males (mean±SE) removed from Brahman (●——●) and Hereford (▲——▲) heifers at monthly intervals.

20.6 TICK POPULATIONS ON YOUNG CALVES

Ticks were removed from calves younger than 14 days, thus the differences in the resistance, if present, could be related to innate resistance, probably genetically controlled.

The data are summarized in Table 20.1. Collections of ticks form 22 calves of each breed showed that the number of ticks on Hereford was similar to that on Brahman. This indicates that the resistance to the common tick species found during this study is probably acquired.

20.7 CONCLUSIONS

Field collections of ticks from two breeds of cattle showed that the common species of ticks were *Amblyomma hebraeum*, *Rhipicephalus appendiculatus* and *R. e. evertsi.*

The number of ticks of all these species was larger on Hereford than on Brahman heifers. Preliminary results also indicated that the density of immature ticks of the above-mentioned species is higher on Hereford than on Brahman.

A positive correlation was found between the number of ticks and the amount of gamma globulins, an indication of the possible production of anti-tick antibodies.

A similar correlation was found between tick burden and the number of

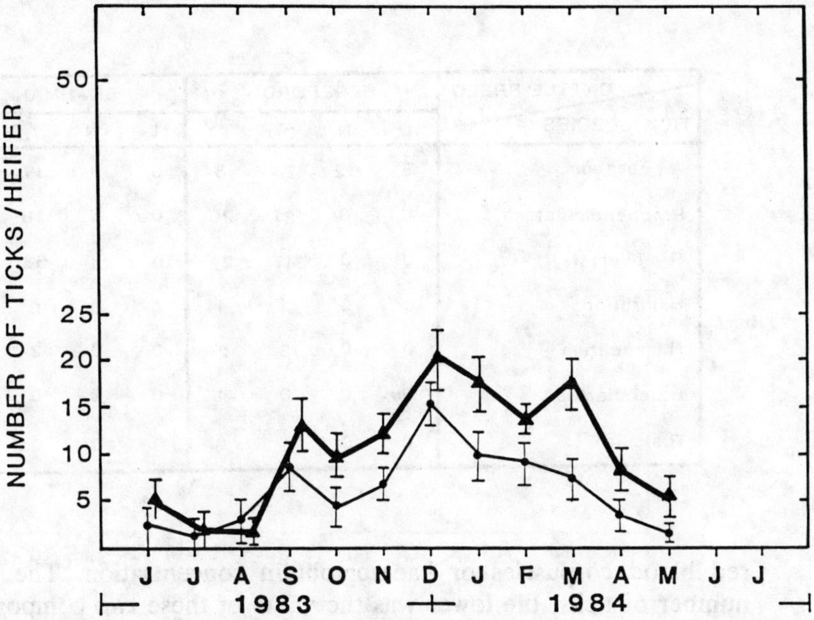

Fig. 20.6 — Number of *Rhipicephalus evertsi evertsi* females (mean±SE) removed from Brahman (●——●) and Hereford (▲——▲) at monthly intervals.

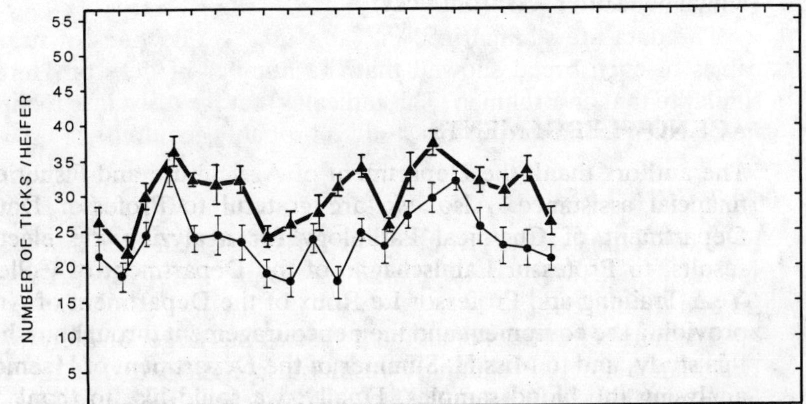

Fig. 20.7 — The amount (g/l) of gamma globulins (mean±SE) separated by electrophoresis from serum of blood taken from Brahman (●——●) and Hereford (▲——▲) heifers at monthly intervals.

Table 20.1 — Number of ticks removed from Hereford and Brahman calves.

CATTLE BREED / TICK SPECIES	HEREFORD				BRAHMAN			
	L	N	♂♂	♀♀	L	N	♂♂	♀♀
A.hebraeum	8	12	24	8	0	1	17	5
R.appendiculatus	0	0	27	20	0	0	18	18
R.e.evertsi	0	0	47	21	0	1	58	32
H.m.rufipes	0	0	1	1	0	0	3	1
H.truncatum	0	0	3	2	0	0	2	1
B.decolartus	0	0	0	1	0	0	0	0
R.simus	0	0	1	0	0	0	0	0

red blood corpuscles or haemoglobulin concentration. The larger the number of ticks, the lower was the value of these two components. The resistance to ticks was probably acquired by the host. However, it appears that the Brahman breed has the ability to acquire this resistance faster, and obviously to a higher degree than the Hereford. This ability to acquire resistance to ticks could be genetically controlled in the Brahman, and is missing in the Hereford.

This information presented here might assist in further experiments which will enable us to understand the mechanisms of acquired immunity. Then we may be able to implement this knowledge in breeding cattle breeds which will suffer less from ticks.

ACKNOWLEDGEMENTS

The authors thank the Department of Agriculture and Fisheries for their financial assistance. Also, we are grateful to Professor Dauth of the Department of Chemical Pathology for analyzing the electrophoresis results, to Professor Labuschagne of the Department of Collective First Year Training and Professor Le Roux of the Department of Anatomy for providing the equipment and their encouragement throughout the course of this study, and to Miss H. Summer of the Department of Haemotology for analysing the blood samples. Finally, we sould like to thank Mr A. M. Spickett of the Onderstepoort Veterinary Research Institute for allowing us to cite some of his unpublished results. Also to be thanked is Mr Tom Knight of the Witpoort farm who not only allowed the use of his farm but also assisted in tick collections.

REFERENCES

Acocks, J. P. H. (1975) Veld types of South Africa. *Memoirs of the Botanical Survey of South Africa*. No. 40. Botanical Research Institute, Pretoria. 128p.

Baker, M. K. & Ducasse F. B. W. (1967) Tick infestation of livestock in Natal. I. The predilection sites and seasonal variations of cattle ticks. *J. S. Afr. Vet. Med. Assoc.* **38**: 447–453.

Bonsma, J. C. (1944) Hereditary Heartwater resistance characteristics in cattle. *Farming in South Africa*, 71–96.

Dicker, R. W. & Sutherst, R. W. (1981) Control of the bush tick (*Haemaphysalis longicornis*) with Zebu × European cattle. *Australian Veterinary Journal* **57**: 66–68.

Drummond, R. O. (1970) Current worldwide research on control of ticks involved in animal diseases. *Misc. Publ. Entomol. Soc. Am.* **6**: 367–372.

Durrand, M. R. E. (1976) Ticks a warning. *Queensland Agric. J.* (Nov–Dec): 541–544.

Garris G. I., Stacey, B. R., Hair, J. A., & McNew, R. W. (1979) A comparison of Lone Star Ticks on Brahman and Hereford cattle. *Entomol. Soc. of Am.* **72**: 869–872.

Gladney, W. J., Ernst, S. E., & Grabbe, R. R. (1974) The aggregation response of the Gulf Coast tick on cattle. *Ann. Entomol. Soc. Am.* **67**: 750–752.

Heller-Haupt, A., Varma, M. R. G., & Langi, A. O. (1981). Acquired resistance to ixodid ticks in laboratory animals. *Trans. R. Soc. Trop. Med. Hyg.* **75**: 147–148.

Johnston, L. A. Y. & Haydock, K. P. (1969) The effect of cattle tick (*Boophilus microplus*) on production of Brahman-cross and British-breed cattle in Northern Australia. *Austal. Vet. J.* **45**: 175–179.

Kaiser, M. N., Sutherst, R. W., & Bourne, A. S. (1982) Relationship between ticks and Zebu cattle in Southern Uganda. *Trop. Anim. Hlth. Prod.* **14**: 63–74.

Latif, A. A. (1984). Resistance to natural tick infestations in different breeds of cattle in Sudan. *Insect Sci. Applic.* **5**: 95–97.

Norval, R. A. I. (1978). Repeated feeding of *Amblyomma hebraeum* (Acarina: Ixodidae) immatures on laboratory hosts. Hosts effects on tick yield, engorgement weight and engorgement period. *J. Parasitol.* **64**: 910–917.

Rechav, Y. (1982) Dynamics of tick populations (Acari: Ixodidae) in the Eastern Cape Province of South Africa. *J. Med. Entomol.* **19**: 679–700.

Rechav, Y. & Whitehead, G. B. (1978). Field trials with pheromone-acaricide mixtures for the control of *Amblyomma hebraeum*. *J. Econ. Entomol.* **71**: 149–151.

Rechav, Y., Kuhn, H. G., & Knight, M. M. (1980) The effects of the tick *Amblyomma hebraeum* (Acari: Ixodidae) on blood of rabbits. *J. Med. Entomol.* **17**: 555–560.

Steelman, C. D. (1976) Effects of external and internal arthropod parasites on domestic livestock production. *Ann Rev. Ent.* **21**: 155–178.

Strother, G. R., Burns, E. C., & Smart, E. I. (1974) Resistance of pure bred Brahman, Hereford and Brahman–Hereford cross bred cattle to the Lone Star Tick, *Amblyomma americanum* (Acarina: Ixodidae) *J. Med. Entomol.* **11**: 559–563.

Waters, K. S., Round, J., & Bond. J. (1982) Controlling cattle ticks by breeding and selection in North Queensland. *Queensland Agric. J.* (Jul–Aug): 200–202.

Whitehead, G. B. (1965) Resistance in the Acarina: ticks. In: Naegle J. A. (ed) *Advances in acarology*, Vol. 2. Ithaca, N.Y. Cornell Univ. Press. p. 53–70.

Whitehead, G. B. (1973) Resistance to acaricides in ticks in the Eastern Cape Province. *S. Afr. Med. J.* **47**: 342–344.

Wikel, S. K. & Allen, J. R. (1982) Immunological basis of host resistance to ticks. In: Obenchain, F. D. & Galun, R. (ed.) *Physiology of ticks.* London, Pergamon Press, pp. 169–196.

Willadsen, P. (1980) Immunity of ticks. In: Lumsden, W. H. R. (ed) *Advances in parasitology*. **18**: 293–313.

Woodham, C. B., Gonzalez Origel, A., Lopez Leon, A., & Guerena Morales, R. (1983) Progress in the eradication of *Boophilus* ticks in Mexico 1960–80. *World animal review* **48**: 18–24.

21

Secretory and digest cells of female *Boophilus microplus:* Invasion and development of *Babesia bovis;* light and electron microscope studies

R. I. S. Agbede, D. H. Kemp, and **H. M. D. Hoyte**

21.1 INTRODUCTION

Babesia undergoes dramatic changes in the tissues of the cattle tick *Boophilus microplus* during its development from the erythrocytic stages ingested by the females to the final forms found in the salivary glands of the larvae. However, details of *Babesia* development in the tick gut are not yet clear, owing in part to our scanty knowledge of the structure and function of gut. Many studies on tick fever of cattle have focused on the complex life cycle of *Babesia bigemina* and *Babesia bovis* without correlating this with the physiological and chemical developments within the tick.

Agbede & Kemp (1985) showed that during feeding, the gut of female *B. microplus* had five cell types, the stem cell, which differentiated into any one of the other cell types, two secretory cells (s_1, s_2), a basophilic cell (b), and two subtypes of digest cells (dhl, dhv) (Fig. 21.1). By the time of rapid tick engorgement just prior to dropping from the host, that is, the time of *B. bovis* infection, only two cell types remained in the gut, the digest cells (dhl, dhv) and the basophilic cells (b). The digest cells were involved with intracellular digestion of the bloodmeal, while the basophilic cells had a variety of functions including, possibly, the synthesis of vitellogenins (egg proteins) (Coons, *et al.* 1982, Agbede 1984).

These histological studies, together with the accumulating information on bovine *Babesia* in ticks (Fig. 21.2, Friedhoff 1981) could provide a basis for answering some of the still outstanding questions on the life cycle of *B. bovis* in *B. microplus*. Are Strahlenkörper (Friedhoff 1981) the gametes of

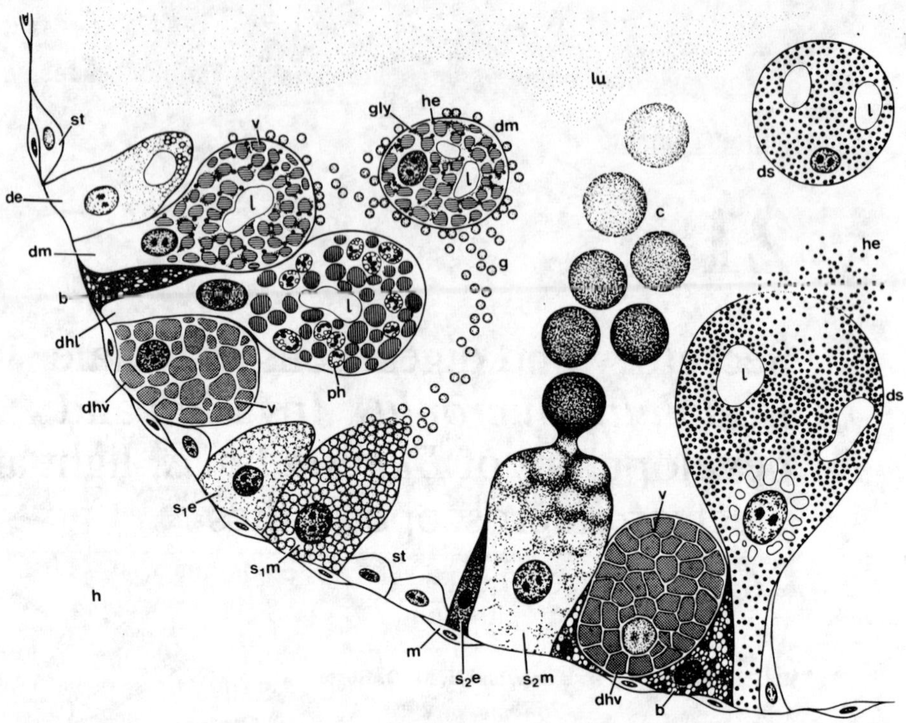

Fig. 21.1 — Suggested development of cells in the gut of *B. microplus* from newly moulted to fully engorged female. b=basophilic cell; c=colloidal mass, secretory cell s_2m; de=digest cell, early; dhl=digest cell with haemoglobin and leucocytes (nymph and 3rd generation female); dm=digest cell, mature (1st and 2nd generation); dhv=digest cell with haemoglobin only; ds=digest cell, spent; g=secretory cell s_1m granules; gly=glycogen; h=haemolymph space; he=haematin; l=lipid; lu=gut lumen; m=muscle cell; ph=phagosome with host leucocyte; s_1e=secretory cell 1, early; s_1m=secretory cell 1, mature; s_2e=secretory cell 2, early (females); s_2m=secretory cell 2, mature (females); st=stem cell; v=vacuole with haemoglobin.

B. bovis? Which of the cell types mentioned above does *B. bovis* penetrate and in which does it develop? By what mechanism do vermicules recognize and penetrate the ovaries? In this chapter we provide new information on the development of *Babesia* in the tick's gut cells.

21.2 MATERIALS AND METHODS

The Yeerongpilly strain of *Boophilus microplus*, free of *Babesia,* was fed on calves with no previous experience of this tick. When female ticks had attained a length of 4.5 to 6 mm they were picked from the donor calf and fed *B. bovis* infected blood using an *in vitro* technique (Willadsen *et al.* 1984). The ticks were taped to a glass slide heated to 35 ± 0.5 °C on an aluminium plate and the infected blood was presented in a glass capillary tube fitted

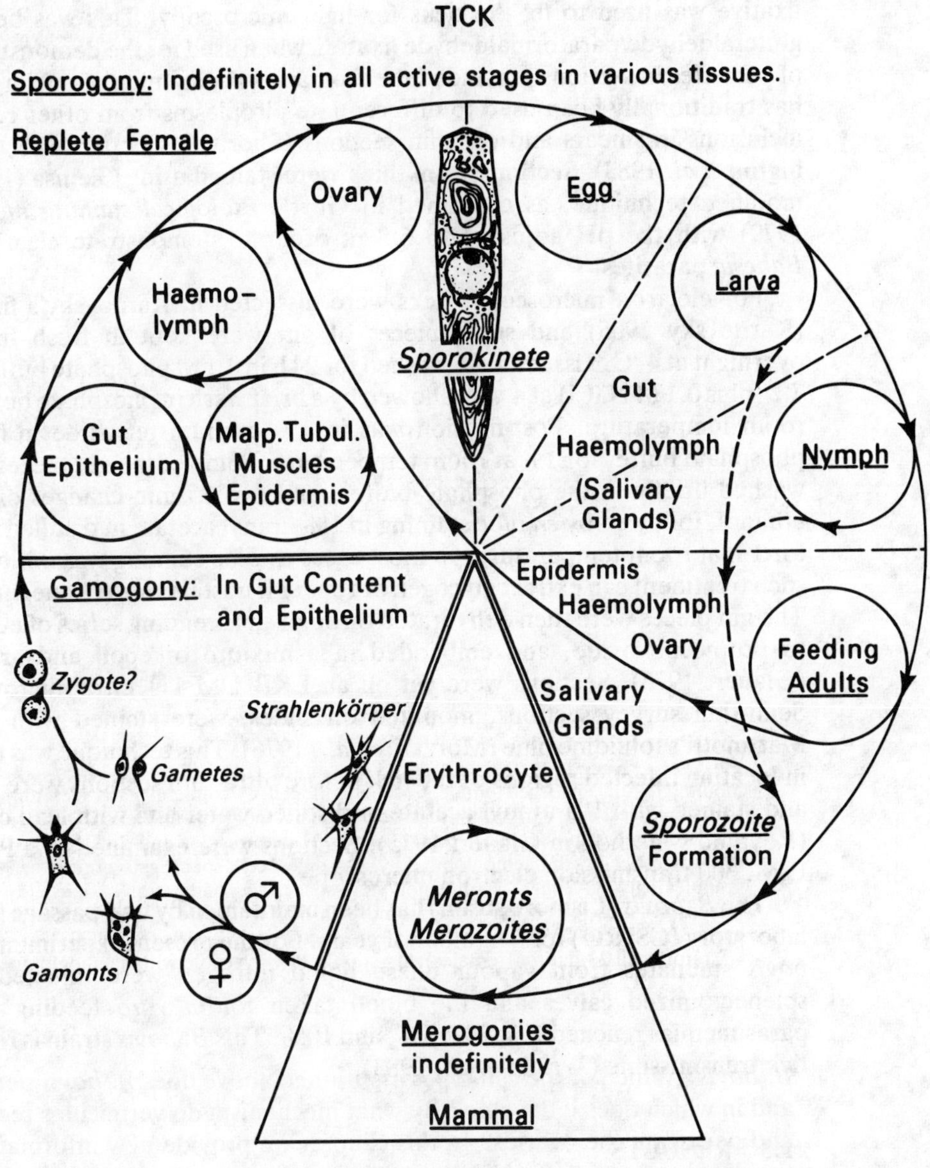

Fig. 21.2 — A hypothetical life cycle for *Babesia*, adapted from Friedhoff (1981).

over the tick's mouthparts. The anticoagulant was preservative-free mucosal heparin (Glaxo) at a final concentration of 10 units/ml. For comparison, ticks were also fed on calves with *B. bovis* parasitaemias of not more than 10%. All engorged females were incubated at 28° ± 0.5 °C, and greater than 90% relative humidity. Three ticks were taken at intervals after incubation for processing to methacrylate embedding (Ruddell 1967a,b). Carnoy's fixative was used to fix the ticks for light microscopy. This was because gluteraldehyde/paraformaldehyde fixation when used for the demonstration of parasites in sections precluded the Romanowsky staining of nuclei which has traditionally been used to differentiate piroplasms from other cellular inclusions in smears and paraffin sections (Shortt & Cooper 1948, Binnington *et al.* 1983). Sections 1 μm thick were stained using Giemsa (Hogg's modified technique, as contained in Gurr's *Biological staining methods* 1979) with the pH adjusted to 6.8 in order to demonstrate clearly the *Babesia* parasites.

For electron microscopy, ticks were dissected in Karnovsky's fixative (Karnovsky 1965) and small pieces of gut were kept in fresh fixative overnight at 4 °C. Tissues were washed for 24 h in 0.1 M phosphate buffer pH 7.2 (plus 0.1 M NaCl) at 4 °C, followed by a brief wash in phosphate buffer at room temperature. Post-fixation was in 1% osmium tetroxide in 0.1 M phosphate buffer for 1 h at room temperature. Some of the gut pieces were washed briefly in the phosphate buffer, and two 15 min changes of 30% ethanol, followed by *en bloc* staining in 1% uranyl acetate in distilled water for 1 h at room temperature. In most cases *en bloc* staining was omitted as such treatment can extract glycogen or render it unstainable (Glauert 1975). The gut pieces were then dehydrated through an ascending series of ethanol to propylene oxide, and embedded in a mixture of epon and araldite (Glauert 1975). Sections were cut on an LKB UM 111 ultra microtome. Semi-thin survey sections, mounted on slides, were stained with 0.2% Martinotti's toluidine blue (Morzaria *et al.*, 1976). This technique was useful in locating infected regions of the tick before ultra-thin sections were made and stained with 1% uranyl acetate in distilled water and with lead citrate (Reynold's method in Glauert 1975). Sections were examined in a Philips E.M. 300 transmission electron microscope.

The *B. bovis* (Lismore strain) has been maintained by tick-passage in this laboratory (CSIRO) for a number of years. For the present experiments, *B. bovis* stabilates from vapour phase liquid nitrogen were injected into splenectomized calves and the blood taken for *in vitro* feeding when parasitaemias reached 1%, 3%, 5%, and 10%. This *B. bovis* strain is readily tick transmissible (Wright *et al.*, 1983).

21.3 RESULTS

The female ticks from *in vitro* experiments (Table 21.1) which had fed overnight (17 h) increased their weight by an average of 148 mg or 5–6 times their original weight. This was less than the weight gain of ticks on the host and may be explained by the inability of their cuticle to expand when stuck to the slide. *In vivo* data are presented in Table 21.2.

A total of 55 engorged females were examined up to day 4 post-incubation and the gut cells observed in these engorged females are shown in Fig. 21.3a,b. In the newly engorged ticks (day 0; *in vitro* or *in vivo*) a few *Babesia* merozoites appeared to have transformed into Strahlenkörper (Fig. 21.4). The numbers and their lengths varied, but they had long rays or spikes and were similar to those in the first phase of development described by

Table 21.1 — *In vitro* feeding; weight (mg) of female ticks before feeding, weight (mg) of bloodmeal, and time (hours) after feeding began when ticks were taken for processing

1% Parasitaemia: (Total 11 ticks)			3% Parasitaemia: (Total 10 ticks)		
Tick wt	Blood wt	Time (hours)	Tick wt	Blood wt	Time (hours)
32	27	1	28	70	10
33	47	3	Feeding stopped after 17 hours		
29	42	3	34±5	137±23	*
Feeding stopped after 17 hours					
29±5	115±30	*			

5% Parasitaemia: (Total 10 ticks)			10% Parasitaemia: (Total 10 ticks)		
Tick wt	Blood wt	Time (hours)	Tick wt	Blood wt	Time (hours)
38	57	3	38	54	3
Feeding stopped after 24 hours			Feeding stopped after 17 hours		
34±7	153±20	*	34	77	17
			36	75	17
			33	129	24
			35	153	48

*Ticks were removed from *in vitro* feeding at 17 h or 24 h after feeding began and ticks were taken for processing at 17 h, 24 h, 48 h, 96 h, and 168 h.

Table 21.2 — *In vivo* feeding; mean weight (mg) of engorged female ticks at incubation (100 ticks/batch collected daily), and time (hours after engorgement) when ticks were processed.

Tick wt	Time
261	0
261	24
344	48
344	72
291	96
291	120
268	144
268	168

Friedhoff (1981). Most were free in the lumen and appeared larger than the forms still within the red cells. The diameter of the Strahlenkörper body averaged 2.63 ± 0.01 μm, and the longest spike was 4.43 ± 0.5 μm. For comparison, red cells in the lumen had an average diameter of 4.41 μm. The morphological sequence in the development of the Strahlenkörper was not clarified in this study. Strahlenkörper of varying shapes and forms were present in the gut lumen up to two days post-incubation, but there was an

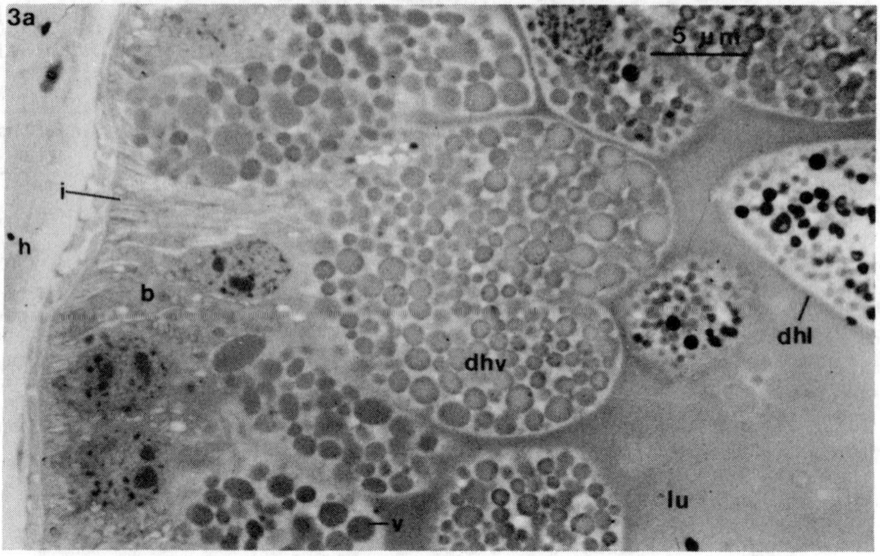

Fig. 21.3 — a and b. Cell types in the gut of *B. microplus* females; basophilic cell, b; basal infolding, i; 3rd generation digest cells with haemoglobin only, dhv; 3rd generation digest cell with haemoglobin and leukocytes, dhl; lumen, lu; haemolymph space, h; vacuole with haemoglobin, v. (Methacrylate section, Haematoxylin, toluidine blue — phyloxine (H & TB-P).

increasing number of the rounded forms without very prominent rays or spikes by 1 day after engorgement. These parasites appeared amoeboid in shape, had dense chromatin at the borders of their nuclei, and were found close to the dhv digest cells (Fig. 21.5). This stage of the parasite appeared to correspond with the fifth phase described by Friedhoff (1981). The width of the body was 4.66 ± 0.11 μm and a few still had long spikes 6.57 ± 0.52 μm in length.

At 2 days after engorgement a new stage was very noticeable inside the dhv digest cells lying in the cell stroma apparently squeezed between the large food vacuoles. Multiplication of this stage in the dhv cells was very marked, giving rise to many star-shaped parasites with a central dense nucleus (Fig. 21.6). In fact, the shape appeared to be determined by the available space between the food vacuoles. The sizes of the parasites at this stage also varied considerably. The largest was 8.87 μm in length by 4.98 μm in width; the smallest was 4.85 μm in length by 3.70 μm in width. Neighbouring parasites appeared to be making contact with their cytoplasmic extensions, and the food vacuoles in cells with *Babesia* were smaller than in non-infected cells (Figs 21.6 and 21.7).

The individual parasites in these conglomerates of four, ten, or more parasites, were larger than the Strahlenkörper and were not fission bodies. By the end of day 3 these parasites had apparently begun to invade the adjacent basophilic cells *en masse,* and by day 4 there were no *Babesia* left in the digest cells (Fig. 21.7 and 21.8). All the ticks processed on day 4 had some basophilic cells infected with *Babesia,* but the number of cells invaded

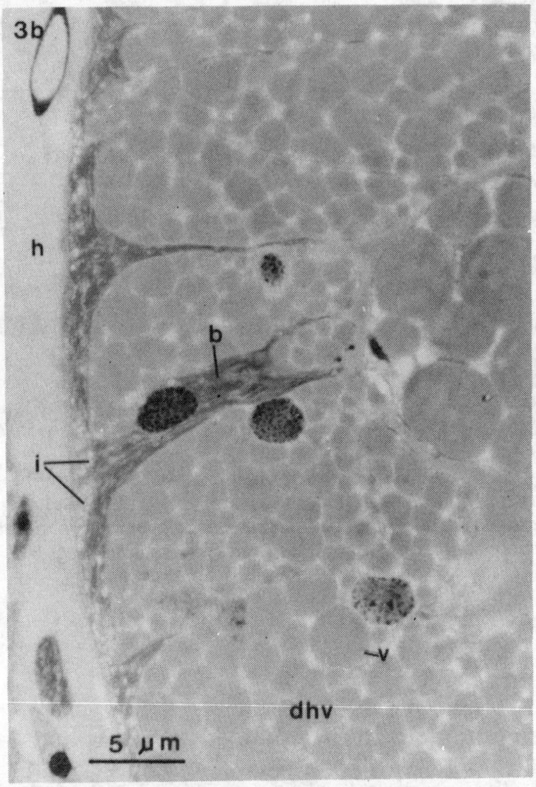

Fig. 21.3 — b. basophilic cell, b; basal infolding, i; 3rd generation digest cells with haemoglobin only, dhv; haemolymph space, h; vacuole with haemoglobin, v. (Methacrylate section, Haematoxylin, toluidine blue — phyloxine (H & TB-P).

was higher in the *in vivo* than the *in vitro* fed ticks. These parasites gave rise to the immature multiple fission bodies (Figs 21.7, 21.8), and then mature multiple fission bodies (kinetes, Fig. 21.9) in the basophilic cells. Development in these cells was completed when fission bodies developed into kinetes (vermicules) which were the motile stages commonly found in the haemolymph by day 5 and later.

Hypertrophy of the infected basophilic cell nuclei was observed. The diameter of the nucleus in non-infected cells was 10.30 ± 0.31 μm and in infected cells 16.36 ± 0.90 μm. As the fission bodies matured the parasitized cells became less basophilic in their staining properties than non-parasitized cells, and showed signs of degeneration.

By day 6, basophilic cell infection was no longer detectable, as the vermicules had escaped into the haemolymph, and by day 7 they had invaded the developing ova and other internal organs. Further development only occurred in the ova where the vermicules rounded up and multiplied within the cell cytoplasm between the yolk granules (Fig. 21.10). Few ova

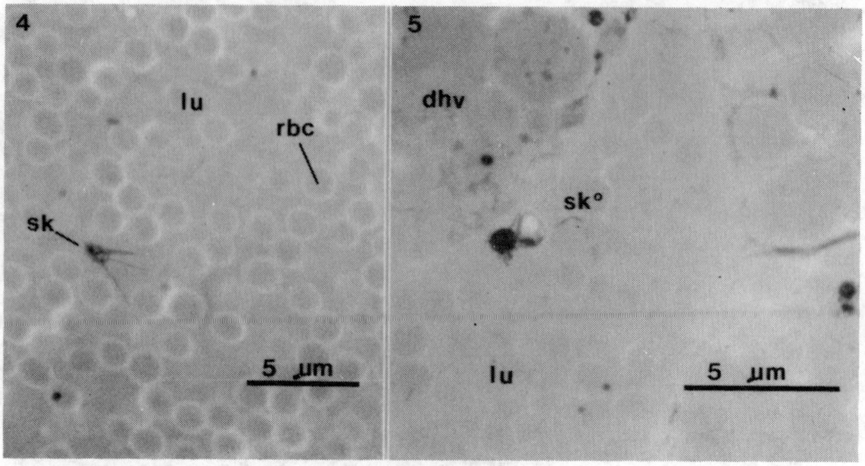

Fig. 21.4 — Gut lumen of newly engorged female; red blood cells, rbc; Strahlenkörper, sk; lumen, lu. (Methacrylate section, Giemsa). Fig. 21.5 — Gut lumen of engorged female 24 h post incubation; 3rd generation digest cell with haemoglobin vacuoles only, dhv; lumen, lu; rounded Strahlenkörper, sk°. (Methacrylate section, Giemsa).

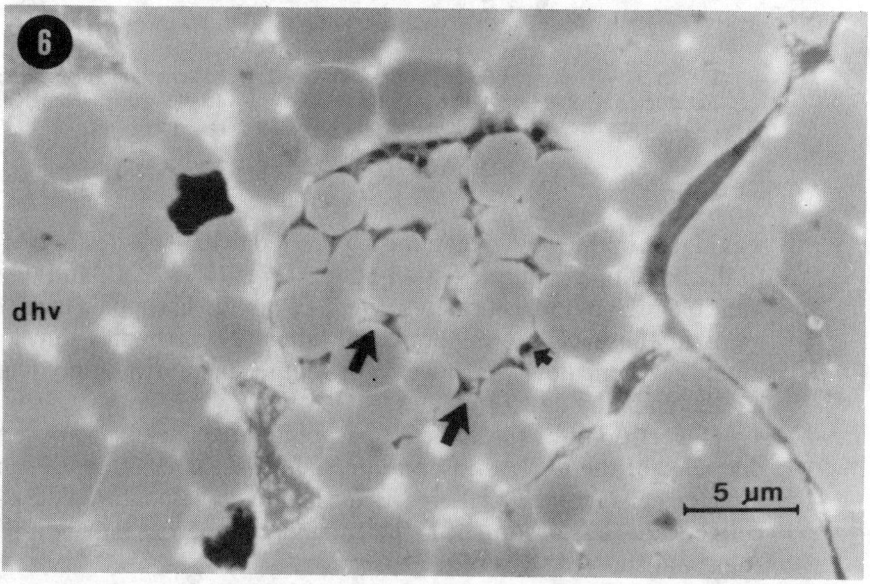

Fig. 21.6 — Digest cell bodies in infected 3rd generation digest cell with haemoglobin only; large arrows indicate star-shaped parasites between the food vacuoles (endosomes); small arrow indicates dark red parasite nucleus; uninfected 3rd generation digest cell, dhv. (Methacrylate section, Giemsa).

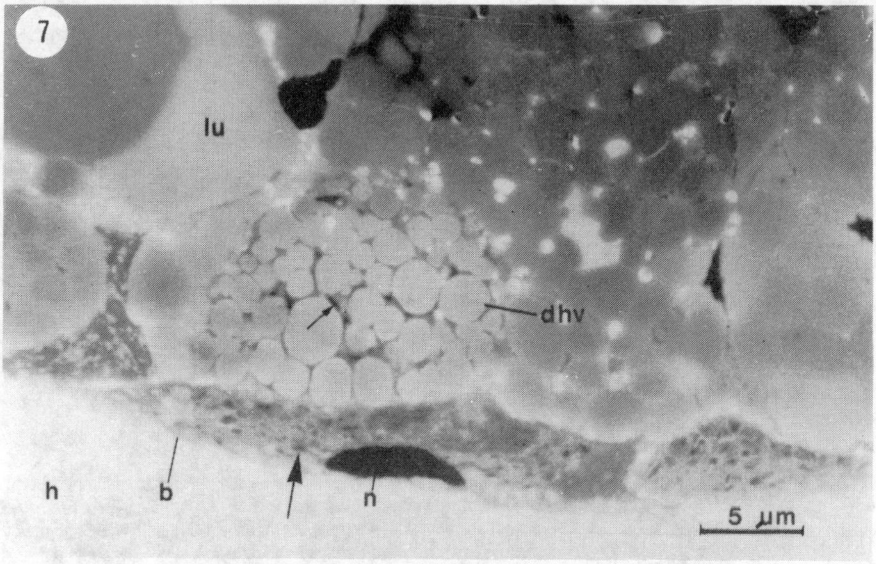

Fig. 21.7 — Digest cell bodies in 3rd generation digest cell, dhv; and invading basophilic cell, b; small arrow indicates parasites in digest cell; large arrow indicates parasites in basophilic cell; nucleus of basophilic cell, n; haemolymph space, h; lumen, lu. (Methacrylate section, Giemsa).

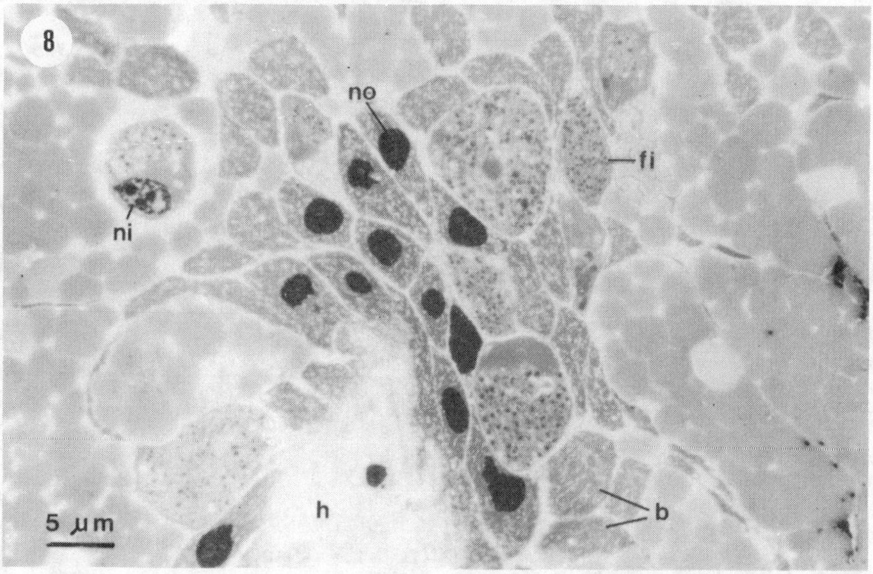

Fig. 21.8 — Immature multiple fission bodies, fi, in infected basophilic cell; uninfected basophilic cell, b; enlarged nucleus of infected cell, ni; nucleus of uninfected basophilic cell, n_0; haemolymph space, h. (Methacrylate section, Giemsa).

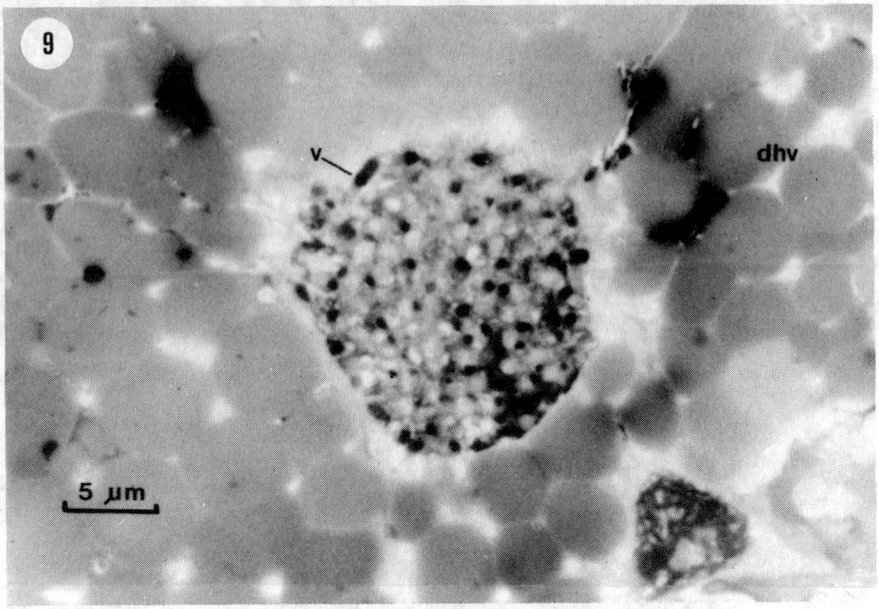

Fig. 21.9 — Mature multiple fission bodies (vermicule), in a basophilic cell; digest cells, dhv; vermicule, v. (Methacrylate section, Giemsa).

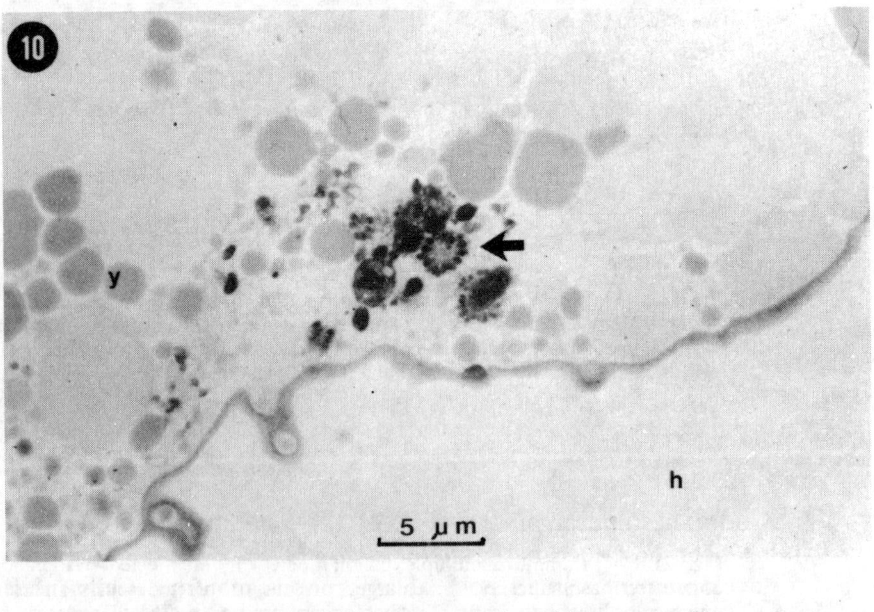

Fig. 21.10 — *B. bovis* parasites in ovum indicated by arrow; yolk granules, y; haemolymph space, h. (Methacrylate section, Giemsa).

were infected (less than 5%), but infected ova were heavily parasitized, resulting in localized degeneration of the ovum. Hypertrophy of the nucleus in the infected cells of the developing ovum was not observed, and the parasite divided in a specific pattern different from the gut stages. The events as they occurred in the ovum appeared to be exactly the same as those reported in detail for *Babesia bovis* in *Rhipicephalus bursa* (Moltmann *et al.*, 1982) and *Babesia major* in *Haemaphysalis punctata* (Morzaria *et al.*, 1976) and therefore were not investigated further.

In ultrastructure, a total of 12 ticks were examined, and some 84 gut cells were infected with fission bodies of *B. bovis*; all were basophilic cells. Two main stages of the parasite were found: the immature fission body in which only the limiting plasma membranes were clearly distinguishable (Fig. 21.11) and the maturing fission body (kinete) which had well defined

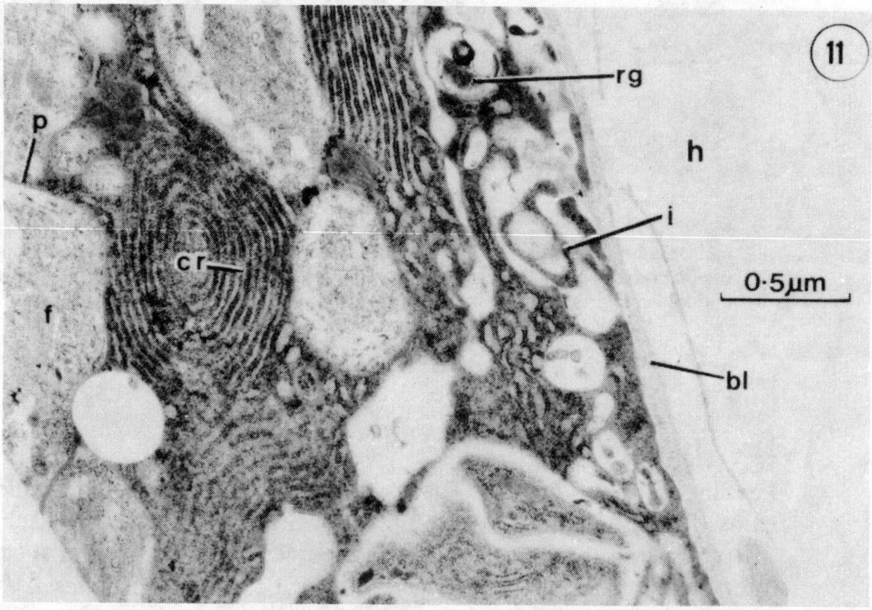

Fig. 21.11 — T.E.M. of immature fission bodies in the gut basophilic cell of *B. microplus* females on the 3rd day after engorgement. Immature fission bodies, f; poorly defined plasma membrane of fission body, p. cisternae of rough endoplasmic reticulum, cr; remnant secretory granule, rg; basal lamina, bl; haemolymph space, h; basal infolding, i.

membranes and organelles (Fig. 21.12). The ultrastructure of the kinetes has been described by Potgieter *et al.* (1976). Both stages appeared to be developing in a small pool of amorphous material. Cells infected with immature fission bodies were well endowed with rough endoplasmic reticulum, but this was much reduced in cells with kinetes (Figs 21.11 and 21.12).

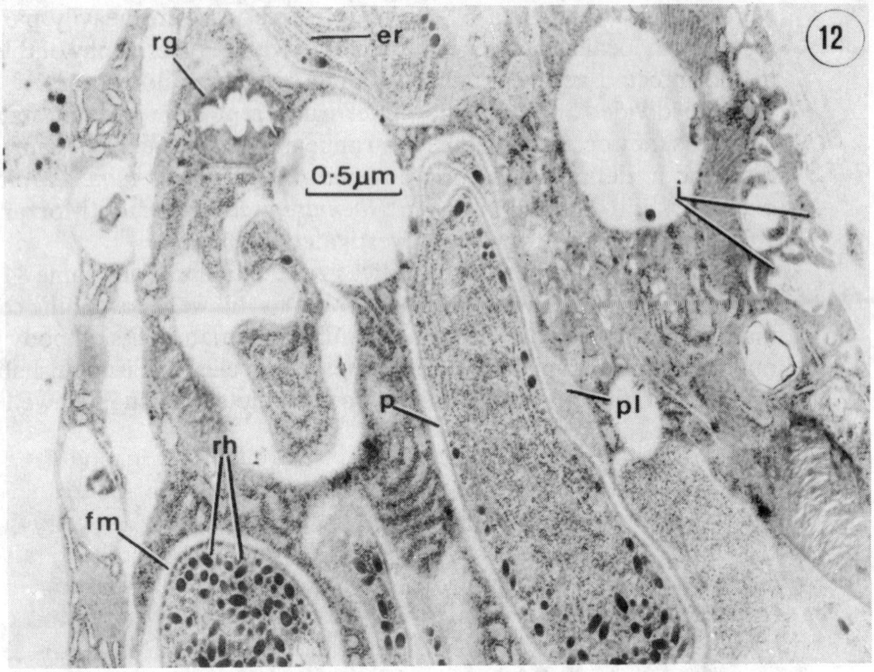

Fig. 21.12 — T.E.M. of mature fission bodies in the gut basophilic cell of *B. microplus* on the 4th day after engorgement (fm); plasma membrane of mature fission body, p; remnant secretory granule, rg; basal infolding, i; rhoptries, rh; rough endoplasmic reticulum of fission body, er; pool of amorphous material surrounding fission body, pl.

21.4 DISCUSSION

Feeding ticks *in vitro* was successful, and most females had good weight gains. This should make it easier to study the early morphological changes of the parasite in the tick gut because the ticks could be fixed at specific time intervals after ingestion of the erythrocytic stages of *Babesia*. It is also possible that *Babesia* development would be more uniform than with *in vivo* infection. More basophilic cells were invaded by *Babesia* in the *in vivo* fed ticks. This might indicate greater infectivity of the *Babesia* under natural conditions, but these ticks also imbibed more infected blood than the *in vitro* fed ticks. The infectivity of the *B. bovis* strain under *in vitro* conditions was confirmed by gut infection in all ticks examined after 2 days.

In the past, the nomenclature of various *Babesia* stages in the tick has been confusing, and to avoid this, the nomenclature of Friedhoff (1981) has been adopted. The term 'kinete' (vermicule) has been used to describe the mature products of the multiple fission which occurred in the basophilic cells of the gut.

In this preliminary study of *B. bovis* in the gut, gametes were not identified at either the light or E.M. level. Friedhoff (1981) reported that sexual fusion of gametes has not been seen in ovine and bovine *Babesia*, though Rudzinska *et al.* (1982, 1983a, 1983b) have distinguished male and female gametes and demonstrated gamete fusion of *Babesia microti* in the gut of *Ixodes dammini.* The issue of gamete fusion has also not been resolved in this study of *B. bovis.* The evidence suggested that the later rounded forms of Strahlenkörper present in the lumen initiated development of parasites in the dhv digest cells, but it was not possible to identify these forms as ookinetes (fertilized macrogametes).

All the earlier stages of Strahlenkörper that were taken into the dhl digest cells appeared to be digested, and only a few parasites left in the lumen presumably managed to penetrate the right digest cells (dhv). This later stage of Strahlenkörper in the lumen was similar to the fifth phase described by Friedhoff (1981) as giving rise to multiple fission bodies in the epithelial cells of the gut (Friedhoff 1969). In the present study it has been shown that although Strahlenkörper penetrated the dhv cells, they did not form fission bodies, but they did divide within these cells, giving rise to a large number of star-shaped stages. For *B. bovis* in *B. miroplus,* this stage is morphologically different from the Strahlenkörper in the lumen or the fission bodies in the basophilic cells. Apart from the difference in location of the parasite, the stage in the digest cell occurs prior to the appearance of fission bodies in the basophilic cells, and it is possibly the precursor of the fission bodies. It is referred to as the digest cell body (dcb). Friedhoff (1969) described a stage of *Babesia bovis* in the gut cells of *Rhipicephalus bursa* (his Figs 3 and 9), but did not identify two stages of development in different gut cell types.

The food vacuoles in the dhv cells increased in size by the accretion of small vacuoles (Agbede & Kemp 1985). The fact that the dcb stages of *Babesia* were not incorporated into the food vacuoles, meant that they were in no danger of being digested with the bloodmeal. The shape and form of these stages (dcb) appeared to be determined by the available space between the food vacuoles in the dhv cells, and in their turn the *Babesia* affected the formation of the food vacuoles. Also the presence of the dcb stages outside the food vacuoles suggested that entry into the cell was an active process and did not take place passively with the food. Wright *et al.* (1981) found that a strain of *B. bovis* parasites without a particular esterase were non-tick transmissible, suggesting that this enzyme had some role in establishing infection in the tick vector. Electron microscopy of parasites penetrating and developing in the digest cells is required to resolve these problems.

Another interesting point is the development of the basophilic cell in relation to *Babesia* development. This cell appeared to originate from the secretory cell type s_2 (Fig. 21.1) present only in the partially fed female (Agbede & Kemp 1985, Agbede, *et al.* 1986). The cells began their second phase of activity after the ticks had engorged. They synthesized new material believed to be egg proteins (vitellogenins) which were presumably secreted into the haemolymph and thence into the eggs (Coons *et al.* 1982, Agbede 1984). The timing of these events suggests that in the evolution of the *Babesia: Boophilus* relationship the *Babesia* has developed an ability to recognize and penetrate this tick gut cell which has the machinery for active synthesis of egg proteins. Fission bodies were all concentrated within the basophilic cells, and hypertrophy of the host cell nucleus suggested a close association of this parasite and the cells. The membranes of the parasitophorous vacuole in which *Babesia bigemina* developed, were found to be continuous with the endoplasmic reticulum of the host cell (Friedhoff & Scholtyseck 1969), suggesting a direct secretion of host cell proteins into the vacuole around the parasite.

Selection of the basophilic cells for development would also make it easier for the liberation of mature vermicules into the haemolymph, because these cells have a much larger surface area in contact with the haemolymph than the digest cells.

Babesia kinetes (vermicules) have been demonstrated in various internal organs of the tick, and there is abundant opportunity for vermicules to be led astray. However, vitellogenin must pass whole into the ova, and a speculative explanation for selection of the ova by the *Babesia* vermicules is that they recognize receptors for these proteins on the surface of the ovarian cells. No studies on the behaviour of the vermicules are available.

ACKNOWLEDGEMENTS

We thank Drs K. C. Bremner, K. C. Binnington, N. Stewart, and I. G. Wright for comments on the manuscript and S. Fiske for some of the figures.

REFERENCES

Agbede, R. I. S. (1984) A study of the gut cells of *Boophilus microplus* in relation to *Babesia bovis* development and host immunity, PhD thesis, University of Queensland.

Agbede, R. I. S. & Kemp, D. H. (1985) Digestion in the cattle-tick *Boophilus microplus*. Light microscope study of the gut cells in nymphs and females, *Int. J. Parasitol.* **15** 147–157.

Agbede, R. I. S., Kemp, D. H. & Hoyte, H. M. D. (1986) *Babesia bovis* development in secretory cells of *Boophilus microplus, Int. J. Parasitol.* **16** 109–114.

Binnington, K. C., Young, A. S. & Obenchain, F. D. (1983) Morphology of normal and *Theileria parva*-infected salivary glands of *Rhipicephalus appendiculatus* (Acari: Ixodoidea), *J. Parasitol.* **69** 421–424.

Coons, L. B., Tarnowski, B. & Ourth, D. D. (1982) *Rhipicephalus sanguineus:* localization of vitellogenin synthesis by immunological methods and electron microscopy, *Exp. Parasitol.* **54** 331–339.

Friedhoff, K. T. (1969) Lichtmikroskopische Untersuchungen über die Entwicklung von *Babesia ovis* (Piroplasmidea) in *Rhipicephalus bursa* (Ixodoidea). I. Die Entwicklung in weiblichen Zecken nach der Repletion, *Z. Parasitenkd.* **32** 191–219.

Friedhoff, K. T. (1981) Morphologic aspects of *Babesia* in the tick. In:

Babesiosis, (eds Ristic, M. & Kreier, J. P.) Academic Press, New York p. 143–169.

Friedhoff, K. T. & Scholtyseck, E. (1969) Feinstrukturen der Merozoiten von *Babesia bigemina* im Ovar von *Boophilus microplus* und *Boophilus decoloratus, Z. Parasitkde.* **32** 266–283.

Glauert, A. M. (1975) *Fixation, dehydration and embedding of biological specimens.* North–Holland, Amsterdam.

Karnovsky, M. J. (1965) A formaldehyde-glutaraldehyde fixative of high osmolality for use in electron microscopy, *J. Cell Biol.* **27** 137A.

Moltmann, U. G., Mehlhorn, H. & Friedhoff, K. T. (1982) Ultrastructural study of the development of *Babesia ovis* (Piroplasmia) in the ovary of the vector tick *Rhipicephalus bursa, J. Protozool.* **29** 30–38.

Morzaria, S. P., Bland, P. & Brocklesby, D. W. (1976) Ultrastructure of *Babesia major* in the tick *Haemaphysalis punctata, Res. Vet. Sci.* **21** 1–11.

Potgieter, F. T., Els, H. J. & Van Vuuren, A. S. (1976) The fine structure of merozoites of *Babesia bovis* in the gut epithelium of *Boophilus microplus, Onderstepoort J. vet. Res.* **43** 1–10.

Ruddell, C. L. (1967a) Hydroxyethyl methacrylate combined with polyethylene glycol 400 and water; an embedding medium for routine 1–2 micron sectioning, *Stain Technol.* **42** 119–123.

Ruddell, C. L. (1967b) Embedding media for 1–2 micron sectioning. 2. Hydroxyethyl methacrylate combined with 2–butoxyethanol, *Stain Technol.* **42** 253–255.

Rudzinska, M. A., Spielman, A., Lewengrub, S., Piesman, J. & Karakashian, S. (1982) Penetration of the peritrophic membrane of the tick *(Ixodes dammini)* by *Babesia microti, Cell Tissue Res.* **221** 471–481.

Rudzinska, M. A., Spielman, A., Lewengrub, S., Trager, W. & Piesman, J. (1983a) Sexuality in piroplasms as revealed by electron microscopy in *Babesia microti, Proc. Natl. Acad. Sci. USA* **80** 2966–2970.

Rudzinska, M. A., Lewengrub, S., Spielman, A. & Piesman, J. (1983b) Invasion of *Babesia microti* into epithelial cells of the tick gut, *J. Protozool.* **30** 338–346.

Shortt, H. E. & Cooper, W. (1948) Staining of microscopical sections containing protozoal parasites by modification of McNamara's method, *Trans. Roy. Soc. Trop. Med. Hyg.* **41** 427–428.

Willadsen, P., Kemp, D. H. & McKenna, R. V. (1984) Bloodmeal ingestion and utilization as a component of host specificity in the tick, *Boophilus microplus, Z. Parasitenkd.* **70** 415–420.

Wright, I. G., Goodger, B. V. & Mahoney, D. F. (1981) Virulent and avirulent strains of *Babesia bovis:* the relationship between parasite protease content and pathophysiological effect of the strain, *J. Protozool.* **28** 118–120.

Wright, I. G., Mirrie, G. B., Mahoney, D. F. & Goodger, B. V. (1983) Failure of *Boophilus microplus* to transmit irradiated *Babesia bovis, Res. Vet. Sci.* **34** 124–125.

22

Development of *Anaplasma marginale* Theiler in ixodid ticks: Coordinated development of a rickettsial organism and its tick host

Katherine M. Kocan

22.1 INTRODUCTION

22.1.1 Ticks as vectors of disease

Ticks have been well documented as efficient vectors of human and animal diseases and are known to transmit organisms from various taxonomic groups — including the protozoans, rickettsiae, viruses, and bacteria — among vertebrate hosts (Hoogstraal 1970, Bertram 1962, Philip & Burgdorfer 1961, Philip 1963, Uilenberg 1980). Of notable economic importance to the cattle industry are four tick-borne diseases: two of these diseases are protozoal (babesiosis, caused by *Babesia bigemina,* and East Coast fever, caused by *Theileria parva*) and two are rickettsial (heartwater disease, caused by *Cowdria ruminantium,* and anaplasmosis, caused by *Anaplasma marginale*) (Bram 1975). None of these agents are known to cause harm to the arthropod host, but all are pathogenic in the associated vertebrate host (Rěhàcěk 1965).

Ixodid ticks are especially suitable vectors of disease agents, in part because they attach securely to hosts and feed for extended periods, thus maximizing the opportunity for exposure to and ingestion of infectious organisms. They are also adaptable to environmental changes and long-lived. Ixodid ticks and the organisms they harbor often overwinter (Philip &

Burgdorfer 1961, Philip 1963). Disease agents may persist through several developmental stages of the tick (transstadial) or may be passed from one generation to the next via the ovaries (transovarial) (Burgdorfer & Varma 1968). In most cases, initial infection of ticks occurs by infection of midgut epithelial cells with organisms ingested in the blood meal (Burgdorfer & Varma 1968). After infecting the ticks, the organisms must be able to remain viable until a host is found on which to take a bloodmeal — often longer than a year. The mechanism by which ticks persist for long periods without feeding is complex. Likewise, the survival of a parasitic organism under these conditions must also be complex, and little is known about the specific factors that facilitate this longevity.

Transmission of disease agents to vertebrate hosts has been documented to occur when ticks feed (Burgdorfer & Varma 1968, Mehlhorn & Schein 1984). Organisms can be passed by contamination of body fluids via the bite or by contamination of the bite lesions with feces or coxal fluid. Most commonly, disease-causing organisms develop in and are released from the salivary glands of feeding ticks, entering the vertebrate host via tick saliva. However, with many tick-borne parasites, especially rickettsial organisms, the exact mechanism of development and transmission is not yet known.

The concept of a developmental cycle of organisms transmitted by ticks has been well documented for the protozoan parasites, *Babesia* sp. and *Theileria* sp. (Mehlhorn & Schein 1984). The developmental cycle is complex and specifically coordinated with the development of the tick. Initial development in gut cells involves sexual reproduction, and reproduction and formation of zygotes. Stages (kinetes) are produced from the zygote which leave the gut and penetrate the salivary gland acini (specifically the type III acinus). Upon attachment of the subsequently-molted tick to a susceptible host, stages already present in the salivary glands begin to develop into sporozoites and are transmitted via the saliva as feeding progresses. Thus, the parasite's development is precisely coordinated with that of its host.

Host factors that modulate parasite development in ticks have not been characterized. Some investigators have suggested modulation by molting hormones. Other studies have shown that simply placing infected ticks at an elevated temperature initiated factors that caused development of infective forms in the gut cells and salivary glands (Burgdorfer & Varma 1968, Hayes & Burgdorfer 1982, Samish 1977, Dalgliesh & Stewart 1976, 1979, Kocan *et al.* 1982, 1983b). However, specific factors controlling parasite development remain largely undescribed.

The ability to understand, classify, and control any parasitic disease is greatly enhanced by knowledge of the development of the etiologic agent in all of its hosts. An understanding of this development and the nutritional requirements of organisms developing in ticks would influence chances of cultivating tick-transmitted organisms *in vitro*. The transmitted stage is a likely point in the life cycle to interrupt; therefore, definition of this stage is important and may lead to new approaches for disease control or prevention.

22.1.2 Rickettsiae and their development in ticks

Current rickettsial classification is outlined in Table 22.1.The rickettsiae encompass a broad group of vector-borne parasites that can only grow and survive within living cells (Weiss & Moulder 1984). Tick-borne rickettsiae can be grouped into two broad categories: (1) those organisms that live and multiply freely within the cytoplasm of host cells, and (2) those that occur only within membrane-bound inclusions within the host cell cytoplasm. Organisms in the first category include species of the genera *Rickettsia* and *Wolbachia*. Organisms in the second category that are found only within inclusions include parasites assigned to the genera *Ehrlichia, Cowdria, Coxiella,* and *Anaplasma.*

The development of rickettsial organisms in ticks has not been as well documented as it has for the protozoans. Perhaps this is because the type genus for the order Rickettsiales, *Rickettsia,* has been reported to divide only by binary fission and infect all tick tissues. The infection has been referred to as 'generalized', and does not involve a specific sequence of development. However, evidence reported for other rickettsial organisms (*Ehrlichia* (Nyindo *et al.* 1971), *Cowdria* (Uilenberg 1983), and *Coxiella* (Baca & Paretsky 1983)) suggests that different stages occur in vertebrate hosts or in cell culture, and these stages may represent a developmental sequence involving reproduction by means other than binary fission. The development of these inclusion-forming organisms in ticks has not been well characterized. In addition, two of these organisms (*Anaplasma* and *Cowdria*) have not been successfully cultivated *in vitro* , nor have the associated diseases been effectively controlled by vaccines. It is likely that the developmental cycles of these organisms in both vertebrate and invertebrate hosts are more complex than once hypothesized.

22.1.3 *Anaplasma marginale* and its transmission by ticks

Anaplasma marginale Theiler is an intraerythrocytic parasite that is the causative agent of bovine anaplasmosis, a disease characterized by severe anemia. The scientific name is based on its staining characteristics and location within the host cell 'anaplasma' referring to the lack of stained cytoplasm and 'marginale' denoting the peripheral location of the organism in the host erythrocyte (Figs 22.1, 22.2a) (Ristic 1977). *A. marginale* is presently classified in the family Anaplasmataceae, order Rickettsiales (Weiss & Moulder 1984). A separate species, *A. caudatum,* was established more recently to accommodate similar organisms that possess an appendage-like structure in association with the marginal body in bovine erythrocytes. The nature of this appendage is not well understood. It can only be seen with light microscopy by a fluorescent antibody staining technique (Scott *et al.* 1961, Kreier & Ristic 1963) and a new methylene blue staining (Schalm *et al.* 1962) of infected erythrocytes, and by electron microscopy of lysed erythrocytes (the electron-dense hemoglobin masks the structure)

Table 22.1 — Classification of the order Rickettsiales according to the 9th edition of Bergey's Manual

ORDER: Rickettsiales		
FAMILY I: Rickettsiaceae		
Tribe I: Rickettsieae		
	Genus	I. *Rickettsia*
	Genus	II. *Rochalimaea*
	Genus	III. *Coxiella*
Tribe II: Ehrlichieae		
	Genus	IV. *Ehrlichia*
	Genus	V. *Cowdria*
	Genus	VI. *Neorickettsia*
Tribe III: Wolbachieae		
	Genus	VII. *Wolbachia*
	Genus	VIII. *Rickettsiella*
FAMILY II: Bartonellaceae		
	Genus	I. *Bartonella*
	Genus	II. *Grahamella*
FAMILY III: Anaplasmataceae		
	Genus	I. *Anaplasma*
	Genus	II. *Aegyptianella*
	Genus	III. *Haemobartonella*
	Genus	IV. *Eperythrozoon*

(Fig. 22.2b) (Espana *et al.* 1959, Ritchie 1962, Simpson *et al.* 1965, Simpson 1977, Kocan *et al.* 1978a,b).

Anaplasma organisms occur in membrane-bound inclusions within erythrocytes (Figs 22.2, 22.3) (Kocan *et al.* 1978a,b). A hypothetical cycle of reproduction of *A. marginale* in erythrocytes was proposed in which the infective agent enters the erythrocyte by endocytosis, with the erythrocyte plasmalemma forming an inclusion vacuole (Ristic & Watrach 1963). Ferritin-conjugated antibody studies have supported this hypothesis by determining that the inclusion membrane is of erythrocyte origin (Kocan *et al.* 1978b, Francis *et al.* 1979). The organism envelope and appendage were recognized by ferritin-labeled anti-*Anaplasma* serum (Fig. 22.3), but the inclusion membrane was only recognized by labeled antierythrocyte serum. However, transfer of the parasite from one erythrocyte to another has not been demonstrated clearly, and the complete developmental sequence within the vertebrate host remains unknown.

Ticks have been established as vectors of anaplasmosis (Dikmans 1950, Ewing 1981, Anthony 1968). Although *A. marginale* can be transmitted mechanically by horseflies and mosquitos, ticks appear principally respon-

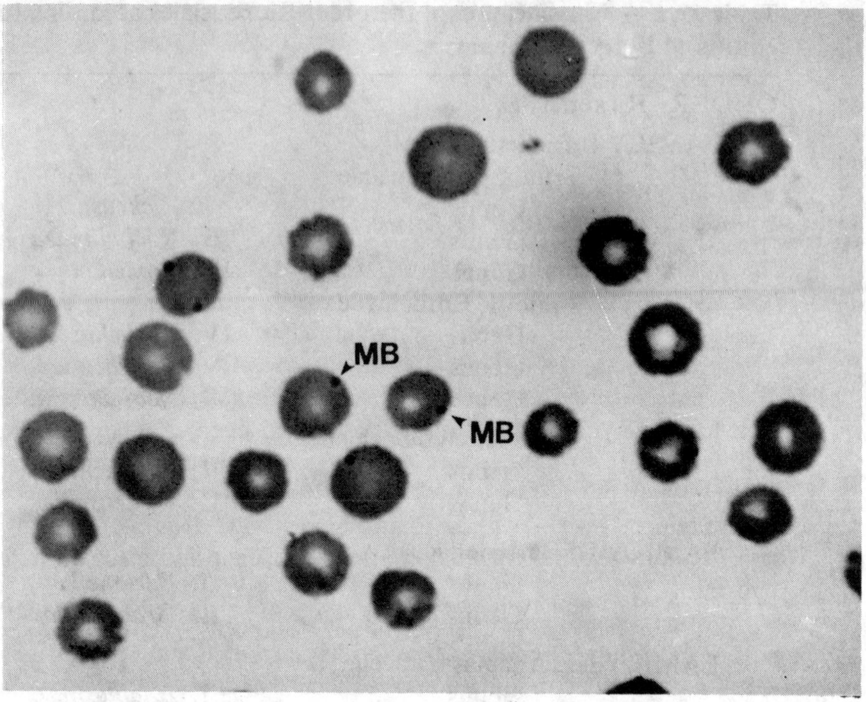

Fig. 22.1 — A Wright's-stained smear of parasitized bovine erythrocytes. Anaplasmal marginal bodies (MB) are at the periphery of erythrocytes. ×488.

sible for the biological transmission of this agent. Tick transmission of enzootic anaplasmosis has been documented in several geographic areas of the world including the United States, South America, and Africa (Bram 1975).

Experimental studies have identified at least 20 species of ticks that are capable of transmitting anaplasmosis (Dikmans 1950, Ewing 1981). Several investigators have verified transstadial transmission, but only one study (of approximately 50 attempts) claims transovarial transmission (Howell *et al.* 1941). In other laboratory studies, transstadial transmission was demonstrated by injection of macerated, feeding infected adult ticks (Anthony 1968). Transmission studies of repeat feedings by hibernating male ticks have demonstrated long-term infectivity in field-collected and laboratory-reared ticks (Sanborn *et al.* 1938, Renshaw *et al.* 1977, Peterson *et al.* 1977, Christensen & Howarth 1966, Howarth & Hokama 1973).

Morphological and fluorescent antibody studies of *A. marginale* in the tick, *Dermacentor andersoni,* were undertaken by several investigators (Anthony & Roby 1962, Friedhoff & Ristic 1966, Anthony *et al.* 1964, Bram & Romanowski 1970). *A. marginale* was demonstrated in the gut contents, excreta, and Malpighian tubules of nymphs and adults for a limited time after feeding on an infected animal, but until work in our laboratory the organism had not been identifiable after host erythrocytes were digested.

22.2 DEVELOPMENT OF *ANAPLASMA MARGINALE* IN *DERMACENTOR ANDERSONI*

The studies from our laboratory reviewed herein were undertaken to demonstrate *A. marginale* in *Dermacentor andersoni* and to describe its development from the infection of nymphal ticks on a calf with anaplasmosis through transmission of the organism to cattle by subsequently-molted adults. For morphological studies, samples of tick tissues were collected on each day of nymphal feeding (6 days) and every 5 days from repletion through molting to the adult stage. At approximately one month post-molting, groups of ticks were placed on susceptible calves to test for transmission. Gut and salivary gland tissues were collected on each of the 9 days of adult feeding for examination with light and electron microscopy (Fig. 22.4).

In our morphological studies, *A. marginale* was found to occur within membrane-bound inclusions that we have termed 'colonies'. The colonies were first described in midgut epithelial cells of adult *D. andersoni* that were infected as nymphs (Kocan *et al.* 1984c). The colonies at various stages of tick development were found, by light microscopy, to have different morphological characteristics and to contain several types of organisms — representing part of a developmental sequence of the organism in the tick. This review of the development of *A. marginale* in ticks will be presented chronologically, beginning from infection of nymphs (while feeding on a calf with anaplasmosis) and continuing through development in and transmission by the adult ticks.

22.2.1 Infection and development in nymphs

Infection of nymphs

Although the specific mechanism of infection of tick gut cells by *A. marginale* is not known, nymphal ticks were infected soon after beginning to feed on a calf with anaplasmosis. Nymphs that fed for as little as 24 hours on an infected calf and were then transferred to an uninfected calf to complete feeding were able to transmit the disease as adults (Kocan 1982). However, when tissues from nymphs collected on each of the 6 days of feeding were examined by light and electron microscopy, colonies of *A. marginale* were not observed (Kocan *et al.* 1984c). Furthermore, no stages of *A. marginale* were seen with electron microscopy either within the midgut lumen or within the epithelial cells of feeding nymphs. The gut contents of feeding nymphs were difficult to evaluate with electron microscopy because the midgut was filled with blood, obscuring parasite morphology. Once concentrated by the tick, the bloodmeal was very electron-dense. As a consequence, the concen-

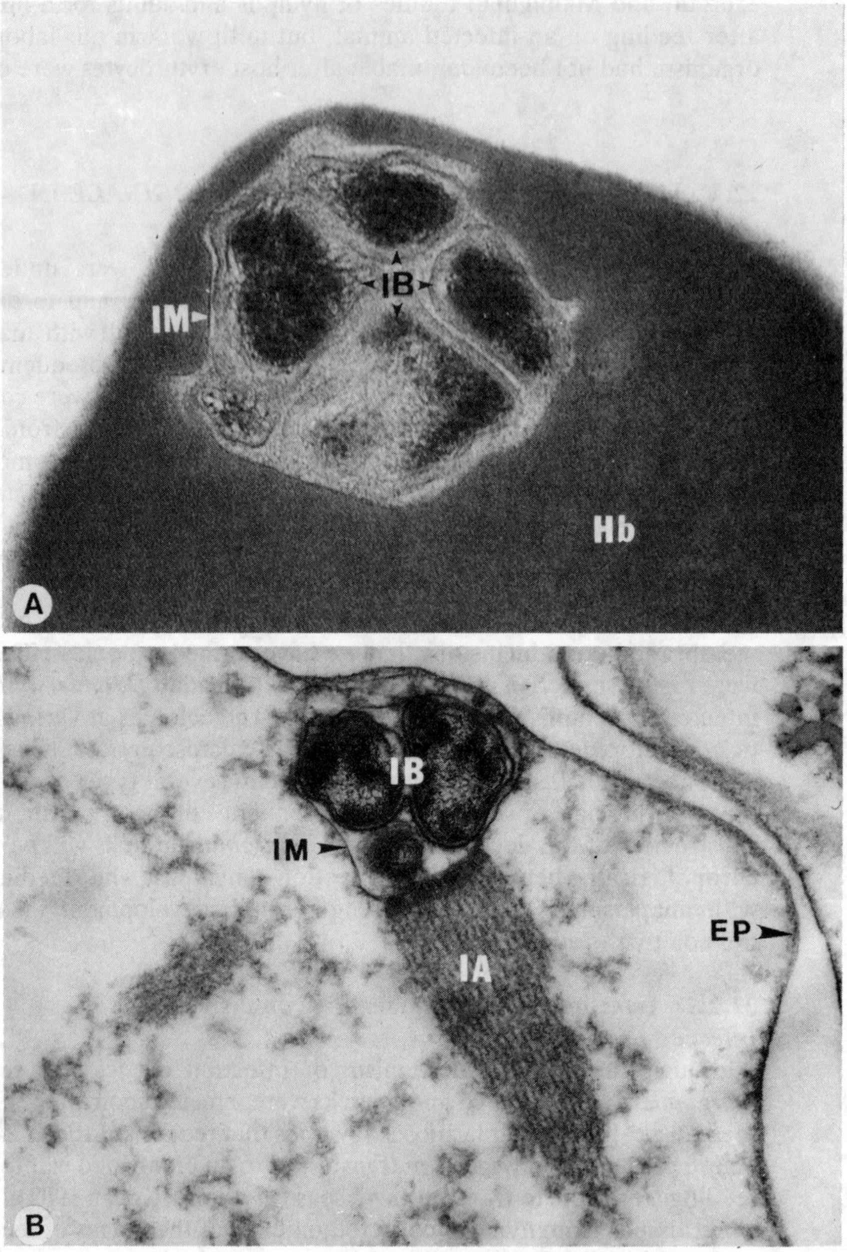

Fig. 22.2 — Inclusion body in hemolyzed bovine erythrocyte. a. The inclusion membrane (IM) surrounds four initial bodies (IB). The erythrocyte contains electron-dense hemoglobin (Hb). ×51,520. b. Inclusion body and inclusion appendage in a hemolyzed bovine erythrocyte. The inclusion appendage (IA) is attached to the inclusion membrane (IM) that surrounds 3 initial bodies (IB). The inclusion is within the erythrocyte plasmalemma (EP). ×33,120.

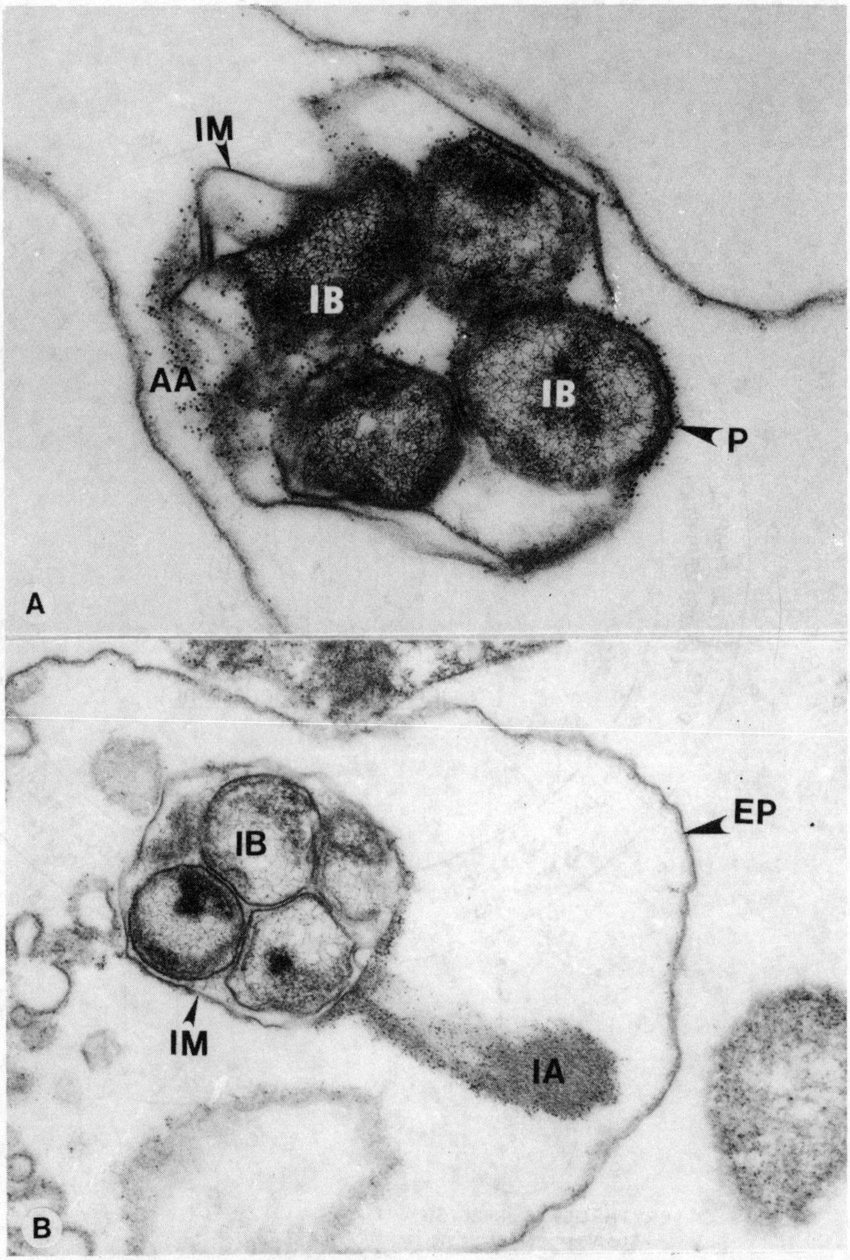

Fig. 22.3 — Anaplasmal inclusion in bovine erythrocytes that have been exposed to ferritin-labeled anti-*A. marginale* serum. a. The outer pellicle (P) of the initial bodies (IB) labeled with ferritin when the inclusion membrane (IM) ruptured. The inclusion membrane was unlabeled except at the junction of the inclusion appendage (AA). ×44,160. b. Damage of the erythrocytic plasmalemma (P) has allowed entrance of the conjugate to label the inclusion appendage (IA). The inclusion membrane (IM) remained intact, thus the initial bodies (IB) did not label. The erythrocytic plasmalemma (EP) and inclusion membrane (IM) did not label. ×44,160.

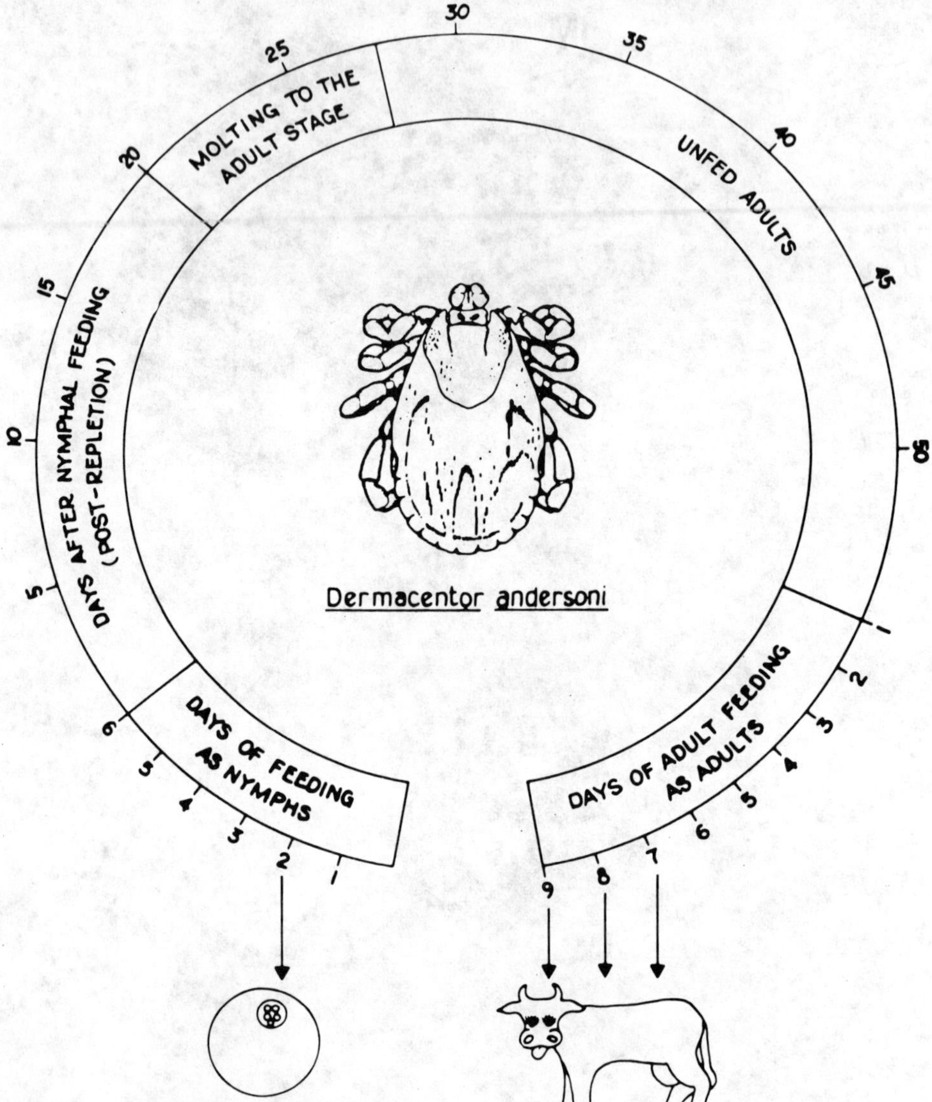

Fig. 22.4 — Experimental protocol for studying development of *A. marginale* in *D. andersoni*. Nymphs were infected by feeding on calves with acute anaplasmosis and allowed to molt to the adult stage where they were studied as feeding adults. Tissue samples were collected on each day of nymphal feeding (total, 6 days) and every 5 days from replete nymphs to 1-month-old adults, as well as on each day of adult feeding (total, 9 days).

trated hemoglobin masked parasite structures in the lumen of the tick's gut to an even greater extent than it did in intact infected bovine erythrocytes (Kocan *et al.* 1978a,b, Kocan *et al.* 1984c).

Replete nymphs
Colonies of *A. marginale* were observed in midgut epithelial cells of replete nymphs throughout development to the adult stage (approximately 20–30 days) (Kocan *et al.* 1984c). Two distinct morphological types of colonies were observed and categorized by light microscopy as nymphal types 1 and 2 (Fig. 22.5). Colonies that were morphologically indistinct, with characteristics in common with both types, were termed transitional nymphal (TsN) colonies. Nymphal type 1 (Ny1) colonies were first observed at 5 days post-repletion, and nymphal type 2 (Ny2) colonies were first observed at 20 days after feeding. The first evidence of colony formation at about 5 days post-repletion appeared more or less concurrently with the rapid formation of midgut epithelial cells and digestion of the blood meal. At completion of feeding, the nymphal tick midgut was essentially a tubular structure filled with concentrated blood, and the gut was lined by a thin layer of epithelial cells. In contrast, at 5 days post-repletion the midgut lumen was not evident and the gut was filled with epithelial cells that had taken up most of the bloodmeal. Therefore, the formation of colonies appeared to have been concurrent with the rapid digestion of the bloodmeal by the tick and attendant cellular proliferation.

When representatives of the 3 nymphal colony types were examined by electron microscopy, organisms of different morphological types were observed within the colonies (Kocan *et al.* 1984c). Ny1 colonies contained small particles and large reticulated forms that were round, and some of which appeared to be dividing by binary fission. (Fig. 22.6). TsN colonies contained reticulated organisms that were more electron-dense, and the colonies were less densely packed with organisms (Fig. 22.7). Ny2 colonies also contained reticulated organisms but they were rod-like in shape, and there was no morphological evidence of binary fission (Fig. 22.8). Organisms within Ny2 colonies appeared to be surrounded by two double-layered membranes. Electron-dense forms, commonly observed in adult ticks, were not seen in any of the three colony types (Ny1, Ny2, or TsN) found in nymphal ticks (Fig. 22.8).

Studies on the infectivity of stages in nymphal ticks revealed that organisms from Ny1 and TsN colonies were not infective for cattle (Kocan 1984). Homogenates of tick gut collected from nymphs that had fed for 4 days on an infected calf or were newly replete were infective. However, homogenates collected from this same group of ticks at 5, 10, and 15 days after feeding to repletion were not infective for susceptible calves. However, gut tissues collected from 20-day repleted nymphs were infective. It was during the 5 to 15 day period after repletion that the Ny1 and TsN colonies were observed and the homogenates were not infective.

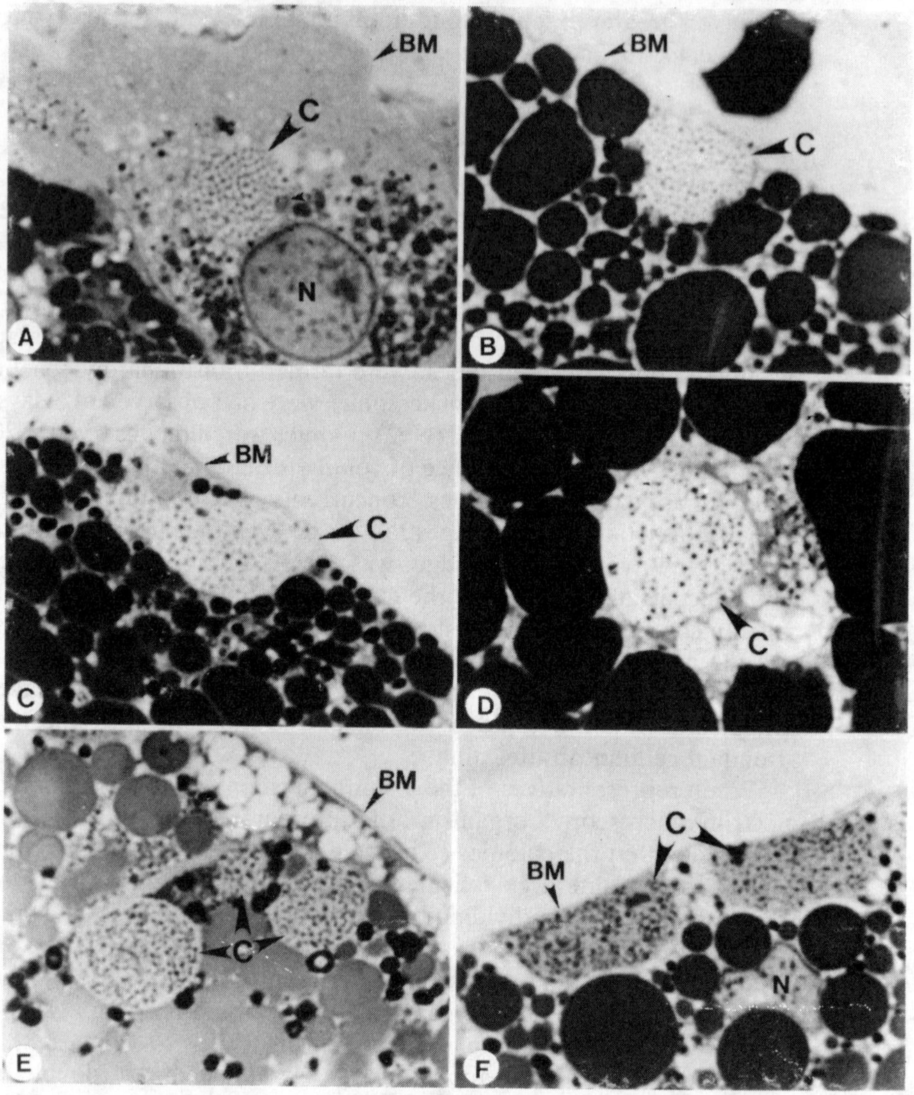

Fig. 22.5 — Photomicrographs of *A. marginale* colony types Ny1, Ny2 and TsN in cross sections of midgut epithelial cells of nymphal ticks that fed on an infected calf. The 1 μm sections were stained with Mallory's stain. ×1573. a. and b. Ny1 colonies (C) next to a nucleus (N) and near the basement membrane (BM) in the midgut epithelial cells. Some Ny1 have a large body at the margin of the colony (small arrow). c and d. TsN colonies (C) located near the basement membrane (BM). e. and f. Ny2 colonies (C) located near or attached to the basement membrane (BM) are often close to host cell nuclei (N).

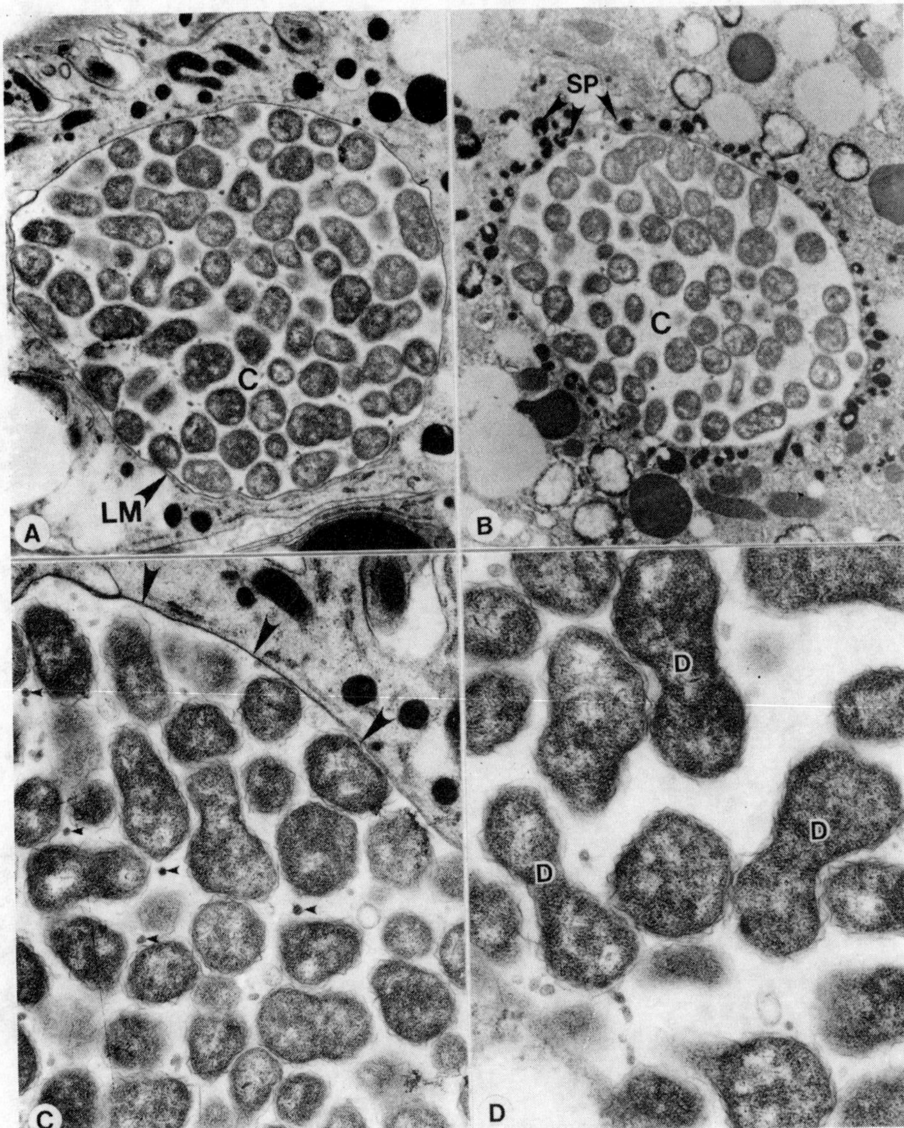

Fig. 22.6 — Electron micrographs of Ny1 colonies in midgut epithelial cells of nymphal ticks at 5 days after feeding to repletion on an infected calf. a. Ny1 colony (C) contains large reticulated organisms. The colony has a well-defined limiting membrane (LM). ×6842. b. Ny1 colony (C) surrounded by small particles (SP). ×6842. c. High magnification of an area of Ny1 colony from Fig. 22.6a. Small particles (small arrows) are between the large reticulated forms. A darkly staining limiting membrane (large arrows) is apparent. ×16,794. d. Higher magnification of the reticulated forms that appear to be dividing by binary fission (D). ×24,320.

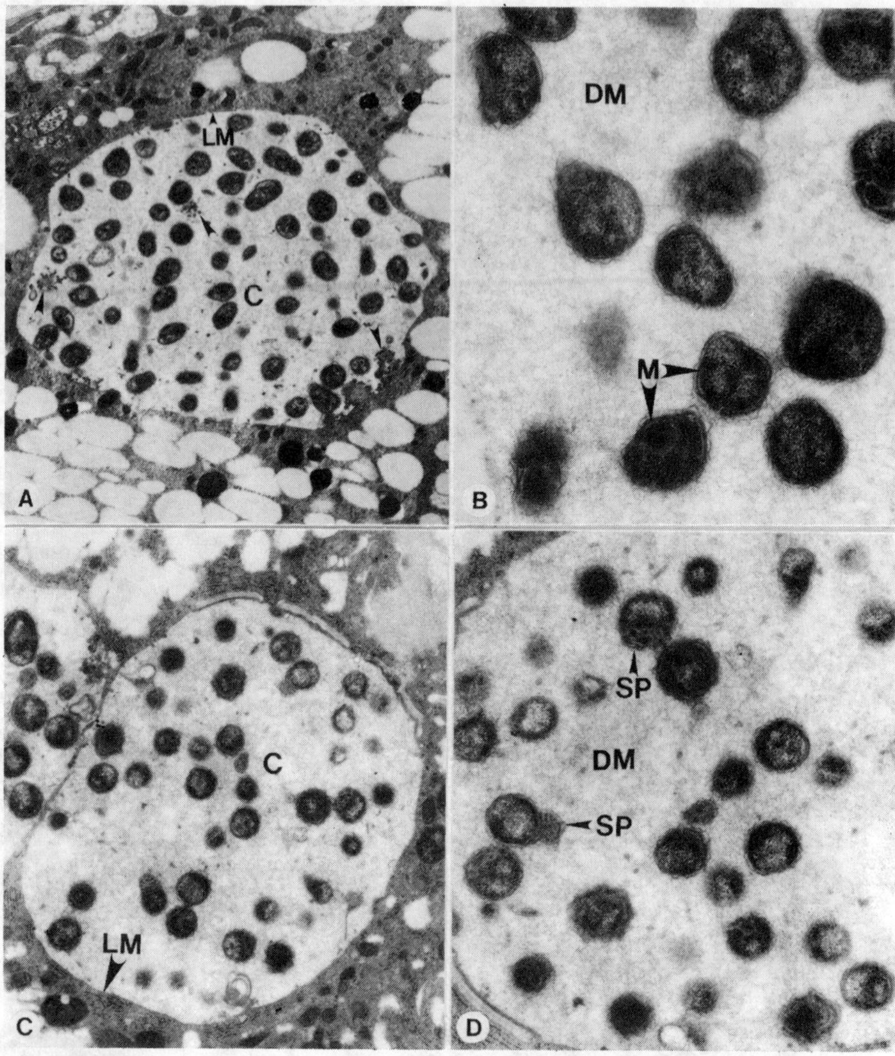

Fig. 22.7 — Electron micrographs of TsN colonies in midgut epithelial cells of nymphal ticks at 10 days (a and b) and 15 days (c and d) after feeding to repletion on an infected calf. a. TsN colony at 10 days after repletion, containing reticulated forms within a limiting membrane (LM). Clusters of small particles (arrows) are evident in the colony. ×6556. b. Higher magnification of the reticulated forms. The organisms vary in electron density and sometimes appear to be surrounded by 2 double membranes (M). A dense matrix (DM) is present. ×22,052. c. TsN colonies at 15 days after repletion. A limiting membrane (LM) is present around the colony (C), but is not as well defined as in the nymphal type 1 colonies. ×8344. d. High magnification from Fig. 22.7c. Organisms are within a dense matrix (DM). Some organisms contain small particles (SP) within the limiting membrane of the organisms. ×13,708.

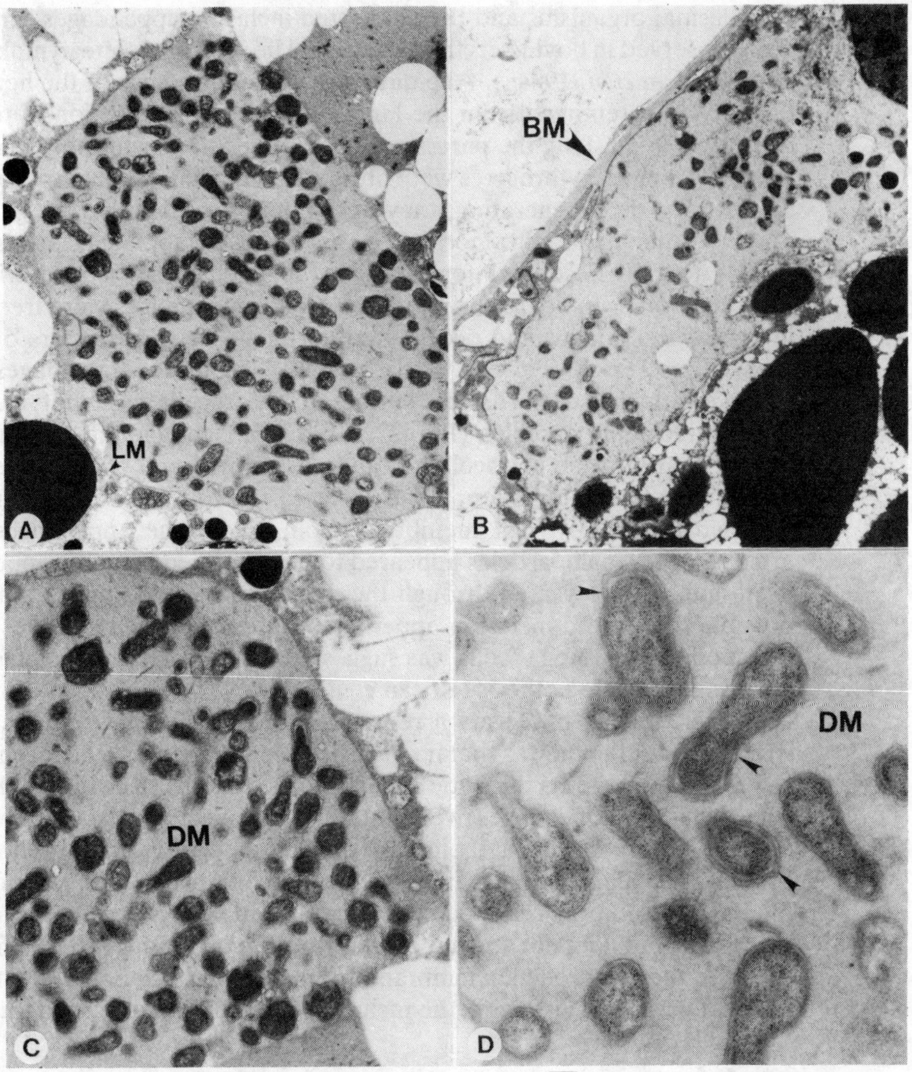

Fig. 22.8 — Electron micrographs of Ny2 colonies in midgut epithelial cells of nymphal ticks at 20 days after feeding to repletion on an infected calf. a. Ny2 colony surrounded by a limiting membrane (LM). Reticulated organisms are within the colonies. ×4263. b. Ny2 colony adjacent to the basement membrane (BM). The colony is irregular in shape. ×5786. c. High magnification of Fig. 22.8a. The reticulated organisms within the colony are round or rod-shaped and are surrounded by a dense matrix (DM). ×8526. d. Higher magnification of the organisms within the Ny2 colonies. The organisms are surrounded by a dense matrix (DM) and appear to have 2 double membranes. ×18,879.

Occurrence of inclusion appendages in nymphal ticks

Anaplasmal organisms and the associated inclusion appendages similar to those observed in bovine erythrocytes could be seen in replete nymphs (Fig. 22.9) (Kocan *et al.* 1984a). At 5 days post-repletion, much of the hemoglobin had been removed from the lumen by the digestive action of tick gut cells, thus unmasking the parasite. Initial bodies similar to those generally found in bovine erythrocytes were observed in the lumen, but most of them appeared to be degenerating; they were not observed within the gut cells. On the other hand, the morphology of inclusion appendages was quite similar to that described for inclusion appendages associated with the marginal bodies in bovine erythrocytes. Some appendages were free in the lumen of the midgut and occurred either alone or with clusters of small vesicular particles (Fig. 22.9). Frequently, inclusion appendages were observed attached to the luminal surface of midgut cells by a blunt, electron-dense attachment complex (Fig. 22.10). The appendage attachment complex appeared to be extracellular, and the pointed end of the appendage extended into the gut lumen. Small particles were often observed immediately across the host cell membrane from where the appendages were attached; the small particles appeared to be generated from the appendage itself and to have passed through the membrane of the midgut cells. The possible role of the inclusion appendage and of the small particles in the infection of tick midgut cells was suggested. Further research is needed to substantiate the hypothesis (Kocan *et al.* 1984a).

An interesting observation regarding the inclusion appendages found within the gut lumen was that no inclusion membrane was evident, nor was any remnant of the erythrocytic membrane identifiable (Kocan *et al.* 1984a). The inclusion appendages were intact, but the other structures of the parasite (e.g., inclusion membrane, initial bodies) were absent or, if present, were degenerating. It is possible that some aspect of digestion of the bloodmeal in the tick is specifically well suited for lysing the erythrocyte membrane but does not affect the integrity of the appendage. It is difficult to separate the erythrocytic membrane from *A. marginale* in the laboratory, especially without distorting the inclusion and associated appendage.

22.2.2 Development in unfed adults

Initial studies in our laboratory demonstrated that adult ticks, infected as nymphs, readily transmitted *A. marginale* to susceptible cattle (Kocan *et al.* 1980c). Furthermore, homogenates of tick gut tissue from infected adults caused infection when inoculated intravenously into susceptible calves (Kocan *et al.* 1981). Adult ticks were infective when they had fed as nymphs on either carrier calves with no apparent *A. marginale* parasitemia, or on calves with acute anaplasmosis with high parasitemia (Kocan *et al.* 1981). However, nymphs that fed on calves with higher parasitemias induced anaplasmosis more rapidly when they fed as adults on susceptible calves than those infected with low parasitemias.

Colonies of *A. marginale* were found in midgut epithelial cells of adult *D. andersoni* experimentally infected as nymphs from groups of ticks that had proved to be infected by animal transmission studies (Kocan *et al.* 1980a). No colonies were observed in associated control ticks, and the identity of the colonial organisms was confirmed as *A. marginale* with fluorescein- and ferritin-labeled antibodies (Kocan *et al.* 1980b, Oberst *et al.* 1981), and with peroxidase-antiperoxidase technique (Staats *et al.* 1983).

Initial studies with the light microscope revealed colonies that varied in staining characteristics and were different from those seen in nymphs. The colonies contained very pleomorphic organisms (Kocan *et al.* 1980a). Through examination of hundreds of colonies, we confirmed that the staining characteristics of the colonies, as discerned with the light microscope, varied considerably, and that the colonies could be categorized into 5 morphological types (Fig. 22.11) (Kocan *et al.* 1983a). Types 1 and 2 colonies contained densely staining masses of organisms. Type 3 colonies contained clumps of organisms, and Types 4 and 5 contained distinctly separate organisms along with large amorphous particles. The mean diameter of the various colony types ranged from 5.6 μm for Type 1 to 10.5 μm for Type 5 colonies. It was hypothesized that the small Type 1 colonies develop into the larger Types 4 and 5 colonies, and that the five colony types represent a developmental sequence of *A. marginale* in midgut epithelial cells of adult *D. andersoni* (Kocan *et al.* 1983a).

The structure of *A. marginale* appeared to vary within the five colony types, and further suggested the occurrence of a developmental sequence of the organism in gut cells (Kocan *et al.* 1983a). Representatives of each colony type were selected by light microscopy and were sectioned for examination by electron microscopy. The ultrastructural features of individual *A. marginale* organisms within colonies varied and included: (1) small electron-dense forms, (2) larger reticulated forms, (3) pleomorphic reticulated forms, and (4) small particles. Types 1 and 2 colonies contained small, electron-dense forms (Fig. 22.12a,b). Type 3 colonies contained electron-dense forms, reticulated forms, and small particles that were often outside of the parasite's limiting membrane (Fig. 22.12c,d). Type 4 colonies contained many reticulated forms that often had small particles within cell membranes (Fig. 22.13). Type 5 colonies contained fewer well-formed reticulated forms than did Type 4 and, in some cases, larger masses of pleomorphic reticulated forms (Fig. 22.14).

The various morphological forms of *A. marginale* within colonies were unique in our experience. They were more similar to stages described in the developmental cycle of *Chlamydia* than to any other organism with which we are familiar. During the course of a chlamydial life cycle, the organism changes from small electron-dense elementary bodies that can adsorb to and infect cells, to larger reticulated bodies that divide by binary fission and subsequently reorganize into elementary bodies (Storz & Page 1971, Storz & Spears 1977). Although the morphological evidence found in our study of *A. marginale* suggests a developmental sequence leading from electron-

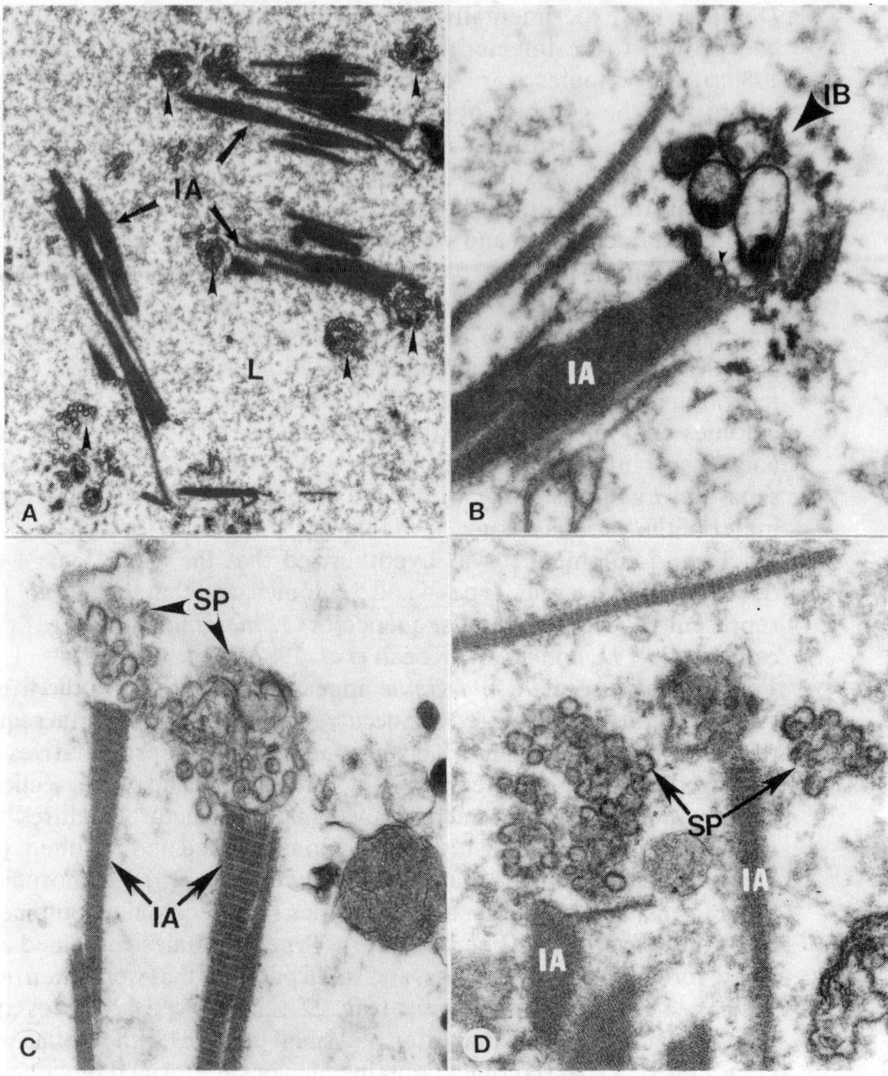

Fig. 22.9 — Electron micrographs of the inclusion appendages of *A. marginale* in the lumen of the midgut of nymphal *D. andersoni*. a. Low magnification of inclusion appendages (IA) in the midgut lumen (L) with associated clusters of small particles (small arrows). ×7688. b. Higher magnification of an inclusion appendage (IA) with 4 anaplasmal initial bodies (IB) typical of those described from bovine erythrocytes. Small particles (small arrows) appear to be arising at the junction of the inclusion membrane and inclusion appendage. ×25,523. c. and d. Inclusion appendages (IA) are often associated with clusters of small particles (SP). The clusters of particles are adjacent to the junction of the inclusion membrane and appendage. c-× 27,060; d-× 24,600.

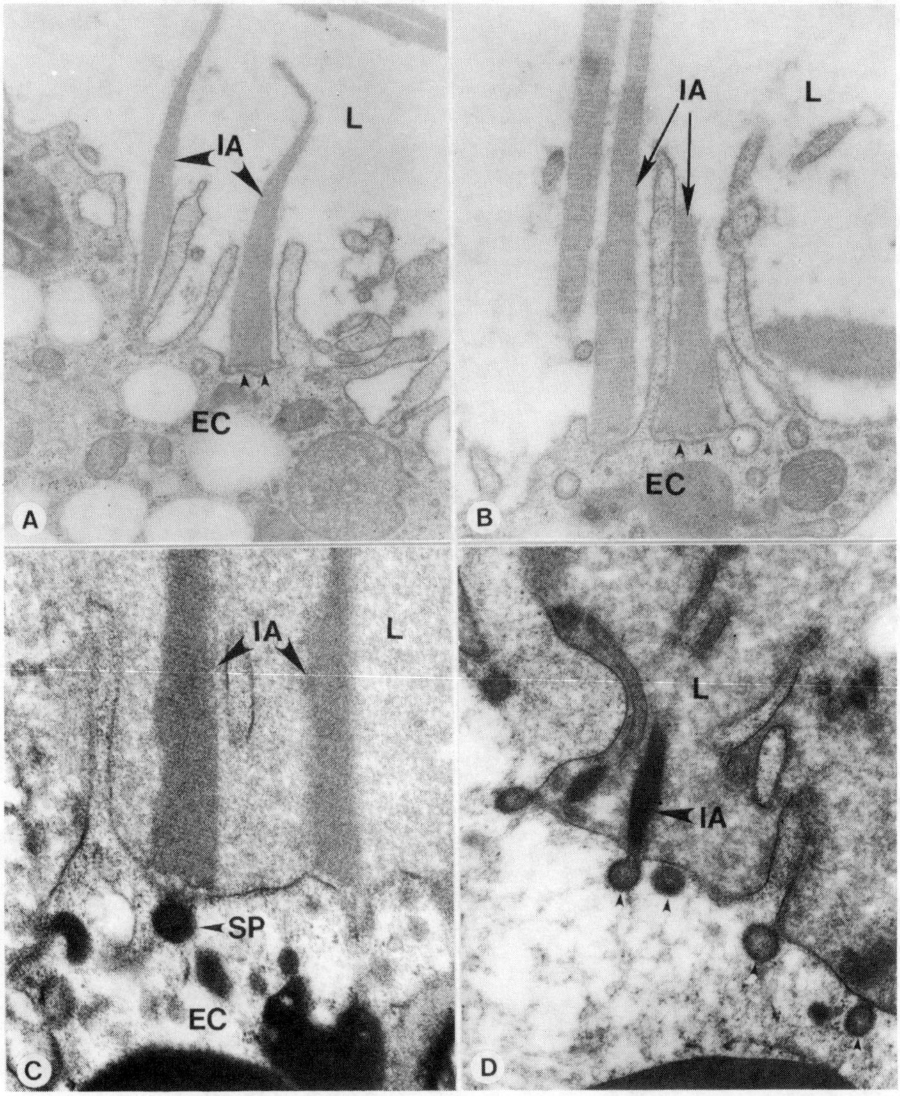

Fig. 22.10 — Electron micrographs of inclusion appendages of *A. marginale* attached to the cell membranes of midgut epithelial cells of nymphal *D. andersoni*. a. and b. Inclusion appendages (IA) attached to midgut epithelial cells (EC) and extending into the lumen (L). The attachment complex is adhered (small arrows) to the midgut cell membrane. The periodicity of the inclusion appendage is clearly visible. c→a ×18,038; d→b ×28,612. c and d. Inclusion appendages (IA) attached to epithelial cells (EC) and extending into the lumen (L) of the midgut. Small particles (small arrows and SP) appear to be arising from the inclusion appendage and passing through the pithelial cell membrane. c-× 29,856; d-× 25,813.

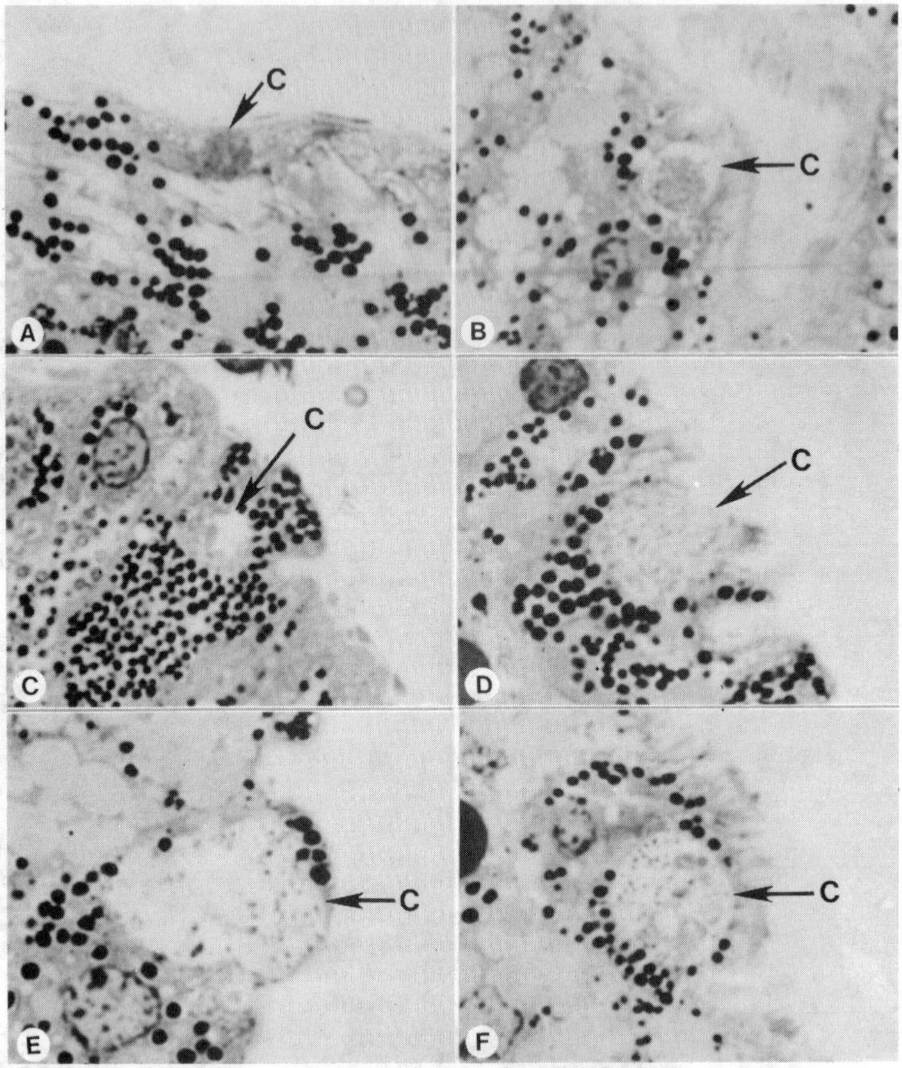

Fig. 22.11 — Photomicrographs of *A. marginale* colony types 1 to 5 in cross-sections of gut epithelial cells of unfed adult ticks that were infected as nymphs. The 1-μm plastic sections are stained with Mallory's stain. ×1523. a. Type 1 colony (C). b. Type 2 colony (C). c. Type 3 colony (C). d. Type 4 colony (C). e. and f. Type 5 colony (C).

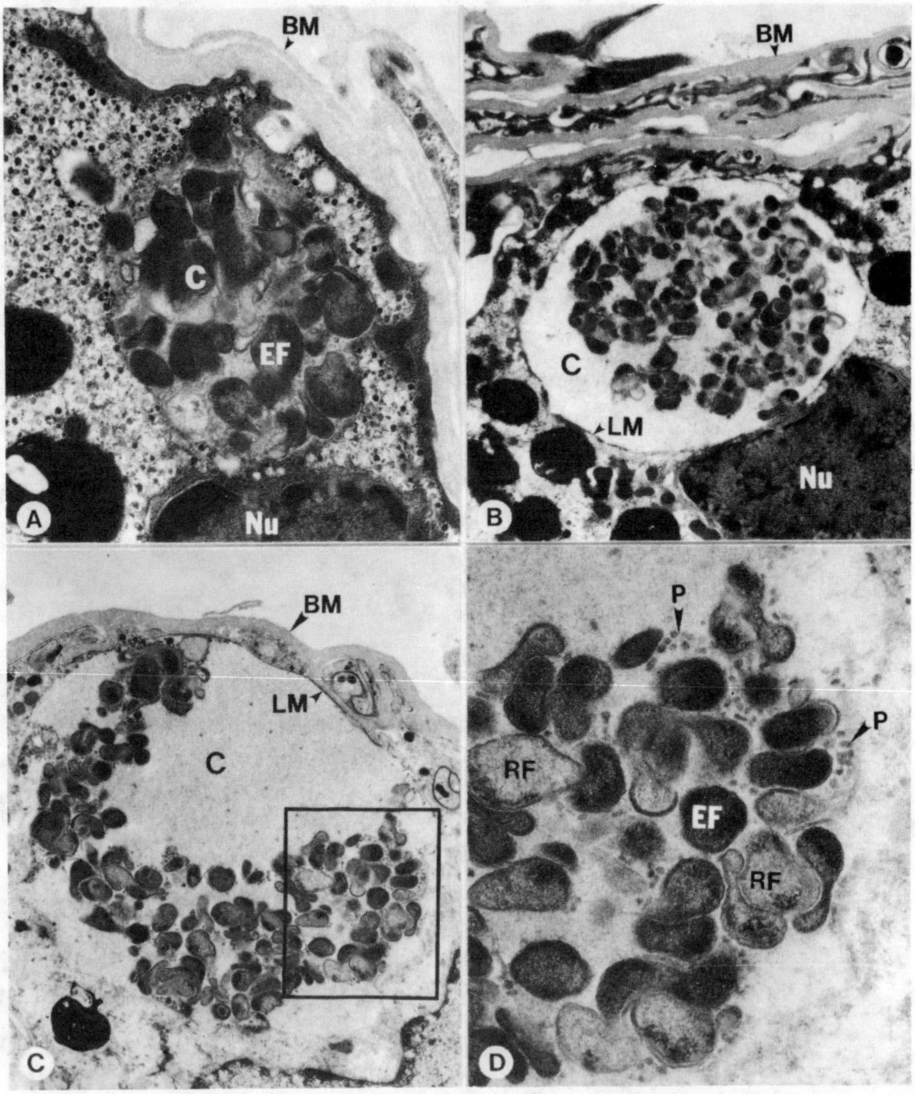

Fig. 22.12 — Electron micrographs of colony types 1 to 3 in midgut cells of adult ticks that were infected as nymphs. a. Type 1 colony (C) containing electron-dense forms (EF), located near the basement membrane (BM) and host cell nucleus (Nu). ×13,995. b. Type 2 colony (C) located near the basement membrane (BM). The colony has a well defined limiting membrane (LM) and is near the host cell nucleus (Nu). ×6606. c. Type 3 colony (C) next to the basement membrane (BM). A limiting membrane (LM) is evident in a portion of the colony. ×7464. The enclosed area is magnified in Fig. 22.12d. d. High magnification of the enclosed area of type 3 colony from Fig. 22.12c. The colony contains electron-dense forms (EF), reticulated forms (RF), and small particles (P). ×23,325.

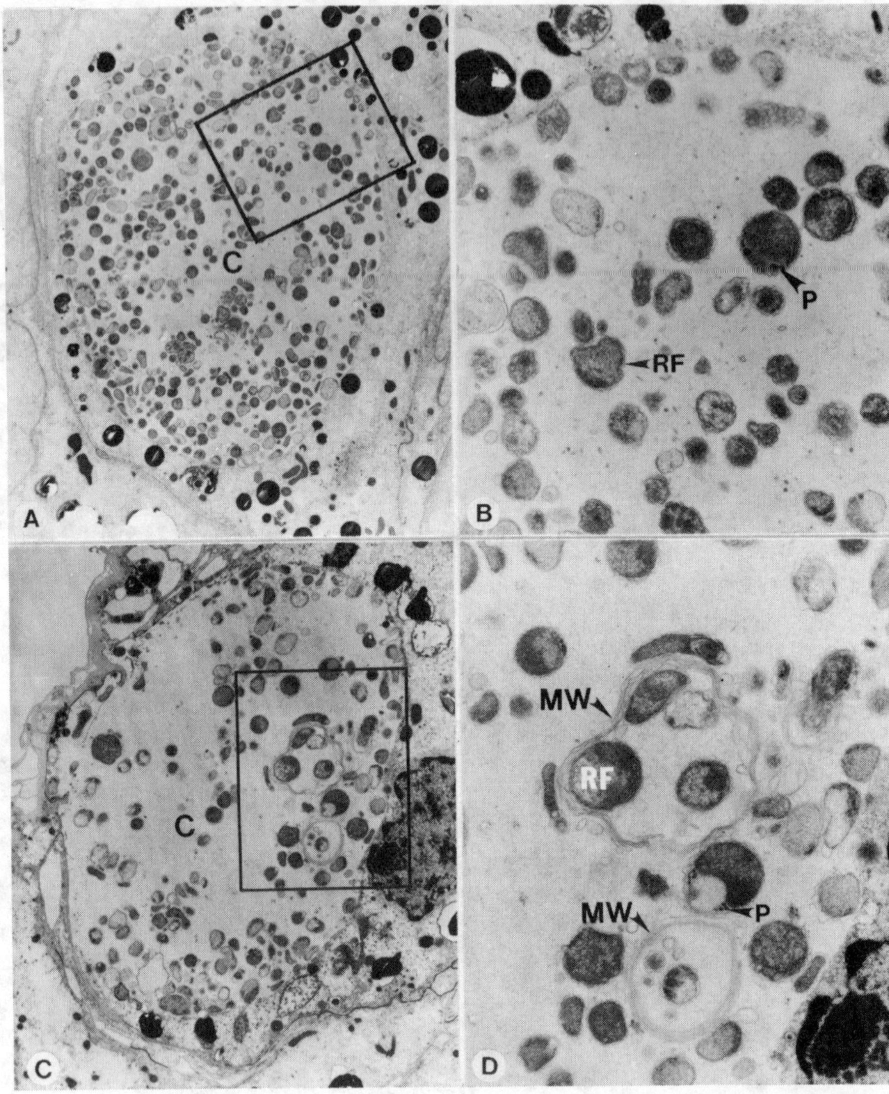

Fig. 22.13 — Electron micrographs of colony type 4 in midgut epithelial cells of adult ticks that were infected as nymphs. a. Low magnification of type 4 colony (C). ×2923. The enclosed area is magnified in Fig. 22.13b. b. Higher magnification of enclosed area of Fig. 22.13a. Reticulated forms (RF) are present and a few contain small particles (P) within the limiting membranes of the reticulated forms. ×11,196. c. Electron micrograph of type 4 colony (C). ×4031. The enclosed area is magnified in Fig. 22.13d. d. Higher magnification of the enclosed area in Fig. 22.13c. Colony contains reticulated forms (RF), a few of which appear to be subparticulating into small particles (P). Membrane whorls (MW) surround some pleomorphic reticulated forms. ×11,663.

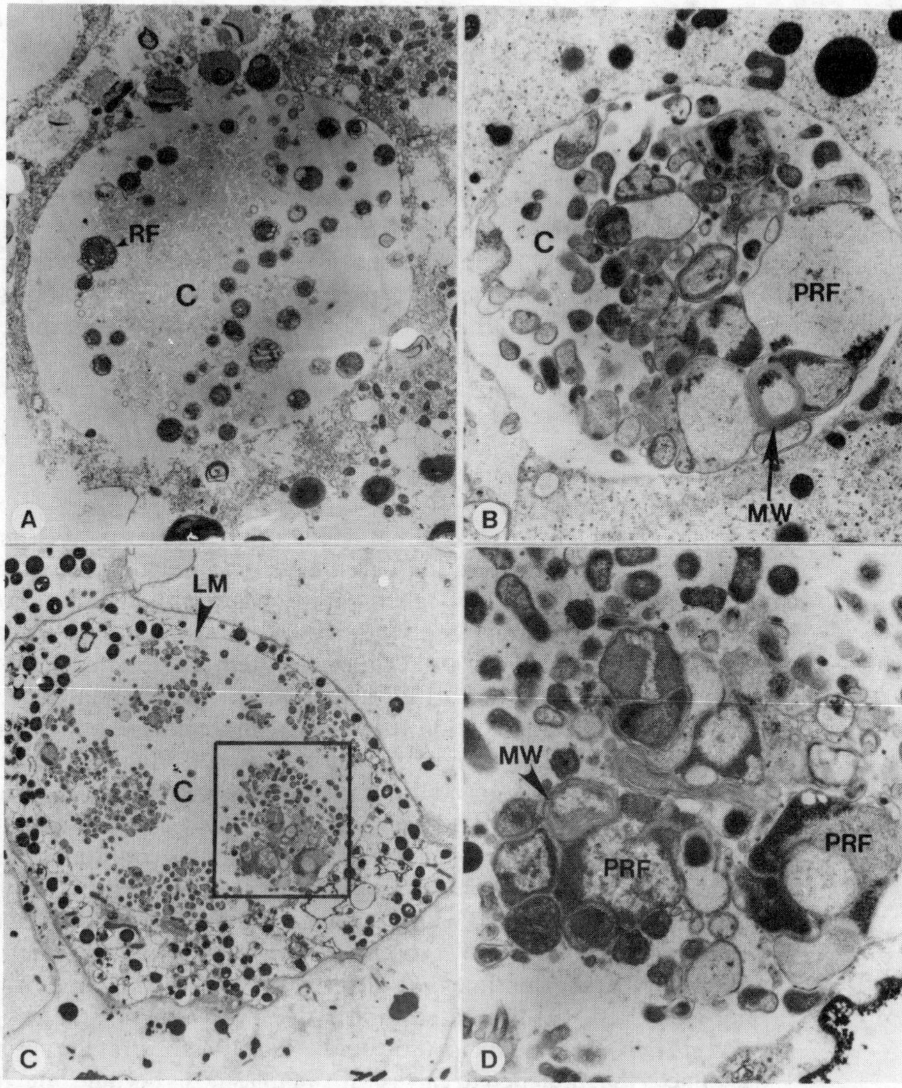

Fig. 22.14 — Electron micrographs of colony type 5 in midgut epithelial cells of adult ticks that were infected as nymphs. a. Type 5 colony (C) that contains several reticulated forms (RF). ×4758. b. Type 5 colony (C) containing pleomorphic reticulated forms (PRF). Membrane whorls (MW) surround some of the pleomorphic reticulated forms. ×5505. c. Type 5 colony (C) with limiting membrane (LM). ×1866. Enclosed area is magnified in Fig. 22.14d. d. Higher magnification of the enclosed area of Fig. 22.14c. Notice numerous pleomorphic reticulated forms (PRF); some reticulated forms are surrounded by membrane whorls (MW). ×9852.

dense form to reticulated form, it is not known whether the reticulated forms reorganize into electron-dense forms as is known to occur in *Chlamydia*. What similarities, if any, exist between the forms of *A. marginale* we have described and the Chlamydiae is not known at present.

Colonies of *A. marginale in Rhipicephalus simus*

Similar colonies of *A. marginale* were described in *Rhipicephalus simus,* a common tick from the Republic of South Africa, confirming a pattern of development of the parasite in a second species of tick (Potgieter *et al.* 1983). As with *D. andersoni,* five morphological types of colonies were present in the midgut epithelial cells of adult ticks that were infected as nymphs. The colonies were also studied by electron microscopy and found to contain organisms similar to those described in *D. andersoni* in North America.

Colony density studies

The density of *A. marginale* colonies (number of colonies per 0.1 mm^2† of tick gut tissue examined) in adult *D. andersoni* infected as nymphs was compared in ticks infected on calves with low and with high parasitemias. These values, in turn, were compared with those from unfed infected *D. andersoni* that had been exposed to an elevated temperature (37° C) for various periods (Kocan *et al.* 1982). Histological and animal inoculation studies confirmed that infection levels of *A. marginale* were significantly higher in ticks infected as nymphs on calves with higher *A. marginale* parasitemias. Incubation of adults also affected colony density. Densities were greatest in adults that had been incubated at 37° C for 2.5 days (Kocan *et al.* 1983b). Corresponding tick transmission studies confirmed that ticks with the greater colony density also had a greater potential to produce infection. Calves inoculated with gut homogenates from *D. andersoni* that were incubated 2.5 days at 37° C (and had high colony density) developed anaplasmosis in a shorter time than did calves exposed to material from ticks (a) not incubated, (b) incubated for 1.5 days, or (c) incubated for 7 days (Kocan *et al.* 1982). The incubation studies also demonstrated that near to 100% of ticks that attached to calves with anaplasmosis became infected and that colonies of *A. marginale* developed almost uniformly in them (Kocan *et al.* 1983b).

Survival of *A. marginale* in unfed, adult ticks

Survival of *A. marginale* in unfed *D. andersoni* decreased rapidly in adults that were over 3 months old (Kocan *et al.* 1984b). Six-month-old ticks were still able to transmit the disease in some cases. However, homogenates from unfed ticks of this age group proved not be infective when inoculated into susceptible cattle. It was interesting to note in dealing with 6-month-old ticks

† Colony densities are the number of colonies per 0.1 mm^2 gut tissue examined. The unit of gut tissue was incorrectly reported as 0.001 mm^2 in Kocan *et al.* (1983b) and Potgieter *et al.* (1983).

that incubation (37° C for 2.5 days) of ticks before tissues were collected restored the infectiveness of the homogenates. In histological studies on the same groups of ticks, colony densities were significantly lower at 6 months than at 3 months. However, samples collected at 6 months from unfed ticks contained colonies even though the corresponding homogenates were not infective. The presence of colonies, therefore, was not synonymous with infectivity. In incubated, 6-month-old ticks, where infectivity was restored, colony densities were significantly increased. It is not known what changes in unfed ticks resulted in the decrease of *A. marginale* within gut epithelial cells. Gut samples from the starved ticks contained greatly reduced amounts of hemoglobin inclusions. Decrease in hemoglobin inclusions may be at least one factor that influenced the survival of anaplasmal organisms.

Studies of hemocytes from ticks infected with *A. marginale*
Hemolymph of infected ticks was studied because hemocytes are migratory cells and are a likely mechanism of transport of orgnisms from one tissue to another. In infections of ticks by *Rickettsia rickettsia,* hemocytes are credited with the spread of infection throughout the tick (Brinton & Burgdorfer 1971).

Hemolymph from adult *D. andersoni* that were infected as nymphs was studied for the presence of *A. marginale* by two methods, viz., animal transmission and fluorescent antibody studies (Kocan *et al.* 1983c). The work was done with unfed ticks that were incubated (37° C for 2.5 days) and with feeding (for 6 days) adults. In both types of ticks, hemolymph was found to be another site of growth of *A. marginale,* but the infectivity was more frequently demonstrated in ticks that had fed for 6 days. Hemolymph collected from incubated ticks caused anaplasmosis in two out of four trials. Hemolymph from feeding ticks caused the disease in four out of four trials. Moderately fluorescing bodies were observed in some hemocytes from incubated ticks, whereas hemocytes from feeding ticks contained numerous clusters of more brightly fluorescing bodies. Fluorescing bodies were not observed in the hemocytes from control ticks. However, when similar collections of hemocytes were examined with electron microscopy, organisms were not seen. This finding was puzzling, warranting a more detailed study of the involvement of hemocytes in the development of *A. marginale*.

The presence of *A. marginale* in hemolymph is an interesting observation, but hemocytes are not a likely mechanism for the transfer of the parasite from the invertebrate to vertebrate host. Hemocytes apparently circulate freely within ticks, and could be important in transferring *A. marginale* within them. However, they have not been described to exit from ticks except when the acarines are injured.

22.2.3 Development and transmission by feeding adults

Although a developmental sequence has been described for *A. marginale* in midgut epithelial cells, growth of the organism in the tick's gut does not imply that this is a likely location from which the organism can transfer to the vertebrate host. Gut regurgitation has been suggested as the mode of transmission of *Cowdria ruminantium* by *Amblyomma* ticks (Uilenberg 1983). However, this mechanism of transmission has not been demonstrated conclusively. In most ixodid ticks, salivary glands are the main route by which fluid, electrolytes, and disease-causing agents pass to the vertebrate host. For this reason, further experiments were designed to examine other possible sites of growth of *A. marginale*.

Studies were designed to determine specifically when, during tick feeding, *A. marginale* was passed from ticks to susceptible cattle. Transmission studies were conducted in which adult ticks, known to be infected, were allowed to feed on calves for 1 to 9 days (Kocan *et al.* 1985). Anaplasmosis developed only in calves on which ticks fed for at least 6 days. Furthermore, as feeding continued, the prepatent periods of the disease shortened, suggesting that increased numbers of infective organisms were being released from feeding ticks beyond the 7th day. The feeding period required to effect transmission was longer than that required by the tick-borne protozoans, *Babesia* sp. and *Theileria* sp. (Mehlhorn & Schein 1984). Ticks infected with these protozoans must feed for 4 to 5 days before stages that are already present in the salivary glands of unfed adults can complete the final developmental sequence that results in sporozoites infective for the vertebrate host.

Studies on gut tissue of feeding adults

The long feeding time required for ticks to transmit *A. marginale* suggests a progressive development of the organism and points to the salivary glands as a site of development and transmission. If regurgitation of gut contents were the mode of transfer of *A. marginale,* transmission would most likely occur before the 6th day of tick feeding. Studies on gut tissues from adult ticks collected on each day of adult feeding (total, 9 days) also did not support the concept of transmission by gut regurgitation (Kocan 1984). Colonies were present in low numbers or not observed in gut tissues collected from ticks late in the feeding cycle (days 7–9) when transmission of the parasite was shown to occur. Another interesting feature noted with light microscopy was that many colonies seen in gut tissues on days 3 to 5 post-attachment were on the hemocoel side of the basement membrane (Fig. 22.15a,b) (Kocan 1984). When these colonies were examined with electron microscopy, it was further discovered that the organisms had entered and developed within myoepithelial cells located on the hemocoel side of the gut basement membrane (Fig. 22.15c,d). Development within the muscle cells indicated an additional site of growth.

Studies on salivary glands of feeding adults

Salivary glands were also collected from male and female ticks on each day of feeding (total, 9 days) (Kocan 1984). Homogenates of salivary glands caused infection when pooled and inoculated into a susceptible calf (Kocan 1984). Colonies of organisms were demonstrated with electron microscopy

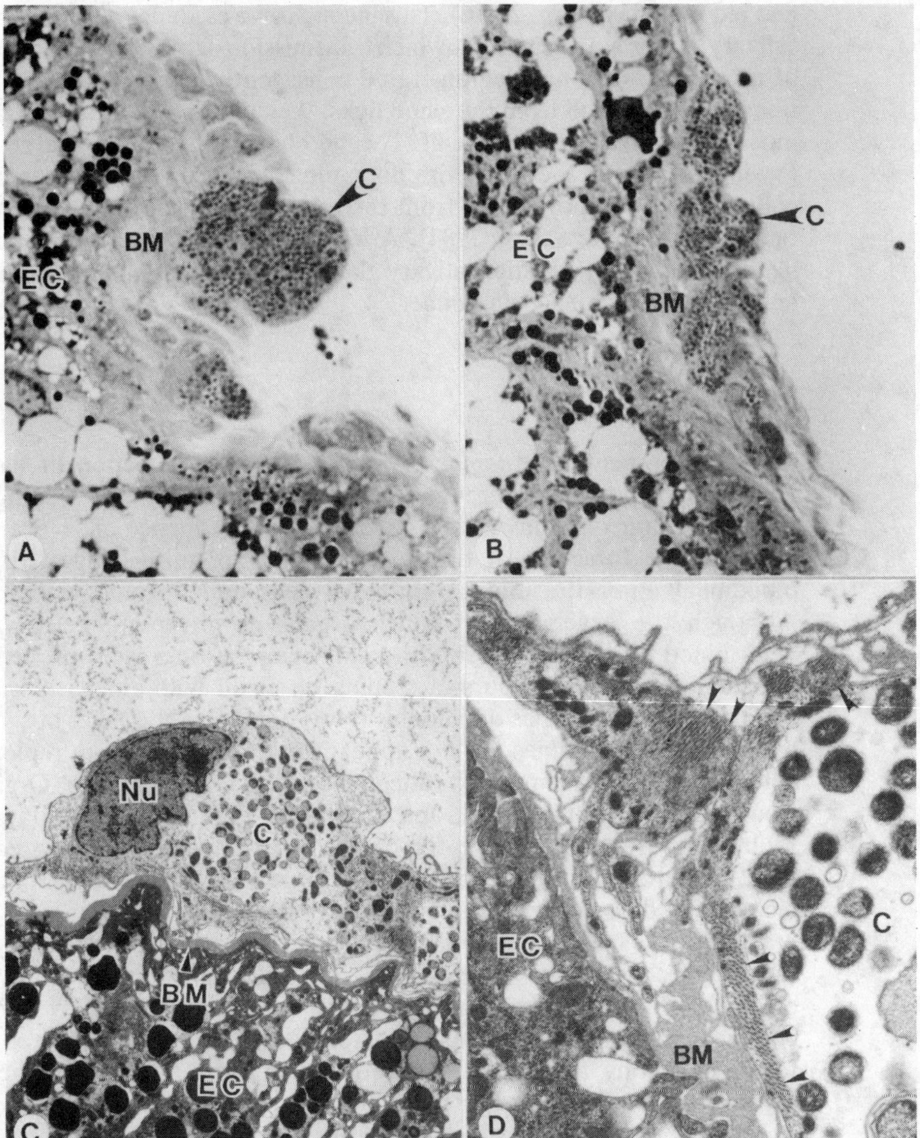

Fig. 22.15 — Colonies of *A. marginale* from adult, feeding ticks that were infected as nymphs. The colonies were located on the hemocoel side of the basement membrane. a. and b. Photomicrographs of colonies (C) located in the hemocoel of the basement (BM) of the tick epithelial cell (EC). ×1648. c. and d. Electron micrographs of colonies (C) on the hemocoel side of the basement membrane (BM) of the tick epithelial cell (EC) with a nucleus (Nu). In a higher magnification of the colony in Fig. 22.14d, muscle fibers (small arrows) confirm the cells as myoepithelial. c-× 3295; d-× 9885.

in the salivary glands of only two (out of many samples) of male ticks that had fed for 7 days (Fig. 22.16). This inconclusive evidence also points to the salivary glands as the site from which transmission occurs; however, colonies of organisms were not demonstrated consistently. Also, when oral secretions were collected from the same ticks, they did not cause infection when inoculated into a susceptible calf. The method of collection of oral secretion (by inoculation of the ticks with dopamine) has been used successfully to collect *Theileria* sporozoites from the salivary glands of infected, pre-fed ticks (Mehlhorn & Schein 1984). Additional experimental approaches are needed to confirm whether salivary glands are involved in the development and transmission of *A. marginale.*

22.3 SUMMARY

The development of *Anaplasma marginale,* from infection of nymphal *Dermacentor andersoni* through its transmission to cattle by feeding adult ticks, is complex and involves a developmental sequence of several stages (Fig. 22.17). Infection of ticks occurs when organisms ingested in the bloodmeal enter the midgut epithelial cells and develop within them. Although the exact way in which organisms infect gut cells was not determined by studies reviewed herein, nymphal ticks were proved to be infected after only 24 hours of feeding on a calf with anaplasmosis. The formation of three types of colonies (Ny1, TsN, and Ny2) was observed in nymphs. These colonies did not appear until after ticks were replete. All colonies in nymphs contained reticulated forms of the parasite. Organisms from two colony types (Ny1 and TsN) seen at 5, 10, and 15 days post-repletion were not infective for cattle, but the later-appearing colonies (Ny2) apparently contained infective organisms because homogenates collected at 20 days post-repletion were infective for cattle.

After ticks had molted to the adult stage, several different colony types (adult types 1–5) were present in the gut cells, and, unlike the nymphal colonies, all colony types were present concurrently. The five types of colonies contained, among them two, forms of *A. marginale*: electron-dense forms in colony types 1–3, and reticulated forms in colony types 4–5. Homogenates made from guts of adult, infected ticks caused anaplasmosis in cattle. It is not known which organisms were infective. When unfed ticks were incubated at 37° C for 2.5 days, colony numbers increased and prepatent periods of the disease induced by the homogenates from those ticks were shorter.

Transmission of *A. marginale* to susceptible cattle by adult ticks did not occur until ticks had fed for 6 days (total feeding time of 9 days). The feeding time required for transmission of this rickettsial organism was longer than that required for transmission of the protozoan parasites *Theileria* and *Babesia* (4 to 5 days). Although the infective stages were not identified or the mechanism of transmission substantially documented, development in salivary glands is suspected. Gut samples collected from feeding adult ticks

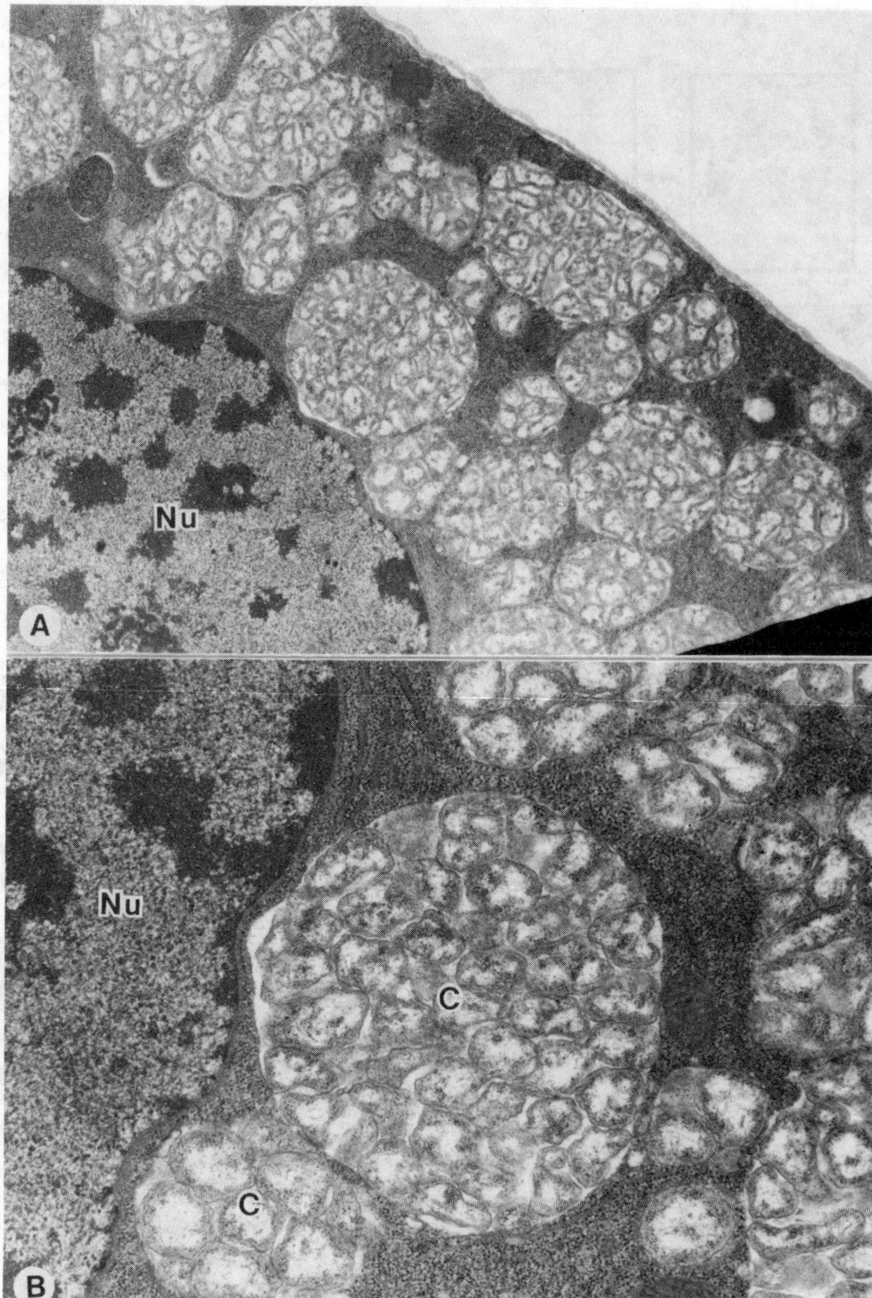

Fig. 22.16 — Electron micrographs of *A. marginale* colonies in salivary gland acini of adult male ticks collected on day 7 of feeding. a and b. Colonies (C) in salivary gland acini next to the cell nucleus (Nu). a-× 4888; b-× 14,288.

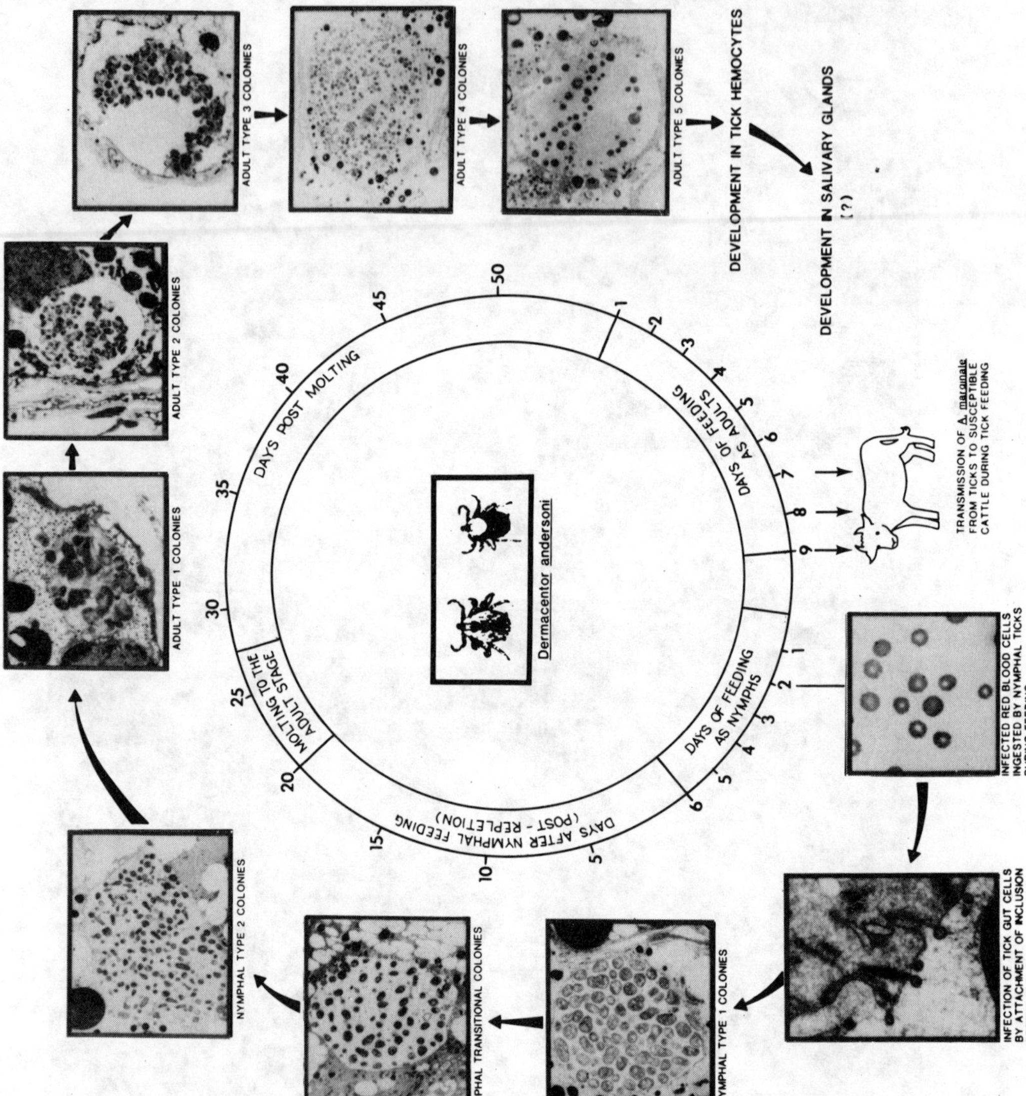

Fig. 22.17 — Summary of the developmental cycle of *A. marginale* in *D. andersoni* from infection of nymphal ticks by feeding on an infected cow through transmission by feeding adults. The developmental cycle involves 3 types of colonies in nymphs (Ny1, TsN and Ny2) and 5 types of colonies in adults (adult colony Types 1–5).

during the time when the parasites were transmitted (days 7–9 of feeding) contained few or no colonies; thus, transmission by gut regurgitation was unlikely. Colonies of *Anaplasma*-like organisms were demonstrated in a few salivary gland acini of male ticks that had fed for 7 days. Further research is needed to confirm whether salivary glands are involved in the development and transmission of *A. marginale*.

The research reviewed in this chapter has documented the existence of a complex developmental sequence for the rickettsial parasite, *Anaplasma marginale,* in an invertebrate host. Host factors that modulate coordinated parasite development have not been described or studied. Discovery of such factors is fundamental to the understanding the growth and pathogenesis of this organism.

ACKNOWLEDGEMENTS

The research reviewed herein was supported by Oklahoma Agricultural Experiment Station Project No. 2-1-11441; USDA–SEA Animal Health Program 83-CRSR-2-2171; USDA-SEA Special Grants No. 5-523820 and 5-5-23874 and the Rockefeller Foundation Grant No. GA AS 8323. Portions of the photographic plates and text reproduced herein appear with the permission of the *American Journal of Veterinary Research.* This manuscript is journal article J4795 of the Oklahoma Agricultural Experiment Station, Stillwater, OK 74078. The author thanks Cynthia McDonald and Marilyn French for clerical assistance, Suzanne Spears for graphic arts, Bob Doyle for photography, and Drs S.A. Ewing and Jakie Hair for reviewing the manuscript.

REFERENCES

Anthony D. W. (1968) *Dermacentor andersoni* Stiles as a vector of bovine anaplasmosis. In: *Proc. 5th National Anaplasmosis Conference, Stillwater, OK,* p. 167–171.

Anthony, D. W.& Roby, T. O. (1962) Anaplasmosis transmission studies with *Dermacentor variabilis* (Say) and *Dermacentor andersoni* (=*D. venustus* Marx) as experimental vectors. In: *Proc. 4th National Anaplasmosis Conference, Reno, Nevada,* p. 78–81.

Anthony, D. W., Madden, P. A., & Gates, D. W. (1964) *Anaplasma marginale* Theiler observed in gut and excreta of *Dermacentor andersoni* Stiles (*Dermacentor venustus* Marx). *Am. J. Vet. Res.* **25**:1464–1471.

Baca, O. G. & Paretsky, D. (1983) Q fever and *Coxiella burnetti*: a model for host-parasite interactions. *Microbiolog. Rev.* **47**:127–149.

Bertram, D. S. (1962) Rickettsial infections and ticks. *Sym. Zool. Soc. London* **6**:179–197.

Bram, R. A. (1975) Tick-borne livestock diseases and their vectors, 1. the global problem. *World Animal Review* **6**:1–5.

Bram, R. A. & Romanowski, R. D. (1970) Recognition of *Anaplasma marginale* Theiler in *Dermacentor andersoni* Stiles by fluorescent antibody method. *J. Parasit.* **56**:32–38.

Brinton, L. P. & Burgdorfer, W. (1971) The structure of *Rickettsia canada* in tissues of *Dermacentor andersoni* Stiles. *J. Bacteriol* **105**:1149–1159.

Burgdorfer, W. & Varma, M. G. R. (1968) Trans-stadial and transovarial development of disease agents in arthropods. *Ann. Rev. Ent.* **12**:347–376.

Christensen J. F., Howarth, J. A. (1966) Anaplasmosis transmission by *Dermacentor occidentalis* taken from cattle in Santa Barbara, California. *Am. J. Vet. Res.* **27**:1473–1475.

Dalgliesh, R. J. & Stewart, N. P. (1976) Stimulation of the development of infective *Babesia bigemina* (= *B argentina*) in unfed *Boophilus microplus* larvae. *Aust. Vet. J.* **52**:543.

Dalgliesh, R. J. & Stewart, N. P. (1979). Observations on the morphology and infectivity of cattle of *Babesia bovis* parasites in unfed *Boophilus microplus* larvae after incubation at various temperatures. *Int. J. Parasitol.* **9**:115–120.

Dikmans, G. (1950) The transmission of anaplasmosis. *Am. J. Vet. Res.* **38**:5–16.

Espana, C., Espana, E. M., & Gonzalez, D. (1959) *Anaplasma marginale.* I. Studies with phase contrast and electron microscopy. *Am. J. Vet. Res.* **20**:795–805.

Ewing, S. A. (1981) Transmission of *Anaplasma marginale* by arthropods. In: *Proc. 7th National Anaplasmosis Conference, Mississippi State, MS.* p. 395–423.

Francis, D. M., Kinden, D. A., & Buening, G. H. (1979) Characterization of the inclusion limiting membrane of *Anaplasma marginale* by immunoferritin labeling. *Am. J. Vet. Res.* **40**:777–782.

Friedhoff K. T. & Ristic, M. (1966) Anaplasmosis XIX. A preliminary study of *Anaplasma marginale* in *Dermacentor andersoni* Stiles by fluorescent antibody technique. *Am. J. Vet. Res.* **27**:643–646.

Hayes, S. F. & Burgdorfer, W. (1982) Reactivation of *Rickettsia rickettsii* in *Dermacentor andersoni* ticks: an ultrastructural analysis. *Infect. Immun.* **37**:779–785.

Hoogstraal, H. (1970) Symposium on current world research on tick-disease relationships and control of tick vectors of disease. *Mis. Publ. Ent. Soc. Am.* **6**:325–372.

Howarth, J. A. & Hokama, Y. (1973) Tick transmission of anaplasmosis under laboratory conditions. In: *Proc. 6th National Anaplasmosis Conference, Las Vegas, Nevada,* pp. 227–230.

Howell, D. E., Stiles, G. W., & Moe, L. H. (1941) The hereditary transmission of anaplasmosis by *Dermacentor andersoni* Stiles. *Am. J. Vet. Res.* **2**:165–166.

Kocan, K. M. (1982) Unpublished data, Veterinary Research, Oklahoma State University, Stillwater, OK 74078.

Kocan, K. M. (1984) Unpublished data, Veterinary Research, Oklahoma State University, Stillwater, OK 74078

Kocan, K. M., Venable, J. H., & Brock, W. E. (1978a) Ultrastructure of anaplasmal inclusions (Pawhuska isolate) and their appendages in intact and hemolyzed erythrocytes and in complement-fixation antigen. *Am. J. Vet. Res.* **39**:1123–1130.

Kocan, K. M., Venable, J. H., Hsu K. C., & Brock, W. E. (1978b) Ultrastructural localization of anaplasmal antigens (Pawhuska isolate) with ferritin-conjugated antibody. *Am. J. Vet. Res.* **39**:1131–1135.

Kocan, K. M., Hair, J. A., & Ewing S. A. (1980a) Ultrastructure of *Anaplasma marginale* Theiler in *Dermacentor andersoni* Stiles and *Dermacentor variabilis* (Say). *Am. J. Vet. Res.* **41**:1966–1976.

Kocan, K. M., Hsu, K. C., Hair, J. A., & Ewing, S. A. (1980b) Demonstration of *Anaplasma marginale* Theiler in *Dermacentor variabilis* (Say) by ferritin-conjugated antibody technique. *Am. J. Vet. Res.* **41**:1977–1981.

Kocan, K. M., Teel, P. D., & Hair, J. A. (1980c) Demonstration of *Anaplasma marginale* Theiler in ticks by tick transmission, animal inoculation and fluorescent antibody studies. *Am. J. Vet. Res.* **41**:183–186.

Kocan, K. M., Hair, J. A., Ewing, S. A., & Stratton, L. G. (1981) Transmission of *Anaplasma marginale* Theiler by *Dermacentor andersoni* Stiles and *Dermacentor variabilis* (Say). *Am. J. Vet. Res.* **42**:15–18.

Kocan, K. M., Barron, S. J., Holbert D., Ewing, S. A., & Hair, J. A. (1982) Influence of increased temperature on *Anaplasma marginale* Theiler in the gut of *Dermacentor andersoni* Stiles. *Am. J. Vet. Res.* **43**:32–35.

Kocan, K. M., Ewing, S. A., Holbert, D., Hair, J. A. (1983a) Morphologic characteristics of colonies of *Anaplasma marginale* Theiler in midgut epithelial cells of *Dermacentor andersoni* Stiles. *Am. J. Vet. Res.* **43**:586–593.

Kocan, K. M., Holbert, D., Ewing, S. A., Hair, J. A., & Barron, S. J. (1983b) Development of colonies of *Anaplasma marginale* in the gut of incubated *Dermacentor andersoni. Am. J. Vet. Res.* **44**:1617–1620.

Kocan, K. M., Oberst, R. D., Ewing, S. A., Hair, J. A., Barron, S. J. (1983c) Demonstration of *Anaplasma marginale* in hemolymph of *Dermacentor andersoni* by animal inoculation and by fluorescent-antibody technique. *Am. J. Vet. Res.* **44**:798–802.

Kocan, K. M., Ewing, S. A., Hair, J. A., & Barron, S. J. (1984a) Demonstration of the inclusion appendage of *Anaplasma marginale* in nymphal *Dermacentor andersoni. Am. J. Vet. Res.* **456**:1800–1807.

Kocan, K. M., Holbert, D., Edwards, W., Ewing, S. A., Barron, S. J., & Hair, J. A. (1984b) Longevity of colonies of *Anaplasma marginale* in midgut epithelial cells of *Dermacentor andersoni.Am. J. Vet. Res.* (in press).

Kocan, K. M., Yelling, T. N., Ewing, S. A., Hair, J. A., & Barron, S. J. (1984c) Morphology of colonies of *Anaplasma marginale* in nymphal *Dermacentor andersoni*. *Am. J. Vet. Res.* **45**:1434–1440.

Kocan, K. M., Barron, S. J., Ewing, S. A., Hair, J. A. (1985) Transmission of *Anaplasma marginale* by adult *Dermacentor andersoni* during feeding calves. *Am. J. Vet. Res.* **46**:1565–1567.

Krier, J. & Ristic, M. (1963) Anaplasmosis. X. Morphologic characteristics of the parasite present in blood of calves infected with the Oregon strain of *Anaplasma marginale*. *Am. J. Vet. Res.* **24**:676–687.

Mehlhorn, H. & Schein, E. (1984) The piroplasms: life cycle and sexual stages. *Ad. Parasit.* **23**:37–103

Nyindo, M. B. A., Ristic, M., Huxsoll, D. L., & Smith, R. D. (1971) Tropical pancytopenia: *in vitro* cultivation of the causative agent—*Ehrlichia canis*. *Am. J. Vet. Res.* **32**:1651–1658.

Oberst, R. D., Kocan, K. M., Hair, J. A., & Ewing, S. A. (1981) Staining characteristics of colonies of *Anaplasma marginale* Theiler in *Dermacentor andersoni* Stiles. *Am. J. Vet. Res.* **432**:2006–2009

Peterson, K. J., Raleigh, R. J., Stroud, R. K., & Goulding, R. L. (1977) Bovine anaplasmosis transmission studies conducted under controlled natural exposure in a *Dermacentor andersoni* (*venustus*) indigenous area of eastern Oregon. *Am. J. Vet. Res.* **38**:351–354.

Philip, C. B. (1963) Ticks as purveyors of animal ailments: A review of pertinent data and of recent contributions. *Ad. acarol.* **1**:285–325.

Philip, C. B. & Burgdorfer, W. (1961) Arthropod vectors as reservoirs of microbial disease agents. *Ann. Rev. Ent.* **6**:391–412.

Potgieter, F. T., Kocan, K. M., McNew, R. W., & Ewing, S. A. (1983) Demonstration of colonies of *Anaplasma marginale* in the midgut of *Rhipicephalus simus*. *Am. J. Vet. Res.* **44**:2256–2261.

Price, W. H. (1954) Variation in virulence of *Rickettsia rickettsii* under natural and experimental conditions. In: *The dynamics of virus and rickettsial infections,* Blakiston Co., Inc., New York, p. 164–183.

Řěhàcěk, J. (1965) Development of animal viruses and rickettsiae in ticks and mites. *Ann. Rev. Entomol.* **10**:1–24.

Renshaw, H. W., Vaughn, H. W., Magonigle, R. A., Davis, W. C., Stauber, E. H., & Frank, R. W. (1977) Evaluation of free-roaming mule deer as carriers of anaplasmosis in an area of Idaho where bovine anaplasmosis is enzootic. *Am. J. Vet. Res.* **170**:334–339.

Ristic, M. (1977) Bovine anaplasmosis. In: Kreier, J. (ed): *Parasitic protozoa* Vol. 4, Academic Press, New York, p. 235–249.

Ristic M. & Watrach, A. M. (1963) Anaplasmosis VI. Studies and a hypothesis concerning the cycle of development of the causative agent. *Am. J. Vet. Res.* **24**:267–277.

Ritchie, A. E. (1962) Simplified specimen preparation for electron microscopic studies of an intra-erythrocytic parasite, *Anaplasma marginale*. *Proc. Soc. Exp. Biol. Med.* **110**:532–535.

Samish, M. (1977) Infective *Theileria annulata* in the tick without a blood meal stimulus. *Nature* **270**:51–52.

Sanborn, C. E., Stiles, G. W., & Moe, L. H. (1938) Anaplasmosis transmission by naturally infected *Dermacentor andersoni* male and female ticks. *N. Am. Vet.* **19**:21–32.

Schalm, O. W., Osebold, J. W., & Murphy, F. A. (1962) Observations with new methylene blue stain as applied to *Anaplasma marginale*. *Calif. Vet.* **16**:36–27.

Scott, W. L., Greer, J. C., & Foote, L. E. (1961) Electron microscopy of *Anaplasma marginale* in the bovine erythrocyte. *Am. J. Vet. Res.* **22**:877–880.

Simpson, C. F. (1977) Morphologic alteration of *Paranaplasma caudata* following treatment of calves with oxytetraceycline. *Am. J. Vet. Res.* **43**:1903–1906.

Simpson, C. F., Kling, J. M., & Love, J. N. (1965) The nature of bands in parasitized bovine erythrocytes. *J. Cell. Biol.* **27**:225–235.

Staats, J. J., Kocan, K. M., Hair, J. A., & Ewing, S. A. (1982) Immunocytochemical labeling of *Anaplasma marginale* Theiler in *Dermacentor andersoni* Stiles with peroxidase-antiperoxidase technique. *Am. J. Vet. Res.* **43**:979–983.

Storz, J. & Page, L. A. (1971) Taxonomy of the Chlamydiae: Reasons for classifying organisms of the genus *Chlamydia,* family Chlamydiaceae, in a separate order, Chlamydiales ord nov. *Int. J. Syst. Bacteriol.* **21**:322–334.

Storz, J. & Spears, P. (1977) Chlamydiales: Properties, cycle of development and effect on eukaryotic host cells. *Curr. Top. Microbiol. Immunol.* **77**:167–214.

Uilenberg, G. (1980) Ticks and tick-borne diseases of veterinary interest in Europe and Africa: prospects for control and eradication, In: Johnson L. A. Y., Coper, M. G. (eds): *Ticks and tick-borne diseases,* Aust. Vet. Assoc., Artarmon, Aust., pp. 1–3.

Uilenberg, G. (1983) Heartwater (*Cowdria ruminantium* infection): Current status. *Ad. Vet. Sci. Comp. Med.* **27**:427–480

Weiss, E. & Moulder, J. W. (1984) The Rickettsias and Chlamydias. In: Krieg, N. R., Holt, J. G. (eds) *Bergey's manual of systemic bacteriology,* Volume 1, Williams and Wilkins, Baltimore, p. 687–739.

Index